AF717933

Environmental Science and Engineering
Subseries: Environmental Science

Series Editors: R. Allan • U. Förstner • W. Salomons

Teja Tscharntke, Christoph Leuschner,
Manfred Zeller, Edi Guhardja,
Arifuddin Bidin
(Eds.)

Stability of Tropical Rainforest Margins

Linking Ecological, Economic and Social Constraints of Land Use and Conservation

With 89 Figures, 1 in colour

Springer

Editors:

Prof. Dr. Teja Tscharntke
Agroecology Group
University of Göttingen
Waldweg 26,
37073 Göttingen, Germany
E-mail: ttschar@gwdg.de

prof. Dr. Manfred Zeller
Institute of Agricultural Economics
and Social Sciences in the Tropics and
Subtropics, University of Hohenheim
(490)
70593 Stuttgart, Germany
E-mail:
manfred.zeller@uni-hohenheim.de

Prof. Dr. Christoph Leuschner
Department of Plant Sciences
University of Göttingen
Untere Karspüle 2
37073 Göttingen,
Germany
E-mail: cleusch@gwdg.de

Prof. Dr. Edi Guhardja
Center for Tropical Forest Margin,
Institut Pertanian Bogor (IPB), Jl.
Raya Gunung Gede, Bogor 16153,
Indonesia
E-mail: storma-ipb@indo.net.id

Dr. Arifuddin Bidin
Universitas Tadulako, Kampus
Bumi Tondo, Palu 94118,
Indonesia
E-mail: storma-palu@gwdg.de

ISSN 1863-5520
ISBN 10 3-540-30289-1 **Springer Berlin Heidelberg New York**
ISBN 13 978-3-540-30289-6 **Springer Berlin Heidelberg New York**

Library of Congress Control Number: 2006936494

This work is subject to copyright. All rights are reserved, whether the whole or part of the material is concerned, specifically the rights of translation, reprinting, reuse of illustrations, recitation, broadcasting, reproduction on microfilm or in any other way, and storage in data banks. Duplication of this publication or parts thereof is permitted only under the provisions of the German Copyright Law of September 9, 1965, in its current version, and permission for use must always be obtained from Springer-Verlag. Violations are liable to prosecution under the German Copyright Law.

Springer is a part of Springer Science+Business Media
springeronline.com
© Springer-Verlag Berlin Heidelberg 2007

The use of general descriptive names, registered names, trademarks, etc. in this publication does not imply, even in the absence of a specific statement, that such names are exempt from the relevant protective laws and regulations and therefore free for general use.

Cover design: E. Kirchner, Heidelberg
Production: A. Oelschläger
Typesetting: Camera-ready by Dr. Stella Aspelmeier

Printed on acid-free paper 30/2132/AO 543210

Contents

List of Contributors

Melani Abdulkadir-Sunito
Department of Communication and Community Development, Faculty of Human Ecology, Bogor Agricultural University, Indonesia

Tatiana Deane de Abreu Sá
Empresa Brasileira de Pesquisa Agropecuária, Embrapa Sede, Parque Estação Biológica - PqEB s/n., Brasília-DF, - Brazil - CEP 70770-901

Diógenes S. Alves
Instituto Nacional de Pesquisas Espaciais (INPE), Av dos Astronautas 1758, CEP 12227-010 São José dos Campos SP, Brazil

M. Amran Amir
Institute of Forest Management, Faculty of Agriculture, Bumi Tondo Campus, Tadulako University, Palu 94118, Indonesia

Iswandi Anas
Laboratory of Soil Biology, Soil Department, Faculty of Agriculture, Institut
Pertanian Bogor (IPB), Darmaga Campus, Bogor 16680, Indonesia

Muhammad Ardiansyah
Institute of Soil Sciences, Bogor Agricultural University (IPB), Bogor, Indonesia

Hadi Susilo Arifin
Department of Landscape Architecture, Faculty of Agriculture, IPB, Bogor, Indonesia

Inge Armbrecht
Departamento de Biología, Universidad de Valle, Apartado Aéreo 25360, Cali, Colombia

Jan Barkmann
Junior Scientist Group "Economic Valuation of Ecosystem Functions and Services" and Environmental and Resource Economics, Department of Agricultural Economics and Rural Development, University of Göttingen, Platz der Göttinger Sieben 7, D-37073 Göttingen, Germany

Pablo C. Benítez
Department of Economics, University of Victoria, PO Box 1700 STN CSC, Victoria BC V8W 2Y2, Canada

Arifuddin Bidin
Universitas Tadulako, Kampus Bumi Tondo, Palu 94118, Indonesia (Sulteng)

Regina Birner
International Food Policy Research Institute (IFPRI), Washington, D.C., USA

Jan Börner
Center for Development Research (ZEF), University of Bonn, Walter-Flex-Str. 3, 53113 Bonn, Germany

Merijn M. Bos
Institute of Agroecology, University of Göttingen, Waldweg 26, D-37073 Göttingen, Germany

Gunnar Brehm
Institut für Spezielle Zoologie und Evolutionsbiologie mit Phyletischem Museum, University of Jena, Erbertstrasse 1, 07745 Jena, Germany

Damayanti Buchori
Department of Pest and Plant Diseases, Faculty of Agriculture IPB, Bogor Agricultural University, Jl. Kamper Kampus, IPB Darmaga, Bogor 16680, Indonesia

Boris Büche
Agroecology, University of Göttingen, Waldweg 26, D-37073 Göttingen, Germany

Manfred Denich
Center for Development Research (ZEF), University of Bonn, Walter-Flex-Str. 3, 53113 Bonn, Germany

Thomas V. Dietsch
Institute of the Environment, Center for Tropical Research, University of California, Los Angeles, La Kretz Hall, Suite 300, Box 951496, Los Angeles, California 90095-1496, USA

Johannes Dietz
Tropical Silviculture, Institute of Silviculture, University of Göttingen, Büsgenweg 1, 37077 Göttingen, Germany and
Plant Ecology, Albrecht-von-Haller-Institute of Plant Sciences, University of Göttingen, Untere Karspüle 2, 37073 Göttingen, Germany

Stefan Erasmi
Institute of Geography, Department of Cartography, GIS and Remote Sensing, University of Göttingen, Goldschmidtstr. 5, 37077 Göttingen, Germany

Raffael Ernst
Department of Animal Ecology and Tropical Biology, Biocenter University of Würzburg, Am Hubland 97074 Würzburg, Germany

Ulrike Falk
Institute of Bioclimatology, University of Göttingen, Büsgenweg 2, Göttingen, Germany

Heiko Faust
Department of Cultural and Social Geography, Institute of Geography, University of Göttingen, Goldschmidtstr. 5, D-37077 Göttingen, Germany

Konrad Fiedler
Department of Population Ecology, University of Vienna, Althanstrasse 14, 1090 Vienna, Austria

Klaus Glenk
Junior Scientist Group "Economic Valuation of Ecosystem Functions and Services" and Environmental and Resource Economics, Department of Agricultural Economics and Rural Development, University of Göttingen, Platz der Göttinger Sieben 7, D-37073 Göttingen, Germany

S. Robbert Gradstein
Institute of Plant Sciences, University of Göttingen, Untere Karspüle 2, 37073 Göttingen, Germany

Gode Gravenhorst
Institute of Bioclimatology, University of Göttingen, Büsgenweg 2, Göttingen, Germany

Stefan Grote
Institute of Bioclimatology, University of Göttingen, Büsgenweg 2, Göttingen, Germany

Edi Guhardja
Kampus IPB Baranang Siang, Ex-Aula Kantor Pusat, Institut Pertanian Bogor, JL Raya Pajajaran, Bogor 16144, Indonesia (Jabar)

Handian Handi
Laboratory for Social Forestry, Faculty of Forestry, Institute Pertanian Bogor, Bogor, Indonesia

Marieke Harteveld
Plant Ecology, Albrecht-von-Haller-Institute of Plant Sciences, University of Göttingen, Untere Karspüle 2, 37073 Göttingen, Germany

Bettina Hedden-Dunkhorst
Federal Agency of Nature Conservation (BfN), Konstantinstr. 110, 53179 Bonn, Germany

Dietrich Hertel
Plant Ecology, Albrecht-von-Haller-Institute of Plant Sciences, University of Göttingen, Untere Karspüle 2, 37073 Göttingen, Germany

Nadine Hilt
Chair of Animal Ecology I, University of Bayreuth, 95440 Bayreuth, Germany

Patrick Höhn
Agroecology, University of Göttingen, Waldweg 26, D-37073 Göttingen, Germany

Dirk Hölscher
Tropical Silviculture, Institute of Silviculture, University of Göttingen, Büsgenweg 1, 37077 Göttingen, Germany

Andreas Ibrom
Institute of Bioclimatology, University of Göttingen, Büsgenweg 2, Göttingen, Germany and
Risø National Laboratory, Frederiksborgvej 399, DK-4000 Roskilde, Denmark

Tania June
Southeast Asian Ministers of Education Organization Regional Centre for Tropical Biology (SEAMEO BIOTROP), BTIC Building, Jl. Raya Tajur Km.6, Bogor, Indonesia and
Faculty of Mathematics and Natural Resources, Bogor Agricultural University/SEAMEO BIOTROP, Bogor, Indonesia

Martin Kappas
Institute of Geography, Department of Cartography, GIS and Remote Sensing, University of Göttingen, Goldschmidtstr. 5, 37077 Göttingen, Germany

Katja Kehlenbeck
Department of Crop Sciences: Tropical Crops, University of Göttingen, Grisebachstr. 6, D-37077 Göttingen, Germany

Alwin Keil
Institute of Agricultural Economics and Social Sciences in the Tropics and Subtropics, University of Hohenheim (490), D-70593 Stuttgart, Germany

Michael Kessler
Institute of Plant Sciences, University of Göttingen, Untere Karspüle 2, 37073
Göttingen, Germany

Alexandra-Maria Klein
Agroecology, University of Göttingen, Waldweg 26, D-37073 Göttingen, Germany

Heiner Kreilein
Institute of Bioclimatology, University of Göttingen, Büsgenweg 2, Göttingen, Germany

Werner Kreisel
Department of Cultural and Social Geography, Institute of Geography, University of Göttingen, Goldschmidtstr. 5, D-37077 Göttingen, Germany

William F. Laurance
Smithsonian Tropical Research Institute, Apartado 2072, Balboa, Republic of Panamá and
Biological Dynamics of Forest Fragments Project, National Institute for Amazonian Research (INPA), C.P. 478, Manaus, AM 69011-970, Brazil

Christoph Leuschner
Plant Ecology, Albrecht-von-Haller-Institute of Plant Sciences, University of Göttingen, Untere Karspüle 2, 37073 Göttingen, Germany

K. Eduard Linsenmair
Department of Animal Ecology and Tropical Biology, Biocenter University of Würzburg, Am Hubland 97074 Würzburg, Germany

Brigitte L. Maass
Department of Crop Sciences: Tropical Crops,
University of Göttingen, Grisebachstr. 6, D-37077 Göttingen, Germany

Adam Malik
Institute of Forest Management, Faculty of Agriculture, Bumi Tondo Campus, Tadulako University, Palu 94118, Indonesia

Sulaiman Mamar
Department of Anthropology, Universitas Tadulako, Kampus Bumi Tondo, Palu 94118, Indonesia

Rainer Marggraf
Environmental and Resource Economics, Department of Agricultural Economics and Rural Development, University of Göttingen, Platz der Göttinger Sieben 7,
D-37073 Göttingen, Germany

Arisbe Mendoza-Escalante
Center for Development Research (ZEF), University of Bonn, Walter-Flex-Str. 3, 53113 Bonn, Germany

Johannes Merklein
Institute of Bioclimatology, University of Göttingen, Büsgenweg 2, Göttingen, Germany

Sonja Migge-Kleian
Ecology Group, Institute of Zoology, University of Göttingen, Berliner Strasse 28, 37073 Göttingen, Germany

Roland Olschewski
Institute of Forest Economics, University of Göttingen, Büsgenweg 5, D-37077 Göttingen, Germany

Alexander Oltchev
Institute of Bioclimatology, University of Göttingen, Büsgenweg 2, Göttingen, Germany and A.N.Severtsov Institute of Ecology and Evolution of the Russiona Acadamy of Sciences (RAS), 33 Leninskiy Prosp., Moscow 119071, Russia

Elise Pendall
Department of Botany, 1000 E. University Ave., University of Wyoming, Laramie, WY 82071, USA

Ivette Perfecto
School of Natural Resources and Environment, Dana Building, University of Michigan, 440 Church St. Ann Arbor, MI 48109, USA

Stacy M. Philpott
Smithsonian Migratory Bird Center, National Zoological Park, 3001 Connecticut Avenue NW, Washington, DC 20008, USA

Ramadhanil Pitopang
Department of Forest Managament and Herbarium Celebense, Tadulako University, Palu, Indonesia

Catherine Potvin
Department of Biology, McGill University, 1205 Docteur Penfield, Montreal, Quebec H3A 1B1, Canada

Golam Rakkibu
Institute of Bioclimatology, University of Göttingen, Büsgenweg 2, Göttingen, Germany

Abdul Rauf
Tadulaku University (UNTAD), Kampus Bumi Tadulako Tondo, Palu - 94118, Indonesia

Mark-Oliver Rödel
Department of Animal Ecology and Tropical Biology, Biocenter University of Würzburg, Am Hubland 97074 Würzburg, Germany

Thomas Ross
Institute of Bioclimatology, University of Göttingen, Büsgenweg 2, Göttingen, Germany

Shahabuddin Saleh
Faculty of Agriculture, University of Tadulako, Kampus Bumi Tadulako Tondo, Palu-Indonesia

Klas Sander
Forest Economist, Environment Department, The World Bank, 1818 H Street NW, Washington DC, USA

Matthias Schaefer
Ecology Group, Institute of Zoology, University of Göttingen, Berliner Strasse 28, 37073 Göttingen, Germany

Björn Schippers
Department of Cultural and Social Geography, Institute of Geography, University of Göttingen, Goldschmidtstr. 5, D-37077 Göttingen, Germany

Wenke Schulz
Ecology Group, Institute of Zoology, University of Göttingen, Berliner Strasse 28, 37073 Göttingen, Germany

Christian H. Schulze
Department of Population Ecology, University of Vienna, Althanstrasse 14, 1090 Vienna, Austria

Stefan Schwarze
Institute of Rural Development, University of Göttingen, Waldweg 26, D-37073 Göttingen, Germany

Luitgard Schwendenmann
Tropical Silviculture, Institute of Silviculture, University of Göttingen, Büsgenweg 1, 37077 Göttingen, Germany

M.T.Felix Sitorus
Department of Communication and Community Development, Faculty of Human Ecology, Bogor Agricultural University, Indonesia

Lorena Soto-Pinto
El Colegio de la Frontera Sur, Unidad San Cristóbal, Carretera Panamericana y Periférico Sur, Apartado Postal 63. C.P. 29290, San Cristóbal de las Casas, Chiapas, México

Ingolf Steffan-Dewenter
Agroecology, University of Göttingen, Waldweg 26, D-37073 Göttingen, Germany

Andrea Steingrebe
Ecology Group, Institute of Zoology, University of Göttingen, Berliner Strasse 28, 37073 Göttingen, Germany

Leti Sundawati
Laboratory for Social Forestry, Faculty of Forestry, Institute Pertanian Bogor, Bogor, Indonesia

Endriatmo Sutarto
Department of Agricultural Socio-Economic Sciences, Institut Pertanian Bogor, Jl. Raya Pajajaran, Bogor 16143, Indonesia

Raquel Thomas
Iwokrama International Centre for Rain Forest Conservation and Development, 77 High Street, Kingston, Georgetown, Guyana

Teja Tscharntke
Agroecology, University of Göttingen, Waldweg 26, D-37073 Göttingen, Germany

André Twele
Institute of Geography, Department of Cartography, GIS and Remote Sensing, University of Göttingen, Goldschmidtstr. 5, 37077 Göttingen, Germany

Adhitya Wardhono
Institute of Rural Development, University of Göttingen, Waldweg 26, D-37073 Göttingen, Germany

Robert Weber
Department of Cultural and Social Geography, Institute of Geography, University of Göttingen, Goldschmidtstr. 5, D-37077 Göttingen, Germany

Maria Wiens
Plant Ecology, Albrecht-von-Haller-Institute of Plant Sciences, University of Göttingen, Untere Karspüle 2, 37073 Göttingen, Germany

Jan-Patrick Witte
Landscape Ecology, Institute of Geography, University of Göttingen, Goldschmidtstr. 5, D-37077 Göttingen, Germany and Environmental and Resource Economics, Department of Agricultural Economics and Rural Development, University of Göttingen, Platz der Göttinger Sieben 7, D-37073 Göttingen, Germany

Lars Woltmann
Ecology Group, Institute of Zoology, University of Göttingen, Berliner Strasse 28, 37073 Göttingen, Germany

Manfred Zeller
Institute of Agricultural Economics and Social Sciences in the Tropics and Subtropics, University of Hohenheim (490), D-70593 Stuttgart, Germany

The stability of tropical rainforest margins, linking ecological, economic and social constraints of land use and conservation – an introduction

Teja Tscharntke[1], Christoph Leuschner[2], Manfred Zeller[3], Edi Guhardja[4], and Arifuddin Bidin[5]

[1] Agroecology, University of Göttingen, Waldweg 26, D-37073 Göttingen, Germany
[2] Plant Ecology, Albrecht-von-Haller-Institute for Plant Sciences, University of Göttingen, Untere Karspüle 2, 37073 Göttingen, Germany
[3] Rural Development Theory and Policy, Institute of Agricultural Economics and Social Sciences in the Tropics and Subtropics, University of Hohenheim (490a), 70593 Stuttgart, Germany
[4] Kampus IPB Baranang Siang, Ex-Aula Kantor Pusat, Institut Pertanian Bogor, JL Raya Pajajaran, Bogor 16144, Indonesia (Jabar)
[5] Universitas Tadulako, Kampus Bumi Tondo, Palu 94118, Indonesia (Sulteng)

Tropical rainforests disappear at an alarming rate causing unprecedented losses in biodiversity and ecosystem services (Hughes et al. 1997, Noble & Dirzo 1997, Tilman et al. 2001, Achard et al. 2002) with Southeast Asia showing the highest rates of deforestation of any major tropical region (Sodhi et al. 2004). Despite an increased recognition of the value of these goods at national and international levels, rainforests continue to be seriously threatened by various forms of encroachments such as low-intensity harvesting of non-timber forest products by the rural poor, large-scale plantation forestry by the state or private actors, and the conversion of forested land by smallholder farmers. Transformation of ecosystems and changes in land use affect important ecosystem services and ultimatively human well-being (Robertson & Swinton 2005).

The stability of rainforest margin areas has been identified as a critical factor in the preservation of tropical forests (Müller & Zeller 2002, Maertens et al. 2006). These areas include the forest edge zone and the belt of agricultural land use systems, which is usually surrounding the forests. Stability has an ecological, social and economic dimension, and understanding the multidisciplinary nature of land use change in tropical rainforest margins is the key to identify more suitable development objectives, such as mitigation of climate change effects and biodiversity loss, poverty reduction, and economic development of rural areas (Vosti & Reardon 1997, Balmford et al. 2002, Totten et

Tscharntke T, Leuschner C, Zeller M, Guhardja E, Bidin A (eds), The stability of tropical rainforest margins, linking ecological, economic and social constraints of land use and conservation, Springer Berlin 2007, pp 1-8

al. 2003, Bawa et al. 2004, Steffan-Dewenter et al. 2006). Rainforest margins are far from stable with respect to social, economic and ecological properties. The challenge is to provide policy-relevant knowledge, based on an integrated scientific approach, to counteract destabilizing processes at different spatial and temporal scales, from the intra-household, household and village level to landscapes and regions.

The global human population is expected to increase further in the near future with the inevitable consequence of an expansion and intensification of agriculture and the associated destructive environmental impacts (Tilman et al. 2001, 2002, Foley et al. 2005). These impacts will particularly happen in regions where agricultural intensification is a relatively recent process such as in the humid tropics (Matson et al. 1997). Currently, too little is known about how to control this process and to identify trade-offs as well as synergies of land use and conservation (Robertson & Swinton 2005). Land use systems need management decisions based on the assessment of environmental costs and benefits and a thorough knowledge of ecosystem properties. These include the evaluation of local land-use practices in the forest (Sodhi et al. 2004), intensification in agroecosystems (Perfecto et al. 2005) and the assessment of alternative land-use options with concepts and data-driven models of large-scale land-use change in tropical, human-dominated landscapes (Laurance 2001). A particular challenge is to show how nature reserves contribute to the adaptive capacity of regional land-use matrices (du Toit et al. 2004) and to develop management schemes for tropical landscapes under the consideration of multiple ecosystem services, not just marketable commodities (Robertson & Swinton 2005).

This edited book provides a platform for scientists from socio-economic and natural sciences interested in the use and conservation of tropical rainforest resources. The research papers aim to contribute to an improved understanding of the processes that have stabilizing or destabilizing effects on ecological and socio-economic systems of tropical rain forest margins and include interdisciplinary papers that strive to integrate environmental, technological and socio-economic issues. The book features three interconnected thematic foci of interdisciplinary research, i.e. forest use and forest disturbance, agroforestry management and integrated land-use concepts.

Part 1: Ecological and socio-economic impacts of different forest use intensities

Forest use intensities greatly differ with respect to ecological and socio-economic benefits and costs. The income and utility derived from different types of forest use varies by socio-economic group differentiated by wealth, ethnicity, education, and other criteria. Responses of biodiversity and ecosystems functions greatly differ with organism group considered, making generalizations difficult.

William F. Laurance summarizes key findings of the Biological Dynamics of Forest Fragments Project in Amazonia, the world's largest and longest-running experimental study on habitat fragmentation. Edge effects play a key role in fragment dynamics and the surrounding matrix has a major influence on fragment connectivity and functioning. Many Amazonian species avoid even small clearings.

Konrad Fiedler et al. focus on tropical forest moths and their response to forest disturbance and recovery. Using large data sets from Borneo and Ecuador, they found strong responses of species composition to disturbance, but not always of species diversity. Beta, rather than alpha, diversity appeared to be meaningful to assess land use impacts.

Raffael Ernst et al. present results on amphibian communities in disturbed forests of the Neo- and Afrotropics. Patterns in community composition, and hence beta diversity, is a key in understanding impacts of human disturbance. These authors argue that alterations in the functional diversity of amphibians are a good predictor of habitat change.

Dietrich Hertel et al. address the effects of forest use and forest conversion on the below-ground compartment in the wet tropical forests of Sulawesi (Indonesia). The authors review the response of fine root biomass to forest disturbance intensity and interpret results with respect to carbon storage in the root system.

Luitgard Schwendenmann et al. present data on how forest conversion affects soil carbon pools, carbon mineralization rates and soil respiration in Central Panama. The active soil organic carbon pool was found to be a sensitive indicator of soil respiration and may indicate land-use changes.

Johannes Dietz et al. analyze how low-intensity forest use in Sulawesi (Indonesia) influences forest structure. They found a higher interception loss in taller stands. Forest management, even at low or moderate intensities, had a significant impact on forest structural parameters and associated ecosystem functioning such as rainfall partitioning in the canopy.

Sonja Migge-Kleian et al. review the driving factors for soil and litter arthropod communities that are influenced by land-use intensification in rainforest margins. Changes in soil/litter moisture and temperature become very important, leading to reduced richness in drier environments. In an experimental study in Sulawesi (Indonesia), the diversity of soil-dwelling species showed a variable response to land-use changes. Disturbance-mediated changes in diversity also include functional changes of the soil/litter communities.

Melani Abdulkadir-Sunito and M. T. Felix Sitorus relate land-use and forest encroachment patterns to ethnicity in Sulawesi (Indonesia). This case study shows social constraints of inter-ethnic relations and land-use systems. Forest encroachment is not only perceived as an economic, but also as an ethno-political action.

Barkmann et al. assess economic preferences for biodiversity and ecosystems services, including the endemic dwarf buffalo "anoa", the preponderance of shading trees in cacao plantations, and the availability of water and rattan.

The local demand for these environmental non-market services, established by choice experiments, does not suffice to compensate financial benefits from rainforest conversion or agroforestry intensification that accrue to individual households.

Stefan Schwarze et al. quantify the contribution of forest products to household incomes in Sulawesi (Indonesia) and investigate links between poverty, livelihood systems and extraction of forest products. Forest products, especially rattan, were a main source of income for the poorest households, while a village agreement could be shown to have a strong negative influence on the likelihood of selling forest products.

Part 2: Sustainable management of agroforestry systems

Most tropical landscapes are highly fragmented and harbor a mosaic of different land use systems. Low-intensity agroforestry may support high biodiversity stabilizing ecosystem functioning, in particular when shaded by natural trees and neighbored by natural forest (Tscharntke et al. 2005). In contrast, high-intensity agroforestry with mono-dominant, planted shade trees and in a monotonous landscape matrix may be characterized by less environmental benefits and high agrochemical inputs. In this second part of the book, the ecological and socio-economic benefits of different agroforest management practices will be compared and related to patterns and processes in natural forests.

Ivette Perfecto et al. evaluate the contribution of traditionally grown shade coffee to the social and ecological stability of Northern Latin America. These agroforestry systems maintain a high biodiversity that in turn contributes to agroecosystem functioning. Coffee certification programs help to sustain important ecological functions such as biological control and carbon sequestration.

Roland Olschewski et al. present an example of how economic valuations of ecosystem services may contribute to the stability of rainforest margins. In a study of coffee bee pollination and pest control from Sulawesi (Indonesia), they quantify the dependence of the farmers' net revenues on the distance of agroforests to natural forests, which provide nesting sites for bees. The promotion of certified "biodiversity-friendly" coffee may enhance rainforest preservation within a mosaic of competing land-use options.

Merijn Bos et al. provide data on the response of insect diversity to rainforest conversion in Sulawesi (Indonesia). Shaded agroforestry systems may still support high diversity levels of beetles, ants, bees, wasps, and dung beetles, and agroforests should be included in conservation plans at the landscape level.

Katja Kehlenbeck et al. describe the plant diversity in the traditional, complex Indonesian homegardens, which are important for subsistence production in rural areas. Homegardens are found to integrate socioeconomic and ecological advantages.

S. Robbert Gradstein et al. studied in Sulawesi (Indonesia) how tree diversity is related to human land use. Roughly one third of tree species in the rainforest (51-63 per 0.25 ha) were important as commercial timber trees, while tree diversity was little affected by moderate land use, but greatly reduced in cacao agroforestry.

Jan Börner et al. evaluate alternatives to slash-and-burn in forest-based fallow systems of eastern Amazonia. These include mulching for the management of soil organic matter and fire-free land clearing with bush choppers and enrichment plantings using fast-growing leguminous tree species. However, costs of mulching are high and taxes are promising policy options to promote fire-free land preparation techniques.

Klas Sander and Manfred Zeller conclude from a case study in Madagascar that the poorest households suffer most from a strict forest conservation approach, while better-off households with more irrigated agricultural land benefit more due to an improved provision of indirect forest services, particularly watershed protection. Furthermore, they discuss the implementation of incentives for afforestation at the household level and management strategies to meet the increasing direct demand on forest resources.

Part 3: Integrated concepts of land use in tropical forest margins

Rain forest margin areas comprise a wealth of land-use systems with different levels of agricultural intensification, from simplified annual crops to highly complex agroforestry systems (Schulze et al. 2005). This part of the book aims to contribute to a better understanding of the dynamics of land use change and related changes in resource availability under various policy scenarios to create improved strategies to reduce and potentially reverse degradation processes.

Alwin Keil et al. analyze the potentials to reduce deforestation by enhancing the technical efficiency of crop production in agroecosystems of the forest margin area. Based on data from Sulawesi (Indonesia), the authors conclude that improvement of technical efficiency may reduce deforestation by increasing incomes on already converted land, but there is also the need to control the influx of migrants attracted by the profitability of higher productivity.

Robert Weber et al. relate ethnicity and migration as cultural impact factors to land-use change in rainforest margins of Sulawesi (Indonesia). The results show a change from a food-first to a cash crop-first strategy, driven by a distinct, immigrating ethnic group (the Bugis) introducing improved knowledge of agricultural intensification.

Stefan Erasmi et al. discuss the state of the art of data analyses for the assessment of land-cover changes in the tropics. Satellite image-based land cover classifications uncover a large inconsistency regarding land cover. The authors suggest improvements for better technical and conceptual monitoring systems.

Andreas Ibrom et al. apply a series of models to predict net primary production from regional meteorological and remote sensing data of a rainforest margin area of Sulawesi (Indonesia). Scenarios suggest that carbon dioxide sequestration and total evaporation increase and sensible heat fluxes and surface temperatures decrease with land-use intensification.

Diogenes S. Alves analyzes deforestation processes in the Brazilian Amazon. Based on the results of four major research programs, he concludes that the tasks of reducing and mitigating deforestation impacts and fostering sustainable land use are not to be engineered but, rather, negotiated, which is a major challenge for science and technology.

Most chapters of this book were presented at an international symposium held at the Georg-August-University of Goettingen in September 2005. The editors thank the four coordinators of the Collaborative Research Center SFB 552 "STORMA" ("Stability of Rainforest Margins in Indonesia"), Daniel Stietenroth, Wolfram Lorenz, Surya Tarigan and Adam Malik, for their invaluable support, Jan Barkmann for the handling of several manuscripts and many peers for their thoughtful reviews. Technical assistance during the editorial process was provided by Stella Aspelmeier. The financial support by the German Science Foundation (the Deutsche Forschungsgemeinschaft DFG) is gratefully acknowledged.

References

Achard F, Eva HD, Stibig H-J, Mayaux P, Gallego J, Richards T, Malingreau J-P (2002) Determination of deforestation rates of the world's humid tropical forests. Science 297: 999-1002

Balmford A, Bruner A, Cooper P, Costanza R, Farber S, Green RE, Jenkins M, Jefferiss P, Jessamy V, Madden J, Munro K, Myers N, Naeem S, Paavola J, Rayment M, Rosendo S, Roughgarden J, Trumper K, Turner RK (2002) Economic reasons for conserving wild nature. Science 297: 950-953

Bawa KS, Kress WJ, Nadkarni NM, Lele S, Raven PH, Janzen DH, Lugo AE, Ashton PS, Lovejoy TE (2004) Tropical Ecosystems into the 21st century. Science 306: 227-228

du Toit JT, Walker BH, Campbell B M (2004) Conserving tropical nature: current challenges for ecologists. Trends in Ecology and Evolution 19: 12-17

Foley JA, DeFries R, Asner GP, Barford C, Bonan G, Carpenter SR, Chapin FS, Coe MT, Daily GC, Gibbs HK, Helkowski JH, Holloway T, Howard

EA, Kucharik CJ, Monfreda C, Patz JA, Prentice IC, Ramankutty N, Snyder PK (2005) Global consequences of land use. Science 309: 570-574

Hughes JB, Daily GC, Ehrlich PR, (1997) Population diversity: Its extent and extinction. Science 278(5338): 689-692

Laurance WF (2001) Future shock: forecasting a grim fate for the earth. Trends in Ecology and Evolution 16: 531-533

Maertens M, Zeller M, Birner R (2006) Sustainable agricultural intensification in forest frontier areas. Agricultural Economics 34 (2): 197-206

Matson PA, Parton WJ, Power AG, Swift MJ (1997) Agricultural intensification and ecosystem properties. Science 277: 504-509

Müller D, Zeller M (2002) Land use dynamics in the Central Highlands of Vietnam: a spatial model combining village survey data and satellite imagery data. Agricultural Economics 27 (3): 333-354

Noble IR, Dirzo R (1997) Forests as human-dominated ecosystems. Science 277: 522-525

Perfecto I, Vandermeer J, Mas A, Pinto LS (2005) Biodiversity, yield, and shade coffee certification. Ecological Economics 54: 435-446

Robertson GP, Swinton SM (2005) Reconciling agricultural productivity and environmental integrity: a grand challenge for agriculture. Frontiers in Ecology and the Environment 3: 38-46

Schulze CH, Waltert M, Kessler PJA, Pitopang R, Shahabuddin, Veddeler D, Mühlenberg M, Gradstein SR, Leuschner C, Steffan-Dewenter I, Tscharntke T (2004) Biodiversity indicator groups of tropical land-use systems: comparing plants, birds, and insects. Ecological Applications 14: 1321-1333

Sodhi NJ, Koh LP, Brook BW, Ng PKL (2004) Southeast Asian biodiversity: an impending disaster. Trends in Ecology and Evolution 19: 654-660

Steffan-Dewenter I, Kessler M, Barkmann J, Bos M, Buchori D, Erasmi S, Faust H, Gerold G, Glenk K, Gradstein SR, Guhardja E, Harteveld M, Hertel D, Höhn P, Kappas M, Köhler S, Leuschner C, Maertens M, Marggraf R, Migge-Kleian S, Mogea J, Pitopang R, Schaefer M, Schwarze S, Sporn SG, Steingrebe A, Tjitrosoedirdjo SS, Tjitrosoemito S, Twele A, Weber R, Woltmann L, Zeller M, Tscharntke T (2006) Socioeconomic context and ecological consequences of rainforest conversion and agricultural intensification. (submitted)

Tilman D, Cassman KG, Matson PA, Naylor R, Polasky S (2002) Agricultural sustainability and intensive production practices. Nature 418: 671-677

Tilman D, Fargione J, Wolff B, D'Antonio C, Dobson A, Howarth R, Schindler D, Schlesinger WH, Simberloff D, Swackhamer D (2001) Forecasting agriculturally driven global environmental change. Science 292: 281-284

Totten M, Pandya SI, Janson-Smith T (2003) Biodiversity, climate, and the Kyoto protocol: risks and opportunities. Frontiers in Ecology and Environment 1: 262-270

Tscharntke T, Klein AM, Kruess A, Steffan-Dewenter I, Thies C (2005) Landscape perspectives on agricultural intensification and biodiversity-ecosystem service management. Ecology Letters 8: 857-874

Vosti SA, Reardon T (eds) (1997) Sustainability, growth and poverty alleviation: a policy and agroecological perspective. The John Hopkins University Press, Baltimore and London

Part I

Ecological and socio-economic impacts of different forest use intensities

Ecosystem decay of Amazonian forest fragments: implications for conservation

William F. Laurance[1,2]

[1] Smithsonian Tropical Research Institute, Apartado 2072, Balboa, Republic of Panamá Email: laurancew@si.edu
[2] Biological Dynamics of Forest Fragments Project, National Institute for Amazonian Research (INPA), C.P. 478, Manaus, AM 69011-970, Brazil

Summary

I summarize key findings from the Biological Dynamics of Forest Fragments Project, the world's largest and longest-running experimental study of habitat fragmentation. Although initially designed to assess the influence of fragment area on Amazonian biotas, the project has yielded insights that go far beyond the original scope of the study. Results suggest that edge effects play a key role in fragment dynamics, that the matrix has a major influence on fragment connectivity and functioning, and that many Amazonian species avoid even small (<100 m wide) clearings. The effects of fragmentation are highly eclectic, altering species richness and abundances, species invasions, forest dynamics, the trophic structure of communities, and a variety of ecological and ecosystem processes. Moreover, forest fragmentation appears to interact synergistically with ecological changes such as hunting, fires, and logging, collectively posing an even greater threat to the rainforest biota.

Keywords: Amazon, edge effects, forest dynamics, habitat fragmentation, hunting, logging, microhabitat changes, rainforest, surface fires.

1 Introduction

The Amazon basin contains over half of Earth's remaining tropical rainforests and is experiencing high rates of deforestation that have accelerated substantially over the past 15 years (Laurance et al. 2001a, 2004, INPE 2005). Because rapid forest conversion is causing widespread habitat fragmentation (Skole and Tucker 1993, Laurance 1998, 2005a), the fates of many Amazonian species will ultimately depend on their capacity to persist in fragmented landscapes or isolated nature reserves.

Tscharntke T, Leuschner C, Zeller M, Guhardja E, Bidin A (eds), The stability of tropical rainforest margins, linking ecological, economic and social constraints of land use and conservation, Springer Berlin 2007, pp 11-37

The Biological Dynamics of Forest Fragments Project (BDFFP) was initiated in 1979 as a large-scale experiment to assess the effects of fragmentation on Amazonian biotas (Lovejoy et al. 1983, 1986, Bierregaard et al. 1992, Laurance et al. 2002). It is the world's largest and longest-running experimental study of habitat fragmentation (cf. Debinski and Holt 2000). Originally, the project's main goals were to assess the influence of fragment size on Amazonian animal and plant communities, to identify a minimum critical size for rainforest reserves, and to help resolve the heated SLOSS (single large versus several small reserves) debate (e.g., Simberloff and Abele 1976, Wilcox and Murphy 1985). Over time, however, many additional research aims have been added as new insights have developed.

A key feature of the BDFFP is that standardized abundance data were collected for trees, understory birds, mammals, amphibians, and various invertebrate groups prior to experimental isolation of the forest fragments. This permitted a far more rigorous assessment of fragmentation effects than would have been possible using only comparisons of fragmented versus intact forest. In addition, the long-term nature of the BDFFP and its synthetic approach, integrating studies of many taxa and numerous ecological and ecosystem processes, have provided insights that are impossible in most other fragmentation studies.

Here I synthesize key BDFFP findings from the past 26 years, based on a survey of over 400 publications and theses, and I also highlight their implications for forest conservation. The first part of this review focuses on extrinsic factors that influence fragment biotas—particularly area, edge, matrix, isolation, and sample effects. The second part identifies key community- and ecosystem-level effects of fragmentation on tropical forests.

1.1 Study area

The 1000-km^2 study site is located 80 km north of Manaus, Brazil in central Amazonia (2° 30' S, 60° W) at 50-100 m elevation (Figure 1). Local soils are nutrient-poor sandy or clay-rich ferrasols, which are widespread in the Amazon basin (Brown 1987). As is typical of the basin, the topography is relatively flat but dissected by many stream gullies. Rainfall ranges from 1,900-3,500 mm annually with a dry season from June to October (Laurance 2001). The climate regime is intermediate between that of the very wet western Amazon and drier, more-seasonal areas in the southeastern and north-central basin. The forest canopy is 30-37 m tall, with emergents reaching to 55 m. Species richness of trees is very high and can exceed 280 species ($\geq$10 cm diameter) per ha (Oliveira and Mori 1999).

The study area is surrounded by large expanses (>200 km) of continuous forest to the west, north, and east. In the early 1980s, five 1-ha fragments, four 10-ha fragments, and two 100-ha fragments were isolated by distances of 80-650 m from surrounding forest by clearing the intervening vegetation to establish cattle pastures. Fragments were fenced to prevent encroachment

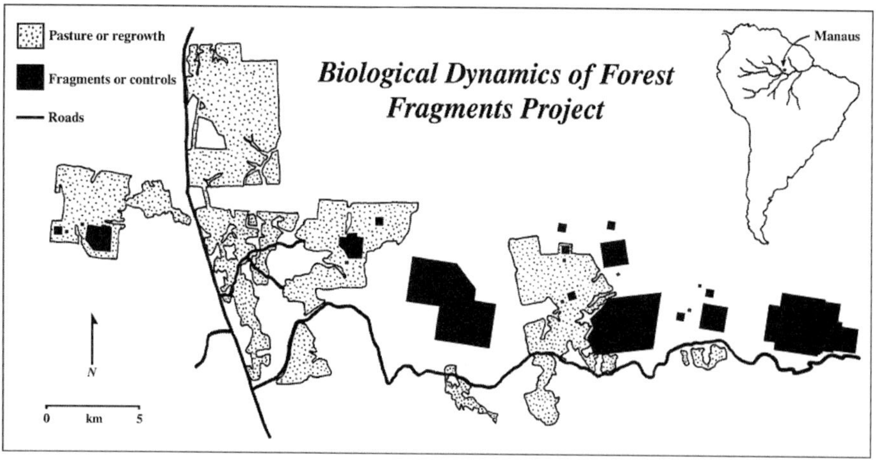

Fig. 1. The BDFFP study area in central Amazonia, showing locations of forest fragments and control sites in intact forest.

by cattle. Twelve reserves ranging from 1-1000 ha in area (three of 1-ha, four of 10 ha, two of 100 ha, and three of 1000 ha) were delineated in nearby continuous forest to serve as experimental controls. Because of low pasture productivity, the ranches were gradually abandoned, and 3-15 m tall secondary forests (dominated by *Cecropia* spp. or *Vismia* spp.) proliferated in many formerly cleared areas. To help maintain fragment isolation, 100 m-wide strips of regrowth were cleared and burned around each fragment on 2-3 occasions. Detailed descriptions of the study area, including the history of each fragment and its surrounding vegetation, are provided elsewhere (Lovejoy et al. 1986, Bierregaard and Stouffer 1997).

2 Extrinsic Factors Affecting Fragment Biotas

2.1 Sample effects

Forest fragments contain a limited subset of any regional biota, in part because small patches inevitably sample fewer species and less habitat diversity than do larger patches (e.g., Wilcox and Murphy 1985, Haila et al. 1993). Results from the BDFFP suggest that such sample effects could be especially important for Amazonian species, which often have patchy distributions at varying spatial scales and complex patterns of endemism (e.g., Zimmerman and Bierregaard 1986, Vasconcelos 1988, Allmon 1991, Rankin-de Merona et al. 1992, Souza and Brown 1994, Didham et al. 1998a, Laurance et al. 1998a, Peres 2005). Pronounced clumping means many species will be missing from any particular fragment or reserve simply because they never occurred there in the first place.

Another key factor is that, in tropical rainforests, most species are locally rare throughout all or much of their geographic range (Hubbell and Foster 1986, Pittman et al. 1999). The acidic, nutrient-poor soils prevalent in much of Amazonia (Brown 1987) appear to promote animal rarity by limiting fruit and flower production and reducing the nutrient content of foliage (reviewed in Laurance 2001). As a result, many invertebrates (Vasconcelos 1988, Becker et al. 1991) and vertebrates (Emmons 1984, Rylands and Keuroghlian 1988, Kalko 1998, Stouffer and Bierregaaard 1995a, Spironello 2001) are considerably less abundant in forests overlaying nutrient-poor Amazonian soils than they are in more-productive areas of the Neotropics. Intrinsic rarity is a critical feature, as demonstrated by studies of Amazonian trees. Even if a species is present when a fragment is initially isolated, its population may be so small that it has little chance of persisting in the long term (Laurance et al. 1998a).

2.2 Area effects

As is generally expected based on other investigations in the tropics (e.g. Laurance and Bierregaard 1997 and references therein, Harcourt and Doherty 2005), BDFFP researchers have often found that species richness is positively correlated with fragment size, and that intact forest contains more species per unit area than fragments (e.g. Figure 2). This arises because many large mammals (Lovejoy et al. 1986), primates (Rylands and Keuroghlian 1988, Schwartzkopf and Rylands 1989, Gilbert and Setz 2001), understory birds (Stouffer and Bierregaard 1995b, Stratford and Stouffer 1999, Ferraz et al. 2003), and even certain beetle, ant, bee, termite, and butterfly species (Powell and Powell 1987, Vasconcelos 1988, Klein 1989, Souza and Brown 1994, Brown and Hutchings 1997, Didham 1997a) are highly sensitive to fragment area. A number of these species have disappeared from even the largest (100 ha) fragments in the study area.

The prediction that extinction rates will be negatively correlated with fragment area (MacArthur and Wilson 1967) is also supported by the BDFFP results. Once isolated, small (1-10 ha) fragments initially lose species at a remarkably high rate; for example, dung and carrion beetle assemblages were markedly altered only 2-6 years after fragment isolation (Klein 1989). Local extinctions of birds (Harper 1989, Stouffer and Bierregaard 1995b, Stratford and Stouffer 1999), primates (Lovejoy et al. 1986, Schwartzkopf and Rylands 1989, Gilbert and Setz 2001), and butterflies (Brown and Hutchings 1997) have also occurred more rapidly in small (1-10 ha) than in large (100 ha) fragments.

In contrast, a few taxa have remained stable or even increased in species richness after fragment isolation. Frog richness increased because of an apparent resilience of most rainforest frogs to area and edge effects and an influx of non-rainforest species from the surrounding matrix (Gascon 1993, Tocher et al. 1997). Butterfly richness also rose after fragment isolation, largely from an invasion of generalist matrix species at the expense of forest-interior butterflies

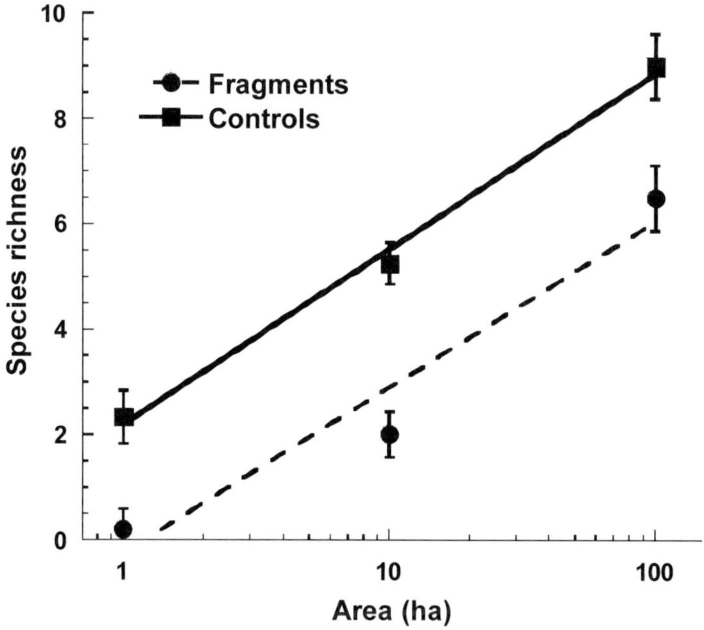

Fig. 2. Species-area relationships for nine species of terrestrial insectivorous birds (mean ± S.E.) in the BDFFP study area. Regression lines are fitted separately for fragments (R^2=94.3%) and control sites (R^2=99.4%) (after Stratford and Stouffer 1999).

(Brown and Hutchings 1997). Small mammal richness has not declined in the BDFFP fragments because most species readily use edge and regrowth habitats (Malcolm 1997). Collectively, BDFFP results reveal that the responses of different species and taxonomic groups to fragmentation are highly individualistic, and suggest that species with small area needs that tolerate matrix and edge habitats are the least vulnerable (e.g., Offerman et al. 1995, Stouffer and Bierregaard 1995b, Didham et al. 1998a, Gascon et al. 1999).

2.3 Edge effects

Edge effects can be a major structuring force in fragmented ecosystems (e.g. Laurance 1997, Didham 1997a, Ries et al. 2004). The BDFFP has helped to reveal the remarkable diversity of edge effects in fragmented rainforests, which alter physical gradients, species distributions, and many ecological and ecosystem processes (Figure 3). Microclimatic changes near edges, such as reduced humidity, increased light, and greater temperature variability, penetrate up to 60 m into fragment interiors (Kapos 1989) and can negatively affect species adapted for humid, dark forest interiors (Lovejoy et al. 1986, Benitez-Malvido 1998). Leaf litter accumulates near edges (Carvalho and Vasconcelos 1999,

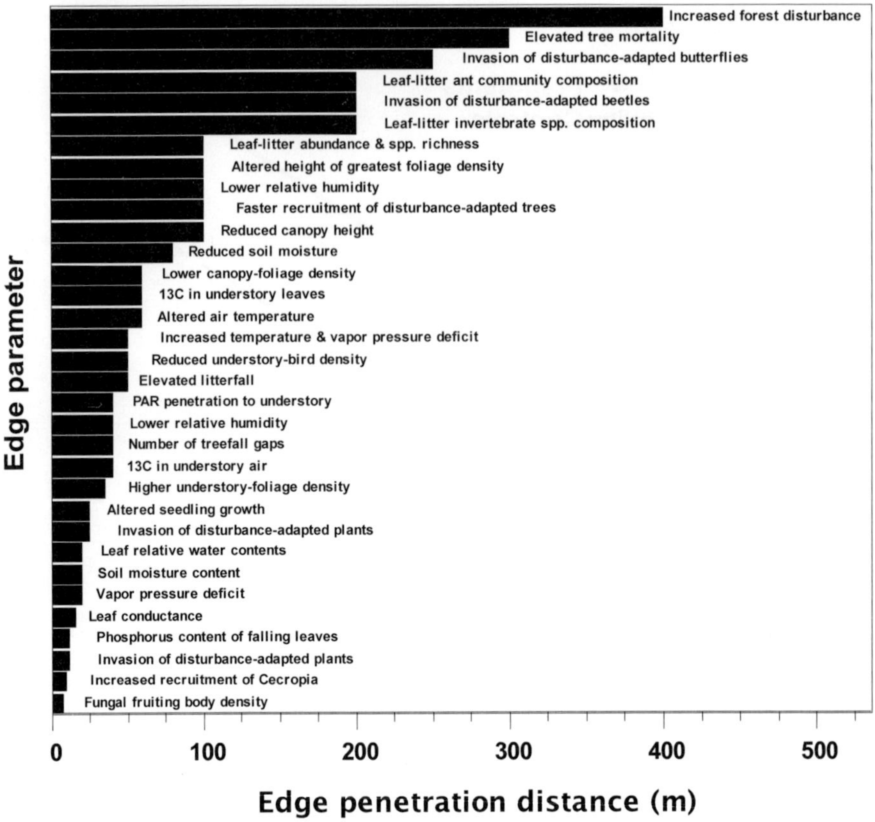

Fig. 3. Penetration distances of different edge effects into the BDFFP forest remnants.

Didham and Lawton 1999) because drought-stressed trees shed leaves and possibly because drier edge conditions slow litter decomposition (Kapos et al. 1993, Didham 1998, Vasconcelos and Laurance 2005). Accumulating litter may negatively affect seed germination (Bruna 1999) and seedling survival (Scariot 2001) and makes forest edges vulnerable to surface fires during droughts (Cochrane et al. 1999).

One of the most striking edge effects is a sharp increase in rates of tree mortality and damage (Ferreira and Laurance 1997, Laurance et al. 1998b). When an edge is created, some trees simply drop their leaves and die standing (Lovejoy et al. 1986), apparently because abrupt changes in light, temperature, or moisture exceed their physiological tolerances. Other trees are snapped or felled by winds, which accelerate over cleared land and then strike forest edges, creating strong turbulence (Laurance 1997). Finally, lianas (woody vines)—important structural parasites that reduce tree growth, survival, and reproduction—increase markedly near edges and may further elevate tree mortality (Laurance et al. 2001b).

The abrupt rise in tree mortality fundamentally alters canopy-gap dynamics (Ferreira and Laurance 1997, Laurance et al. 1998b), which can influence forest structure, composition, and diversity (Brokaw 1985, Hubbell and Foster 1986, Denslow 1987). Smaller fragments often become hyper-disturbed, leading to progressive changes in floristic composition. New trees regenerating within 100 m of forest edges are significantly biased toward disturbance-loving pioneer and secondary species and against old-growth, forest-interior species (Laurance et al. 1998c, in press). The pioneer tree *Cecropia sciadophylla*, for example, has increased 33-fold in density since the BDFFP fragments were isolated (Laurance et al. 2001b).

Some animals respond positively to edges. Certain termites, leafhoppers, scale insects, aphids, aphid-tending ants (Fowler et al. 1993), and light-loving butterflies (Brown and Hutchings 1997) increase near edges. Birds that forage in treefall gaps, such as some arboreal insectivores, hummingbirds, and habitat generalists, often become abundant near edges (Bierregaard and Lovejoy 1989, Bierregaard 1990, Stouffer and Bierregaard 1995a, 1995b). Frugivorous bats increase near edges, probably because such areas have higher fruit abundance than forest interiors (Kalko 1998). The insectivorous marsupial *Metachirus nudicaudatus* apparently increased in fragments because dead trees and ground cover, which provide favored foraging microhabitats, increased near edges (Malcolm 1991).

Many other animal species respond negatively to edges and thus are likely to be vulnerable to fragmentation. Numerous flies, bees, wasps (Fowler et al. 1993), beetles (Didham et al. 1998a, 1998b), ants (Carvalho and Vasconcelos 1999), and butterflies (Brown and Hutchings 1997) decline in abundance near edges. A number of insectivorous understory birds avoid edges (Quintela 1985), particularly solitary species, obligatory ant followers, and those that forage in mixed-species flocks (S.G. Laurance 2004). Some frog species use breeding habitat independent of its proximity to edges (Gascon 1993), whereby others may be edge avoiders (e.g., Pearman 1997).

Edge effects in fragmented forests are evidently additive, whereby forest adjoined by two or more nearby edges suffers greater edge effects than does forest adjoined by just a single edge (Malcolm 1994, 1998, Ries et al. 2004, Fletcher 2005). In the BDFFP study area, an additive-edge model better predicts structural changes to forest fragments than does a single-edge model (Malcolm 1994). In addition, the population density of disturbance-adapted successional trees increased far more in edge plots with four nearby edges (658±850%) than in those with two (264±353%) or one (129±225%) nearby edges (nearby edges were defined as those within 100 m of the plot center). Species richness of successional trees and stand-level tree mortality were also much (ca. 50-120%) higher in plots with 2-4 nearby edges than in those with a single nearby edge (Laurance et al. 2006).

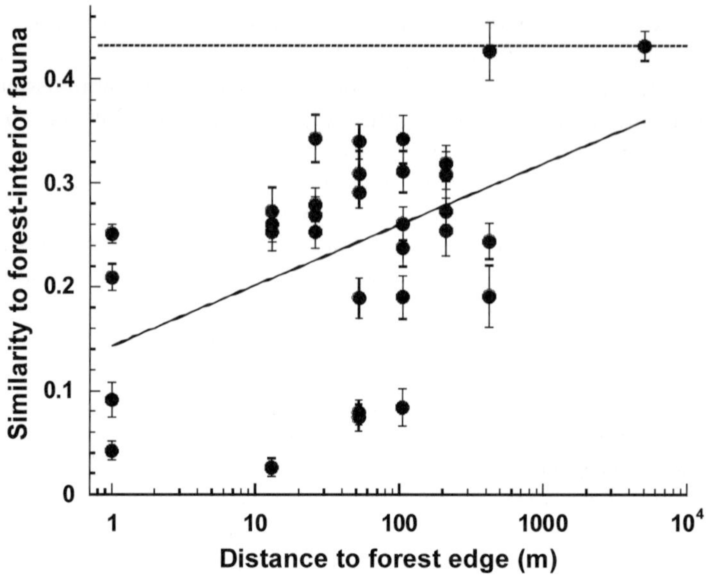

Fig. 4. Changes in the composition of leaf-litter beetle assemblages as a function of distance from forest edge. For each sample, the mean percentage similarity (± SE) to forest-interior samples (ca. 5000 m from edge) is shown. Dotted line shows the average background level of similarity among different forest-interior samples. The regression was highly significant (R^2=23.2%, p=0.005) (after Didham 1997b).

The most striking edge effects in the BDFFP study area occur within 100 m of forest edges (Figure 3). However, wind damage to forests can penetrate up to 300-400 m from edges (Laurance et al. 1998b, 2000, Lewis 1998), and changes in beetle, ant, and butterfly communities can be detected as far as 200-400 m from edges (Figure 4, Brown and Hutchings 1997, Didham 1997b, Carvalho and Vasconcelos 1999). Notably, some edge effects occur over even larger spatial scales in more-seasonal areas of the Amazon: ground fires in two fragmented landscapes of eastern Amazonia were sharply elevated in frequency within at least 2400 m of forest edges (Cochrane and Laurance 2002).

2.4 Edge evolution

Another important finding is that rapid changes in the physical permeability of edges occur in the initial years after fragmentation. Newly created edges are structurally open and thereby permeable to lateral light penetration and hot, dry winds from adjoining cattle pastures. After a few years, these microclimatic alterations decline in intensity as edges are partially sealed by a profusion of secondary growth (Kapos 1989, Camargo and Kapos 1995, Kapos et al. 1997). Desiccation-related plant mortality may also decline over time because of an increase in drought-tolerant species or physiological acclimation of plants near edges. Unlike microclimatic changes, however, wind damage to forests is unlikely to lessen as fragment edges become older and less permeable because downwind turbulence usually increases as edge permeability is reduced (Savill 1983). In terms of edge permeability, three phases of edge evolution can be identified: initial isolation, edge-closure, and post-closure.

In the initial isolation phase (<1 year after edge formation), the gradient between the forest interior and edge is steepest, with hot, dry conditions and increased light and wind penetrating into the fragment. There is a dramatic pulse in tree mortality; many trees die standing (Laurance et al. 1998b). Leaf-litter accumulates as drought-stressed trees shed leaves to conserve water, or replace shade-adapted leaves with sun-adapted leaves (Didham 1998). Abundances of many animals fluctuate sharply. The most sensitive species decline almost immediately.

During the edge-closure phase (1-5 years after edge formation), a proliferation of secondary vegetation and lateral branching by edge trees progressively seals the edge. Edge gradients in microclimate become more complex but do not disappear entirely (Kapos et al. 1997). Plants near the edge die or become physiologically acclimated to edge conditions. Treefall gaps proliferate within the first 100-300 m of edges, partly as a result of increased windthrow. Additional animal species disappear from fragments. Edge-favoring plants and animals sometimes increase dramatically in abundance (Laurance and Bierregaard 1997).

In the post-closure phase (>5 years after edge formation), edge-related changes are largely stabilized, although external land-use changes (such as fires or the development of adjoining regrowth) can disrupt this equilibrium (Gascon et al. 2000). Windthrow remains elevated near edges, despite the fact that the edge is partially sealed by secondary growth. Proliferating lianas near edges probably contribute to increased tree mortality. Turnover rates of trees increase near edges because of elevated tree mortality and recruitment, and increasing numbers of short-lived pioneer species. Pioneer plants have rapid leaf replacement, contributing to the accumulation of leaf litter near edges. Although edge closure occurs relatively quickly in tropical rainforests because of rapid plant growth, edges are still more dynamic and vulnerable to climatic vicissitudes than are forest interiors (Laurance et al. 2002).

2.5 Matrix effects

An increasing body of evidence suggests that the matrix of modified land surrounding forest fragments can have manifold effects on fragment ecology (e.g. Gustafson and Gardner 1996, Weins 1997, Bender and Fahrig 2005). For example, fragments surrounded by 5-10 m-tall regrowth forest experienced less-intensive changes in microclimate (Didham and Lawton 1999) and had lower edge-related tree mortality (Mesquita et al. 1999) than did similar fragments adjoined by cattle pastures. Edge avoidance by mixed-species bird flocks was also reduced when fragments were surrounded by regrowth rather than cattle pastures (Stouffer and Bierregaard 1995b). Patterns of tree regeneration in the BDFFP fragments are strongly influenced by the species of secondary trees proliferating in the nearby matrix, which are evidently producing a large seed rain into the fragments (Nascimento et al. 2006).

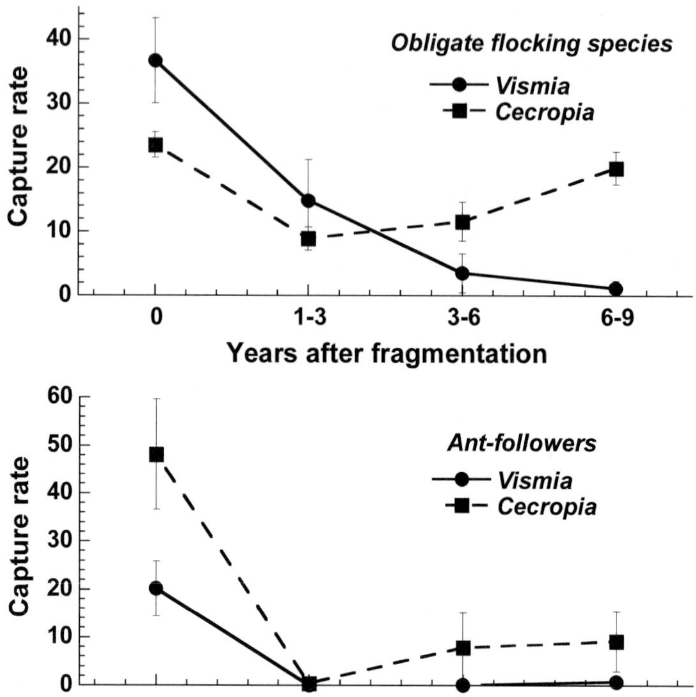

Fig. 5. Changes in capture rates (mean ± SE captures/1000 mistnet hours) over time for two guilds of rainforest birds in 10-ha forest fragments that gradually became surrounded by *Vismia*-dominated and *Cecropia*-dominated regrowth (after Stouffer and Bierregaard 1995b).

Of even more significance is that the matrix influences fragment connectivity (Weins 1997, Bender and Fahrig 2005). Several species of primates (Gilbert and Setz 2001), antbirds, obligate flocking birds (Figure 5, Stouffer and Bierregaard 1995b), and euglossine bees (Becker et al. 1991) that disappeared soon after fragment isolation recolonized fragments when regrowth regenerated in the surrounding landscape. Among rainforest frogs, birds, small mammals, and bats, matrix-avoiding species were much more likely to decline or disappear in the BDFFP fragments than were those that use the matrix (Offerman et al. 1995, Stouffer and Bierregaard 1995a, 1995b, Kalko 1998, Gascon et al.

1999, Borges and Stouffer 1999, Stratford and Stouffer 1999, S.G. Laurance and Gomez 2005).

Some matrix habitats are more suitable for rainforest fauna than others. Regrowth dominated by *Cecropia* trees, which tends to be tall and floristically diverse with a relatively closed canopy (Williamson et al. 1998), is used by more rainforest bird, frog, and ant species than is more-open *Vismia*-dominated regrowth (Stouffer and Bierregaard 1995b, Tocher 1998, Borges and Stouffer 1999, Vasconcelos 1999, Stouffer and Borges 2001). Virtually any kind of regrowth is better than cattle pastures; for example, forest-dependent dung and carrion beetles are far more likely to cross a matrix of regrowth than one that has been completely clearcut (Klein 1989). In general, the more closely the matrix approximates the structure and microclimate of primary forest, the more likely that fragmentation-sensitive species can use it.

The matrix can have both positive and negative effects on fragmented populations. Because game in farmland mosaics is often intensively hunted (Robinson and Redford 1991, Rabinowitz 2000), the matrix can become a population sink for exploited species (Woodroffe and Ginsberg 1998). The matrix can also be a source of fruits, flowers, and other resources that help maintain fragment populations (Bierregaard et al. 1992, Brown and Hutchings 1997). Finally, the matrix supports many nonforest species; for example, from 8-25% of all frog, bird, small mammal, and ant species in the BDFFP study area are exclusively associated with the matrix (Gascon et al. 1999).

2.6 Distance effects

A key finding of the BDFFP is that even small clearings are barriers for many rainforest organisms (as discussed above, such barrier effects often diminish somewhat when fragments are surrounded by regenerating forest rather than pasture). Many terrestrial insectivorous birds have disappeared from the BDFFP fragments and failed to recolonize even those isolated by only 80 m, despite a proliferation of regrowth around many fragments (Stratford and Stouffer 1999). Clearings of just 15-100 m are insurmountable barriers for certain dung and carrion beetles (Klein 1989), euglossine bees (Powell and Powell 1987), and arboreal mammals (Malcolm 1991, Gilbert and Setz 2001). Peccaries (Offerman et al. 1995) and many insect-gleaning bats (Kalko 1998) are also highly reluctant to enter clearings. Even an unpaved road only 30-40 m wide dramatically alters the community structure of understory birds and inhibits the movements of many species (S. G. Laurance et al. 2004, see also Keller et al. 2004).

Some species will cross small clearings but are inhibited by larger expanses of degraded land. Woodcreepers (Dendrocolaptidae) were induced by translocations to move between the BDFFP fragments and nearby areas (80-150 m) of mainland forest (Harper 1989), but have disappeared from slightly more-isolated areas such as Barro Colorado Island in Panama (Robinson 1999). Large predators like jaguars (*Panthera onca*) and pumas (*Puma concolor*) traverse pastures and regrowth in the BDFFP study area, but would likely avoid these areas if hunters were present or human density was higher (Rabinowitz 2000). Some ant-following birds (*Pithys albifrons, Gymnopithys rufigula, Dendrocincla merula*) translocated into forest fragments where army ants are absent will cross clearings of 100-320 m to return to primary forest (Lovejoy et al. 1986, Harper 1989), although clearings of only 100 m preclude their movements under normal circumstances (Bierregaard and Lovejoy 1989, Stouffer and Bierregaard 1995b, S.G. Laurance and Gomez 2005).

Amazonian animals avoid clearings for many reasons. Most understory species have had little reason to traverse clearings in their evolutionary history, so the avoidance of such areas is probably an innate response (Greenberg 1989). Other species are constrained by morphology or physiology; strictly arboreal species, for instance, will find even a small pasture an impenetrable barrier. Specialized habitat needs probably limit yet others; for example, rainforest birds that flip over dead leaves in order to find insects, like the antbird *Myrmornis torquata*, probably can not manipulate the large leaves of *Cecropia* trees, and therefore avoid *Cecropia*-dominated regrowth (Stratford and Stouffer 1999). A final factor that limits inter-fragment movements, at least in Amazonian birds, is that few species are migratory. In temperate forests, even truly isolated fragments can be colonized in the breeding season by migratory species (e.g., Blake and Karr 1987), but Amazonian birds appear less likely to do so.

3 Ecological Changes in Fragmented Communities

3.1 Hyperdynamism

BDFFP results and findings from other studies suggest that, for many organisms, fragmentation alters population and community dynamics (Laurance 2002). At the outset, deforestation causes recurring disturbances. Surface fires, loggers, hunters, miners, fuelwood gatherers, and livestock can all penetrate into forest remnants and cause a diversity of ecological changes (Schelhas and Greenberg 1996, Laurance and Bierregaard 1997, Curran et al. 1999). For instance, smoke from nearby forest burning strongly disturbed butterfly communities in the BDFFP fragments, accelerating the loss of forest-interior species (Brown and Hutchings 1997).

The proliferation of forest edges also has important effects, because edges are intrinsically less stable than forest interiors. For example, insect activity is highly variable near edges and is influenced more strongly than forest interiors by daily weather variation (Fowler et al. 1993). Tree-mortality rates are sharply elevated near edges and vary markedly over time because of periodic windstorms, droughts, and successional changes in edge structure (Laurance et al., 1998b, 2002, Mesquita et al. 1999).

In addition, small populations in fragments may be less stable than those in continuous forest. Bat communities in the BDFFP fragments appear to exhibit an unusually rapid turnover of species, apparently because of high rates of disappearance of forest-interior species coupled by an influx of opportunistic frugivores that feed along forest edges and in nearby regrowth (Sampaio 2000). Population turnover in the social spider *Anelosimus eximius* was much higher near forest edges than in forest interiors, suggesting that small fragment populations are unstable (Ventincinque et al. 1993). Small mammal abundances fluctuated dramatically in the BDFFP fragments, especially in the first few years after isolation, relative to populations in intact forest (Malcolm 1991).

Finally, fluxes of animals and plant propagules to and from the surrounding matrix can sometimes destabilize fragment populations. When the forest surrounding the BDFFP fragments was initially felled, displaced birds flooded into the fragments, leading to sharply elevated densities and increased territorial behavior by resident birds (this increase was temporary; total bird numbers fell to pre-fragmentation levels within 200 days of fragment isolation) (Bierregaard and Lovejoy 1989). Dramatic irruptions of some Heliconine and Ithomiine butterflies occurred in the BDFFP fragments when their weedy food plants (*Passiflora* vines and *Solanum* bushes) proliferated near fragment margins (Brown and Hutchings 1997).

3.2 Hyperabundance

Many species decline or disappear in fragmented forests, but others can increase dramatically, especially if they favor disturbed or edge habitats or readily tolerate the surrounding matrix. Examples of edge- and disturbance-favoring groups include certain rodents and marsupials (Malcolm 1997), gap-favoring and nectarivorous birds (Stouffer and Bierregaard 1995a, 1995b, S.G. Laurance 2004), frugivorous bats (Kalko 1998), understory insects (Malcolm 1991, 1994), pioneer trees (Laurance et al. 1998c), and lianas (Laurance et al. 2001b). Species that thrive in fragments because they can exploit the adjoining matrix include shrub-frugivorous bats (Kalko 1998) and the tamarin *Sanguinus midas* (Rylands and Keuroghlian 1988).

Other species may increase in fragments when their competitors or predators disappear (e.g. Kruess and Tscharntke 1994, Terborgh et al. 1997, 2001), or because they have flexible behavioral repertoires. Howler monkeys (*Alouatta seniculus*), for instance, can achieve high densities in small forest fragments where only a few other monkeys are present (Gilbert and Setz 2001). The woodcreeper *Xiphorhynchus pardalotus* often forages with mixed-species and canopy flocks in intact forest, but in fragments it will forage alone and even on edges abutting pastures (Bierregaard 1990). Some canopy-feeding hummingbirds will also forage along forest edges and in treefall gaps, and thereby increase in fragments (Stouffer and Bierregaard 1995a).

3.3 Species invasions

Species-rich rainforests may be relatively resistant to invasions (Rejmánik 1996, Laurance and Bierregaard 1997), but habitat disturbance (Von Holle 2005) and increasing propagule pressure from invaders (Fine 2002) are both likely to increase invader establishment in fragmented landscapes. Many non-rainforest species have colonized matrix habitats in the BDFFP landscape, although to date incursions into fragments have been more limited. The most conspicuous invaders of fragments are generalist frogs (Tocher et al. 1997) and light-loving butterflies (Brown and Hutchings 1997), although many other taxa have been detected, including open-forest bats (Kalko 1998), exotic and generalist palms (Scariot 1998), Africanized honeybees (Dick 2001), *Glaphrocanthon* beetles (Klein 1989), generalist fruitflies (*Drosophila* spp., Martins 1989), and leaf-cutting ants native to tropical savannas (*Atta laevigata* and *Acromyrmex laticeps*, Vasconcelos and Cherrett 1995). Incursions of non-rainforest birds (e.g., *Troglodytes aedon*, *Ramphocelus carbo*) into the BDFFP fragments have been surprisingly limited, despite widespread local extinctions of many native insectivorous birds (Stouffer and Bierregaard 1995b, Stouffer and Borges 2001). Likewise, exotic lianas are apparently uncommon in the BDFFP fragments (Laurance et al. 2001b), unlike forest remnants in some other tropical regions (reviewed in Laurance 1997).

However, the BDFFP study area is relatively young (forest clearing began only in 1980) and still largely isolated from other human-dominated landscapes. Many non-native species, such as generalist frogs, probably arrived from settled areas by traversing along road verges and powerline clearings (Tocher et al. 1997, Gascon et al. 1999). As encroaching degraded lands draw nearer, the pressure from invading species is likely to increase. In this sense, older Amazonian frontiers are likely to be more severely degraded by invaders than are landscapes that have only recently been colonized and fragmented.

3.4 Changes in trophic structure

Because they have high area and energy requirements, predators and large-bodied species are predicted to decline in habitat fragments whereas smaller species at lower trophic levels—such as generalist herbivores and ominivores—should increase (Holt 1996). Such changes have been hypothesized to cause ecological distortions that help drive the process of species impoverishment in fragments (Terborgh et al. 1997).

Results from the BDFFP partially support these hypotheses. Among birds and forest-interior bats, insectivores have been especially vulnerable to fragmentation whereas many omnivores and nectarivores have remained stable or increased in fragments (Bierregaard and Stouffer 1997, Kalko 1998, Sampaio 2000). There is also a decline of large mammals in fragments, including predators (Lovejoy et al. 1986, Bierregaard et al. 1992), but because the BDFFP landscape is protected from hunting, the reductions are less dramatic than typically occur in other tropical areas (Robinson and Redford 1991).

Patterns among insects are more complex and may partly reflect shifts in resource abundance in fragmented forests. The guild composition of termites is altered in fragments, with lower species richness and an increase in litter-feeders and those intermediate between soil-feeding and wood-feeding types (Souza and Brown 1994); such changes could result to some extent from increased litter and wood debris in fragments. Dung and carrion beetles are less abundant and diverse in fragments, in part because many vertebrates on which they rely have declined or disappeared (Klein 1989). Among leaf-litter beetles, there are proportionally more predator species and fewer wood-boring species in fragments and near edges (Didham et al. 1998b). Of these patterns, the relative increase in fragments of predatory beetle diversity (Didham et al. 1998b) and the declines of decomposer beetle (Klein 1989) and termite (Souza and Brown 1994) diversity seem contrary to the simple prediction that fragments should be biased toward taxa at lower trophic levels.

3.5 Changes in ecological processes

Tropical rainforests are renowned for their ecological complexity (Janzen 1969, Gilbert 1980). Fragmentation clearly alters some ecological processes, but the generality of these effects is not yet known (Harrison and Bruna 1999). For example, fragmentation has a strong positive effect on pollination or fecundity in the emergent tree *Dinezia excelsia* (Dick 2001), but no detectable effect in the understory herb *Heliconia acuminata* (Bruna and Kress 2002). Hypothetically at least, the disappearance of many euglossine bees in the BDFFP fragments could reduce the fecundity of orchids, which rely entirely on euglossines for pollination (Powell and Powell 1987). Likewise, the decline in fragments of dung beetles, which bury dung for their larvae that often contains seeds, might reduce seed survival and germination for some plant species (Klein 1989, Andresen 2001).

Predation intensity is almost certainly altered in Amazonian fragments. Predation on understory and litter arthropods has probably declined because of a collapse of insectivorous bird (Stouffer and Bierregaard 1995b, Stratford and Stouffer 1999), bat (Kalko 1998, Sampaio 2000), and army ant (Harper 1989, Bierregaard et al. 1992) assemblages. It seems plausible that these declines could be partly responsible for increased insect abundance near forest edges (Lovejoy et al. 1986, Fowler et al. 1993) and might even promote increased herbivory in fragments (Benitez-Malvido et al. 1999). The decline of large carnivores may reduce predation on some vertebrates, but there is no indication of mesopredator release (Crooks and Soulé 1999) in the BDFFP fragments (Meyer 1999).

Tropical rainforests sustain myriad species with coevolved interdependencies and may therefore be prone to secondary extinctions (Gilbert 1980), although such losses might be limited by ecological redundancy in many mutualisms (e.g., Horvitz and Schemske 1990). An interesting example from the BDFFP involves several species of obligatory ant-following birds, which accompany marauding swarms of army ants in order to capture fleeing insects. Each ant colony raids over areas of up to 100 ha, and the birds' home ranges must encompass 2-3 colonies because each colony spends several weeks per month in an inactive phase (Harper 1989). Because army ants need such large areas, the ant-followers are highly prone to extinction in fragments (Stouffer and Bierregaard 1995b). In addition, the decline of peccaries in BDFFP fragments has led to reduced abundances of at least four frog species (*Phyllomedusa* spp. and *Colostethus* sp.) that breed only in peccary wallows (Zimmerman and Bierregaard 1986). Understanding the effects of fragmentation on such interdependent species is a priority for future research.

3.6 Changes in ecosystem processes

Tropical forests have a major influence on the global climate, in part by storing large quantities of terrestrial carbon. The rapid destruction of these forests probably accounts for at least a quarter of all greenhouse gas emissions, contributing significantly to global warming (Houghton 1991, Fearnside 2000). An unexpected finding is the degree to which fragmentation alters carbon storage. Elevated tree mortality leads to a decline of living biomass near edges (Figure 6, Laurance et al. 1997, 1998d, Nascimento and Laurance 2004), especially because large canopy and emergent trees, which contain a high proportion of forest biomass, are particularly vulnerable to fragmentation (Laurance et al. 2000). As the biomass from the dead trees decomposes, it is converted into greenhouse gases such as carbon dioxide and methane. This loss of living biomass is not offset by increased numbers of lianas and small successional trees (Laurance et al. 1998d, 2001b), which have lower wood densities and therefore store less carbon than the old-growth species they replace (Laurance et al. 1998d, Nascimento and Laurance 2004). In tropical forests worldwide, millions of tons of atmospheric carbon emissions may be released each year by this process (Laurance et al. 1998e). Edge-related losses of biomass are predicted to increase sharply once fragments fall below 100-400 ha in area, depending on fragment shape (Laurance et al. 1998b).

The rate of carbon cycling is also altered. In intact forests, carbon can be stored for very long periods in large trees, some of which can live for more than a thousand years (Chambers et al. 1998). In fragments, the residence times for carbon surely will decrease as smaller, short-lived plants replace large old-growth trees and rates of litter deposition increase near edges. The dynamics of this cycle can have major effects on carbon storage in vegetation and soils and the rate of input of organic material into tropical rivers and streams (Wissmar et al. 1981).

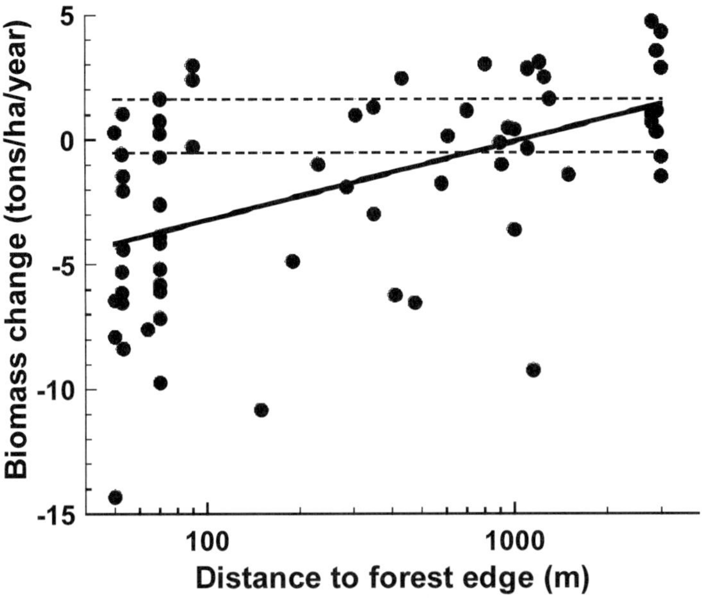

Fig. 6. Annual change in above-ground tree biomass in the BDFFP study area as a function of distance from forest edge. Each data point represents a 1-ha plot that was studied for periods of up to 18 years. The dotted lines show the 95% confidence intervals for forest-interior plots (>500 m from edge) (after Laurance et al. 1997).

4 Conclusions and Outlook

The BDFFP has yielded scores of insights into the effects of habitat fragmentation on rainforest biotas. Results suggest that edge effects and area-related extinctions will rapidly degrade smaller (<100 ha) fragments, which are predominant in anthropogenic landscapes (Laurance and Bierregaard 1997, Gascon et al. 2000). Species' abundances in fragments will differ from those in intact forest, with some species declining and others becoming hyperabundant. Fundamental processes such as canopy-gap dynamics, predation, and carbon storage will be altered or disrupted. Fragments will be strongly influenced by the surrounding matrix, which affects landscape connectivity, the intensity of edge effects, species invasions, and the frequency or intensity of disturbances such as windstorms and fire. Over time, fragmented communities will become increasingly dominated by matrix-tolerant generalists, disturbance-adapted opportunists, and species with small area requirements.

The BDFFP is a controlled experiment, and the ecological effects of fragmentation should be even greater in other tropical landscapes. Firstly, the BDFFP fragments are primarily square, which makes them less vulnerable to edge effects than more-irregularly shaped fragments. Secondly, the BDFFP fragments are located near large tracts of continuous forest, which facilitates rescue effects (Brown and Kodric-Brown 1977) and recolonization for some species, and may help maintain natural rainfall and hydrological cycles (Shulka et al. 1990). Thirdly, many of the BDFFP fragments have become surrounded by regrowth, which increases fragment connectivity while reducing the intensity of some edge effects. Finally, the BDFFP study area is protected from hunters, loggers, miners, and recurring surface fires that have dramatically exacerbated the effects of fragmentation in other tropical landscapes (Curran et al. 1999, Cochrane and Laurance 2002).

The BDFFP findings have not identified a single "minimum critical size" for tropical nature reserves. Results have, however, helped to demonstrate that such reserves should be both large and numerous. The low densities and patchy distributions of most Amazonian species, large spatial scale of some edge effects, irregular shapes of many nature reserves, and synergistic interactions of fragmentation with other human impacts all indicate that Amazonian reserves should be as large as possible—ideally on the order of thousands of square kilometers or more (cf. Peres and Terborgh 1995, Laurance 2000, 2005b, 2006, Cochrane and Laurance 2002, Peres 2005). Moreover, the high turnover of many taxa at regional scales (high gamma diversity) implies that multiple reserves should be stratified along major environmental gradients to capture a large fraction of the regional biota. Finally, the extreme sensitivity of many species to forest clearings and edge effects suggests that relatively wide, continuous corridors of primary forest must be maintained – with limited hunting pressure – to permit faunal movements, plant dispersal, and gene flow among reserves.

Acknowledgements
This chapter has been adapted and updated, with permission, from an earlier article in *Conservation Biology* (Laurance et al. 2002). I thank Teja Tscharntke and an anonymous referee for commenting on the manuscript.

References

Allmon WD (1991) A plot study of forest floor litter frogs, central Amazon, Brazil. J Trop Ecol 7: 503-522

Andresen E (2001) Effects of dung presence, dung amount, and secondary dispersal by dung beetles on the fate of *Micropholis guyanensis* (Sapotaceae) seeds in central Amazonia. J Trop Ecol 17: 61-78

Becker P, Moure JB, Peralta FJA (1991) More about euglossine bees in Amazonian forest fragments. Biotropica 23: 586-591

Benitez-Malvido J (1998) Impact of forest fragmentation on seedling abundance in a tropical rain forest. Conserv Biol 12: 380-389

Benitez-Malvido J, Garcia-Guzman G, Kossman-Ferraz ID (1999) Leaf-fungal incidence and herbivory on tree seedlings in tropical rainforest fragments: anexperimental study. Biol Conserv 91: 143-150

Bierregaard RO (1990) Avian communities in the understory of Amazonian forest fragments. In: Keast A (ed) Biogeography and ecology of forest bird communities. SPB Academic Publishing, The Hague, pp 333-343

Bierregaard RO, Lovejoy TE (1989) Effects of forest fragmentation on Amazonian understory bird communities. Acta Amazônica 19: 215-241

Bierregaard RO, Lovejoy TE, Kapos V, dos Santos AA, Hutchings RW (1992) The biological dynamics of tropical rainforest fragments. BioScience 42: 859-866

Bierregaard, RO, Stouffer PC (1997) Understory birds and dynamic habitat mosaics in Amazonian rainforest. In: Laurance WF, Bierregaard RO (eds) Tropical forest remnants: ecology, management, and conservation of fragmented communities. University of Chicago Press, Chicago, pp 138-155

Blake JG, Karr JR (1987) Breeding birds of isolated woodlots: area and habitat relationships. Ecology 68: 1724-1734

Borges SH, Stouffer PC (1999) Bird communities in two types of anthropogenic successional vegetation in central Amazonia. Condor 101: 529-536

Brokaw NVL (1985) Gap-phase regeneration in a tropical forest. Ecology 66: 682-687

Brown KS Jr (1987) Soils and vegetation. In: Whitmore TC, Prance GT (eds) Biogeography and quaternary history in tropical America. Oxford Monographs in Biogeography 3, pp 19-45

Brown KS Jr, Hutchings RW (1997) Disturbance, fragmentation, and the dynamics of diversity in Amazonian forest butterflies. In: Laurance WF, Bierregaard RO (eds) Tropical forest remnants: ecology, management, and conservation of fragmented communities. University of Chicago Press, Chicago, pp 91-110

Brown JH, Kodric-Brown A (1977) Turnover rates in insular biogeography: effect of immigration on extinction. Ecology 58: 445-449

Bruna EM (1999) Seed germination in rainforest fragments. Nature 402: 139

Bruna EM, Kress WJ (2002) Habitat fragmentation and the demographic structure of an Amazonian understory herb (*Heliconia acuminata*) Conserv Biol 16: 1256-1266

Camargo JLC (1993) Variation in soil moisture and air vapour pressure deficit relative to tropical rain forest edges near Manaus, Brazil. M.Phil. thesis. Cambridge University, Cambridge, U.K.

Camargo JLC, Kapos V (1995) Complex edge effects on soil moisture and microclimate in central Amazonian forest. J Trop Ecol 11: 205-211.

Carvalho KS, Vasconcelos HL (1999) Forest fragmentation in central Amazonia and its effects on litter-dwelling ants. Biol Conserv 91: 151-158

Chambers JQ, Higuchi N, Schimel J (1998) Ancient trees in Amazonia. Nature 391: 135-136.

Cochrane MA, Alencar A, Schulze MD, Souza CM, Nepstad DC, Lefebvre P, Davidson E. (1999) Positive feedbacks in the fire dynamics of closed canopy tropical forests. Science 284: 1832-1835

Cochrane MA, Laurance WF (2002) Fire as a large-scale edge effect in Amazonian forests. J Trop Ecol 18: 311-325

Crooks KR, Soulé ME (1999) Mesopredator release and avifaunal extinctions in a fragmented system. Nature 400: 563-566

Curran LM, Caniago I, Paoli GD, Astianti D, Kusneti M, Leighton M, Nirarita CE, Haeruman H (1999) Impact of El Nino and logging on canopy tree recruitment in Borneo. Science 286: 2184-2188

Debinski DM, Holt RD (2000) A survey and overview of habitat fragmentation experiments. Conserv Biol 14: 342-355

Denslow JS (1987) Tropical rainforest gaps and tree species diversity. Ann Rev Ecol Syst 18: 431-451.

Dick C (2001) Habitat change, African honeybees, and fecundity in the Amazonian tree *Dinizia excelsa* (Fabaceae). In: Bierregaard RO, Gascon C, Lovejoy TE, Mesquita R (eds) Lessons from Amazonia: ecology and conservation of a fragmented forest. Yale University Press, New Haven, pp 146-157

Didham RK (1997a) The influence of edge effects and forest fragmentation on leaf-litter invertebrates in central Amazonia. In: Laurance WF, Bierregaard RO (eds) Tropical forest remnants: ecology, management, and conservation of fragmented communities. University of Chicago Press, Chicago, pp 55-70

Didham RK (1997b) An overview of invertebrate responses to habitat fragmentation. In: Watt AD, Stork NE, Hunter MD (eds) Forests and insects. Chapman and Hall, London, pp 303-320

Didham RK (1998) Altered leaf-litter decomposition rates in tropical forest fragments. Oecologia 116: 397-406

Didham RK, Hammond PM, Lawton JH, Eggleton P, Stork NE (1998a) Beetle species responses to tropical forest fragmentation. Ecol Monogr 68: 295-303

Didham RK, Lawton JH (1999) Edge structure determines the magnitude of changes in microclimate and vegetation structure in tropical forest fragments. Biotropica 31: 17-30

Didham RK, Lawton JH, Hammond PM, Eggleton P (1998b) Trophic structure stability and extinction dynamics of beetles (Coleoptera) in tropical forest fragments. Phil Proc Roy Soc Lond B 353: 437-451

Emmons LH (1984) Geographic variation in densities and diversities of nonflying mammals in Amazonia. Biotropica 16: 210-222

Fearnside PM (2000) Global warming and tropical land-use change: greenhouse gas emissions from biomass burning, decomposition and soils in forest conversion, shifting cultivation and secondary vegetation. Clim Change 46: 115-158

Ferraz G, Russell GJ, Stouffer PC, Bierregaard RO, Pimm SL, Lovejoy TE (2003) Rates of species loss from Amazonian forest fragments. Proc Nat Acad Sci USA 100: 14069-14073

Ferreira LV, Laurance WF (1997) Effects of forest fragmentation on mortality and damage of selected trees in central Amazonia. Conserv Biol 11: 797-801

Fine PVA (2002) The invasibility of tropical forests by exotic plants. J Trop Ecol 18: 687-705

Fletcher RJ (2005) Multiple edge effects and their implications in fragmented landscapes. J Anim Ecol 74: 342-352.

Fowler HG, Silva CA, Ventincinque E (1993) Size, taxonomic and biomass distributions of flying insects in central Amazonia: forest edge vs. understory. Rev Biol Trop 41: 755-760

Gascon C (1993) Breeding habitat use by Amazonian primary-forest frog species at the forest edge. Biodiv Conserv 2: 438-444

Gascon C, Lovejoy TE, Bierregaard RO, Malcolm JR, Stouffer PC, Vasconcelos H, Laurance WF, Zimmerman B, Tocher M, Borges S (1998) Matrix habitat and species persistence in tropical forest remnants. Biol Conserv 91: 223-229

Gascon C, Williamson GB, Fonseca, GAB (2000) Receding edges and vanishing reserves. Science 288: 1356-1358

Gilbert LE (1980) Food web organization and the conservation of Neotropical diversity. In: Soulé, ME, Wilcox BA (eds) Conservation biology: an evolutionary-ecological perspective. Sinauer, Sunderland, Massachusetts, pp 11-33

Gilbert KA, Setz EFZ (2001) Primates in a fragmented landscape: six species in central Amazonia. In: Bierregaard RO, Gascon C, Lovejoy TE, Mesquita R (eds) Lessons from Amazonia: ecology and conservation of a fragmented forest. Yale University Press, New Haven, pp 262-270

Greenberg R (1989) Neophobia, aversion to open space, and ecological plasticity in Song and Swamp Sparrows. Can J Zool 67: 1194-1199

Gustafson EJ, Gardner RH (1996) The effect of landscape heterogeneity on the probability of patch colonization. Ecology 77: 94-107

Haila Y, Hanski IK, Raivio S (1993) Turnover of breeding birds in small forest fragments: the sampling colonization hypothesis corroborated. Ecology 74: 714-725.

Harcourt AH, Doherty DA (2005) Species-area relationships of primates in tropical forest fragments: a global analysis. J Appl Ecol 42: 630-637

Harper LH (1989) The persistence of ant-following birds in small Amazonian forest fragments. Acta Amazônica 19: 249-263

Harrison S, Bruna EM (1999) Habitat fragmentation and large-scale conservation: what do we know for sure? Ecography 22: 225-232

Holt RD (1996) Food webs in space: an island biogeographic perspective. In: Polis GA, Winemiller KO (eds) Food webs: integration of patterns and dynamics. Chapman and Hall, New York, pp 313-323

Horvitz CC, Schemske DW (1990) Spatiotemporal variation in insect mutualists of a neotropical herb. Ecology 71: 1085-1097

Houghton RA (1991) Tropical deforestation and atmospheric carbon dioxide. Clim Change 19: 99-118

Hubbell SP, Foster RB (1986) Commonness and rarity in a Neotropical forest: implications for tropical tree conservation. In: Soulé ME (ed) Conservation biology: the science of scarcity and diversity, Sinaur, Sunderland,Massachusetts, pp 205-232

INPE (2005) Deforestation estimates for the Brazilian Amazon. Instituto Nacional de Pesquisas Espaciais, São Jose dos Campos, Brazil Janzen D H (1969) Allelopathy by myrmecophytes: the ant *Azteca* as an allelopathic agent of *Cecropia*. Ecology 50: 147-153

Kalko EKV (1998) Organization and diversity of tropical bat communities through space and time. Zool: Anal Complex Syst 101: 281-297

Kapos V (1989) Effects of isolation on the water status of forest patches in the Brazilian Amazon. J Trop Ecol 5: 173-185.

Kapos V, Ganade G, Matsui E, Victoria RL (1993) ^{13}C as an indicator of edge effects in tropical rainforest reserves. J Ecol 81: 425-432

Kapos V, Wandelli E, Camargo JL, Ganade G (1997) Edge-related changes in environment and plant responses due to forest fragmentation in central Amazonia. In: Laurance WF, Bierregaard RO (eds) Tropical forest remnants: ecology, management, and conservation of fragmented communities. University of Chicago Press, Chicago, pp 33-44

Keller I, Nentwig W, Largiader CR (2004) Recent habitat fragmentation due to roads can lead to significant genetic differentiation in an abundant flightless ground beetle. Molecular Ecol 13: 2983-2994

Klein BC (1989) Effects of forest fragmentation on dung and carrion beetle communities in central Amazonia. Ecology 70: 1715-1725

Kruess A, Tscharntke T (1994) Habitat fragmentation, species loss, and biological control. Science 264: 1581-1584

Laurance SG (2004) Responses of understory rainforest birds to road edges in central Amazonia. Ecol Applic 14: 1344-1357

Laurance SG, Gomez M (2005) Clearing width and movements of understory rainforest birds. Biotropica 37: 149-152

Laurance SG, Stouffer PC, Laurance WF (2004) Effects of road clearings on movement patterns of understory rainforest birds in central Amazonia. Conserv Biol 18: 1099-1109

Laurance WF (1997) Hyper-disturbed parks: edge effects and the ecology of isolated rainforest reserves in tropical Australia. In: Laurance WF, Bierregaard RO (eds) Tropical forest remnants: ecology, management, and conservation of fragmented communities. University of Chicago Press, Chicago, pp 71-83

Laurance WF (1998) A crisis in the making: responses of Amazonian forests to land use and climate change. Trends Ecol Evol 13: 411-415

Laurance WF (2000) Do edge effects occur over large spatial scales? Trends Ecol Evol 15: 134-135

Laurance WF (2001) The hyper-diverse flora of the central Amazon: an overview. In: Bierregaard RO, Gascon C, Lovejoy TE, Mesquita R (eds) Lessons from Amazonia: ecology and conservation of a fragmented forest. Yale University Press, New Haven, pp 47-53.

Laurance WF (2002) Hyperdynamism in fragmented habitats. J Veg Sci 13: 595-602

Laurance WF (2005a) Amazon burning. New Scientist, 15 October, pp 34-39

Laurance WF (2005b) When bigger is better: The need for Amazonian megareserves. Trends Ecol Evol 20: 645-648.

Laurance WF (2006) More reasons for megareserves in Amazonia. Natureza e Conservação 4: 96-106

Laurance WF, Bierregaard RO (eds) (1997) Tropical forest remnants: ecology, management, and conservation of fragmented communities. University of Chicago Press, Chicago.

Laurance WF, Cochrane MA, Bergen S, Fearnside PM, Delamonica P, Barber C, D'Angelo S, Fernandes T (2001a) The future of the Brazilian Amazon. Science 291: 438-439

Laurance WF, Delamonica P, Laurance SG, Vasconcelos HL, Lovejoy TE (2000) Rainforest fragmentation kills big trees. Nature 404: 836

Laurance WF, Albernaz AKM, Fearnside PM, Vasconcelos HL, Ferreira LV (2004) Deforestation in Amazonia. Science 304: 1109

Laurance WF, Ferreira LV, Gascon C, Lovejoy TE (1998d) Biomass loss in Amazonian forest fragments. Science 282: 1611a

Laurance WF, Ferreira LV, Rankin-de Merona JM, Laurance SG (1998b) Rain forest fragmentation and the dynamics of Amazonian tree communities. Ecology 79: 2032-2040

Laurance WF, Ferreira LV, Rankin-de Merona JM, Laurance SG, Hutchings R, Lovejoy TE (1998c) Effects of forest fragmentation on recruitment patterns in Amazonian tree communities. Conserv Biol 12: 460-464.

Laurance WF, Gascon C, Rankin-de Merona JM (1998a) Predicting effects of habitat destruction on plant communities: a test of a model using Amazonian trees. Ecol Applic 9: 548-554

Laurance WF, Laurance SG, Delamonica P (1998e) Tropical forest fragmentation and greenhouse gas emissions. For Ecol Manage 110: 173-180

Laurance WF, Laurance SG, Ferreira LV, Rankin-de Merona JM, Gascon C, Lovejoy TE (1997) Biomass collapse in Amazonian forest fragments. Science 278: 1117-1118

Laurance WF, Lovejoy TE, Vasconcelos HL, Bruna EM, Didham RK, Stouffer PC, Gascon C, Bierregaard RO, Laurance SG, Sampiao E (2002) Ecosystem decay of Amazonian forest ragments: a 22-year investigation. Conserv Biol 16: 605-618

Laurance WF, Nascimento HEM, Laurance SG, Andrade A, Fearnside PM, Ribeiro JELS (2006) Rain forest fragmentation and the proliferation of successional trees. Ecology (in press)

Laurance WF, Perez-Salicrup D, Delamonica P, Fearnside PM, D'Angelo S, Jerozolinski A, Pohl L, Lovejoy TE (2001b) Rain forest fragmentation and the structure of Amazonian liana communities. Ecology 82: 105-116

Laurance WF, Williamson GB, Delamonica P, Olivera AA, Gascon C, Lovejoy TE, Pohl L (2002) Effects of a severe drought on Amazonian forest fragments and edges. J Trop Ecol 17: 771-785

Lewis S (1998) Treefall gaps and regeneration: a comparison of continuous forest and fragmented forest in central Amazonia. Ph.D. thesis, University of Cambridge, Cambridge

Lovejoy TE, Bierregaard RO, Rankin JM, Schubart HOR (1983) Ecological dynamics of forest fragments. In: Sutton SL, Whitmore TC, Chadwick AC (eds) Tropical rain forest: ecology and management. Blackwell Scientific, Oxford, pp 377-384

Lovejoy TE, Bierregaard RO, Rylands AB, Malcolm JR, Quintela CE, Harper LH, Brown KS, Powell AH, Powell GVN, Schubart HOR, Hays MB (1986) Edge and other effects of isolation on Amazon forest fragments. In: Soulé ME (ed) Conservation biology: the science of scarcity and diversity. Sinauer, Sunderland, Massachusetts, pp 257-285

MacArthur RO, Wilson EO (1967) The theory of island biogeography. Princeton University Press, Princeton

Malcolm JR (1991) The small mammals of Amazonian forest fragments: pattern and process. Ph.D. thesis, University of Florida, Gainesville

Malcolm JR (1994) Edge effects in central Amazonian forest fragments. Ecology 75: 2438-2445

Malcolm JR (1997) Biomass and diversity of small mammals in Amazonian forest fragments. In: Laurance WF and Bierregaard RO (eds) Tropical forest remnants: ecology, management, and conservation of fragmented communities. University of Chicago Press, Chicago, pp 207-221

Malcolm JR (1998) A model of conductive heat flow in forest edges and fragmented landscapes. Climatic Change 39: 487-502.

Martins MB (1989) Invasão de fragmentos florestais por especies de *Drosophila* (Diptera, Drosophilidae). Acta Amazônica 19: 265-271

Mesquita R, Delamônica P, Laurance WF (1999) Effects of surrounding vegetation on edge-related tree mortality in Amazonian forest fragments. Biol Conserv 91: 129-134

Meyer J (1999) Efeitos da fragmentação de floresta sobre predação em ovos de aves e remoção de sementes. M.Sc. thesis, National Institute for Amazonian Research (INPA), Manaus, Brazil

Nascimento HEM, Laurance WF (2004) Biomass dynamics in Amazonian forest fragments. Ecol Applic 14: S127-S138

Nascimento HEM, Andrade A, Camargo JLC, Laurance WF, Laurance SG, Ribeiro JELS (In press) Effects of the surrounding matrix on tree recruitment in Amazonian forest fragments. Conserv Biol.

Offerman HL, Dale VH, Pearson SM, Bierregaard RO, O'Neill RV (1995) Effects of forest fragmentation on neotropical fauna: current research and data availability. Environ Rev 3: 191-211

Oliveira AA, Mori S (1999) Central Amazonian terra-firme forests: high tree species richness. Biodiv Conserv 8: 1219-1244

Pearman PB (1997) Correlates of amphibian diversity in an altered landscape of Amazonian Ecuador. Conserv Biol 11: 1211-1225

Peres CA (2005) Why we need megareserves in Amazonia. Conserv Biol 19: 728-733

Peres CA, Terborgh JW (1995) Amazonian nature reserves: an analysis of the defensibility status of existing conservation units and design criteria for the future. Conserv Biol 9: 34-46

Pittman NCA, Terborgh JW, Silman MR, Nuñez P (1999) Tree species distributions in an upper Amazonian forest. Ecology 80: 2651-2661

Powell AH, Powell GVN (1987) Population dynamics of male euglossine bees in Amazonian forest fragments. Biotropica 19: 176-179

Quintela CE (1985) Forest fragmentation and differential use of natural and man-made edges by understory birds in central Amazonia. M.Sc. thesis, University of Chicago, Chicago.

Rabinowitz A (2000) Jaguar. Island Press, Washington, D.C.

Rankin-de Merona JM, Prance JM, Hutchings RW, Silva MF, Rodrigues MA, Uehling ME (1992) Preliminary results of a large-scale inventory of upland rain forest in the central Amazon. Acta Amazônica 22: 493-534

Rejmánek M (1996) Species richness and resistance to invasions. In: Orians G, Dirzo R, Cushman H (eds) Biodiversity and ecosystem processes in tropical forests. Springer-Verlag, Berlin, Germany, pp 153-172

Ries L, Fletcher RJ, Battin J, Sisk TD (2004) Ecological responses to habitat edges: mechanisms, models, and variability explained. Ann Rev Ecol Syst 35: 491-522

Robinson JG, Redford KH (eds) (1991) Neotropical wildlife use and conservation. University of Chicago Press, Chicago

Robinson WD (1999) Long-term changes in the avifauna of Barro Colorado Island, Panama, a tropical forest isolate. Conserv Biol 13: 85-97

Rylands AB, Keuroghlian A (1998) Primate populations in continuous forest and forest fragments in central Amazonia. Acta Amazônica 18: 291-307

Sampaio, E (2000) The effects of fragmentation on structure and diversity of bat communties in a central Amazonian tropical rain forest. Ph.D. thesis, University of Tuebingen, Tuebingen, Germany

Savill PS (1983) Silviculture in windy climates. Forestry Abstr 44: 473-488

Scariot A (1998) Forest fragmentation: the effects on palm diversity in central Amazonia. J Ecol 87: 66-76

Scariot A (2001) Seedling mortality by litterfall in Amazonian forest fragments. Biotropica 32: 662-669

Schelhas J, Greenberg R (eds) (1996) Forest patches in tropical landscapes. Island Press, Washington, D.C.

Schwartzkopf L, Rylands AB (1989) Primate species richness in relation to habitat structure in Amazonian rainforest fragments. Biol Conserv 48: 1-12

Shukla J, Nobre CA, Sellers P (1990) Amazon deforestation and climate change. Science 247: 1322-1325

Sizer N, Tanner EVJ (1999) Responses of woody plant seedlings to edge formation in a lowland tropical rainforest, Amazonia. Biol Conserv 91: 135-142

Simberloff DS, Abele LG (1976) Island biogeographic theory and conservation practice. Science 191: 285-286

Skole D, Tucker CJ (1993) Tropical deforestation and habitat fragmentation in the Amazon: satellite data from 1978 to 1988. Science 260: 1905-1910

Souza, OFF, Brown VK (1994) Effects of habitat fragmentation on Amazonian termite communities. J Trop Ecol 10: 197-206

Spironello WR (2001) The brown capuchin monkey (*Cebus apella*): ecology and home range requirements in central Amazonia. In: Bierregaard RO, Gascon C, Lovejoy TE, Mesquita R (eds) Lessons from Amazonia: ecology and conservation of a fragmented forest. Yale University Press, New Haven, pp 271-283

Stouffer PC, Bierregaard RO (1995a) Effects of forest fragmentation on understory hummingbirds in Amazonian Brazil. Conserv Biol 9: 1085-1094

Stouffer PC, Bierregaard RO (1995b) Use of Amazonian forest fragments by understory insectivorous birds. Ecology 76: 2429-2445

Stouffer PC, Borges SH (2001) Conservation recommendations for understory birds in Amazonian forest fragments and secondary areas. In: Bierregaard RO, Gascon C, Lovejoy TE, Mesquita R (eds) Lessons from Amazonia: ecology and conservation of a fragmented forest. Yale University Press, New Haven, pp 248-261

Stratford JA, Stouffer PC (1999) Local extinctions of terrestrial insectivorous birds in a fragmented landscape near Manaus, Brazil. Conserv Biol 13: 1416-1423

Terborgh JW, Lopez L, Tello J, Yu D, Bruni AR (1997) Transitory states in relaxing ecosystems of tropical land-bridge islands. In: Laurance WF, Bierregaard RO (eds) Tropical forest remnants: ecology, management, and conservation of fragmented communities. University of Chicago Press, Chicago, pp 256-274

Terborgh J, Lopez L, Nuñez PV, Rao M, Shahabuddin G, Orihuela G, Riveros M, Ascanio R, Adler G, Lambert T, Balbas L (2001) Ecological meltdown in predator-free forest fragments. Science 294: 1923-1926

Tocher M (1998) A comunidade de anfíbios da Amazônia central: diferenças na composição específica entre a mata primária e pastagens. In: Gascon C, Moutinho P (eds) Floresta Amazônica: dinâmica, regeneração e manejo. National Institute for Amazonian Research (INPA), Manaus, Brazil, pp 219-232

Tocher M, Gascon C, Zimmerman BL (1997) Fragmentation effects on a central Amazonian frog community: a ten-year study. In: Laurance WF, Bierregaard RO (eds) Tropical forest remnants: ecology, management, and conservation of fragmented communities. University of Chicago Press, Chicago, pp 124-137

Vasconcelos HL (1988) Distribution of *Atta* (Hymenoptera, Formicidae) in "terra-firme" rain forest of central Amazonia: density, species composition and preliminary results on effects of forest fragmentation. Acta Amazônica 18: 309-315

Vasconcelos HL (1999) Effects of forest disturbance on the structure of ground-foraging ant communities in central Amazonia. Biodiv Conserv 8: 409-420

Vasconcelos HL, Cherrett JM (1995) Changes in leaf-cutting ant populations (Formicidae: Attini) after the clearing of mature forest in Brazilian Amazonia. Stud Neotrop Fauna Environ 30: 107-113

Vasconcelos HL, Laurance WF (2005) Influence of habitat, litter type, and soil invertebrates on leaf-litter decomposition in a fragmented Amazonian landscape. Oecologia, DOI: 10.1007/s00442-005-0117-1

Ventincinque EM, Fowler HG, Silva C (1993) Modes and frequencies of colonization and its relation to extinctions, habitat and seasonality in the social spider *Anelosimus eximius* in the Amazon (Araneidae: Theriinae). Psyche 100: 35-41

Von Holle B (2005) Biotic resistance to invader establishment of a southern Appalachian plant community is determined by environmental conditions. J Ecol 93: 16-26

Weins JA (1997) Metapopulation dynamics and landscape ecology. In: Gilpin M, Hanski I (eds) Metapopulation biology. Academic Press, San Diego, California, pp 43-62

Wilcox BA, Murphy DD (1985) Conservation strategy: the effects of fragmentation on extinction. Am Nat 125: 897-887

Williamson GB, Mesquita R, Ickes K, Ganade G (1998) Estratégias de árvores pioneiras nos Neotrópicos. In: Gascon C, Moutinho P (eds) Floresta Amazônica: dinâmica, regeneração e manejo. National Institute for Amazonian Research (INPA), Manaus, Brazil, pp 131-144

Wissmar RC, Richey JE, Stallard RF, Edmond JM (1981) Plankton metabolism and carbon processes in the Amazon River, its tributaries, and floodplain waters, Peru-Brazil, May-June 1977. Ecology 62: 1622-1633

Woodroffe R, Ginsberg JR (1998) Edge effects and the extinction of populations inside protected areas. Science 280: 2126-2128

Zimmerman BL, Bierregaard RO (1986) Relevance of the equilibrium theory of island biogeography and species-area relations to conservation with a case from Amazonia. J Biogeogr 13: 133-143

Moths at tropical forest margins – how mega-diverse insect assemblages respond to forest disturbance and recovery

Konrad Fiedler[1*], Nadine Hilt[2], Gunnar Brehm[3], and Christian H. Schulze[1]

[1] Department of Population Ecology, University of Vienna, Althanstrasse 14, 1090 Vienna, Austria
[2] Chair of Animal Ecology I, University of Bayreuth, 95440 Bayreuth, Germany
[3] Institut für Spezielle Zoologie und Evolutionsbiologie mit Phyletischem Museum, University of Jena, Erbertstrasse 1, 07745 Jena, Germany

*corresponding author: Konrad Fiedler, Email: konrad.fiedler@univie.ac.at

Summary

Tropical forest moth ensembles are often extraordinarily rich in species and complex in structure. There is an increasing number of quantitative studies from all major tropical realms which now allows for a more rigorous assessment of the ways how such ensembles change along habitat transects from the natural forest towards the edge of large conservation areas, or in the course of forest recovery at such margins. Such knowledge is also essential for deciding if moths can be used for monitoring the biotic effects of forest disturbance on mega-diverse insect faunas at all, and which moth groups are the most suitable targets for this purpose. Using very large data sets from low (Mt. Kinabalu National Park, Borneo) and high elevations (Podocarpus National Park, Ecuador), we investigate how species richness and species composition of moths change at small spatial scales along gradients of disturbance at the edge of natural forest reserves. Local species diversity did not always decline, and sometimes even increased, along the gradients. Ensembles of moths from the families Arctiidae or Sphingidae were as rich as in natural forest, or were even more diverse in habitats close to the forest margin. Geometridae and Pyraloidea ensembles, in contrast, tended to be impoverished. Correlations of alpha diversity measures between moth taxa were often poor and thus did not allow for using one group as a 'biodiversity indicator' of others. Estimates of local diversity also depended on the temporal scale of assessment. Species composition was remarkably sensitive to habitat alterations at forest edges even at very small spatial scales, despite the high dispersal potential of many

Tscharntke T, Leuschner C, Zeller M, Guhardja E, Bidin A (eds), The stability of tropical rainforest margins, linking ecological, economic and social constraints of land use and conservation, Springer Berlin 2007, pp 39-60

moths. Patterns of beta diversity were highly concordant across unrelated moth taxa with very different life history syndromes. Geometridae ensembles responded more sensitively to changes in canopy openness than arctiid moths and yielded more robust results with regard to sampling effects. Therefore geometrids are particularly promising candidates for environmental monitoring in tropical landscapes. Much of the faunal differentiation between forest and margin habitats was due to shifts in abundance relationships, and not to the presence or absence of 'indicator species'. Therefore, abundance-based monitoring appears more appropriate to detect effects of environmental change on mega-diverse moth ensembles. Our results suggest that patterns of beta (rather than alpha) diversity are generally more meaningful to assess the impact of processes at forest edges relevant to nature conservation and landscape planning.

Keywords: alpha diversity, Andes, Arctiidae, beta diversity, Borneo, community composition, disturbance gradient, Ecuador, forest recovery, Geometridae, herbivorous insects, moths, Mount Kinabalu, Pyraloidea, succession

1 Introduction

Tropical forests harbour the largest fraction of terrestrial biodiversity (Sayer et al. 2000, Wright 2005). Decline and deterioration of forests still progress at an alarming rate in most tropical countries (Wright 2005). Hence the reduction of tropical forests is one of the most pressing challenges to the preservation of global biodiversity. Apart from the loss of primary forest areas due to clear-cutting and conversion into agroforests, farmland or settlements, another suite of major threats is related to fragmentation issues (Laurance 1998, Laurance et al. 2002, Armenteras et al. 2003, Wright 2005). Most remaining tropical forests are islands embedded in a non-forested, anthropogenically transformed landscape matrix. Accordingly, principles of island biogeography and metapopulation dynamics (e.g. size and quality of forest remnants, or distances between fragments) determine the long-term fate of populations, species, and communities of organisms that thrive in these fragments. One special topic related to the island nature of today's tropical forests are edge effects (Murcia 1995, Laurance 1998): forest edges are ecological boundaries characterized by very steep gradients from the forest interior to the surrounding open vegetation. Such gradients are well documented for abiotic factors (light, temperature, humidity, wind speed, rainfall; e.g. Camargo and Kapos 1995, Davies-Colley et al. 2000, Newmark 2001) and have manifold effects on the vegetation (Fox et al. 1997, Williams-Linera et al. 1998, Mesquita et al. 1999). Consequences for organisms at higher trophic levels are also evident. In phytophagous animals, such consequences may be mediated indirectly through the dependence on particular plant species whose presence or abundance changes as a consequence of altered abiotic conditions. Alternatively, abiotic conditions may directly affect behaviour or impose physiological constraints (e.g. inability to tolerate intense solar radiation or low humidity outside the forest shade).

Edge effects are of special importance in small habitat fragments, where they may penetrate so far that the entire fragment must be rather viewed as an edge habitat (Fox et al. 1997). However, for large and rather well protected conservation areas understanding the processes at forest margins is also of paramount importance. These margins form the interface between forest and cultivated landscape, and they are usually characterized by a dynamic mosaic of succession habitats (villages, farmland, abandoned fields and pastures, shrub land, secondary forest). This habitat heterogeneity may even increase the species richness relative to forest areas of comparable size. On the other hand, disturbance regimes at forest margins may facilitate the establishment of invasive or weedy species in remaining natural habitats. Therefore, the study of tropical forest margins is an important issue in conservation biology.

In contrast to the effects of land-use practices such as selective logging (Meijaard et al. 2005), or the irreversible conversion of forest (establishment of agroforests and plantations, agriculture, settlement: Holloway et al. 1992), rather little is known as to how animal communities respond to the altered conditions at tropical forest margins. This is especially obvious when it comes to those organisms that account for the largest fraction of terrestrial species diversity, viz. the arthropods. Available studies often focus on single species or small sets of species. Here we review the response of very large insect ensembles to environmental changes at the margin of large tropical forest reserves. We selected moths as a model group for this approach. A growing number of studies in all tropical realms has been conducted in recent decades dealing with disturbance and/or edge effects on moth diversity (e.g. Holloway 1985, 1998, Holloway et al. 1992, Chey et al. 1997, Intachat et al. 1997, 1999, Willott 1999, Kitching et al. 2000, Basset et al. 2001, Lepš et al. 2001, Ricketts et al. 2001, Axmacher et al. 2004). We integrate our own findings with these studies to examine whether generalized trends can now be formulated.

Four reasons render moths suitable targets for this research approach.

1. Moths are very rich in species, thus allowing for the assessment of effects on 'megadiverse' ensembles (terminology: Fauth et al. 1996). At the same time, moths comprise many families that widely differ in ecological requirements (e.g. hostplant affiliations) and morphological traits (size, development of wings or mouthparts: Scoble 1995, Kristensen 1999, Holloway et al. 2001). Thus, cross-taxon comparisons are possible to test the generality of findings, but still within the boundaries of a common ground plan.
2. Moths are taxonomically far better known than any other insect group of comparable diversity (Scoble 1999, Holloway et al. 2001). Hence species identifications are possible for a large fraction of samples, with all the advantages precise identifications provide over mere parataxonomic sorting (Krell 2004).

3. Moths can be readily sampled in large numbers by attraction to light (Muirhead-Thomson 1991), resulting in samples amenable to thorough statistical evaluation with high power. In contrast to a commonly held misconception, weak artificial light sources have a small attraction radius (usually below 30 m: Beck and Linsenmair 2006, Wirooks 2005). Thus, samples are drawn from the local community around the trap site and allow for ecological characterization of moth ensembles at small spatial scales.
4. Adult moths and caterpillars differ widely in their resource requirements and dispersal capacities. Moth larvae are largely sedentary organisms, the majority being herbivores with often restricted host plant ranges. Adult moths, in contrast, are usually capable of flight (wing reductions occur in less than 1% species worldwide and are rare in tropical forests: Sattler 1991) and feed on nectar. Specialization to certain nectar sources is an exception, and many moths even do not feed at all ('capital breeders': Tammaru and Haukioja 1996). Moth communities are shaped by the balance between availability of resources (such as larval hostplants) and adult dispersal. If moth communities reflect habitat differences on small spatial scales (in the range some dozens to hundreds of meters distance between sites) it can be concluded that local resource availability is the major determinant of community composition. In contrast, if differences between communities largely depend on spatial distance, dispersal has to be regarded as the key to understand diversity patterns.

We use very large, quantitative samples of moths collated across habitat gradients at the margins of two large conservation areas (situated in Borneo and southern Ecuador) to address the following questions:

- How does moth diversity change along the gradient from the natural forest to the more disturbed habitats situated directly at forest margins? Are these changes uniform across regions or moth families with different life habits?
- How do species compositions change across forest margins? Is there an abrupt or continuous change? Are different moth taxa differentially sensitive? Is there a high species turnover, or do communities rather differ due to shifts in relative abundance of their component species?

To answer these questions we attracted moths to light-towers (gauze cylinders, height 1.6 m, diameter 0.6 m). These were placed at ground level and equipped with weak light sources (15 W blacklight tube in Borneo, 2 x 15 W tubes in Ecuador) to minimize attraction of moths from adjacent habitats. Sampling sites were selected to ensure that (a) in a radius of 50 m around the light source the vegetation was homogeneous and (b) that the light tower was not visible from the nearest other sampling site. All moth individuals of the selected target taxa which were attracted to the light source were sampled manually during the first 3 h after dusk. Manual sampling is more effective than automatic traps in the case of Geometridae and the resulting specimens can be more easily identified due to their better condition (Axmacher and Fiedler 2004). All moth samples were subsequently set, sorted and identified using taxonomic literature and reference collections in large research museums (see acknowledgments). Further details on procedures for moth collecting and processing of samples can be found in Schulze (2000), Beck et al. (2002), and Fiedler and Schulze (2004) for the Bornean studies, and in Hilt and Fiedler (2005, 2006) and Hilt et al. (2006) for the sampling in Ecuador.

2 Moths at the margin of Mount Kinabalu National Park (Sabah, North East Borneo)

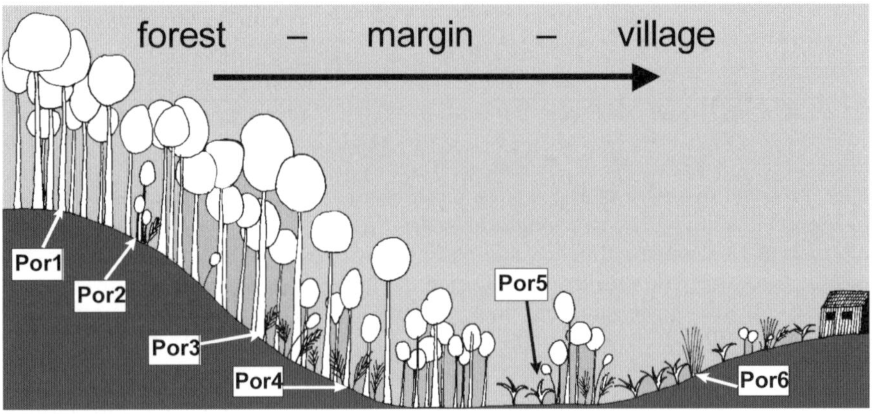

Fig. 1. Schematic illustration of the sampled transect from natural forest to the forest margin at Poring (Mt. Kinabalu National Park, Borneo) showing the locations of light trapping sites (modified from Fiedler and Schulze 2004, see Beck et al. 2002 for more information about sampling sites). Por1: primary forest without signs of recent anthropogenic disturbance; Por2: disturbed primary forest; canopy layer opened due to a large old tree fall gap promoting dense understory vegetation with pioneer plants; Por3: old secondary forest with large gaps caused by selective logging; Por4: young secondary forest approx. 20m high with dense upper canopy layer and understory vegetation; Por5: secondary forest fragment situated in cultivated area adjacent to a banana plantation; Por6: cultivated area with banana plantation and dense bamboo thicket. Distance between Por1 and Por6 was less than 2 km.

Mount Kinabalu National Park is one of the last remaining large forest areas in NE Borneo (Wong and Phillipps 1996). We sampled moth ensembles along two habitat gradients from primary forest to farmland, one situated at the north eastern (Serinsim village) and the other at the south eastern margin (Poring village; Figure 1) of the park (Schulze 2000, Beck et al. 2002, Fiedler and Schulze 2004). Sampling sites were situated at low elevations (170-630 m a.s.l.) and represented forest margin ecotones in the lowland to hill forest zone. In one large family, the Geometridae (500 observed species, 4585 individuals), species diversity decreased along both gradients, but not in a uniform manner. Species diversity (measured as Fisher's alpha to alleviate the strong sample size dependence of recorded species numbers: Rosenzweig 1995) at disturbed forest sites towards the margin was as high, or partially even higher, than in natural forest (Beck et al. 2002). Maximum local diversity was not observed

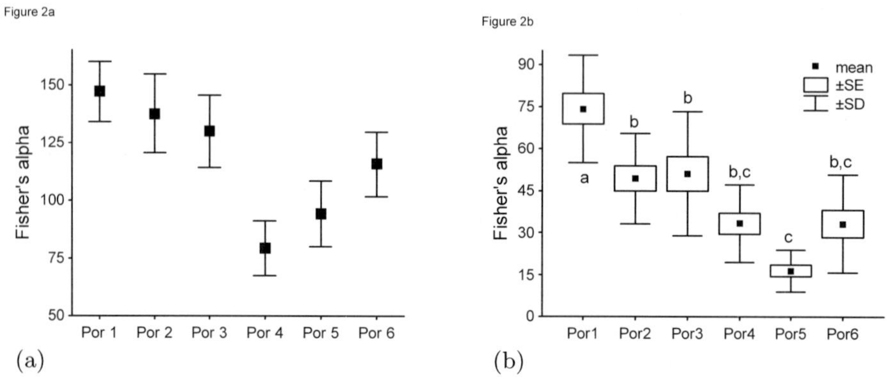

Fig. 2. Local diversity of Pyraloidea moth ensembles along the transect depicted in Figure 1, analysed at two temporal scales. (a) Fisher's alpha ± 95% confidence intervals of 13 aggregated nightly samples per site. (b) Fisher's alpha (filled squares: means, boxes: ±1SE, whiskers: ±1SD) for the same data set, based on averages of diversity estimates per sampling night. Means marked with different letters differ significantly (Scheffé test, $P<0.05$, following ANOVA: $F_{5;72}=18.47$, $P<0.0001$). The 13 nightly samples were taken in two collecting periods in the year 1997 (see Fiedler and Schulze 2004).

in natural forest, but in secondary forest sites at both forest margins. Margin habitats without closed canopy held an impoverished geometrid fauna, even though the natural forest was only a few hundred meters away. In another very large ensemble, the superfamily Pyraloidea (680 species observed, 7724 individuals), the pattern was somewhat different. Forest sites with old-grown trees were characterized by exceptionally high species diversity. Margin habitats where no old-grown trees were left, in contrast, had a far less diverse (but still rich) pyraloid fauna (Figure 2a). Interestingly, a slightly different pattern emerged when viewed at higher temporal resolution (Figure 2b). Pyraloids are the single most abundant moth group in SE Asia to be attracted to artificial light sources. Therefore, even samples taken during single nights are large enough for the calculation of meaningful diversity measures. We resampled pyraloid ensembles at 6 sites near Poring 13 times within one year (Fiedler and Schulze 2004). At this small temporal scale primary forest indeed ranked as the most diverse habitat. Weakly disturbed forest sites held significantly lower pyraloid diversity, and the latter decreased further towards the forest margin. However, losses of diversity towards the forest margin of Mt. Kinabalu National Park were not observed in all moths. Tiger moth diversity (Arctiidae, 29 observed species, 420 individuals) was practically constant from forest to farmland; this applied to the subfamilies Arctiinae and Syntominae (Schulze 2000) as well as to combined samples (K. Fiedler and C.H. Schulze, unpublished data). Hawkmoth ensembles also turned out to be rather insensitive (Schulze and Fiedler 2003). Only with large samples from multiple sites a subtle disturbance effect on hawkmoth diversity could be detected (Beck et al. 2006).

Table 1. Spearman rank correlations (r_S and P values, 1000 permutations for matrix correlations, Clarke and Gorley 2001) of alpha and beta diversity in three moth groups at the margin of Mt. Kinabalu National Park (Borneo). Alpha diversity was measured as Fisher's alpha at six sites near Poring (Beck et al. 2002, Fiedler and Schulze 2004). Matrix correlations are based on pairwise NESS similarities between samples (sample size parameter set as m=10). Note that correlation coefficients are high and positive in sign. Due to the low number of sites (n=6) and hence small statistical power, none of the reported relationships is significant if controlled for a table-wide false discovery rate (P<0.05) according to the method of Benjamini and Hochberg (1995).

		Pyraloidea	Arctiidae
alpha diversity	Geometridae	r_S=0.77 P=0.07	r_S=0.83 P=0.04
	Pyraloidea		r_S=0.60 P=0.21
beta diversity	Geometridae	r_S=0.52 P=0.05	r_S=0.42 P=0.07
	Pyraloidea		r_S=0.84 P=0.02

Changes in species composition across the forest margin were pronounced in all studied moth taxa. Ordinations of geometrid samples (using nonmetric multidimensional scaling based on NESS or CNESS indices: Brehm and Fiedler 2004) depicted the gradient from the forest interior to the margin habitats, yet with slight differences between the two study areas (Beck et al. 2002). In Pyraloidea, a major break was observed between three sites with and three sites without old-grown trees (Fiedler and Schulze 2004). A similar pattern recurred within the Arctiidae. Matrix rank correlation tests reflect the conclusions drawn from ordination analyses (Table 1): the pattern of faunal change was more concordant between Arctiidae and Pyraloidea than both families were to the Geometridae. Within Pyraloidea, temporal dynamics and between-site differences affected ensemble composition almost equally (two-way crossed ANOSIM, 1000 permutations, statistics based on Spearman's rank correlation as implemented in Clarke and Gorley 2001; site differences: $r_S = 0.436$, sampling time: $r_S = 0.485$, both $P<0.001$). Overall, dynamics of pyraloid ensembles were concordant across the forest margin and were not more pronounced at the more open sites than in the forest interior (Fiedler and Schulze 2004).

3 Moths at the margin of Podocarpus National Park (Andes, southern Ecuador)

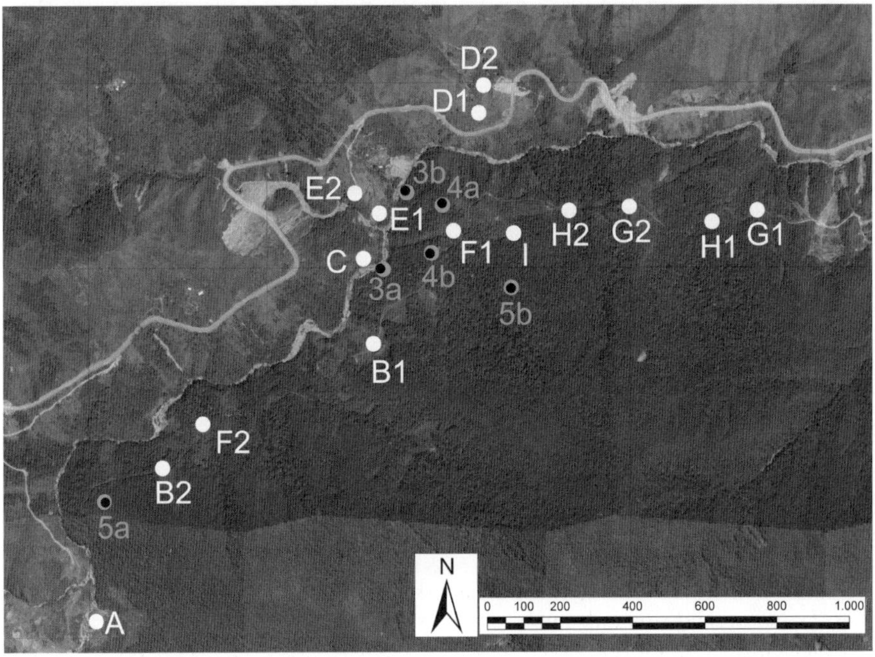

Fig. 3. Composite aerial photograph of the study area at the margin of Podocarpus National Park, southern Ecuador (after Hilt and Fiedler 2005). Moth ensembles were monitored at 21 sites representing a succession gradient, with 4-9 samples taken per site (Hilt and Fiedler 2005, Hilt et al. 2006). Sites A, B, C and E are early succession sites (without woody vegetation) situated near the margin of closed natural forest; sites D are two abandoned cattle pastures; sites G, H and I are late succession habitats (with shrub or tree cover); and sites 3a to 5b are located inside natural forest. Note the small spatial scale of the study. Photograph courtesy of E. Jordan.

Podocarpus National Park is one of the last remaining large blocks of continuous montane forest in southern Ecuador (Madsen and Øllgaard 1994).

We sampled moth ensembles (Geometridae, Arctiidae) at 21 sites situated in various types of succession habitats and in natural forest along the boundary of this conservation area (Figure 3; see Hilt and Fiedler 2005, 2006 for further details about sites). All sites were situated in a narrow altitudinal belt (1800-2000 m a.s.l.) where montane forest is the potential natural vegetation (Paulsch 2002, Homeier 2004). Light-trapping sites were close to the margin of continuous natural forest, with a maximum distance of 445 m (Hilt and Fiedler 2005). The study area is part of a global biodiversity hotspot (e.g. Myers et al. 2000, Brooks et al. 2002), and this is also true for moths (Brehm et al. 2005).

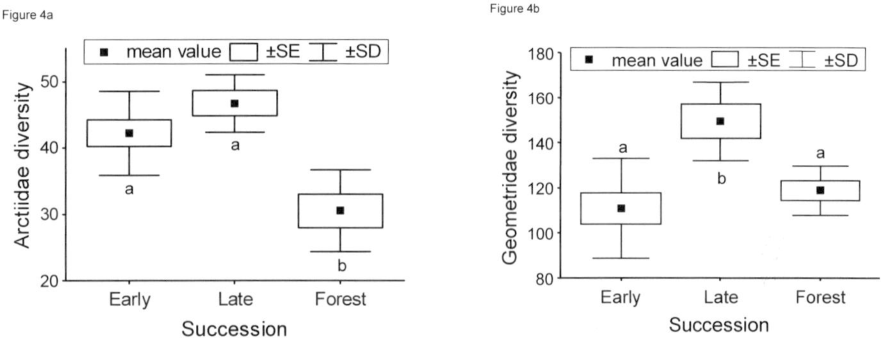

(a) (b)

Fig. 4. Local diversity of Arctiidae (a) and Geometridae (b) at the margin of Podocarpus National Park, measured using Fisher's alpha for aggregated samples from each site. Sites were grouped into three categories: natural forest understorey (N=6), late succession sites (N=5), and early succession sites plus abandoned cattle pastures (i.e. margin habitats without woody vegetation, N=10). Filled squares: means, boxes: ±1SE, whiskers: ±1SD. Means marked with different letters differ significantly (Scheffé test, P<0.05, following ANOVA: $F_{2;18}$ >7.30, P<0.005).

Diversity patterns varied at the local scale between the studied moth families. Local arctiid diversity, measured again as Fisher's alpha (recorded totals: 287 species, 9211 individuals), was significantly higher in late and early succession habitats close to the forest edge than in natural forest understorey (Figure 4a). Geometrid diversity (recorded totals: 868 species, 23720 individuals) was clearly highest in late successional habitats again, but was markedly low in open early succession habitats (Figure 4b; see Brehm and Fiedler 2005, Hilt et al. 2006). The differences between the two moth families in their response to habitat differences became even more obvious when examining the covariance of local diversity measures (Table 2). Correlations were never significant, and sometimes even negative in sign, in all comparisons between arctiids and geometrids. Correlations were significantly positive within Geometridae and also mostly within Arctiidae (Hilt and Fiedler 2006). In the latter family, only the subfamily Lithosiinae (whose larvae are suspected to feed frequently

Table 2. Concordance between local diversity of moth ensembles from the margin of Podocarpus National Park (Ecuador) across different taxonomic levels. Given are Pearson's correlation coefficients r between alpha diversity of Arctiidae, Geometridae and various subordinated taxa (measured as Fisher's alpha of the log-series distribution). A: Arctiidae, G: Geometridae. Ennominae and Larentiinae are subfamilies of Geometridae; Phaegopterini and Ctenuchini (here viewed as including Euchromiini; see Jacobson and Weller 2002) are clades in the Arctiidae-Arctiinae; Lithosiinae are another subfamily of Arctiidae. * $P<0.01$, ** $P<0.005$, *** $P<0.001$, n.s. not significant ($P>0.05$). Bold figures: results significant after controlling for a table-wide false discovery rate ($P<0.05$) according to the method of Benjamini and Hochberg (1995).

	Arctiidae total	Ennominae (G)	Larentiinae (G)	Phaegopterini (A)	Ctenuchini/ Euchromiini (A)	Lithosiinae (A)
Geometridae total	0.18, n.s.	**0.94*****	**0.73*****	-0.24, n.s.	-0.14, n.s.	0.12, n.s.
Arctiidae total		0.10, n.s.	-0.01, n.s.	**0.63****	**0.65****	**0.74*****
Ennominae (G)			0.55*	-0.27, n.s.	-0.19, n.s.	-0.002, n.s.
Larentiinae (G)				-0.26, n.s.	-0.25, n.s.	0.02, n.s.
Phaegopterini (A)					**0.68*****	0.15, n.s.
Ctenuchini/ Euchromiini (A)						0.21, n.s.

on lichens instead of vascular plants: Weller et al. 1999, Kitching and Rawlins 1999) revealed no concordance with other arctiid clades.

The poor concordance of diversity patterns across more distantly related moth clades indicates that differences in the habitat requirements of these insects are so strong that they lead to differential success in colonizing microhabitats in the heterogeneous mosaic at the forest margin. In addition, this finding undermines the potential to use one such moth group to predict diversity of others ('biodiversity indication': Lawton et al. 1998, Brehm and Fiedler 2003, Schulze et al. 2004). Even at the small spatial scale of the sampling design adopted here, idiosyncrasies of clades prevented the emergence of common ecological trends in local species diversity. Some moth groups, like Arctiidae, obviously do better in margin habitats than others do, such as the Geometridae. This finding is in line with earlier observations in SE Asia and Australia that arctiid ensembles tend to be less sensitive to anthropogenic habitat disturbance and may even benefit from opening of the forest canopy (Holloway 1985, 1998, Kitching et al. 2000, Schulze 2000).

With regard to beta diversity the picture changes remarkably. Matrix rank correlations between CNESS distance matrices of Geometridae and Arctiidae were all highly significant, irrespective of the weight given to rare or abundant species ($0.576 < r_S < 0.902$, all $P << 0.001$: Hilt 2005). Also within the Arctiidae, matrix rank correlations were significant for most comparisons between subfamilies and tribes (Hilt and Fiedler 2006). Hence, while various moth taxa obviously respond quite discordantly with regard to their potential to form diverse or poor ensembles in different microhabitats along the forest margin, the pattern of species turnover turned out to be unexpectedly concordant. It should be noted at this point that site-to-site variation of moth ensembles within natural forest does of course exist. However, species composition was remarkably similar between natural forest sites at comparable altitude in the Ecuadorean study area (Brehm et al. 2003, Süßenbach 2003), and this also extends to forest sites after moderate experimental manipulation (K. Fiedler and M. Zimmermann, unpublished results).

4 Moth ensembles in environmental gradients: do taxa differ in sensitivity?

The above results already indicate that not only individual species, but even entire moth taxa differ in their overall sensitivity to environmental change. This is an important issue for practical conservation measures: the selection of target groups that are to be included in biodiversity assessment or monitoring programmes should be based on solid evidence as to the suitability of the chosen 'indicators'. From what has been shown above, and in line with the pertinent literature, it would appear that Arctiidae comprise a greater fraction of 'weedy' species (Holloway 1985, 1998, Kitching et al. 2000), i.e. species whose polyphagous larvae thrive on cultivated plants and especially on pioneer vegetation. In contrast, Old World tropical Geometridae ensembles respond very sensitively to the loss or reduction in canopy cover of forest habitats (Intachat et al. 1997, 1999, Beck et al. 2002, but see Axmacher et al. 2004). We here apply two analytical approaches to our very large New World (i.e. Ecuadorian) data set to test this hypothesis.

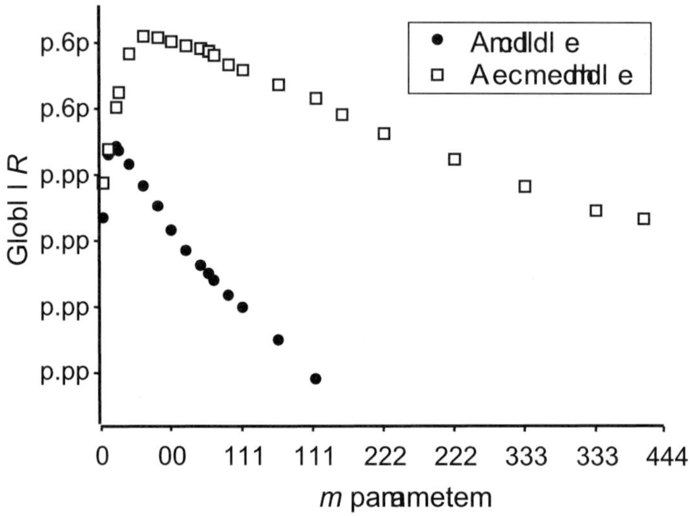

Fig. 5. Discriminatory power of Analyses of Similarities (expressed as values of the global R statistics, one-way ANOSIM, 1000 permutations) as a function of the sample size parameter m (of CNESS indexes as a measure of faunal distances) for two moth families. Basis for the analyses were the samples from 21 sites at the margin of Podocarpus National Park, Ecuador, separated into four habitat categories (natural forest understorey, late succession habitats with woody vegetation, early succession habitats without woody vegetation, abandoned pastures). Highest resolution is obtained at m=10 for the Arctiidae and m=30 for the Geometridae. Resolution decays more strongly with increasing emphasis on rare species (= high m values) in the Arctiidae. Due to lower samples sizes, no values were available in Arctiidae beyond m=152.

As a first approach, we explored how well faunal distance matrices for Arctiidae and Geometridae reflect habitat groupings according to their vegetation and successional status. The 21 sites at the margin of Podocarpus National Park were *a priori* grouped into abandoned cattle pastures, early and late succession habitats, and mature forest understorey. We then used the R statistics of ANOSIM (Clarke 1993, Legendre and Legendre 1998) as a measure of the sensitivity of a given moth group to reflect habitat differences. For this purpose, we calculated all possible pairwise distances between moth samples from the 21 sites. Since samples from mega-diverse tropical moth communities are never 'complete' and prone to sampling effects (Brehm and Fiedler 2004), the NESS index or its metric analogue CNESS (Trueblood et al. 1994) are the most suitable measures of faunal distances in such cases. In this comparison, Geometridae clearly emerged as the environmentally more sensitive moth group with higher R values throughout (Figure 5). Since this difference was observed at all values of the sample size parameter m, the higher resolution of analyses based on Geometridae samples cannot be regarded as merely an effect of the larger catch sizes that were achieved with the more abundant and more speciose geometrids. Arctiidae achieved relatively good discriminatory power only in a small range of values of the sample size parameter m, while ecological resolution was heavily reduced when emphasizing dominant (m=1) or rare species (high m values). Geometridae were superior in that regard since high segregation between samples was obtained over a wide range of m values. In practice this means that in a monitoring study robust results can be expected with Geometridae as target groups.

ANOSIM, like unconstrained ordination techniques such as non-metric multidimensional scaling (Clarke 1993), draws exclusively on faunal data and does not include environmental variables *a priori*. We therefore also applied a constrained ordination technique (Økland 1996) to our data, namely Redundancy Analysis (RDA: Legendre and Legendre 1998), to explore differential sensitivity of Arctiidae and Geometridae. As environmental variables we used canopy openness (measured from hemispherical photographs) as a proxy of integrity of the forest, and altitude as a proxy of overall habitat temperature regimes. Studies in natural forest in Podocarpus National Park revealed how sensitive moth ensembles respond to even minor altitudinal differences in species composition (Brehm et al. 2003). Geometridae responded very sensitively to canopy openness (Figure 6a). Altitude had a measurable effect as well, which was almost independent of canopy openness. In the Arctiidae, in contrast, the effects of these two environmental variables were less pronounced, and more difficult to separate from each other (Figure 6b). Therefore, also according to the RDA, geometrid moths were more indicative of environmental gradients than arctiids. Notably, arctiids appeared not to be as suitable for indication purposes as has been found for these moths in Nearctic forests (Summerville et al. 2004). Our analyses underscore how different the responses of moth ensembles at the community level are. They also support the view that Geometridae make up excellent targets for environmental monitoring of insect faunas of tropical forests, even at small spatial scales, where the flight capacity of all species should be large enough to allow them to reach any site when dispersing from the remaining natural forest.

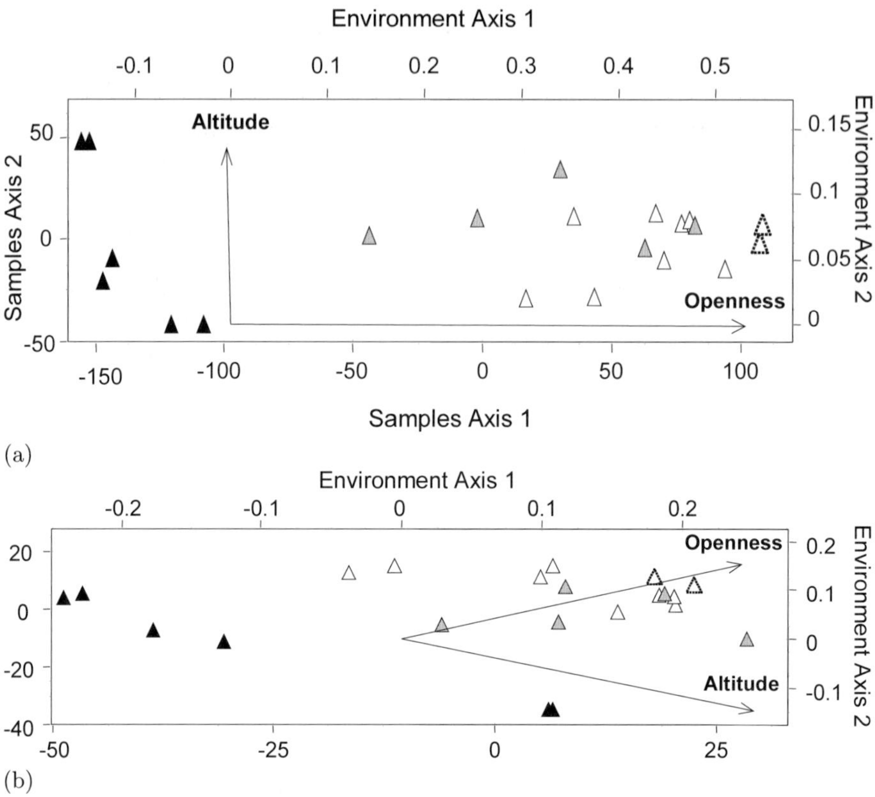

Fig. 6. Explanatory power of two environmental variables (canopy openness, measured from hemispherical photographs; and altitude) for variation in ensemble composition of moths at the margin of Podocarpus National Park (Ecuador). Graphs are ordination plots of samples with vectors of predictor variables overlaid according to a Redundancy Analysis (RDA). Arrows: environmental vectors, symbols: 21 sites at the forest margin. Filled symbols: natural forest understory, shaded symbols: late woody succession vegetation, open symbols: early succession stages, dotted triangles: abandoned pastures. (a) Geometridae. Canonical axis 1 explains 28.2%, axis 2 explains 1.9% of data variance (total: 30.1%). Correlations with altitude are –0.004 (axis 1) and 0.136 (axis 2), and those with canopy openness are 0.531 (axis 1) and –0.002 (axis 2). (b) Arctiidae. Canonical axis 1 explains 12.2%, axis 2 explains 4.4% of data variance (total: 16.8%). Correlations with altitude are 0.251 (axis 1) and –0.147 (axis 2), and those with canopy openness are 0.242 (axis 1) and 0.152 (axis 2). Environmental variables explain differences between Geometridae ensembles far better than those in Arctiidae ensembles. Moreover, the two environmental variables have almost independent effects in Geometridae, but are less well separable in their influence on Arctiidae.

5 Moth ensembles in margin habitats: how different from forest?

This leads to the issue as to how different moth ensembles are across the habitats of a forest margin. For species that can thrive equally well inside and outside the forest this distinction is obviously irrelevant. Very few, if any, of the moth species covered by our studies in Borneo and Ecuador appear to have such generalist habitat requirements. One might rather expect that different species live in either the forest or the surrounding landscape matrix. If so, there should be a sharp boundary and little overlap between ensembles inside and outside the forest. Alternatively, habitats outside the natural forest might be so poor in quality for practically all moth species (for example due to frequent human intervention) that they are just sinks which rely on frequent immigration of dispersers from nearby forest. If this were the case, the further away from the source forest, the more depauperate the moth ensembles should be. Moreover, such ensembles from margin habitats should consist of impoverished subsets of the forest fauna, which can hierarchically be nested in the forest fauna with increasing distance from the source area.

Our studies in Borneo and Ecuador reveal that neither of these extremes is the case. While some moth species turned out to be 'true' forest dwellers, and others exclusively occur in the landscape matrix around closed forest, most species were found on both sides of the forest margin, but at different abundances. Generally, the segregation of moth ensembles was more robust when abundances (i.e. CNESS or NESS index measures) rather than incidences (e.g. the Soerensen index or CNESS with very high m values) were used to express faunal similarities. This indicates that shifts in the local abundance relations, rather than the mere presence or absence of species, generate the gradients in beta diversity across forest margins. On the other hand, our data also allow to refute a strict source-sink concept. If moth ensembles from a habitat in the surrounding landscape matrix were really just impoverished subsets of forest ensembles (e.g. would contain only the more dispersive species, but in roughly the same abundance relationships as they have in the forest), then CNESS distances would become negligible at least at high m values. This is because at high m values the differences caused by 'sampling effects' (here: viewing the ensemble outside the forest as a small sub-sample from the source community) should vanish. This was, however, never the case (e.g. Hilt and Fiedler 2005, 2006, Hilt et al. 2006).

We therefore conclude that moth ensembles at forest margins are governed by a balance of both of these processes. There is considerable dispersal of mobile moths (described by Ricketts et al. 2001 as 'halos' of forest moths around natural forest remnants), and certainly not all of these stray individuals will find suitable resources at all sites to successfully reproduce. In line with that conclusion we observed high proportions of singletons among arctiid and geometrid samples from early succession habitats and especially on abandoned pastures (Hilt and Fiedler 2005, Hilt et al 2006). Such habitats are likely to represent sinks for the majority of arctiid and geometrid moth species in Ecuador since most potential larval host plants are not available. However, at the same time there are abundance shifts of species that are a permanent component of communities inside as well as outside the forests. These shifts, caused by the matching of habitat requirements and local resource availability for each individual species, allow some moths to be more abundant in the margin habitats (e.g. species that otherwise live in tree-fall gaps, landslide areas or forest canopies; see also Schulze et al. 2001), whereas others are more abundant in closed forest, without being 'exclusive' species of any of these habitat types.

6 Conclusions and Outlook

Very species rich moth ensembles at the margins of tropical conservation areas revealed a high sensitivity towards environmental change at surprisingly small spatial scales. Hence, despite the substantial dispersal potential of most moths, these organisms are perfectly suitable to monitor effects of environment on the diversity and dynamics of mega-diverse herbivore communities. A common pattern emerging from our studies is that local alpha diversity of different taxonomic groups shows quite discordant idiosyncratic responses. For example, Geometridae ensembles are distinctly impoverished just a few dozen meters away from the natural forest, while this is never the case in Arctiidae moths. Moreover, it is a matter of scale whether ensembles from disturbed habitats appear as more, or less, diverse than the respective ensembles in 'undisturbed' natural forest. Such a scale dependency is not only restricted to the spatial dimension (Hill and Hamer 2000, 2004, Rahbek 2005), but may also extend to the temporal dimension, as shown by our results on Pyraloidea from Borneo. In contrast to these less predictable patterns of alpha diversity, the patterns of species turnover (beta diversity) were surprisingly invariant across taxonomic groups with widely diverging life history syndromes. They were also rather stable with regard to reductions in sample size (i.e. when comparing smaller samples of subordinated clades with each other or the entire moth family: Brehm et al. 2003, Hilt and Fiedler 2006). We therefore suggest that for the assessment of environmental effects on rich insect communities the monitoring and analysis on the community level is a much more promising and powerful approach than the selection of a few 'indicator' species, the choice of which is by necessity arbitrary.

Our analyses also underline that in mobile organisms the differentiation between local communities is a matter of shifts in relative abundance composition rather than species composition. Hence, analyses using abundance information are far more sensitive than those relying on presence-absence data. This problem is particularly obvious in speciose tropical communities, where sampling effects (tourists, singletons, under-recording of many rare species in finite samples: Novotny and Basset 2000, Gotelli and Colwell 2001, Brehm and Fiedler 2004) pose great challenges to the modes of data collection and analysis.

Finally, our results show that margin faunas are different from those seen in the 'natural' forest as well as in the open landscape matrix dominated by human interference. The analysis of ensemble structures suggests that margin communities are neither simple subsets of the forest fauna, nor comprised predominantly of 'weedy' non-forest species. Such ensembles are rather a unique blend of species, reflecting local habitat conditions (e.g. resource availability and microclimate) in the habitat mosaic of the forest margin. Further research, especially on larger temporal scales, will be needed to assess how unique and stable such margin ensembles really are, and how they depend on the varying influx from source populations inside and outside the forest. For tropical insects, such a deeper mechanistic understanding will only become feasible once more is known about the habitat requirements and life histories of the component species. Given the poor knowledge of even most basic traits such as host plant affiliations and adult food requirements of insects in tropical realms, this is a long way to go – and one which needs to be pursued immediately in the light of ongoing losses of tropical near-natural habitats and their unique biodiversity.

Acknowledgements
We thank the Economic Planning Unit at the Prime Minister's Department and Sabah Parks for granting research permits in Malaysia; and the Ministerio del Medio Ambiente as well as the foundation Nature and Culture International for granting permits and access to study sites in Ecuador. We heartily appreciate the support by many taxonomists and museum curators who assisted with identifications and provided access to research collections: Christoph Häuser, Daniel Bartsch (Stuttgart); Jeremy Holloway, Martin Honey, Linda Pitkin, Malcolm Scoble, Michael Shaffer (London); Axel Hausmann (Munich); Bernard Landry (Geneva); Matthias Nuss (Dresden); Wolfgang Speidel (Bonn); Suzanne Rab-Green (New York); Benoît Vincent (Paris); Wolfgang Nässig (Frankfurt). Jan Beck, Dirk Süßenbach and many student helpers contributed to the collection and processing of moth samples. The Deutsche Forschungsgemeinschaft financially supported our work (Fi 547/4; Fi 547/6; FOR 402/1, FOR 402/2).

References

Armenteras D, Gast F, Villareal H (2003) Andean forest fragmentation and the representativeness of protected natural areas in the eastern Andes, Colombia. Biol Conserv 113: 245-256

Axmacher JC, Fiedler K (2004) Manual versus automatic moth sampling at equal light sources – a comparison of catches from Mt. Kilimanjaro. J Lepid Soc 58: 196-202

Axmacher JC, Tünte H, Schrumpf M, Müller-Hohenstein K, Lyaruu HVM, Fiedler K (2004) Diverging diversity patterns of vascular plants and geometrid moths during forest regeneration on Mt. Kilimanjaro, Tanzania. J Biogeogr 31: 895-904

Basset Y, Charles E, Hammond DS, Brown VK (2001) Short-term effects of canopy openness on insect herbivores in a rain forest in Guyana. J Appl Ecol 38: 1045-1058

Beck J, Kitching IJ, Linsenmair KE (2006) Effects of habitat disturbance can be subtle yet significant: biodiversity of hawkmoth assemblages (Lepidoptera: Sphingidae) in Southeast-Asia. Biodiv Conserv 15: 465-486

Beck J, Linsenmair KE (2006) Feasibility of light-trapping in community research on moths: attraction radius of light, completeness of samples, nightly flight times and seasonality of Southeast-Asian hawkmoths (Lepidoptera: Sphingidae). J Res Lepid 39 (in press)

Beck J, Schulze CH, Linsenmair KE, Fiedler K (2002) From forest to farmland: diversity of geometrid moths along two habitat gradients on Borneo. J Trop Ecol 18: 33-51

Benjamini Y, Hochberg Y (1995) Controlling the false discovery rate: a practical and powerful approach to multiple testing. J R Statist Soc B57: 289-300

Brehm G, Fiedler K (2003) Faunal composition of geometrid moths changes with altitude in an Andean montane rain forest. J Biogeogr 30: 431-440

Brehm G, Fiedler K (2004) Ordinating tropical moth ensembles from an elevational gradient: a comparison of common methods. J Trop Ecol 20: 165-172

Brehm G, Fiedler K (2005) Diversity and community structure of geometrid moths of disturbed habitat in a montane area in the Ecuadorian Andes. J Res Lepid 38: 1-14

Brehm G, Homeier J, Fiedler K (2003) Beta diversity of geometrid moths (Lepidoptera: Geometridae) in an Andean montane rainforest. Diversity Distrib 9: 351-366

Brehm G, Pitkin LM, Hilt N, Fiedler K (2005) Montane Andean rain forests are a global diversity hotspot of geometrid moths. J Biogeogr 32: 1621-1627

Brooks TM, Mittermeier RA, Mittermeier CG, da Fonseca GAB, Rylands AB, Konstant WR, Flick P, Pilgrim J, Oldfield S, Magin G, Hilton-Taylor C (2002) Habitat loss and extinction in the hotspots of biodiversity. Conserv Biol 16: 909-923

Camargo JLC, Kapos V (1995) Complex edge effects on soil moisture and microclimate in central Amazonian forest. J Trop Ecol 11: 205-221

Chey VK, Holloway JD, Speight MR (1997) Diversity of moths in forest plantations and natural forests in Sabah. Bull Entomol Res 87: 371-385

Clarke KR (1993) Non-parametric multivariate analyses of changes in community structure. Austral J Ecol 18: 117-143

Clarke KR, Gorley RN (2001) PIMER v5: user manual/tutorial. Primer-E, Plymouth

Davies-Colley RJ, Payne GW, van Elswijk M (2000) Microclimate gradients across a forest edge. New Zealand J Ecol 24: 111-121

Fauth JE, Bernardo J, Camara M, Resetarits WJ, van Buskirk J, McCollum SA (1996) Simplifying the jargon of community ecology: a conceptual approach. Am Nat 147: 282-286

Fiedler K, Schulze CH (2004) Forest modification affects diversity (but not dynamics) of speciose tropical pyraloid moth communities. Biotropica 36: 615-627

Fox BJ, Taylor JE, Fox MD, Williams C (1997) Vegetation changes across edges of rainforest remnants. Biol Conserv 82: 1-13

Gotelli NJ, Colwell RK (2001) Quantifying biodiversity: procedures and pitfalls in the measurement and comparison of species richness. Ecol Letts 4: 379-391

Hamer KC, Hill JK (2000) Scale-dependent effects of habitat disturbance on species richness in tropical forests. Conserv Biol 14: 1435-1440

Hill JK, Hamer KC (2004) Determining impacts of habitat modification on diversity of tropical forest fauna: the importance of spatial scale. J Appl Ecol 41: 744-754

Hilt N (2005) Diversity and species composition of two different moth families (Lepidoptera: Arctiidae vs. Geometridae) along a successional gradient in the Ecuadorian Andes. PhD thesis, Univ Bayreuth. URL: http://opus.ub.uni-bayreuth.de/volltexte/2006/201

Hilt N, Fiedler K (2005) Diversity and composition of Arctiidae moth ensembles along a successional gradient in the Ecuadorian Andes. Diversity Distrib 11: 387-398

Hilt N, Fiedler K (2006) Arctiid moth ensembles along a successional gradient in the Ecuadorian montane rain forest zone: how different are subfamilies and tribes? J Biogeogr 33: 108-120

Hilt N, Brehm G, Fiedler K (2006) Diversity and ensemble composition of geometrid moths along a successional gradient in the Ecuadorian Andes. J Trop Ecol 22: 155-166

Holloway JD (1985) Moths as indicator organisms for categorising rain forest and monitoring changes and regenerating processes. In: Chadwick AC, Sutton SL (eds) Tropical rain forest: the Leeds symposium. Philosophical and Literary Society, Leeds, pp 235-242

Holloway JD (1998) The impact of traditional and modern cultivation practices, including forestry, on lepidopteran diversity in Malaysia and Indonesia. In: Newbery DM, Prins HHT, Brown ND (eds) Dynamics of tropical ecosystems. Blackwell, Oxford, pp 567-597

Holloway JD, Kirk-Spriggs AH, Chey VK (1992) The response of some rain forest insect groups to logging and conversion to plantation. Phil Trans R Soc Lond B 335: 425-436

Holloway JD, Kibby G, Peggie D (2001) The families of Malesian moths and butterflies. Brill, Leiden

Homeier J (2004) Baumdiversität, Waldstruktur und Wachstumsdynamik zweier tropischer Bergregenwälder in Ecuador und Costa Rica. Diss Bot 391: 1-207

Intachat J, Holloway JD, Speight MR (1997) The effect of different forest management practices on geometroid moth populations and their diversity in Peninsular Malaysia. J Trop Forest Sci 9: 411-430

Intachat J, Holloway JD, Speight MR (1999) The impact of logging on geometroid moth populations and their diversity in lowland forests of Peninsular Malaysia. J Trop Forest Sci 11: 61-78

Jacobson NL, Weller SJ (2002) A cladistic study of the Arctiidae (Lepidoptera) by using characters of immatures and adults. Thomas Say Publications in Entomology: Monographs. Entomological Society of America, Lanham

Kitching RL, Orr AG, Thalib L, Mitchell H, Hopkins MS, Graham AW (2000) Moth assemblages as indicators of environmental quality in remnants of upland Australian rain forest. J Appl Ecol 37: 284-297

Kitching IJ, Rawlins JE (1999) The Noctuoidea. In: Kristensen NP (ed) Handbook of zoology, vol 1, part 1, Lepidoptera, moths and butterflies. De Gruyter, Berlin, New York, pp 355-401

Krell F-T (2004) Parataxonomy vs. taxonomy in biodiversity studies – pitfalls and applicability of 'morphospecies' sorting. Biodiv Conserv 13: 795-812

Kristensen NP (1999) Lepidoptera, moths and butterflies. In: Fischer M (ed) Handbook of zoology, vol 1, part 1. De Gruyter, Berlin, New York

Laurance WF (1998) A crisis in the making: responses of Amazonian forests to land use and climate change. Trends Ecol Evol 13: 411-415

Laurance WF, Lovejoy TE, Vasconcelos HL, Bruna EM, Didham RK, Stouffer PC, Gascon C, Bierregaard RO, Laurance SG, Sampaio E (2002) Ecosystem decay of Amazonian forest fragments: a 22-year investigation. Conserv Biol 16: 605-618

Lawton JH, Bignell DE, Bolton B, Bloemers GF, Eggleton P, Hammond PM, Hodda M, Holt RD, Larsen TB, Mawdsley NA, Stork NE, Srivastava DS, Watt AD (1998) Biodiversity inventories, indicator taxa and effects of habitat modification in tropical forest. Nature 391: 72-75

Legendre P, Legendre L (1998) Numerical ecology, 2nd ed. Elsevier, Amsterdam

Lepš J, Novotný V, Basset Y (2001) Habitat and successional status of plants in relation to the communities of their leaf-chewing herbivores in Papua New Guinea. J Ecol 89: 186-199

Madsen JE, Øllgaard B (1994) Floristic composition, structure, and dynamics of an upper montane rain forest in Southern Ecuador. Nordic J Bot 14: 403-423

Meijaard E, Sheil D, Nasi R, Augeri D, Rosenbaum B, Iskandar D, Setyawati T, Lammertink M, Rachmatika I, Wong A, Soehartono T, Stanley S,

O'Brien T (2005) Life after logging: reconciling wildlife conservation and production forestry in Indonesian Borneo. Center for International Forestry Research, Bogor

Mesquita RCG, Delamônica P, Laurance WF (1999) Effect of surrounding vegetation on edge-related tree mortality in Amazonian forest fragments. Biol Conserv 91: 129-134

Muirhead-Thomson RC (1991) Trap responses of flying insects. Academic Press, London

Murcia C (1995) Edge effects in fragmented forests: implications for conservation. Trends Ecol Evol 10: 58-62

Myers N, Mittermeier RA, Mittermeier CG, da Fonseca GAB, Kent J (2000) Biodiversity hotspots for conservation priorities. Nature 403: 853-858

Newmark WD (2001) Tanzanian forest edge microclimatic gradients: dynamic patterns. Biotropica 33: 2-11

Novotný V, Basset Y (2000) Rare species in communities of tropical insect herbivores: pondering the mystery of singletons. Oikos 89: 564-572

Økland RH (1996) Are ordination and constrained ordination alternative or complementary strategies in general ecological studies? J Veget Sci 7: 289-292

Paulsch A (2002) Development and application of a classification system for undisturbed and disturbed tropical montane forest based on vegetation structure. PhD thesis, Univ Bayreuth. URL: http://opus.ub.uni-bayreuth.de/volltexte/2002/1

Rahbek C (2005) The role of spatial scale and the perception of large-scale species-richness patterns. Ecol Letts 8: 224-239

Ricketts TH, Daily GC, Ehrlich PR, Fay JP (2001) Countryside biogeography of moths in a fragmented landscape: biodiversity in native and agricultural habitats. Conserv Biol 15: 378-388

Rosenzweig ML (1995) Species diversity in space and time. Cambridge Univ Press, Cambridge

Sattler K (1991) A review of wing reduction in Lepidoptera. Bull Br Mus Nat Hist (Ent) 60: 243-288

Sayer J, Ishwaran N, Thorsell J, Sigaty T (2000) Tropical forest biodiversity and the World Heritage Convention. Ambio 29: 302-309

Schulze CH (2000) Auswirkungen anthropogener Störungen auf die Diversität von Herbivoren – Analysen von Nachtfalterzönosen entlang von Habitatgradienten in Ost-Malaysia. PhD thesis, Univ Bayreuth

Schulze CH, Fiedler K (2003) Hawkmoth diversity in northern Borneo does not reflect the influence of anthropogenic habitat disturbance. Ecotropica 9: 99-102

Schulze CH, Linsenmair KE, Fiedler K (2001) Understorey versus canopy: patterns of vertical stratification and diversity among Lepidoptera in a Bornean rain forest. Plant Ecol 153: 133-152

Schulze CH, Waltert M, Kessler PJA, Pitopang R, Shahabuddin, Veddeler D, Mühlenberg M, Gradstein SR, Leuschner C, Steffan-Dewenter I, Tscharntke T (2004) Biodiversity indicator groups of tropical land-use systems: comparing plants, birds, and insects. Ecol Appl 14: 1321-1333

Scoble MJ (1995) The Lepidoptera: form, function, and diversity, 2nd ed. Oxford University Press, Oxford

Scoble MJ (ed) (1999) Geometrid moths of the world – a catalogue (Lepidoptera: Geometridae). CSIRO Publishing, Collingwood

Summerville KS, Ritter LM, Crist TO (2004) Forest moth taxa as indicators of lepidopteran richness and habitat disturbance: a preliminary assessment. Biol Conserv 116: 9-18

Süßenbach D (2003) Diversität von Nachtfaltergemeinschaften entlang eines Höhengradienten in Südecuador (Lepidoptera: Pyraloidea, Arctiidae). PhD thesis, Univ Bayreuth. URL: http://opus.ub.uni-bayreuth.de/volltexte/2003/33

Tammaru T, Haukioja E (1996) Capital breeders and income breeders among Lepidoptera – consequences to population dynamics. Oikos 77: 561-564

Trueblood DD, Gallagher ED, Gould DM (1994) Three stages of seasonal succession on the Savin Hill Cove mudflat, Boston Harbor. Limnol Oceanogr 39: 1440-1454

Weller SJ, Jacobson NL, Conner WE (1999) The evolution of chemical defences and mating systems in tiger moths (Lepidoptera: Arctiidae). Biol J Linn Soc 68: 557-578

Williams-Linera G, Domínguez-Gastelú V, García-Zurita ME (1998) Microenvironment and floristics of different edges in a fragmented tropical rainforest. Conserv Biol 12: 1091-1102

Willott SJ (1999) The effects of selective logging on the distribution of moths in a Bornean rainforest. Phil Trans R Soc Lond B 354: 1783-1790

Wirooks L (2005) Die ökologische Aussagekraft des Lichtfangs – eine Studie zur Habitatbindung und kleinräumigen Verteilung von Nachtfaltern und ihren Raupen. Wolf & Kreuels, Havixbeck-Hohenholte

Wong KM, Phillipps A (eds) (1996) Kinabalu: Summit of Borneo. Sabah Society with Sabah Parks, Kota Kinabalu

Wright SJ (2005) Tropical forests in a changing environment. Trends Ecol Evol 20: 553-560

Amphibian communities in disturbed forests: lessons from the Neo- and Afrotropics

Raffael Ernst[1*], K. Eduard Linsenmair[1], Raquel Thomas[2], and Mark-Oliver Rödel[1]

[1] Department of Animal Ecology and Tropical Biology, Biocenter University of Würzburg, Am Hubland 97074 Würzburg, Germany
[2] Iwokrama International Centre for Rain Forest Conservation and Development, 77 High Street, Kingston, Georgetown, Guyana (formerly, Guyana Forestry Commission, Planning and Research Development Division)

*corresponding author: Raffael Ernst, present address: Department of Zoology, Staatliches Museum für Naturkunde Stuttgart, Forschungsmuseum Schloss Rosenstein, Rosenstein 1, 70191 Stuttgart, Germany
phone: +49 (0)711-8936 284 Email: ernst@biozentrum.uni-wuerzburg.de

Summary

Timber harvesting is currently the most common commercial utilisation activity in tropical forests. Assessing the effects of logging on different aspects of biodiversity and general ecosystem properties is hence of prime importance if the few remaining areas of intact tropical forest are to be protected effectively and efficiently.

As we will point out in this chapter, tropical amphibian communities are an appropriate model system for studies on the impacts of human-induced environmental changes on the dynamics of complex biological systems. Here, we elaborate on patterns of diversity changes in tropical forest amphibian communities facing habitat alterations associated with selective logging in two globally important eco-regions (Côte d'Ivoire, Upper Guinea, West Africa and Guyana, the Guiana Shield, northern South America), and discuss findings from other previously conducted studies with similar focus.

A key statement that we stress on in this chapter is the fact that common measures of diversity, such as species richness and α-diversity only inadequately reflect processes of diversity change following anthropogenic disturbance. They also fail to describe actual impacts on the dynamics of complex biological systems. We argue that commonly used measures produce an incoherent and insufficient picture of diversity patterns and the underlying processes that shape these patterns. Thus, an understanding of higher levels

Tscharntke T, Leuschner C, Zeller M, Guhardja E, Bidin A (eds), The stability of tropical rainforest margins, linking ecological, economic and social constraints of land use and conservation, Springer Berlin 2007, pp 61-87

of diversity, such as β-diversity and functional diversity (and hence compositional patterns) appears to be the key to effectively mitigating the impacts of human-induced disturbance on amphibian communities.

We show that the predictability of amphibian community composition depends on the respective level of anthropogenic disturbance imposed on a particular habitat. Hence, human activities that lead to changes in the structure of a forest, such as logging, not only alter simple system descriptors, such as the number of species in a given community, but rather alter the dynamics of the entire system. In this context, functional diversity is shown to be an important aspect underlying the actual mechanism that leads to the observed change of predictability patterns. Functional differences between species, rather than number of species per se appear to be the decisive factor in sustaining desirable ecosystem states and thus in maintaining important ecosystem services.

Because biological diversity appears to play a substantial role in ecosystem resilience required to safeguard essential ecosystem functions in the face of environmental change, we call for a critical revision of common diversity assessments approaches. We advocate the reconsideration of the uncritical use of widespread measures and descriptors of biodiversity on grounds of inconsistent patterns found throughout numerous studies, including those presented herein.

The complex nature of natural communities and the multiple aspects of biodiversity at different levels, make it necessary to incorporate processes acting on different organizational and spatial scales. When investigating the impacts of human-induced environmental changes on diverse vertebrate communities in the tropics, we should hence address compositional changes, as well as β-, and functional aspects of biodiversity. Special attention should also be drawn to the particular disturbance history of a given site and large scale cross-regional comparisons should be given priority. The consideration of these approaches in future studies would likely provide deeper insight in ecosystem processes at large scales and improve the effectiveness of current ecosystem management strategies.

Key words: anthropogenic disturbance, habitat alteration, logging, communities, Amphibia, diversity, predictability patterns, community dynamics, conservation, West Africa, northern South America

1 Introduction

Tropical forests cover 6 % of the planet. However, as early as in the late 1980s, an area of 142,000 km^2 was estimated to be lost every year, equalling 1 % of the total tropical forest area (Myers 1989). Deforestation rates have been increasing ever since (Lambin et al. 2003). The major threats to tropical rain forests are clearance and fragmentation of the forest, and the overexploitation of organisms in the remaining area (Laurance et al. 2001, Primack 2002).

Tropical amphibians are a diverse vertebrate group that is threatened by multiple factors. Since their first appearance approximately 299-251 million years ago, this group has evolved into what could be the most diverse group of tetrapod vertebrates (Köhler et al. 2005). Currently, 5,743 amphibian species are recognized and given the rate of new descriptions and species estimates, amphibians could possibly outnumber bird and reptile species within the next years (white papers of Amphibian Conservation Summit, Conservation International and IUCN 2005). Despite this alleged evolutionary success story, many amphibian species throughout the world are facing severe population declines or even extinction. The Global Amphibian Assessment (http://www.globalamphibians.org/) revealed that almost one third (32%) of all amphibian species are threatened and around 120 species have gone extinct in recent times. This percentage is far higher than in all other known groups of organisms (Stuart et al. 2004, Young et al. 2004). These facts have prompted many amphibian population and community studies acknowledging an alarming phenomenon known as the Global Amphibian Decline (Lips et al. 2005), which has affected amphibian populations in various areas throughout the world. Although many factors, such as increased ultraviolet-B radiation, climatic change associated with global warming, and environmental contaminations adversely affect amphibians (e.g. Blaustein et al 1994, 1996, Pechman and Wilbur 1994, Pounds et al. 1999), habitat loss and alteration are probably the most serious causes of amphibian population declines, currently affecting almost 70 % of the species (http://www.globalamphibians.org). Many amphibian species of which populations have previously been weakened or experienced declines due to habitat degradation and destruction, ultimately appear to fall victim to a previously unknown disease, the pathogenic chytrid fungus (*Batrachochytrium dendrobatidis*). This pathogen seems to be responsible for the loss of entire populations previously facing environmental stress (e.g. Weldon and du Preez 2004). New evidence supports the possibility that the fungus rapidly spreads and wipes out amphibian population in connection with global warming (Cary and Alexander 2003, Pounds and Puschendorf 2004, Pounds et al. 2006), which in turn is substantially promoted by rapidly increasing deforestation. Given the current rate of deforestation and habitat destruction, it cannot be overstated that habitat loss is the principle factor determining the uncertain future of many amphibian populations and furthermore the general level of species extinctions in tropical forests.

Due to their physiology, mostly short generation times, and predominantly biphasic lifecycle, amphibians are particularly sensible with respect to habitat changes and altered microclimatic conditions that accompany these changes. There is a common misperception concerning amphibian habitat requirements (Dodd and Smith 2003). They are usually thought to be exclusively associated with large wetland habitats and it is assumed that preventing the net loss of wetland habitats is an adequate measure to protect entire amphibian communities. However, amphibian habitat requirements, especially in the tropics, are very diverse and far more complex, not only concerning different species, but also with regard to different life history stages of a particular species. So far little is known about differences between sexes or size classes.

Most amphibians have much shorter reproductive cycles than, for example large mammals or trees, resulting in faster detectable alterations in population size and structure with changing environmental conditions. Many species reflect landscape differences and exhibit a high sensitivity to habitat modification (Pineda et al. 2005, Ernst and Rödel 2005, Ernst et al. in press). Amphibians in general also comprise a significant portion of the vertebrate fauna of tropical forests throughout the world, where they are important, both as predators and as prey (Inger 1980a, b, Duellman 1990). Standardized quantitative methods exist for estimating their species richness and abundance (Heyer et al. 1994, Rödel and Ernst 2004, Veith et al. 2004). For these reasons amphibian communities are an appropriate model system for studies on the impacts of human-induced environmental changes on the dynamics of complex biological systems.

This chapter focuses on the effects of forest disturbance due to timber harvesting on amphibian communities. Timber harvesting is currently the most common and lucrative utilisation of tropical forest resources (Johns 1997), and it has been identified as one of the major threats to global vertebrate diversity (e.g. Thiollay 1992, Lauck 2005).

Fig. 1. Location of field studies (arrows). From west to east: i) Mabura Hill Forest Reserve (MHFR), central Guyana - 05° 09.322' N, 058° 41.983' W -; ii) Taï National Park (TNP), south-western Côte d'Ivoire - 05° 50.003' N, 007° 20.536' W -.

Rain forests and their fauna are not identical and just as the forests in the major rain forest regions differ, so does the percentage of forest remaining and the nature of threats to its continued survival (Primack and Corlett 2005). Despite the urgent need for comparable studies on actual effects of logging on diverse vertebrate communities, across various regions, there is still a lack of knowledge and generalizable results. In an attempt to further narrow this gap we summarize some of our work on amphibian communities in two geographically distinct eco-regions, Côte d'Ivoire (Upper Guinea, West Africa) and Guyana (Guiana Shield, northern South America); see Figure 1; for detailed study site descriptions see ter Steege et al. 1996, Riezebos et al. 1994). We discuss our findings in relation to other previously conducted studies with a similar focus. In the following sections we specifically elaborate on patterns of diversity changes in tropical forest amphibian communities facing human-induced environmental changes. We examine observed pattern changes at different levels of diversity (α-, β-, functional diversity) and discuss their respective significance and suitability in the context of the study of ecosystem dynamics and functioning. We address problems arising from the uncritical use of common diversity descriptors and finish with a discussion of the management implications.

2 Forest disturbance and logging methods – From single tree harvesting to clear felling

The creation of forest margins, which is the leading theme of this book, is intrinsically linked with commercial timber harvesting. This is because most medium and large scale logging operations move into pristine forest areas from a surrounding matrix, which in most cases is already highly anthropogenically altered. Logging operations thereby increase edge area and create margin situations in the remaining forest stand. Even small scale logging creates margins that are often big enough to foster negative effects. This may be especially true with regard to very sensitive and relatively immobile organisms, such as amphibians.

Most extant tropical rain forests have been logged or will be logged in the near future and huge areas are damaged by commercial logging every year (Whitmore 1997).

The most common form of timber extraction from tropical forest is selective logging. This kind of logging aims at the extraction of a limited number of tree species with high commercial value, which generally comprise only a small proportion of the trees present in a given area. Technological advances and an increasing demand for wood products, however, have resulted in a substantial increase in the number and intensity of logging operations world-wide (Dawkins and Philip 1998). A recent study revealed that selective logging, which in the past has often been neglected in deforestation studies, doubles previous estimates of anthropogenically induced forest degradation (Asner et al. 2005). The term 'selective logging' has hence become misleading, because, even though only a small percentage of the trees are felled for timber today, a much larger percentage is killed or damaged during logging operations (e.g. Jonsson and Lindgren 1990). Logging damage may appear to be random, mimicking natural disturbance by creating canopy openings and debris piles comparable to those found in natural Chablis of mature rainforests. Yet, the scale, level of selectivity, and intensity of the disturbance caused by logging operations often differs. Estimated annual rates of gap formation in natural forests range between 1-2 % (Hartshorn 1990, van der Meer and Bongers 1996). Low-dynamical forests, such as the tropical rain forest of Guyana (ter Steege and Hammond 2001), fall well below these rates (0.4 %, van Dam and Rose 1997). Gaps created by logging are frequently larger than natural gaps and usually more gaps per area are being created (van der Hout 1999). In addition, large parts of forest landscapes are affected at the same time (due to e.g. road construction), even if logging operations are restricted to small patches, while natural disturbance, with the exception of large scale disturbances, such as extensive fires or hurricanes, will only affect smaller parts at a time.

Logging also has direct impacts on the biotic and abiotic conditions of the forest (Uhl and Vieira 1989, Johns 1997, Vitt et al. 1998). As mentioned above, the direct impacts of selective logging include the removal of harvested trees and associated damage to the residual stand, but it also leads to significant disturbance of the soil and catchment's hydrology caused by the use of heavy machinery. Unlike most forms of natural disturbance, logging creates new habitats through soil compaction that, in combination with increasing solar radiation, a direct result of the newly opened canopy, leads to alterations in the hydro-regime. Forest understoreys are characterized by strong vertical and horizontal gradients in light availability (Poorter and Arets 2003) with irradiance levels of 1-2 % of full sunlight under closed canopy. In tree fall gaps, however, irradiance levels can reach up to 25 % (Chazdon and Fetcher 1984). Although gap formation will locally change a multitude of parameters, increased light availability and hence intensified solar radiation with respective changes in temperature and humidity, is the most obvious change (van Dam 2001).

The sustainable management of tropical forests has been discussed very controversially (e.g. Bowles et al. 1998, Chazdon 1998, Gascon et al.1998) and the potential of forest and forest fauna recovery after logging applying various selective logging systems has been debated (e.g. Dawkins and Philip 1998, Clarke et al. 2005a). Forest recovery after logging depends highly on the severity of the impact. High logging intensities may have a disastrous effect on forest recovery (e.g. Chai and Udarbe 1997) and intensive exploitation may simply cause devastation (Nepstad et al. 1999). Selectivity is yet another problem as the selective removal of large reproductive trees may have severe consequences for regeneration (van Ulft 2004) and hence population development (Tilman et al. 1994). The overall impact on biodiversity and the forest ecosystem is difficult to quantify. Even though the conservation value of secondary and logged forests is recently obtaining more and more recognition (Chazdon 1998, Peña-Claros 2001), the long-term capacity to support biodiversity is still insufficiently known for tropical forests that are managed for timber (Arets 2005). As timber harvesting is currently the most common use of tropical forests (Johns 1997) and since the quality of forest management will be one of the major determinants of global biodiversity in the next

50 years, assessing the effects of logging activities on biodiversity and general ecosystem functions is of primary importance to conservation (compare Fimbel et al. 2001).

3 Potential processes determining amphibian diversity in disturbed forests

The long-term effects of logging on amphibian diversity are only poorly understood. Generally, logging leads to habitat alterations, which in turn may result in species alterations. The effects and the actual impact strongly depend on the scale and type of logging. Extensive logging (e.g. clear-cutting) can lead to drastic climatic changes and loss of forest microhabitats. As a consequence, true forest species may go extinct, whereas open habitat species can invade newly opened areas. Even lower intensity logging (e.g. patchy or selective logging) has the potential of dramatically impacting habitats and associated species. Impacts may be effective on smaller scales, but they may be just as drastic, i.e. the loss of particular species or even entire functional groups due to microclimatic changes and changes in the hydro regime, as a direct consequence of an increased canopy openness (compare Figure 2).

4 Direct effects of disturbance on simple system descriptors

Changes in the natural disturbance regime, e.g. due to logging, can induce changes in species richness, α-diversity and species composition (Clark 1990, Denslow 1995). They are also known to affect relative abundances in arboreal ants (Floren et al. 2001), insect herbivores (Basset et al. 2001), termites (Lima et al. 2000), lizards (Goldingay et al. 1996), birds (Thiollay 1992, Sekercioglu 2002, Laiolo et al. 2003), and mammals (Laurence and Laurence 1996, Clarke et al. 2005b). The direction of the effects of logging on basic system descriptors such as species richness and diversity in forest communities is not consistent with regard to different forests and different species groups. Both increased and decreased diversity after selective logging, as well as no changes in diversity have been reported. Even trends identifiable within particular taxa are highly variable, both within and between studies. (Table 1). Another major concern relates to the fact that only few studies indicate the intensity of logging or even consider it as a factor in the analyses. Most often they are not systematic with regard to temporal disturbance patterns (i.e. considering different forest recovery states or the time that has passed since logging occurred). In many studies, sites with very different disturbance histories are combined and analyzed without distinguishing between differing time spans elapsed since logging occurred. In our studies on the effects of logging on the

Table 1. Response of amphibians to logging. Only results related to logged areas are listed, even if studies addressed additional disturbance gradients (? = not data; D = decrease; I = increase; C = change; NC = no change). Modified after Alzevedo-Ramos et al. 2003.

Focal taxa	Time since logging [years]	Logging intensity [m^3h^{-1}]	Factor	Variables monitored and effect	Location	Source
Amphibia	?	?	logging	richness (I)	Peninsular Malaysia	Heang et al. 1996
Amphibia	?	?	logging	abundance (D)	Kalimantan, Borneo	Lang 2000
Anurans	>20	?	logging to shade coffee plantations	richness (D), abundance (I and D according to species)	Highlands of La Antigua River, Mexico	Pineda et al. 2005
Amphibia	5-25	?	logging to cacao plantation	abundance (from 5 to 25 yrs.D), biomass (D), richness (I), species diversity (I), evenness (I)	Costa Rica	Heinen 1992
Amphibia	5-20	?	logging to pasture	richness (NC) and composition (C) of hylid species (I) and *Eleutherodactylus* species (D)	Amazonia, Ecuador	Pearman 1997
Amphibia	3	25	logging techniques	richness (NC), composition (NC)	eastern Amazonia, Brazil	Azevedo-Ramos et al. unpubl.
Anurans	5-12	clearcut	logging to pasture	richness (I), composition (C), abundance (NC) *Phyllomedusa tarsius*	BDFFP-Project, Amazonia, Brazil	Tocher et al. 2001
Amphibia	12-16	57	logging	richness (Côte d'Ivoire: NC, Guyana: D), species diversity (Côte d'Ivoire: D, Guyana: D, NC from 12 to 16 yrs.), functional diversity (Côte d'Ivoire: D, Guyana: D)	Côte d'Ivoire, Guyana	Ernst et al. in press

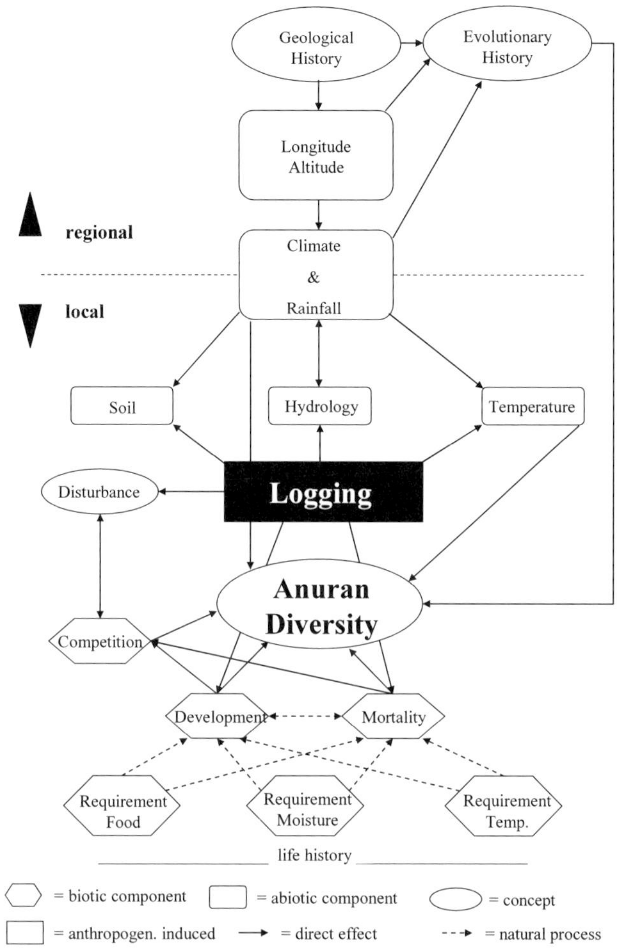

Fig. 2. Simplified diagram of potential direct, and indirect influences of logging on anuran diversity. Note that life history traits and individual responses may vary between different groups or species and across scales. The direction of the respective effects (positive or negative) may thus be species- / group-specific, rather than universal.

diversity of two tropical anuran communities in West Africa and the Guiana Shield we were able to demonstrate that the comparison of commonly used indices and simple system descriptors produces only idiosyncratic and inconsistent patterns with regard to the different geographic sites and disturbance regimes being compared. In case of the West African anuran communities, we found that species richness was identical in both, primary and exploited forest communities, while species richness of the Guyanan anuran communities decreased in exploited sites. We also demonstrated that there were no significant differences in diversity at the α-level in different recovery states. These results show that, given their conceptual basis, the search for general community patterns using traditional α-diversity measures, appears to be questionable. Common measures of biodiversity do not account for qualitative or compositional differences between assemblages. They may thus not adequately reflect actual changes and may hence be unsuitable and inappropriate indicators of forest disturbance as has previously been pointed out by Nummelin and Kaitala (2004).

5 The impacts of disturbance on community dynamics

As has been pointed out in the previous section, there is a substantial number of studies focusing on the direct effects of logging and habitat fragmentation and degradation on common system descriptors, such as species richness, relative abundance, and α-diversity, without providing a coherent and consistent picture of the underlying ecological processes and effects on the systems studied. Higher level diversity aspects and general system properties relating to ecosystem processes and functioning (e.g. compositional predictability patterns, functional diversity) are hardly ever addressed in most studies, even though they may be more likely to yield general patterns essential for drawing generalized conclusions with regard to entire systems. Anthropogenic habitat alteration has long been neglected as a factor in the analysis of system dynamics and compositional predictability patterns in biological communities. Yet, this issue is receiving growing attention as degradation, fragmentation, and destruction of natural ecosystems proceeds with alarming rapidity (Didham et al. 1996, Wardle 1999, Floren and Linsenmair 2001, 2005, Floren et al. 2001). The potential effects of human-induced environmental changes on system dynamics are yet poorly understood.

Results from our studies on anuran leaf litter communities in West Africa (Ernst and Rödel 2005) suggest that the predictability of community composition at least partly depends on anthropogenic disturbance imposed on a particular habitat. We were able to detect a transition from generally unpredictable systems in undisturbed primary forests to predictable systems in highly altered secondary forests, as has previously also been shown for canopy arthropod communities (Floren and Linsenmair 2001, 2005, Floren et al. 2001). Primary forest sites in geographic proximity to one another turned out to be similar with respect to species composition, despite differences in habitat parameters. As tests for spatial autocorrelations have clearly shown (Ernst and Rödel 2005), simple spatial auto-correlative patterns do not provide a sufficient explanation. The compositional resemblance in geographically closer sites can be understood as the result of underlying random walk dynamics, analogous to Hubbell's ecological drift (Hubbell 2001). These dynamics produce otherwise variable species assemblages across the entire range of the habitat complex. Similar communities are hence the product of stochastic ex-

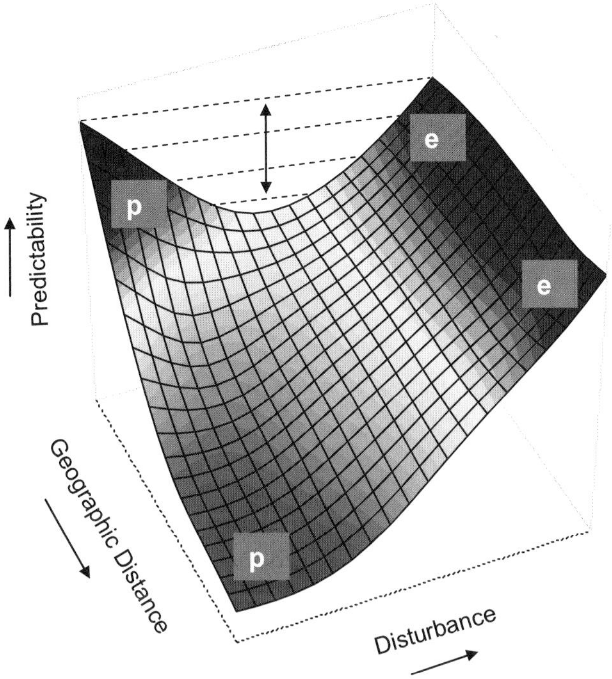

Fig. 3. Hypothetical relationship between geographic inter-site distance, disturbance level, and predictability of community composition. P = primary forest communities; e = exploited forest communities. Note relative indifference to geographic proximity in highly disturbed sites, whereas predictability in undisturbed sites depends strongly on geographic proximity. As a result, β-diversity is generally higher in primary forest than in exploited forest. It remains unsolved whether predictability changes gradually from high to low levels of disturbance, or if the pattern change is the result of a transitional threshold. At small geographic distances the pattern may remain constant. However the underlying process may be different, as indicated by the bend at intermediate disturbance levels. In low disturbance systems geographic proximity may be the key factor, whereas in high disturbance systems species-habitat associations and hence environmental restrictions may be more important. Modified after Ernst and Rödel (2005).

tinction and colonization events occurring in species with high turnover rates, recruited from a common regional species pool (Hecnar and M'Closkey 1996). The upper limit on the species composition of a new community developing in a given place is set by the composition of the regional species pool. Membership in the regional species pool is constrained by, e.g., geological history of respective regions, and the evolutionary processes of speciation within different taxonomic groups. Random walk dynamics may also result in high turnover rates, i.e. high β-diversity between sites, which indeed was the case in the primary forest communities compared. Communities in exploited forests on the other hand have been shown to be largely subjected to the influence of strong environmental site filters. This factor mainly controls membership in the local species pool, the latter hence being constrained by physiological tolerances related to factors such as temperature and humidity. In exploited forests we uncovered correlations of community composition with environmental characteristics, whereas geographic proximity did not seem to play a role. In the past, communities in these sites have been subject to habitat alteration, generally producing more restrictive environments, especially with respect to microclimatic parameters. The existence of a strong site filter has a tremendous effect on the general processes governing recruitment from the regional species pool as outlined above. The remaining set of species thus consists of either highly adapted species, resulting in stronger species-habitat relations than would be detected in primary forest habitats, or it consists of species that exhibit a broad-scale physiological tolerance. As a direct result, turnover rates, i.e. high β-diversity, are lower in these exploited sites, even though, as has been pointed out in the previous section, α-diversity and other simple system descriptors may not differ between communities of differently disturbed habitats. We found the same patterns in leaf litter anuran communities of primary and exploited forest habitats in central Guyana (Ernst et al. 2005, Ernst unpubl. data), a site with a very different evolutionary and geographic history. Figure 3 illustrates the hypothetical relationship between geographic inter-site distance, disturbance level, and predictability of community composition based on the results of our studies. Similar patterns have likewise been reported from tropical moth communities in Borneo, Ecuador, and Tanzania (see Fiedler et al. previous chapter) as well as tropical ant communities in Borneo (Floren et al. 2001, Floren and Linsenmair 2005).

As more studies, explicitly focusing on the impacts of anthropogenic disturbance on complex biological communities, are being conducted, it becomes even more evident that human activities that lead to changes in the structure of a habitat, such as logging, not only alter the composition of organisms that form a particular community but also have the potential of altering the dynamics of the entire system. The anthropogenic disturbance history of an area should hence always be considered when detecting deterministic patterns in the structuring of a community, as general applicability of deterministic explanations according to conventional niche theory may be attenuated. In their seminal work on response diversity, ecosystem change, and resilience, Elmqvist et al. (2003) pointed out that ecosystems can be strikingly conservative in their organization and function, despite differing histories and species compositions (Forys and Allen 2002, Brown et al. 2001). The same may be true for predictability pattern changes after human-induced disturbance in complex biological communities. Even though evidence for direct impacts on system dynamics has generally been scarce because anthropogenic disturbance has largely been neglected as an important factor in previous works, the studies presented in this book, as well as a number of recent studies indicate that amphibian communities are not simply a unique case but that we are very likely dealing with a general pattern. The observed pattern changes illustrate strikingly that anthropogenic disturbance affects the underlying processes that shape diversity patterns at large scales.

6 In search of key factors

It seems tempting to argue that structural simplicity within disturbed forests and therefore a reduction of system parameters per se (the number of occurring species and/or habitats and therefore the degrees of freedom of the entire system) may be essential factors explaining the transition from rather stochastically organized to deterministically organized communities. However, a comparison of merely structural habitat features and simple species numbers neglects qualitative differences, which can be severe. As has repeatedly stressed, simple system descriptor changes may not be representative, as patterns are incoherent and vary among different forests and different species groups, as well as between studies (compare Figure 4). Even if a reduction in species richness or α-diversity after anthropogenic disturbance is recorded, it is important to note that a reduction of species numbers is not necessarily an unbiased reduction of functional trait diversity. Composition and number of species is determined by biogeographical i.e. regional filters, local dispersal filters, and site filters. The diversity that passes these filters does not necessarily represent a random sample with respect to functional traits as has been pointed out in the previous section. At the α-diversity level, two communities, one consisting of 10 different, equally abundant species belonging to a single genus, the second consisting of 10 different, equally abundant species belonging to five different genera, exhibit identical species richness, diversity and evenness, although their composition is clearly distinct. Yet, these differences bear the potential of answering the question of why disturbances such as logging not only affect system descriptors, such as species richness, abundance, and diversity, but also drastically alter the system's dynamics. This has previously been suggested for the relationship between biodiversity and ecosystem functioning in grassland ecosystems (Tilman et al. 1997, Hector 2001). As evidence of links between functional diversity and ecosystem processes is constantly growing (Chapin et al. 2000, Diaz and Capido 2001, Loreau et al. 2001, Kinzig et al. 2001), it appears to be of utmost importance to be able to quantify the loss of functional diversity after human disturbances, such as logging. However, it is difficult to completely separate the effects of species and functional group richness in statistical analyses. Their particular contribution to changes in predictability patterns of biological systems is yet to be tested systematically.

In a recent study we approached this problem by assaying the impact of selective logging on different levels of diversity in two tropical anuran communities.The results of this study illustrated that simple measures of diver-

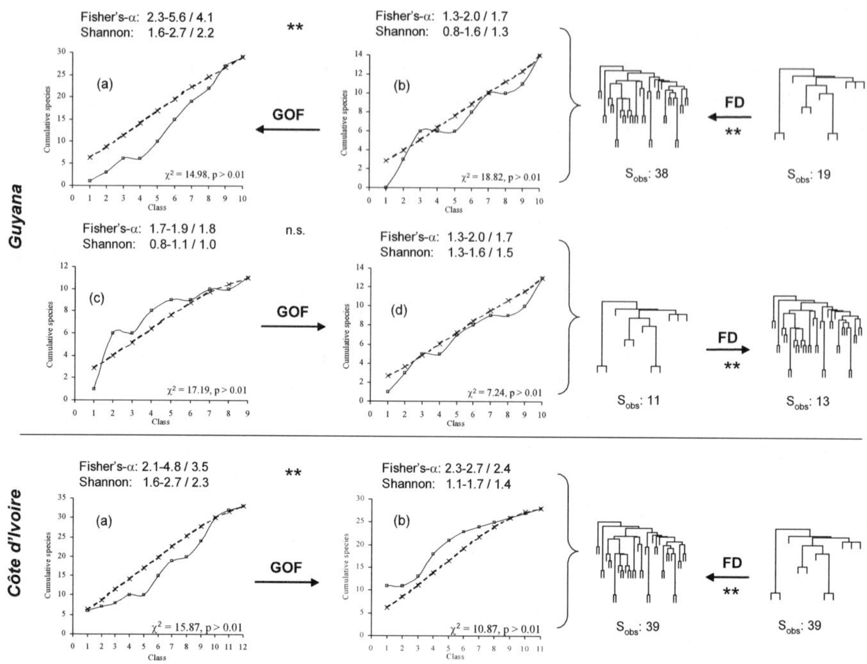

Fig. 4. Comparison of species richness (S_{obs}= species richness given as species observed), α-diversity (represented by the parametric Fisher's α and non-parametric Shannon index, given are range and mean) and functional diversity (FD sensu Petchey and Gaston 2002) between anuran communities of differently disturbed forests in Côte d'Ivoire and Guyana (a) primary forest community, (b) exploited forest community, (c) exploited forest (1992) community, (d) exploited forest (1988) community. ** = statistically significant, n.s. non significant difference in α-diversity and functional diversity, respectively. Goodness-of-fit (GOF) to log-series distribution illustrated as cumulative observed (solid line) and expected (broken line) number of species in relation to abundance class. Direction of arrows indicates increase. Complexity of functional dendrograms symbolizes high/low functional diversity in communities of respective habitats. Data from (Ernst et al. in press).

sity, such as species richness, may not adequately reflect the actual extent of biodiversity loss (e.g. loss of functional trait diversity) after anthropogenic disturbance (Ernst et al. in press). Whereas species richness in the amphibian communities that where studied remained unchanged or only little affected, functional diversity (FD) differed significantly. We found extinctions after disturbance to follow a non-random pattern. They appeared to be trait dependent as has previously been shown to be the case in four natural communities, one being an animal assemblage (Petchey and Gaston 2002). Lea et al. (2005) also report on similar findings in amphibian communities in Nigerian landscapes undergoing long-term degradation. They detected a shift from a predominance of forest specialists to a predominance of generalists, while species diversity remained stable or even increased. In case of the African and South American amphibian communities studied, we found a clear trend with regard to group membership and loss of particular species in exploited sites. Functional groups most affected by habitat alteration associated with logging contained species that depend on lotic or large lentic habitats for reproduction (e.g.: *Aubria subsigillata, Leptopelis macrotis,* and *Kassina lamottei* in Côte d'Ivoire; *Hyalinobatrachium* spp., *Atelopus spumarius,* and *Phyllomedusa bicolor* in Guyana; Figure 5). These habitat features are most often disappearing in disturbed forests due to an altered microclimate. High intensity logging may drastically alter hydrological cycles and enhance erosion and siltation (ter Steege et al. 1996). Very often, previously permanent aquatic habitats become ephemeral and lotic habitats become stagnant or disappear entirely. On the other hand, groups that persisted contained species that have reproductive modes making them less dependent or entirely independent of large aquatic habitats for reproduction as is the case in many direct developing or tree hole breeding species (e.g.: *Arthroleptis* spp., *Phrynobatrachus tokba,* and *Acanthixalus sonjae* in Côte d'Ivoire; *Eleutherodactylus marmoratus* and *Osteocephalus oophagus* in Guyana; Figure 5). These species were able to maintain stable populations or even increased in abundance. Despite the different evolutionary history of the communities in the two geographically separated regions, functional groups that were most affected or least affected showed a high resemblance.

In case of high levels of functional redundancy (sensu Loreau 2004) the loss of a particular species may seem irrelevant to ecosystem functioning. However, under the influence of disturbance, the simplification of the system may in fact be more severe and fundamental as entire functional groups are eliminated from the system, hence leading to tremendous structural impoverishment. Symptomatically, systems that lose entire functional groups as a result of environmental change are characterized by low response diversity (sensu Elmqvist et al. 2003) across scales (Jackson et al. 2001). The removal of species within the same functional group may also have dramatic consequence for evolutionary characteristics (Pauly et al 2002) with additional negative cascading effects (Carr et al. 2002). This kind of diversity erosion beyond the species level may hence potentially affect the stability of ecosystem structure and function tremendously (McCann 2000, Jackson et al. 2001).

Patterns found in different recovery states deviated in terms of differences between changes in α-diversity versus functional diversity, which is alarming from a conservation or management point of view. Whereas no significant differences in diversity of communities in different recovery states were detected at the α-level, significant differences in functional diversity were observed. Anuran communities showed a tendency towards recovery i.e., communities of old recovery states were functionally more diverse than those of young recovery states (Figure 4), yet this recovery potential would have remained concealed if looking at α-diversity exclusively. However, it has to be said that the difference in recovery time (only four years) may be too short to explore

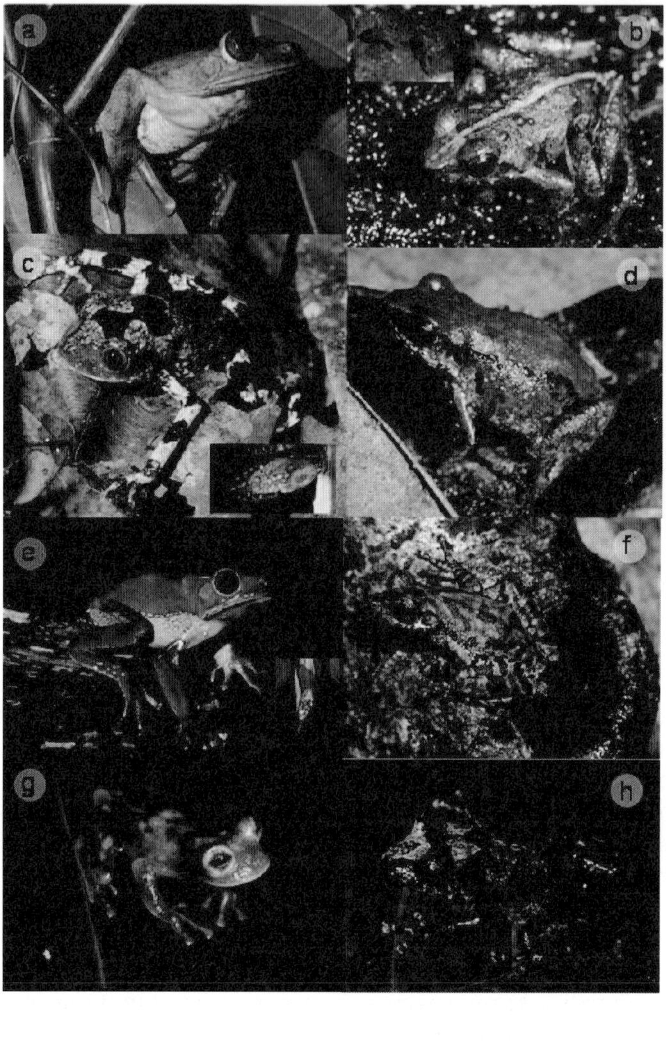

Fig. 5. Representatives of functional groups most (left) / least (right) affected by logging (a-d: West Africa, e-h: South America). a. *Leptopelis macrotis;* b. *Phrynobatrachus tokba*, inlet: froglets undergoing semi-direct development in jelly layer; c. *Kassina lamottei*, inlet: large lentic-nektonic tadpole; d. *Arthroleptis* sp.1; e. *Phyllomedusa bicolor*, inlet: clutch of *P. vaillantii*, a syntopic species, wrapped in leaf; f. *Osteocephalus oophagus* in front of water filled tree hole; g. *Hyalinobatrachium iaspidiense*; h. *Eleutherodactylus marmoratus.*

faunal recovery patterns, even in organisms with high population turnover rates and generation cycles, such as amphibians. Nonetheless, differences were significant and it seems promising to tests whether this pattern proves to be consistent in systems in which differences in recovery time are more distinguished. This would be especially important as the role of disturbed systems in conserving the remaining biodiversity will most likely increase in the near future.

What has previously been pointed out could have a tremendous influence on the planning of ecosystem management and restoration. Overall, patterns at the α-level appeared to be very inconsistent, whereas patterns at higher levels β- or functional diversity were more coherent (compare also Figure 4). Given that these very differences, i.e. species' identity not number per se, seem to influence ecosystem functioning tremendously (see O'Connor and Crowe 2005) it seems necessary to rethink common approaches of diversity assessments aimed at the protection of entire communities or even ecosystems, even more so if essential ecosystem services are to be secured in the long run.

7 Challenges and lessons to be learned

Many of the chapters in this book and the studies and publications associated with them leave little doubt as to the tremendous negative effects of habitat loss and fragmentation on numerous rainforest species, entire communities and the principles that govern their functioning. These factors will ultimately also affect the human population, locally as well as globally, that depends on a vast number of forest resources and the ecosystem services that these forests provide. It has been widely acknowledged that, although forests have been altered by humans for millennia, recent changes have occurred at an unprecedented pace and scale. However, necessary steps towards an effective protection of the remaining tracts of intact tropical forest and the biodiversity contained therein have up to now not been taken. Another common theme of many chapters in this book is the call for a revision of "traditional" diversity perspectives and measurement approaches, which all too often addressed simple α-diversity changes exclusively and neglected sometimes more important and informative patterns, such as changes in β-diversity, functional diversity or changes in predictability patterns of entire communities. Concepts for management and conservation planning were hence very often based on incorrect or at least incomplete assumptions.

8 In a nutshell

There are three major conclusions that can be drawn from our results and that have direct implications for management and conservation even beyond issues innate to amphibian conservation.

1. Scale matters: It is important to monitor different levels of biodiversity (α-, β-, functional diversity) in order to reveal important patterns that influence system dynamics and to be able to assess the actual loss after anthropogenic disturbance. Differential responses of different taxa, functional groups within particular taxa, and hence, particular species to habitat alterations induced by anthropogenic disturbance, complicate generalizations and require a good knowledge of the organism groups studied. Taxonomic and autecological studies are therefore not simply a luxurious addition but rather a prerequisite for the development of integrative coherent concepts.

2. Time matters: The disturbance history of a site is important to estimate its role in the structuring of a given biological community and to assess its impact on the system's dynamics. The analysis of e.g. different chronosequences and hence recovery states is important to detect potential recovery patterns. This implies that there is an urgent need for long-term studies and studies addressing pre- and post-harvesting phenomena, as these provide the only real opportunity for uncovering patterns in time, which are central to long-term conservation planning. The resulting information can then be used to establish sound and ecologically sustainable management concepts not only for the remaining primary forest that is protected in national parks and nature reserves, but also for areas set aside for sustainable use of tropical forests. This is especially true as these areas will very likely constitute the majority of forest habitats in the decades to come and because in reality, only a small fraction of tropical forests is adequately protected in reserved areas.

3. Geographic history matters at the local scale: For generalizations to be meaningful it is essential that predictions have been tested in systems that differ sufficiently, yet share some key factors that make them comparable in the first place. It is important to differentiate between common (in many cases regional or even global) and specific (in many cases site specific or local) phenomena. This again is of course intrinsically a matter of scale. Despite different geographic and evolutionary histories, general patterns at a particular level (here e.g. β-diversity, functional diversity or predictability) are similar in otherwise distinct geographic realms. On the other hand, slight differences may exist, hence possibly limiting assertions and general predictive power. Differences are mainly differences by order of magnitude in factors, such as variations in species composition and functional diversity.

Both aspects, general and specific are, however, important when it comes to sound ecosystem management plans and conservation strategies. Congruence in general patterns determines the overall management agenda, whereas site-specific differences have to be considered in the implementation of the agenda at the local scale.

9 Some specific recommendations with regard to the conservation of amphibians in altered tropical forest

Habitat diversity and habitat size

Although this has not specifically been addressed in this chapter, there is multiple evidence that habitat diversity is a crucial factor in the conservation of amphibian diversity. Even though amphibians may generally require less area than, e.g., large mammals or birds of prey, to maintain stable populations, habitat diversity is likely to increase with habitat size. A minimum habitat size is also essential to maintain ecosystem functions that determine e.g. climatic conditions and thereby directly affect amphibians. An important implication of our findings directly involves habitat size. Amphibian diversity in primary habitats has been shown to be determined by the dynamics of compositionally unpredictable local communities. Local community composition varies across the spatial range as a result of stochastic patterns of extinction and colonization events. Species turnover is hence comparatively high. Conserving overall amphibian diversity and thereby a large proportion of the functional traits of members (species) of a particular community requires conserving a large proportion of all species that make up the community. However, within these highly dynamic systems this is only possible if a range of habitats and hence local communities are being conserved. Obviously, this requires the protection of large tracts of forest. With respect to the protection of diverse habitat features simple measures can be taken when logging operations are being carried out. These are partly addressed and implemented in already existing national legislation (e.g. Guyana Forestry Commission's code of practice for timber harvesting, GFC 2002). The exclusion of diverse and representative parts of the concession from logging activities should be considered. A special focus should be given to the presence of diverse aquatic habitats, such as permanent forest ponds, swamp areas, streams and small creeks. Old emergent trees provide a number of unique microhabitats and should thus be spared from felling. Some species, e.g. the large neotropical hylid *Trachycephalus resinifictrix,* are almost exclusively recorded calling from the crown level of the largest emergent trees, they hardly ever descend to the ground. These old trees usually provide large water filled tree holes on which this species depends for reproduction and should hence be given special consideration for protection as habitat trees.

Protection of interior forest conditions and hence microclimatic and hydrological factors

As has repeatedly been pointed out, logging may drastically alter hydrological cycles and microclimate. Hydrological cycles within exploited sites have been reported to change dramatically, i.e. previously permanent creeks were becoming ephemeral, only being lotic during the peak rainy season. Large lentic ponds were reportedly more prone to desiccation. These changes especially affect species that depend on these habitats for reproduction, e.g. members of the neotropical genera *Atelopus* and *Hyalinobatrachium* or the afrotropical genera *Aubria* and *Kassina*. A significant proportion of any given tropical amphibian community is composed of small, leaf litter inhabiting species (app. 25-30 %; Allmon 1991). These may especially be prone to decreases in abundance or even extinction, as they rely on dense and moist litter for cover and nutrition (small leaf litter arthropods). However, leaf litter quantities frequently decrease as a result of forest stand thinning and formerly moist and shaded patches are opened up and hence prone to desiccation.

Continued logging and hence intensified disturbance along with a tremendous decrease in primary habitat size may thus put the survival of populations of these high priority conservation species at stake. Silvicultural techniques that minimize negative effects on canopy structure and openness and that prevent extensive soil compaction and collateral damage of the remaining stand may represent effective measures in safeguarding local amphibian communities. Among these practices are so called "Reduced Impact Logging" (RIL) methods that incorporate the clipping of lianas before felling to prevent incidental damage on surrounding trees, applying log hauling methods that do not make use of heavy machinery and restricting logging operations to dry periods, thus avoiding excessive soil compaction.

Avoiding access, fragmentation, and the formation of large margin areas

Historically, anthropogenic pressure in areas such as the Guiana Shield has been limited due to the very low population density and minimal access to the area in general. However, road construction accompanying logging operations has facilitated the access to previously uninhabited areas. Locally operating logging companies additionally facilitate access through skid trails and roads linking concessions. New areas are thus frequently opened and the size of forest margins and edges is being increased. Restricting road formation and, thus access to previously intact areas may therefore be essential for the maintenance of forest connectivity. The establishment of large buffer zones may be and efficient additional tool, guaranteeing the protection of interior-forest habitats. Maintaining connectivity through protection of adequate corridors between primary and exploited forest patches is yet another important step to be taken in order to guarantee the persistence of colonization and recolonization dynamics.

Guaranteeing adequate recovery periods: Higher functional diversity in old forest recovery states can be seen as an indication that amphibian communities may have some potential of recovery, given that 1: original logging was conducted in a sustainable manner, 2: remaining forest stands maintain sufficient connectivity among each other to allow recolonization of true forest species, and 3: adequate recovery periods are provided. One important implementation of this finding could be the extension of harvesting cycles i.e. long rotation periods, in combination with reduced impact logging methods and low intensity harvest in order to guarantee recovery of the system in the long run. However, systematic long-term studies are needed in order to specify and to express concretely recommendations with regard to harvesting cycles and recovery potential of the forest communities affected by the disturbance.

Acknowledgements
RE was supported by a doctoral scholarship from the German Academic Exchange Service (DAAD). Iwokrama International Centre for Rain Forest Conservation and Development helped with transportation and various administrative services. Permission to conduct biodiversity research in Guyana was given by the Environmental Protection Agency Guyana. The field study in Africa was supported by the DAAD (post-doc scholarship to MOR in 1999-2000) and the BIOLOG-program of the German Ministry of Education and Science (BMB+F; Project W08 BIOTA-West, FZ 01 LC 00410, 2001-2002). Research and access permission were given by the Ministère de l'Enseignement Supérieur et de la Recherche Scientifique,and the Ministère de la Construction et de l'Environnement of the Republic of Ivory Coast. Two anonymous reviewers provided valuable comments on the manuscript. These supports are gratefully acknowledged.

References

Allmon WD (1991) A plot study of forest floor litter frogs, Central Amazon, Brazil. Journal of Tropical Ecology 7: 503-522

Arets EJMM (2005) Long-term responses of populations and communities of trees to selective logging in tropical rain forests in Guyana. Tropenbos-Guyana Series 13, Georgetown, Guyana (Tropenbos Guyana Programme)

Asner GP, Knapp DE, Broadbent EN, Oliveira PJC, Keller M, Silva JN (2005) Selective logging in the Brazilian Amazon. Science 310: 480-482

Azevedo-Ramos C, De Carvalho Jr. O, Nasi R (2003) Animal indicators, a tool to assess biotic integrity after logging tropical forests? Unpublished report, Núcleo de Altos Estudos Amazônicos (NAEA), Universidade Federal do Pará

Basset Y, Charles E, Hammond DS, Brown VK (2001) Short-term effects of canopy openness on insect herbivores in a rain forest in Guyana. Journal of Applied Ecology 38: 1045-1058

Blaustein AR, Hokit DG, O'Hara RK, Holt RA (1994) Pathogenic fungus contributes to amphibian losses in the Pacific Northwest. Biological Conservation 67: 251-254

Blaustein AR, Wake DB (1996) Das Rätsel des weltweiten Amphibiensterbens. In: Biologische Vielfalt, Spektrum Verlag, Heidelberg

Bowles IA, Rice RE, Mittermeier RA, da Fonseca GAB (1998) Logging and tropical forest conservation. Science 280: 1899-1900

Brown JH, Ernest SKM, Parody JM, Haskell JP (2001) Regulation of diversity: maintenance of species richness in changing environments. Oecologia 126: 321-332

Carey C, Alexander MA (2003) Climate change and amphibian declines: is there a link? Diversity and Distributions 9: 111-121

Chai DNP, Udarbe MP (1997) The effectiveness of current silvicultural practice in Sabah. The Malaysian Forester 40: 27-35

Chapin FS, Zavelata ES, Eviner VT, Naylor RL, Vitousek PM, Reynolds HL, Hooper DU, Lavorel S, Sala OE, Hobbie SE, Mack MC, Díaz S (2000) Consequences of changing biodiversity. Nature 405: 234-242

Chazdon RL (1998) Ecology: Tropical forests - log 'em or leave 'em? Science 281: 1295-1296

Chazdon RL, Fetcher N (1984) Photosynthetic light environments in a lowland tropical rain forest in Costa Rica. Journal of Ecology 72: 553-564

Clark DB (1990) The role of disturbance in the regeneration of neotropical moist forest. In: Bawa, KS, Hadley, M. (eds) Reproductive ecology of tropical forest plants. UNESCO, Paris, France, pp 291-315

Clarke FM, Pio DV, Racey PA (2005b) A comparison of logging systems and bat diversity in the Neotropics. Conservation Biology 19: 1194-1204

Clarke FM, Rostant LV, Racey PA (2005a) Life after logging: post-logging recovery of a neotropical bat community. Journal of Applied Ecology 42: 409-420

Dawkins HC, Philip MS (1998) Tropical moist forest silviculture and management; a history of success and failure. CAB international, Wallingford, UK

Denslow JS (1995) Disturbance and diversity in tropical rain forests: The density effect. Ecological Applications 5: 962-968

Díaz S, Cabido M (2001) Vive la différence: plant functional diversity matters to ecosystem processes. Trends in Ecology and Evolution 16: 646-655

Didham RK, Ghazoul J, Stork NE, Davis A J (1996) Insects in fragmented forests: A functional approach. Trends in Ecology and Evolution 11: 255-260

Dodd CK Jr., Smith LL (2003) Habitat destruction and alteration. Historical trends and future prospects for amphibians. In: Semlitsch RD (ed) Amphibian Conservation. Smithsonian Institution Press, Washington D.C., pp 94-112

Duellman WE (1990) Herpetofauna in neotropical rainforests: comparative composition, history, and resource use. In: Gentry, AH (ed) Four neotropical rainforests. Yale University Press, Yale, pp 455-505

Elmqvist T, Folke C, Nyström M, Peterson G, Bengtsson J, Walker B, Norberg J (2003) Response diversity, ecosystem change, and resilience. Frontiers in Ecology and the Environment 1: 488-494

Ernst R, Rödel M-O (2005) Anthropogenically induced changes of predictability in tropical anuran assemblages. Ecology 86: 3111-3118

Ernst R, Rödel M-O, Arjoon D (2005) On the cutting edge – The anuran fauna of the Mabura Hill Forest Reserve, Central Guyana. Salamandra 41: 179-194

Ernst R, Linsenmair KE, Rödel M-O (in press) Diversity erosion beyond the species level: Dramatic loss of functional diversity after selective logging in two tropical amphibian communities. Biological Conservation

Fimbel RA, Grajal A, Robinson JG (2001) The cutting edge. Conserving wildlife in logged tropical forests. Biology and Resource Management Series. Columbia University Press, New York

Floren A, Linsenmair KE (2001) The influence of anthropogenic disturbance on the structure of arthropod communities. Plant Ecology (special issue) 153: 153-167

Floren A, Linsenmair KE (2005) The importance of primary forest for species diversity: an investigation using arboreal ants as an example. Ecosystems 8: 559-567

Floren A, Freking A, Biehl M, Linsenmair KE (2001) Anthropogenic disturbance changes the structure of arboreal tropical ant communities. Ecography 24: 547-554

Forys EA, Allen CR (2002) Functional group change within and across scales following invasions and extinctions in the Everglades ecosystem. Ecosystems 5: 339-347.

Gascon C, Mesquita R, Higuchi, N (1998) Logging on in the rain forests. Science 281: 1453

GFC (2002) Code of Practice for Timber Harvesting, 2nd edition. Guyana Forestry Commission, Georgetown, November 2002

Goldingay R, Daly G, Lemckert F (1996) Assessing the impacts of selective logging on reptiles and frogs in the montane forest of southern New South Wales. Wildlife Research 23: 495-510

Hartshorn GS (1990) An overview of Neotropical forest dynamics. In: Gentry AH (ed) Four Neotropical rain forests. Yale University Press, pp 585-599

Heang KB, Liat LB, Lambert MRK (1996) To determine the effects of logging and conversion of primary forets to tree crop plantations on herpetofaunal diversity in Peninsular Malaysia. Conservation, Management and Development of Forest Resources 1996: 126-140

Hecnar SJ, M'Closkey RT (1996) Regional dynamics and the status of amphibians. Ecology 77: 2091-2097

Hector A (2001) Biodiversity and the functioning of grassland ecosystems: Multi site comparisons. In: Kinzig AP, Pacala SW, Tilman D (eds) The functional consequences of biodiversity: Empirical progress and theoretical extensions. Monographs in population biology 33. Princton University Press, Princeton and Oxford, pp 71-95

Heinen JT (1992) Comparisons of the leaf litter herpetofauna in abandoned cacao plantations and primary forest in Costa Rica: Some implications for faunal restoration. Biotropica 24: 431-439

Heyer WR, Donnelly MA, McDiarmid RW, Hayek L-AC, Foster MS (1994) Measuring and monitoring biological diversity. Standard methods for amphibians. Smithsonian Institution Press, Washington and London

Hubbell SP (2001) The unified neutral theory of biodiversity and biogeography. Monographs in population biology 32. Princeton University Press, Princeton and Oxford

Inger RF (1980a) Densities of floor-dwelling frogs and lizards in lowland forests of Southeast Asia and Central America. American Naturalist 115: 761-770

Inger RF (1980b) Relative abundances of frogs and lizards in forests of Southeast Asia. Biotropica 12: 14-22

Jackson JBC, Kirby MX, Berger WH, Bjorndal KA, Botsford LW, Bourque BJ, Bradbury RH, Cooke R, Erlandson J, Estes JA, Hughes TP, Kidwell S, Lange CB, Lenihan HS, Pandolfi JM, Peterson CH, Steneck RS, Tegner MJ, Warner RR (2001) Historical overfishing and the recent collaps of coastal ecosystems. Science 293: 629-637

Johns AG (1997) Timber production and biodiversity conservation in tropical rain forest. Cambridge University Press, Cambridge, Australia

Jonsson T, Lindgren P (1990) Logging technology for tropical forests – for or against? Report for ITTO. Forskningsstiftelsen 'Skogsarbeten' (Forest Operation Institute), Stockholm, Sweden

Kinzig AP, Pacala SW, Tilman D (2001) The functional consequences of biodiversity: empirical progress and theoretical extensions. Princton University Press, Princton and Oxford

Köhler J, Vieites DR, Bonett RM, Garcia FH, Glaw F, Steinke D, Vences M (2005) New amphibians and global conservation: a boost in species discoveries in a highly endangered vertebrate group. BioScience 55: 693-696

Laiolo P, Caprio E, Rolando A (2003) Effects of logging and non-native tree proliferation on the birds overwintering in the upland forests of north-western Italy. Forest Ecology and Management 179: 441-454

Lambin EF, Geist HJ, Lepers E (2003) Dynamics of land-use and land-cover change in tropical regions. Annual Review of Environment and Resources 28: 205-241

Lang DA (2000) What is the impact of conventional logging on anuran diversity and abundance in the Bulungan research forest, East Kalimantan? Report for the Center of International Forestry Research, Bogor, Indonesia

Lauck B (2005) Can life history studies contribute to understanding the impacts of clearfell logging on pond breeding anurans? A review. Applied Herpetology 2: 125-137

Laurance WF, Cochrane MA, Bergen S, Fearnside PM, Delamonica P, Barber CV, D'Angelo S, Tito F (2001) The future of the Brazilian Amazon. Science 291: 438-439

Laurence W, Laurence S (1996) Response of five arboreal marsupials to recent selective logging in tropical Australia. Biotropica 28: 310-322

Lea JM, Luiselli L, Politano E (2005) Are there shifts in amphibian faunal composition in Nigerian landscapes undergoing long-term degradation? A case study from a montane environment. Revue d'Écologie (Terre Vie) 60: 65-76

Lima AP, Cordeiro-Duarte AC, Luizão FJ, Higushi N (2000) Effects of selective logging intensity on two termite species of the genus *Syntermes* in Central Amazonia. Forest Ecology and Management 137: 151-154

Lips KR, Burrowes PA, Mendelson JR, Parra-Olea G (2005) Amphibian declines in Latin America: Widespread population declines, extinctions, and impacts. Biotropica 37: 163-165

Loreau M, (2004) Does functional redundancy exist ? Oikos 104: 606-611

Loreau M, Naeem S, Inchausti J, Bengtsson J, Grime JP, Hector A, Hooper DU, Huston MA, Raffaelli D, Schmid B, Tilman D, Wardle DA (2001) Biodiversity and ecosystem functioning: current knowledge and future challenges. Science 294: 804-808

McCann KS (2000) The diversity-stability debate. Nature 405: 228-233

Myers N (1989) Deforestation rates in tropical forests and their climatic implications. Friends of the earth, 26-28 Underwood St. London N1 7JQ, U.K.

Nepstad DC, Verssimo A, Alencar A, Nobre C, Lima E, Lefebvre P, Schlesinger P, Potter C, Moutinho P, Mendoza E, Cochrane M, Brooks V (1999) Large-scale impoverishment of Amazonian forests by logging and fire. Nature 398: 505-508

Nummelin M, Kaitala S (2004) Do species dominance indices indicate rain forest disturbance by logging. Biotropica 36: 628-632

O'Connor NE, Crowe TP (2005) Biodiversity loss and ecosystem functioning: distinguishing between number and identity of species. Ecology 86: 1783-1796

Pearman PB (1997) Correlates of amphibian diversity in an altered landscape of Amazonian Ecuador. Conservation Biology 11: 1211-1225

Pechmann JHK, Wilbur HM (1994) Putting declining amphibian populations in perspective: natural fluctuations and human impacts. Herpetologica 50: 65-84

Peña-Claros M (2001) Secondary forest succession - Processes affecting the regeneration of Bolivian tree species. PhD thesis, Utrecht University, Utrecht, The Netherlands. Promab Scientific Series 3, Riberalta, Bolivia

Petchey OL, Gaston KJ (2002) Extinction and the loss of functional diversity. Proceedings of the Royal Society London B269: 1721-1727

Pineda E, Moreno C, Escobar F, Halffter G (2005) Frog, bat, and dung beetle diversity in the cloud forest and coffee agroecosystems of Veracruz, Mexico. Conservation Biology 19: 400-410

Poorter L, Arets EJMM (2003) Light environment and tree strategies in a Bolivian tropical moist forest: an evaluation of the light partitioning hypothesis. Plant Ecology 166: 295-306

Pounds JA, Fogden MPL, Campbell JH (1999) Biological responses to climate change on a tropical mountain. Nature 398: 611-615

Pounds JA, Puschendorf R (2004) Clouded futures. Nature 427: 107-108

Pounds JA, Bustamante MR, Coloma LA, Consuegra JA, Fogden MPL, Foster PN, La Marca E, Masters KL, Merino-Viteri A, Puschendorf R, Ron SR, Sanchez-Azofeifa GA, Still SJ, Young BE (2006) Widespread amphibian extinctions from epidemic disease driven by global warming. Nature 439: 161-167

Primack RB (2002) Essentials of conservation biology. Sinauer Associates, Sunderland, MA.

Primack R, Corlett R (2005) Tropical rain forests: An ecological and biogeographical comparison. Blackwell Publishers

Riezebos EP, Vooren PA, Guillaumet JL (1994) Le Parc National de Taï, Côte d'Ivoire. Tropenbos Series 8, Wageningen, The Netherlands

Rödel M-O, Ernst R (2004) Measuring and monitoring amphibian diversity in tropical forests. I. An evaluation of methods with recommendations for standardization. Ecotropica 10: 1-14

Sekercioglu CH (2002) Effects of forestry practices on vegetation structure and bird community of Kibale National Park, Uganda. Biological Conservation 107: 229-240

Stuart SN, Chanson JS, Cox NA, Young BE, Rodrigues ASL, Fischmann DL, Waller RW (2004) Status and trends of amphibian declines and extinctions worldwide. Science 306: 1783-1786

ter Steege H, Boot RGA, Brouwer LC, Caesar JC, Ek RC, Hammond DS, Haripersaud PP, van der Hout P, Jetten VG, van Kekem AJ, Kellman MA, Khan Z, Polak AM, Pons TL, Pulles J, Raaimakers D, Rose SA, van der Sanden JJ, Zagt RJ (1996) Ecology and logging in a tropical rain forest in Guyana with recommendations for forest management. Tropenbos Series 14, Wageningen, The Netherlands (Tropenbos Foundation)

ter Steege H, Hammond DS (2001) Character convergence, diversity, and disturbance in tropical rain forest in Guyana. Ecology 82: 3197-3212

Thiollay J-M (1992) Influence of selective logging on bird species diversity in a Guianan rain forest. Conservation Biology 6: 47-63

Tilman D, Knops J, Wedin D, Reich P, Ritchie M, Sieman E (1997) The influence of functional diversity and composition on ecosystem processes. Science 227: 1300-1302

Tilman D, May RM, Clarence LL, Nowak MA (1994) Habitat destruction and the extinction debt. Nature 371: 65-66

Tocher MD, Gascon C, Meyer J (2001) Community composition and breeding success of Amazonian frogs in continuous forest and matrix-habitat aquatic sites. In: Bierregaard Jr. RO, Gascon C, Lovejoy TE, Mesquita R (eds)

Lessons from Amazonia: The ecology and conservation of a fragmented forest. Yale University Press, London, pp 235-247

Uhl C, Vieira IC (1989) Ecological impacts of selective logging in the Brasilian Amazon: A case study from the Paragominas region of the state of Para. Biotropica 21: 98-106

van Dam O (2001) Forest filled with gaps: effects of gap size on water and nutrient cycling in tropical rain forest - A study in Guyana. Ph.D. thesis, Utrecht University. Tropenbos-Guyana Series 10, Tropenbos-Guyana Programme, Georgetown, Guyana

van Dam O, Rose SA (1997) Natural tree fall gaps in the Pibiri gap experiment study area, Guyana. Working document, Tropenbos-Guyana Programme, Georgetown, Guyana

van der Hout P (1999) Reduced impact logging in the tropical rain forest of Guyana - ecological, economic and silvicultural consequences. Ph.D. thesis, Utrecht University. Tropenbos-Guyana Series 6, Tropenbos-Guyana Programme, Georgetown, Guyana

van der Meer PJ, Bongers F (1996) Formation and closure of canopy gap in the rain forest at Nouragues, French Guiana. Vegetatio 126: 167-179

van Ulft LH (2004) The effect of seed mass and gap size on seed fate of tropical rain forest tree species in Guyana. Plant Biology 6: 214-221

Veith M, Lötters S, Andreone F, Rödel M-O (2004) Measuring and monitoring amphibian diversity in tropical forests. II. Estimating species richness from standardised transect census. Ecotropica, 10: 85-99

Vitt LJ, Avila-Pires TCS, Caldwell JP, Oliveira VRL (1998) The impact of individual tree harvesting on thermal environments of lizards in Amazonian rain forest. Conservation Biology 12: 654-664

Wardle DA (1999) Biodiversity, ecosystems and interactions that transcend the interface. Trends in Ecology and Evolution. 14: 125-127

Weldon C, du Preez LH (2004) Decline of the Kihansi Spray Toad, *Nectophrynoides asperginis*, from the Udzungwa Mountains, Tanzania. Froglog 62: 2-3

Whitmore TC (1997) Tropical forest disturbance, disappearance, and species loss. In: Laurance WF, Bierregaard RO Jr. (eds) Tropical forest remnants. Ecology, management, and conservation of fragmented communities. University of Chicago Press, Chicago, pp 3-12

Young BE, Stuart SN, Chanson JS, Cox NA, Boucher TM (2004) Disappearing jewels. The status of New World amphibians. NatureServe, Arlington

Fine root mass, distribution and regeneration in disturbed primary forests and secondary forests of the moist tropics

Dietrich Hertel*, Christoph Leuschner, Marieke Harteveld, and Maria Wiens

Plant Ecology, Albrecht-von-Haller-Institute of Plant Sciences, University of Göttingen, Untere Karspüle 2, 37073 Göttingen, Germany

* corresponding author: Dietrich Hertel, phone +49 (0)551 395708, fax +49 (0)551 392029, Email: dhertel@gwdg.de

Summary

In most tropical countries, rainforests are encroached by the local population and converted into agricultural land or pasture at constantly high rates. An increasing portion of this new agricultural land is abandoned within a few years and left for succession by woody plants. In many regions, already a considerable fraction of the forested area is covered by secondary forests formed by rapidly growing, short-lived pioneer tree species. In the Lore Lindu National Park on Sulawesi (Indonesia), like in other tropical countries, many of the remaining patches of protected primary forest are also offended at their margins by neighbouring villagers, who practise selective logging or local planting of crops inside the forest. Thus, disturbed old-growth forest and regrowing secondary forest after clearcut represent woody vegetation types, which are of rapidly growing importance throughout the tropics. Considerable research has focused on the effects of disturbance – natural as well as human-induced – on above-ground vegetation in primary forests, and on leaf and canopy characteristics of secondary as compared to primary forest trees. Only a few studies, however, have addressed the effects of disturbance and forest conversion on the below-ground compartment in wet tropical forests.

This review surveys the literature on fine root biomass (roots < 2 mm) and its distribution and activity in disturbed old-growth and secondary tropical moist forests and compares the results with non-disturbed old-growth forests. Particular attention is being paid to the relationship between disturbance intensity, above-ground forest structure and fine root system size and structure. Secondly, we focus on the age-dependence of fine root biomass in regrowing secondary forests and the time required for fine root biomass recov-

Tscharntke T, Leuschner C, Zeller M, Guhardja E, Bidin A (eds), The stability of tropical rainforest margins, linking ecological, economic and social constraints of land use and conservation, Springer Berlin 2007, pp 89-108

ery after clearcut. The results are interpreted with respect to carbon storage in the root system of tropical moist forests and its response to disturbance and conversion.

Keywords: fine root mass, root distribution, old-growth forests, secondary forests, natural disturbance, anthropogenic disturbance, disturbance intensity, disturbance type, forest regeneration, hurricane, tree fall, fire, logging, forest conversion, agroforest system, crop species

1 Introduction

Large parts of remaining tropical moist forests are more or less influenced by human disturbance, if they have not been converted to agricultural land or pasture (Nepstad et al. 1999, Achard et al. 2002, Lambin et al. 2003). Thus, forests with a variable degree of disturbance represent an increasingly important ecosystem type in tropical landscapes. On the other hand, forest disturbance caused by natural tree mortality (background or catastrophic) is a typical element of forest dynamics in tropical forests (Tanner et al. 1991, Attiwill 1994, Scatena et al. 2005). In fact, many studies have analysed the effect of forest disturbance on structure, species composition and carbon cycle of tropical rain forests (e.g. Laurance et al. 1998, Richter et al. 1999, Grove et al. 2000, Kwit and Platt 2003, Nascimento and Laurance 2004). However, most of these studies concentrated on above-ground effects of forest disturbance, while little is yet known on the influence of natural or anthropogenic disturbance events on the root system of the forests. Especially the fine root system (i.e. roots smaller than 2 mm in diameter) might be strongly affected by forest disturbance, since these roots are distributed mainly at the upper soil horizons and represent the most dynamic part of the root system (Fitter 1996, Jackson et al. 1996). These roots play an important role for water, carbon, and nutrient cycling in forests (Fitter 1996, Vogt et al. 1996, Cairns et al. 1997).

In the humid tropics, different types of forest disturbance exist. Its intensity, whether anthropogenic or natural, is largely variable with respect to severity, frequency, duration, and spatial extent (Tanner et al. 1991, Lugo and Scatena 1996). The most prevalent natural disturbances in tropical moist forests are canopy gaps caused by senescent trees, epidemic insect herbivory, storm events, landslides, and fire events by lightning (Coley and Barone 1996, Gariguata and Ostertag 2001, Scatena et al. 2005). Widespread anthropogenic forest disturbances are selective timber extraction, fire regime, forest conversion to agroforest systems under remaining forest tree cover, or complete forest conversion to agricultural land or pasture (Lambin et al. 2003, Asner et al. 2005). Large-scale clearcut, on the other hand, leaves extended areas for succession in which secondary forests of variable age and regeneration status replace abandoned land (Houghton et al. 1993, Silver et al. 2000b, Gariguata and Ostertag 2001).

The aim of this review is (i) to study the influence of different types and intensities of natural and anthropogenic disturbance on the fine root system in tropical moist forests, and (ii) to analyse the regeneration process of the fine root system following disturbance events and the age-dependence of fine root mass in regrowing secondary forests. We surveyed the literature on fine root mass and distribution data in disturbed old-growth (primary) forests to compare the records with unaffected old-growth forests. Tropical secondary forests were also included in this survey. Only studies from the tropical moist region were considered. The selection was done according to the global distribution limits of tropical moist forests as defined by Richards (1996) and by using a rainfall limit of 1700 mm yr^{-1} for separating tropical moist from tropical dry forests (Richards 1996). We only selected studies that reported data on fine roots, i.e. roots < 2 mm in diameter (in two cases < 3 mm). In contrast to earlier studies, we attempted to distinguish between live (biomass), dead (necromass), and total (live + dead) fine root mass records in our data set. Further details on the selection criteria of the data base are given in Hertel et al. (2006) and Hertel and Leuschner (2006).

2 Effect of low to severe disturbance in tropical old-growth forests on the fine root system

2.1 Natural disturbances

Only a limited number of studies have investigated the impact of natural disturbance events on fine root systems of tropical moist forests. These studies refer to a broad range of different disturbance events from high to low disturbance intensities and from large- to small-scale disturbance extents. A broad variety of time intervals that elapsed between disturbance event and the date of study is present in these investigations (Table 1). Two different types of disturbances were investigated in these studies: storm events of high or medium intensity and small-scale tree fall gaps caused by natural mortality. In the vast majority of studies, these natural disturbances led to a reduction in standing fine root mass (live or live + dead). The size of reduction was dependent on (i) disturbance intensity and its spatial extension, and (ii) the time elapsed between the disturbance event and the date of investigation. Studies in the Luquillo Experimental Forest on Puerto Rico documented an almost entire loss of the living fine root system one or three months after the severe hurricane Hugo hit the island (Parrotta and Lodge 1991, Silver et al. 1996). On the other hand, hurricane Iniki with a medium intensity led to a reduction of the living fine root mass by only 28 % compared to the pre-hurricane state in Hawaiian *Metrosideros* forests (Herbert et al. 1999). This

Table 2. Compilation of data on anthropogenic disturbance in relation to fine root mass in disturbed and undisturbed old-growth forests of the moist tropics. Classification of the categories for disturbance intensity and area were chosen according to the available information on the disturbance event in the respective data source. *: "l" = live fine root mass, "l+d" = live + dead fine root mass.

Disturbance intensity	Disturbed forest area	Disturbance event	Fine root mass undisturbed (g m^{-2})*	Fine root mass disturbed (g m^{-2})*	Percental change caused by disturbance	Years after disturbance	Locality	Country	Reference
high	large	conversion to agroforest	408 (l)	223 (l)	-45	ca. 12	Sulawesi	Indonesia	Leuschner et al. 2006
high	medium	logging	480 (l+d)	434 (l+d)	-10	ca. 10	W Ghats	India	Sundarapandian & Swamy 1996
medium	large	fire	946 (l)	464 (l)	-51	2.5	Oaxaca	Mexico	Asbjornsen et al. 2005
medium	large	fire	793 (l)	383 (l)	-52	2.5	Oaxaca	Mexico	Asbjornsen et al. 2005
medium	small	logging	408 (l)	225 (l)	-45	ca. 1	Sulawesi	Indonesia	Leuschner et al. 2006
low	medium	fire	480 (l+d)	353 (l+d)	-26	ca. 1	W Ghats	India	Sundarapandian & Swamy 1996
low	small	logging	624 (l)	350 (l)	-44	ca. 1	Meghalaya	India	Upadhaya et al. 2005
low	small	logging	463 (l)	325 (l)	-30	ca. 1	Meghalaya	India	Upadhaya et al. 2005
low	small	logging	408 (l)	313 (l)	-23	ca. 1	Sulawesi	Indionesia	Leuschner et al. 2006

(2006) reported, that the fine root necromass was higher in the undisturbed forest compared to disturbed forest with low- to medium-intense logging or agroforestry systems.

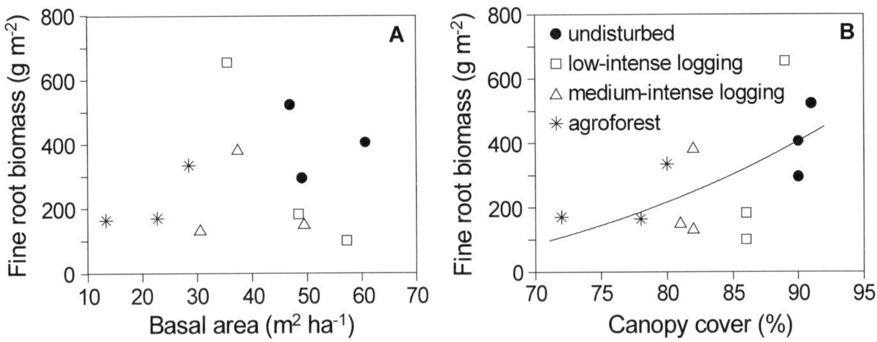

Fig. 1. Relationship between standing fine root biomass (live fine roots < 2 mm, organic layer plus 0-50 cm mineral soil) and basal area (A) or canopy cover (B) of 12 study plots in the four forest-use types differing in forest disturbance intensity on Sulawesi (Indonesia). Relation between fine root biomass and basal area was not significant; relation of fine root biomass and canopy cover: $r^2 = 0.34$, $r^2_{adj.} = 0.27$, $p < 0.05$. The data on basal area were appropriated by J. Dietz.

The relationship between disturbance intensity, forest above-ground structure, and fine root mass was investigated in detail in a tropical lower montane forest in Central Sulawesi, Indonesia. In this study, 12 forest stands representing four different types of forest-use intensity were investigated (with three replicate stands per type): (A) undisturbed natural forest, (B) natural forest with slight disturbance by selective logging, (C) disturbed natural forest with moderate selective logging activity, and (D) heavily disturbed natural forest which had been converted to a cacao-agroforest system under remaining rainforest shade trees. Along the disturbance gradient from A to D, changes in size and vertical distribution of the fine root system were studied as a function of above-ground forest structure and associated use intensity (Leuschner et al. 2006). Various attributes of above-ground structure declined with increasing disturbance intensity, i.e. tree height, dbh (diameter at breast height), basal area, and canopy cover, while stem density showed no directional change along this disturbance gradient. Live fine root mass was highest in the undisturbed forest (A) and decreased with increasing logging or disturbance intensity. In the moderately logged forests (C), a similar amount of tree fine root biomass was found as in the agroforestry systems under shade trees (D). The decrease in fine root biomass coincided roughly with the decrease in stand basal area (Figure 1A). However, the relation of the two parameters was not significant in the data set. Instead, we found a significant positive relation between fine root biomass and canopy cover (Figure 1B).

From these results it appears that tree fine root biomass tends to decrease not only after the harvest of large trees, but also with the cutting of thin understorey timbers, as was practised in the sligthly disturbed forests of the type B, despite of only minor reduction in stand basal area. However, extraction of larger trees (as increasingly happened in the forest-use types B to D) led to a decrease in both, canopy cover and basal area, resulting in a further decrease in fine root biomass.

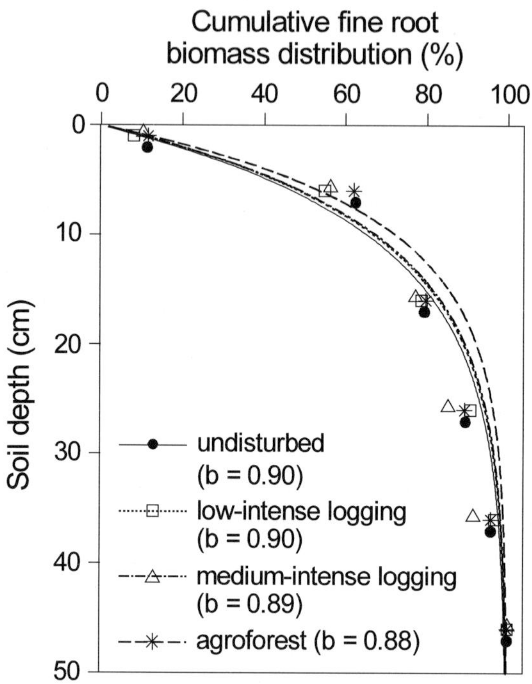

Fig. 2. Cumulative vertical fine root biomass distribution in soil profiles of pre-montane forests differing in forest disturbance intensity on Sulawesi (Indonesia). β-values given in brackets refer to the regression equation (according to Gale & Grigal 1987) of the form $y = 1 - \beta^d$ to describe the decrease in fine root biomass FRB (y = cumulative FRB in g m^{-2} until soil depth d) with soil depth (d in cm) of the 4 forest-use types.

Gale and Grigal (1987) formulated the hypothesis that the vertical fine root distribution pattern should change with forest succession, with late-successional species having a more shallow distribution and early-successional species revealing a more even distribution with soil depth. Consequently, if disturbance leads to the invasion of light-demanding early-successional species into the newly created gaps, one may expect the vertical root distribution patterns to change after selective logging or other major disturbances. In fact, many mature tropical forests have high amounts of fine root biomass near the soil surface (e.g. Sanford and Cuevas 1996, Herbert et al. 1999, Hertel et al. 2003). A rapid exponential decrease of the fine root biomass with soil depth seems to occur irrespective of soil type or moisture regime (Schenk 2005). Most likely, this pattern is mainly a consequence of nitrogen and phosphorus availability in the soil, which typically decreases downwards from the soil surface. Indeed, Yang et al. (2004) showed, that the vertical decrease in fine root biomass was stronger in a pristine subtropical forest in China than in adjacent forest plantations with more light-demanding tree species. In agroforestry systems, the vertical extension of the fine root system is highly dependent on tree species. Root systems with a fine root biomass more evenly distributed over soil depth are recorded frequently (Akinnifesi et al. 1999). In the disturbance gradient of the Sulawesi study, no effect of disturbance intensity on the vertical fine root distribution patterns was visible (Figure 2). Regardless of differences in absolute fine root biomass values (Table 1, Figure 1), fine root biomass decreased strongly with soil depth. This is confirmed by the calculated β values of 0.88 to 0.90, that indicate rather superficial rooting patterns of the trees in all four forest-use types (A-D), comparable to the late-successional tree species group in Gale and Grigal (1987) and Jackson et al. (1996); in contrast, β values of around 0.97 would be expected for early-successional species. One reason for this surprising result could be that after logging, small to medium-sized gaps are not only occupied by pioneer species, but also by late-successional species that play an important role in forest regeneration after the harvest or tree fall of old-growth individuals (Kwit and Platt 2003). Thus, only a minor shift in the composition of tree functional groups may occur after disturbance, which does not result in a change in vertical fine root distribution patterns. Similarly, the conversion of old-growth forest to agroforestry systems does not necessarily lead to a marked change in the vertical distribution of fine roots, since rooting depth of agroforest crops was found to depend largely on tree density (Akinnifesi et al. 1999), soil type and tree species (Lehmann 2003). Especially crops (such as cacao) grown under shade trees were often found to have more shallow fine root systems (Lehmann 2003). We conclude, that reductions in profile totals of fine root biomass as caused by disturbance are not necessarily accompanied by structural changes in the tree fine root systems.

3 Regeneration of tree fine root systems after natural forest disturbance

Studies on how fast the fine root system of tropical old-growth forests recovers after natural disturbance events have been conducted only exceptionally so far. In our literature survey, we only found four investigations on the regeneration of fine root biomass following natural forest disturbance. As indicated above, intensity and spatial extension of the disturbance event determine how much fine root biomass remains after disturbance. The recovery process, therefore, should depend on disturbance intensity because it creates soil patches with low root densities. However, the recovery process also depends on soil fertility (Ostertag 1998) and, thus, may reveal a considerable variability across the sites.

Based on the four studies of regenerating tropical moist forests, we attempted to calculate a relative recovery rate for the fine root biomass after disturbance expressed as percental increase in biomass relative to the initial biomass. This information is based on fine root biomass data recorded at different sampling dates after disturbance, irrespective of the type and severity of disturbance. A number of studies provided no precise biomass data for subsequent sampling dates in the same stand, or did not give a root biomass prior to disturbance. In these cases, we attempted to fill the information gaps by additional data from earlier investigations in the same forest patch, or by extrapolation from adjacent sampling locations in the same forest area. Consequently, our results must be seen as rough estimates of the fine root recovery process. Accordingly, tropical moist forests show recovery rates for their fine root system in the range of 3.1 to 8.2 % per month (mean: 5.7 %) relative to the initial fine root biomass. This is equivalent to recovery periods of 32 to 12 months, respectively, until the fine root biomass has returned to normal state, or an average recovery period of about one and a half year. Surprisingly, we could not detect a clear effect of the disturbance intensity on the regeneration process: recovery rates were 5.4 and 6.0 % month^{-1} in two heavily disturbed forest stands, 8.2 % month^{-1} in a slightly disturbed forest, but only 3.1 % month^{-1} in a forest of medium disturbance intensity (Parrotta and Lodge 1991, Silver et al. 1996, Ostertag 1998, and Herbert et al 1999, respectively). These calculations are based on the assumption that the fine root system recovers at a constant rate, i.e. that the process is independent of time. This may not be true, because a disturbance event causes fine root death and thus lowers fine root density and reduces competition intensity among fine roots. Moreover, dying fine roots should lead to a mineralization pulse shortly after disturbance, which may stimulate fine root growth during the first weeks. By conducting a non-linear regression analysis, we were able to describe the relationship between relative fine root biomass and time elapsed after disturbance by a saturation function (Figure 3). Accordingly, fine root biomass recovery starts immediately after disturbance and proceeds at a high rate during the first 5-6 months until half of the initial fine root biomass is regrown. According to this model, the recovery rate decreases thereafter and may approach the initial biomass state after about two years. This result is not so different from the linear recovery model, which predicted complete recovery after one and a half year. Thus, both calculation methods provide evidence that the fine root system of old-growth forests regenerates rather quickly after natural disturbance events, irrespective of disturbance type and intensity. However, more studies on the recovery process must be conducted to confirm these conclusions.

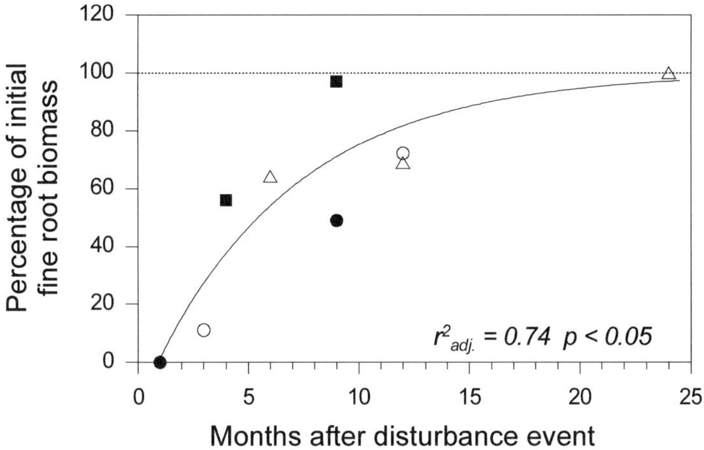

Fig. 3. Percentage of initial fine root biomass in primary forests related to time after natural disturbance event. Probability and $r^2_{adj.}$ values refer to the selected regression model $y = 100 - (114.9 + x)\ (1.2\ x^{-1})^{-1}$. Data references: filled circles: Parrotta & Lodge (1991), empty circles: Silver et al. (1996), filled squares: Ostertag 1998, empty triangles: Herbert et al. (1999). In the case of the data reported in Parrotta & Lodge (1991) the initial fine root biomass was approximated from the data presented in Silver et al. (1996) from the same study area.

4 Size and development of the fine root system in tropical secondary forests

Large areas that were once covered by tropical moist natural forests are nowadays occupied by secondary forests (Silver et al. 2000b, Achard et al. 2002). These areas represent different stages of forest regeneration after anthropogenic or natural disturbance. With increasing human pressure on the remaining natural forests, secondary forests are increasingly important in terms of land area covered, and with respect to carbon sequestration, for the provision of timber, and for biodiversity conservation (Gariguata and Ostertag 2001). Many secondary forests established naturally on clearcut land represent different stages of a secondary succession. Only few investigations on the fine root system of secondary forests in the humid tropics have been conducted so far. Thus, comparison between the fine root systems of secondary forests and undisturbed old-growth forests is hampered by a lack of data. It was stated that the fine root systems of young secondary forests with early-successional species can be as large as those of mature natural forests (Guariguata and Ostertag 2001, Hölscher et al. 2005). This was found in a study of a one-year-old secondary forest, which was compared to a primary forest in Costa Rica (Raich 1980, Raich 1983). On the other hand, there are reports that the fine root mass of secondary forests increases substantially with stand age (Berish

1982, Cavelier et al. 1996). This apparent contradiction may be explained by the fact that in all of the cited investigations, total fine root mass (live + dead) was recorded. Hence, the inclusion of the dead fine root fraction may have obscured the relationship between the live fine root mass and stand age. Our literature survey of studies in which the live fine root fraction was separated from dead fine roots found only one study on a paired comparison of secondary and old-growth forests for the humid tropics (Hertel et al. 2003), and a second one for semi-humid subtropical China (Muthukumar et al. 2003). In both cases, fine root biomass (live fine roots) was higher in the old-growth forest compared to the secondary forests. When considering not only paired secondary / old-growth forest comparisons, but including all available fine root biomass data of secondary and natural old-growth forests in the moist tropics, we found no significant difference between the two forest types (Hertel and Leuschner 2006). Secondary forests of variable age had an average biomass of nearly 400 g m^{-2} as compared to 450 g m^{-2} of live fine root mass in old-growth forests (Figure 4). However, the data on living fine root mass of secondary forests ($n = 4$) is too scarce for more general conclusions. Instead, we used data on total fine root mass (live + dead) of secondary forests based on a larger number of studies ($n = 21$) for our analysis. Comparison of the total fine root mass seems to confirm that the size of the fine root system in secondary forests may not be significantly different from mature old-growth forests (Figure 4).

Regression analysis of our data set showed that total fine root mass increases significantly with the age of the secondary forests (Figure 5A). On the other hand, 68 % of the variability in our data set was explained by other factors. This large part of variance not attributed to the parameter stand age might partly be a result of the inclusion of the dead root fraction in the data set. Fine root necromass of tropical moist forests is often larger in forest stands on acidic soils (Hertel and Leuschner 2006). Since soil pH often decreases during secondary succession on clearcut or burnt sites (Guariguata and Ostertag 2001), the increase in total fine root mass may be due to an increase in necromass with stand development. Additionally, live fine root mass is often larger in tropical forests on acidic soils than in forests on less acidic soils (Gower 1987, Cavelier 1992, Maycock and Congdon 2000). This is supported by the results of our data analysis suggesting that the soil acidity status has a great influence on the total fine root mass of secondary forests (Figure 5B). In fact, Hertel et al. (2003) showed that fine root biomass as well as necromass increased with decreasing soil pH during secondary succession. Thus, there is evidence that not only necromass, but also fine root biomass increases during secondary succession of tropical forests at least at sites with high soil proton concentration.

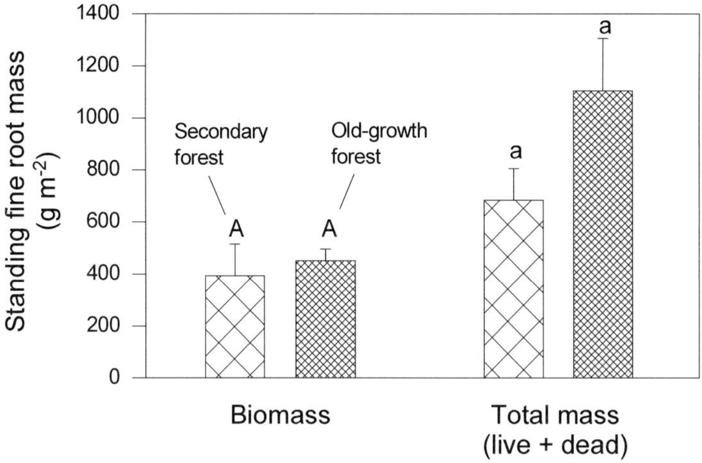

Fig. 4. Fine root biomass and total fine root mass (live + dead) of old growth and secondary forests in the humid tropics. Given are means and standard errors; n = 4 (biomass) or 19 (total mass), and 55 (biomass) or 49 (total mass) for secondary and old-growth forests, respectively. Different capital or small letters indicate significant differences between old-growth and secondary forests; $p < 0.05$, Scheffé's test for unbalanced, parametrically distributed samples. All data represent mass of fine roots < 2 mm in diameter. Note that the respective sampling depths of the studies were not equal. Data references: Klinge (1973), Huttel (1975), Huttel & Bernhard-Reversat (1975), Edwards & Grubb (1977), Stark & Spratt (1977), Tanner (1977), Kira (1978), Jordan & Escalante (1980), Raich (1980), Tanner (1980), Berish (1982), Edwards & Grubb (1982), Raich (1983), Gower (1987), Berish & Ewel (1988), Cuevas & Medina (1988), Sanford (1989), Cavelier (1992), Lugo (1992), Vance & Nadkarni (1992), Silver & Vogt (1993), Cavelier et al. (1996), Sanchez-Gallen & Alvarez-Sanchez (1996), Silver et al. (1996), Sundarapandian & Swamy (1996), Newberry et al. (1997), Denslow et al. (1998), Herbert & Fownes (1999), Malhi et al. (1999), Priess et al. (1999), Chuyong et al. (2000), Maycock & Congdon (2000), McGroddy & Silver (2000), Pavlis & Jenik (2000), Silver et al. (2000), Ostertag (2001), Schuur & Matson (2001), Yavitt & Wright (2001), Ibrahima et al. (2002), Kitayama & Aiba (2002), Nepstad et al. (2002), Hertel et al. (2003), Jaramillo et al. (2003), Moutinho et al. (2003), Davidson et al. (2004), Röderstein et al. (2005), Moser et al. (unpubl.). For more detailed information on the data sources see Hertel & Leuschner (2006).

5 Conclusions

The results of our literature survey show that the fine root systems of tropical moist forests are highly vulnerable to both natural and anthropogenic disturbances. The intensity and spatial extension of the disturbance event determine the decrease in fine root biomass caused by the disturbance. However, even small disturbance events (e.g. small-scale tree fall, slight selective logging) can lead to a noticeable reduction of the fine root system in natu-

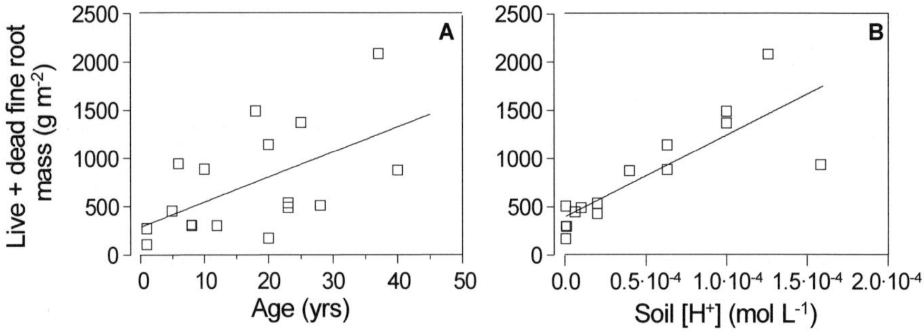

Fig. 5. A. Relationship between total fine root mass (live + dead) and age (1-50 years) for secondary forests in the humid tropics. Linear regression equation: $y = 280 + 26.2 \cdot x$, $r^2 = 0.32$, $r^2_{adj} = 0.27$, $p < 0.05$, $n = 17$. B. Relationship between total fine root mass (live + dead) and soil proton concentration for secondary forests in the humid tropics. Linear regression equation: $y = 397 + 8.5 \cdot 10^6 \cdot x$, $r^2 = 0.68$, $r^2_{adj} = 0.66$, $p < 0.001$, $n = 15$. All data represent mass of fine roots < 2 mm in diameter. Note that the respective sampling depths of the studies were not equal. Data references: Raich (1980) (*A*), Berish (1982) (*A*), Raich (1983) (*A*), Lugo (1992) (*A, B*), Cavelier et al.(1996) (*A, B*), Pavlis & Jenik (2000) (*A, B*), Hertel et al. (2003) (*A, B*), Jaramillo et al. (2003) (*A, B*). For more detailed information on the data sources see Hertel & Leuschner (2006).

ral forests of the moist tropics. Our rough estimation of the recovery time of the fine root biomass in naturally disturbed old-growth forests, on the other hand, suggests a rather quick regeneration of the fine root system after such disturbance events.

Although we did not find a significant difference in total fine root mass (live + dead), and fine root biomass (live fine roots) between tropical moist secondary and old-growth forests on average, our data analysis showed a significant increase in total fine root mass with stand age (0-50 years) of secondary forests. This finding gives evidence that the establishment of the fine root system during secondary succession following forest conversion can last several years to reach the initial biomass state of the natural forest. We conclude that the sensibility of the fine root system in tropical moist forests to natural and anthropogenic disturbance events is an important factor with respect to the carbon sequestration and the stability of these forests.

References

Achard F, Eva HD, Stibig HJ, Mayaux P, Gallego J, Richards T, Malingreau J-P (2002) Determination of deforestation rates of the world's humid tropical forests. Science 297: 999-1002

Akinnifesi FK, Kang BT, Ladipo DO (1999) Structural root form and flue root distribution of some woody species evaluated for agroforestry systems. Agroforest Syst 42: 121-138

Asbjornsen H, Velazquez-Rosas N, Garcia-Soriano R, Gallardo-Hernandez C (2005) Deep ground fires cause massive above- and below-ground biomass losses in tropical montane cloud forests in Oaxaca, Mexico. J Trop Ecol 21: 427-434

Asner GP, Knapp DE, Broandbent EN, Oliveira PJC, Keller M, Silva JN (2005) Selective logging in the Amazon. Science 310: 480-482

Attiwill PM (1994) The disturbance of forest ecosystems – the ecological basis for conservative management. For Ecol Manage 63: 247-300

Berish CW (1982) Root biomass and surface area in three successional tropical forests. Can J For Res 12: 699-704

Berish CW, Ewel JJ (1988) Root development in simple and complex tropical successional ecosystems. Plant Soil 196: 73-84

Cairns MA, Brown S, Helmer EH, Baumgardner GA (1997) Root biomass allocation in the world's upland forests. Oecologia 111: 1-11

Cavelier J (1992) Fine-root biomass and soil properties in a semideciduous and a lower montane rain forest in Panama. Plant Soil 142: 187-201

Cavelier J, Estevez J, Arjona B (1996) Fine root biomass in three successional stages of an Andean cloud forest in Colombia. Biotropica 28: 728-736

Chuyong GB, Newbery DM, Songwe NC (2000) Litter nutrients and retranslocation in a central African rain forest dominated by ectomycorrhizal trees. New Phytol 148: 493-510

Coley PD, Barone JA (1996) Herbivory and plant defenses in tropical forests. Ann Rev Ecol Syst 27: 305-335

Cuevas E, Medina E (1988) Nutrient dynamics within Amazonian forests. II. Fine root growth, nutrient availability and leaf litter decomposition. Oecologia 76: 222-235

Curran LM, Trigg SN, McDonald AK, Astiani D, Hardiono YM, Siregar P, Caniago I, Kasischke E (2004) Lowland forest loss in protected areas of Indonesian Borneo. Science 303: 1000-1003

Davidson EA, Ishida FY, Nepstad DC (2004) Effects of an experimental drought on soil emissions of carbon dioxide, methane, nitrous oxide, and nitric oxide in a moist tropical forest. Global Change Biol 10: 718-730

Denslow JS, Ellison AM, Sanford RE (1998) Treefall gap size effects on above- and below-ground processes in a tropical wet forest. J Ecol 86: 597-609

Edwards PJ, Grubb PJ (1977) Studies of mineral cycling in a montane rainforest in New Guinea. I. The distribution of organic matter in the vegetation and soil. J Ecol 65: 943-969

Edwards PJ, Grubb PJ (1982) Studies of mineral cycling in a montane rainforest in New Guinea. IV. Soil characteristics and the division of mineral elements between the vegetation and soil. J Ecol 70: 649-666

Fitter A (1996) Characteristics and functions of root systems. In: Waisel Y, Eshel A, Kafkafi, U (eds) Plant roots: the hidden half. Dekker, New York, pp 1-20

Gale MR, Grigal DF (1987) Vertical root distributions of northern tree species in relation to successional status. Can J For Res 17: 829-834

Gower ST (1987) Relations between mineral nutrient availability and fine root biomass in two Costa Rican tropical forests: a hypothesis. Biotropica 19: 171-175

Grove SJ, Turton SM, Siegenthaler DT (2000) Mosaics of canopy openness induced by tropical cyclones in lowland rain forests with contrasting management histories in northeastern Australia. J Trop Ecol 16: 883-894

Guariguata MR, Ostertag R (2001) Neotropical secondary forest succession: changes in structural and functional characteristics. For Ecol Manage 148: 185-206

Herbert DA, Fownes JH (1999) Forest productivity and efficiency of ressource use across a chronosequence of tropical montane soils. Ecosystems 2: 242-254

Herbert DA, Fownes JH, Vitousek PM (1999) Hurricane damage to a Hawaiian forest: Nutrient supply rate affects resistance and resilience. Ecology 80: 908-920

Hertel D, Leuschner Ch (2006) Fine root mass and fine root production in tropical moist forests as dependent on soil, climate and elevation. In: Bruijnzeel S, Juvik J (eds) Mountains in the mist: Science for Conserving and Managing tropical montane cloud forests. Hawaii Univ Press pp 283-297

Hertel D, Leuschner Ch, Hölscher D (2003) Size and structure of fine root systems in old-growth and secondary tropical montane forests (Costa Rica). Biotropica 35: 143-153

Hertel D, Hölscher D, Köhler L, Leuschner Ch (2006) Changes in fine root system size and structure during secondary succession in tropical montane forests of the Cordillera de Talamanca, Costa Rica. In: Kappelle M (ed) Ecology and Conservation of Neotropical Montane Oak Forests. Ecological Studies 185, Springer, New York (in press)

Hölscher D, Mackensen J, Roberts J-M (2005) Forest recovery in the humid tropics: changes in vegetation structure, nutrient pools and the hydrological cycle. In: Bonell M, Bruijnzeel S (eds) Forests, water and people in the humid tropics. Cambridge University Press, Cambridge, pp 598-622

Houghton RA, Unruh JD Lefebvre PA (1993) Current land cover in the tropics and its potential for sequestering carbon. Global Biogeochem Cycles 7: 305-320

Huttel Ch (1975) Root distribution and biomass in three Ivory Coast rain forest plots. In: Golley FB, Medina E (eds) Tropical ecological systems. Ecol. Studies 11, Springer, New York, pp 123-130

Huttel Ch, Bernhard-Reversat F (1975) Recherches sur l'écosystème de la forêt subéquatoriale de basse Côte-d'Ivoire. Cycle de la matière organique. Terre Vie 29: 203-228

Ibrahima A, Schmidt P, Ketner P, Mohren GJM (2002) Phytomasse et cycle des nutrients dans la forêt tropicale dense humide du Sud Cameroun. Tropenbos-Cameroon documents 9. The Tropenbos-Cameroon program, Kribi, Cameroon, 84 pp

Jackson RB, Canadell J, Ehleringer JR, Mooney HA, Sala OE, Schulze ED (1996) A global analysis of root distribution for terrestrial biomes. Oecologia 108: 389-411

Jaramillo VJ, Ahedo-Hernández R, Kauffman JB (2003) Root biomass and carbon in a tropical evergreen forest of Mexico: changes with secondary succession and forest conversion to pasture. J Trop Ecol 19: 457-464

Jordan CF, Escalante G (1980) Root productivity in an Amazonian rain forest. Ecology 61: 14-18

Kira T (1978) Community architecture and organic matter dynamics in tropical lowland rain forest of Southeast Asia with special reference to Pasoh Forest, West Malaysia. In: Tomlinson PB, Zimmerman MH (eds) Tropical trees as living systems. Cambridge University Press, New York, pp 561-590

Kitayama K, Aiba S-I (2002) Ecosystem structure and productivity of tropical rain forests along altitudinal gradients with contrasting soil phosphorus pools on Mount Kinabalu, Borneo. J Ecol 90: 37-51

Klinge H (1973) Root mass estimation in lowland tropical rain forests of central Amazonia, Brazil. I. Fine root masses of a pale yellow latosol and a giant humus podzol. Trop Ecol 14: 29-38

Kwit C, Platt WJ (2003) Disturbance history influences regeneration of nonpioneer understory trees. Ecology 84: 2575-2581

Lambin EF, Geist HJ, Lepers E (2003) Dynamics of land-use and land-cover change in tropical regions. Annu Rev Environ Resour 28: 205-241

Laurance WF, Ferreira LV, Rankin-De Merona JM, Laurance SG (1998) Rain forest fragmentation and the dynamics of Amazonian tree communities. Ecology 79: 2032-2040

Lehmann J (2003) Subsoil root activity in tree-based cropping systems. Plant Soil 255: 319-331

Leuschner Ch, Wiens M, Harteveld M, Hertel D, Tjitrosemito S (2006) Patterns of fine root mass and distribution along a disturbance gradient in a tropical montane forest, Central Sulawesi. Plant Soil 283: 163-174

Lugo AE (1992) Comparison of tropical tree plantations with secondary forests of similar age. Ecol Monogr 62: 1-41

Lugo AE, Scatena FN (1996) Background and catastrophic tree mortality in tropical moist, wet, and rain forests. Biotropica 28: 585-599

Malhi Y, Baldocchi DD, Jarvis PG (1999) The carbon balance of tropical, temperate and boreal forests. Plant Cell Environ 22: 715-740

Maycock CR, Congdon RA (2000) Fine root biomass and soil N and P in north Queensland rain forests. Biotropica 32: 185-190

McGroddy M, Silver WL (2000) Variations in belowground carbon storage and soil CO_2 flux rates along a wet tropical climate gradient. Biotropica 32: 614-624

Moutinho P, Nepstad DC, Davidson EA (2003) Influence of leaf-cutting ant nests on secondary forest growth and soil properties in Amazonia. Ecology 84: 1265-1276

Muthukumar T, Sha L, Yang X, Cao M, Tang J, Zheng Z (2003) Distribution of roots and arbuscular mycorrhizal associations in tropical forest types of Xishuangbanna, southwest China. Appl Soil Ecol 22: 241-253

Nascimento HEM, Laurance WF (2004) Biomass dynamics in Amazonian forest fragments. Ecol Appl 14: S127-S138 Suppl

Nepstad DC, 19 others (2002) The effects of partial throughfall exclusion on canopy processes, aboveground production, and biogeochemistry of an Amazon forest. J Geophysical Res 107: 1-18

Nepstad DC, Verissimo A, Alencar A, Nobre C, Lima E, Levebvre P, Schlesinger P, Potter Ch, Moutinho P, Mendoza E, Cochrane M, Brooks V (1999) Large-scale impoverishment of Amazonian forests by logging and fire. Nature 398: 505-508

Newbery DM, Alexander IJ, Rother JA (1997) Phosphorus dynamics in a lowland African rain forest: the influence of ectomycorrhizal trees. Ecol Monogr 67: 367-409

Ostertag R (1998) Belowground effects of canopy gaps in a tropical wet forest. Ecology 79: 1294-1304

Ostertag R (2001) Effects of nitrogen and phosphorus availability on fine-root dynamics in Hawaiian montane forests. Ecology 82: 485-499

Parrotta JA, Lodge DJ (1991) Fine root dynamics in a subtropical wet forest following hurricane disturbance in Puerto Rico. Biotropica 23: 343-347

Pavlis J, Jenik J (2000) Roots of pioneer trees in the Amazonian rain forest. Trees 14: 442-455

Priess T, Then C, Fölster H (1999) Litter and fine-root production in three types of tropical premontane rain forest in Venezuela. Plant Ecol 143: 171-187

Raich JW (1980) Fine root regrow rapidly after forest felling. Biotropica 12: 231-232

Raich JW (1983) Effects of forest conversion on the carbon budget of a tropical soil. Biotropica 15: 177-184

Richards PW (1996) The tropical rain forest. 2nd edn, Cambridge Univ Press, Cambridge, UK

Richter DD, Markewitz D, Trumbore SE, Wells CG (1999) Rapid accumulation and turnover of soil carbon in a re-establishing forest. Nature 400: 56-58

Röderstein M, Hertel D, Leuschner Ch (2005) Above- and below-ground litter production in three tropical mountain forests (Ecuador). J Trop Ecol 21: 483-492

Sánchez-Gallén I, Alvarez-Sánchez J (1996) Root productivity in a lowland tropical rain forest in Mexico. Vegetatio 123: 109-115

Sanford RL, Cuevas E (1996) Root growth and rhizosphere interactions in tropical forests. In: Mulkey SS, Chazdon RL, Smith AP (eds) Tropical forest plant ecophysiology. Chapman and Hall, New York, pp 268-300

Sanford RL (1989) Fine root biomass under a tropical forest light opening in Costa Rica. J Trop Ecol 5: 251-256

Scatena FN, Planos-Gutierrez EO, Schellekens J (2005) Natural disturbances and the hydrology of humid tropical forests. In: Bonnell M, Bruijnzeel LA (eds) Forests-Water-People in the Humid Tropics. Cambridge University Press, Cambridge, pp 489-512

Schenk HJ (2005) Vertical vegetation structure below ground: scaling from root to globe. Progr Bot 66: 341-373

Schuur EAG, Matson PA (2001) Net primary productivity and nutrient cycling across a mesic to wet precipitation gradient in Hawaiian montane forest. Oecologia 128: 431-442

Silver WL, Neff J, McGroddy M, Veldkamp E, Keller M, Cosme R (2000a) Effects of soil texture on belowground carbon and nutrient storage in a lowland Amazonian forest ecosystem. Ecosystems 3: 193-209

Silver WL, Ostertag R Lugo AE (2000b) The potential for carbon sequestration through reforestation of abandoned tropical agricultural and pasture lands. Restor Ecol 8: 394-407

Silver WL, Scatena FN, Johnson AH, Siccama TG, Watt F (1996) At what temporal scales does disturbance affect belowground nutrient pools. Biotropica 28: 441-457

Silver WL, Vogt KA (1993) Fine root dynamics following single and multiple disturbance in a subtropical wet forest ecosystem. J Ecol 81: 729-738

Stark N, Spratt M (1977) Root biomass and nutrient storage in rain forest oxisols near San Carlos de Rio Negro. Trop Ecol 18: 1-9

Sundarapandian SM, Chandrasekaran S, Swamy PS (1996) Influence of disturbance on fine root biomass and productivity in two deciduous forests of western Ghats, Tamil Nadu. Curr Sci India 70: 242-246

Sundarapandian SM, Swamy PS (1996) Fine root biomass distribution and productivity patterns under open and closed canopies of tropical forest ecosystems at Kodayar in Western Ghats, South India. For Ecol Manage 86: 181-192

Tanner EVJ (1977) Four montane rain forests of Jamaica: a quantitative characterization of the floristics, the soils and the foliar mineral levels, and a discussion of the interrelations. J Ecol 65: 883-918

Tanner EVJ (1980) Studies on the biomass and productivity in a series of montane rain forests in Jamaica. J Ecol 68: 573-588

Tanner EVJ, Kapos V, Healey JR (1991) Hurricane effects on forest ecosystems in the Carribbean. Biotropica 23: 513-521

Upadhaya K, Pandey H, Law P, Tripathi R (2005) Dynamics of fine and coarse roots and nitrogen mineralization in a humid subtropical forest ecosystem of northeast India. Biol Fert Soils 41: 144-152

Vance ED, Nadkarni NM (1992) Root biomass distribution in a moist tropical montane forest. Plant Soil 142: 31-39

Vogt KA, Vogt DJ, Palmiotto PA, Boon P, O'Hara J, Asbjornson H (1996) Review of root dynamics in forest ecosystems grouped by climate, climatic forest type and species. Plant Soil 187: 159-219

Yang YS, Chen GS, Lin P, Xie JS, Guo JF (2004) Fine root distribution, seasonal pattern and production in four plantations compared with a natural forest in Subtropical China. Ann For Sci 61: 617-627

Yavitt JB, Wright SJ (2001) Drought and irrigation effects on fine root dynamics in a tropical moist forest, Panama. Biotropica 33: 421-434

Surface soil organic carbon pools, mineralization and CO_2 efflux rates under different land-use types in Central Panama

Luitgard Schwendenmann[1*], Elise Pendall[2], and Catherine Potvin[3]

[1] Tropical Silviculture, Institute of Silviculture, University of Göttingen, Büsgenweg 1, 37077 Göttingen, Germany
[2] Department of Botany, 1000 E. University Ave., University of Wyoming, Laramie, WY 82071, USA
[3] Department of Biology, McGill University, 1205 Docteur Penfield, Montreal, Quebec H3A 1B1, Canada

*corresponding author: Luitgard Schwendenmann, phone: +49 (0)551 3991118
Email: lschwen@gwdg.de

Summary

The global carbon cycle is being perturbed by changes in land-use, especially in the tropics. This chapter compares surface soil organic carbon stocks, carbon mineralization rates and soil CO_2 efflux between an undisturbed forest and a clearing at Barro Colorado Island and between a pasture and plantation at Sardinilla, Central Panama. Our results on C cycling at two study sites with contrasting parent material and soil type were compared with other studies throughout the moist tropics. Differences in soil carbon stocks in the topsoil (0-5 cm) of the clearing (15 Mg C ha^{-1}) and the undisturbed forest site (22 Mg C ha^{-1}) were statistically not significant. Our inventory revealed that the highest carbon stock (29 Mg C ha^{-1}) was found under the native tree plantation, although at least part of this high value is site-related. Thus, no carbon change could be detected two years after the conversion of the site from a pasture into a native tree plantation. Soil CO_2 efflux rates at the pasture site (8 µmol CO_2 m^{-2} s^{-1}) were significantly higher than in forest, clearing and plantation (5-6 µmol CO_2 m^{-2} s^{-1}). Large CO_2 flux rates in the pasture might be explained by high belowground biomass production which leads to high root respiration rates. Our incubation experiment showed that pasture and clearing soil had a higher proportion of active pool carbon than plantation and forest. Higher amounts of active pool C indicate the existence of carbon readily mineralizable by microbes. Our results demonstrate that the active

Tscharntke T, Leuschner C, Zeller M, Guhardja E, Bidin A (eds), The stability of tropical rainforest margins, linking ecological, economic and social constraints of land use and conservation, Springer Berlin 2007, pp 109-131

pool C is a good predictor of soil respiration. Thus, active soil organic carbon is a sensitive indicator for changes in soil organic carbon following land use change.

Keywords: forest, mineralization rate, pasture, plantation, soil carbon stocks, soil CO_2 efflux, stable isotopes, Central Panama

1 Introduction

The current discussion about global change, including land-use change and greenhouse gas emissions, has increased interest in the global carbon cycle (Clark 2004). Tropical forests play a critical role with respect to global carbon pools and fluxes as these forests store about half of the world's biomass (Brown and Lugo 1982) and 20% of the global soil carbon (Jobbagy and Jackson 2000). The global carbon cycle is being altered in response to human interference; for instance land-use changes in the tropics are estimated to contribute about 23% to human-induced CO_2 emissions (Houghton 2003a).

Soil organic matter (SOM) or soil organic carbon (SOC) encompass all organic constitutes and fractions in the mineral soil, including plant and animal tissue in variable stages of decomposition, living biomass of microorganisms, root and microbial exudates and well-decomposed and highly stable organic material. Although SOM consists of many C compounds, it is often divided conceptually into three pools of different magnitude and turnover times (Figure 1). The turnover time represents the time carbon resides in a certain pool and hence is a measure of the stability of carbon pools. The labile, active pool is composed of microbial biomass and easily decomposable compounds from leaf litter and root-derived material with short turnover times (from weeks to years), the slow (also called intermediate) pool (consisting of refractory components of litter, weakly sorbed carbon) has turnover times from 10 to more than 100 years and the passive (also called resistant or inert) pool (composed of highly humified organic compounds, often mineral-stabilized) has a turnover time on the order of 10^3 years (Parton et al. 1987, Trumbore 1997).

Soil organic matter is a major factor in ecosystem functioning and determines whether soils act as sinks or sources of carbon in the global carbon cycle. Carbon input, magnitude of soil organic carbon pools and finally carbon mineralization depend on many factors (Figure 1). Changing patterns of land-use and land-use management practices can have significant direct and indirect effects on soil organic pools, due to changes in plant species, primary productivity, litter quantity and quality and soil structure. However, the impact of land-use changes on organic carbon pools in the mineral soil depends also on long-term site-specific factors (e.g. climate, topography and parent material) and is often overridden by the high spatial heterogeneity of soil organic carbon. Consequently, effects on SOC pools are evident only with the most intensive practices. For example, forest clear-cuttings for pasture in the humid

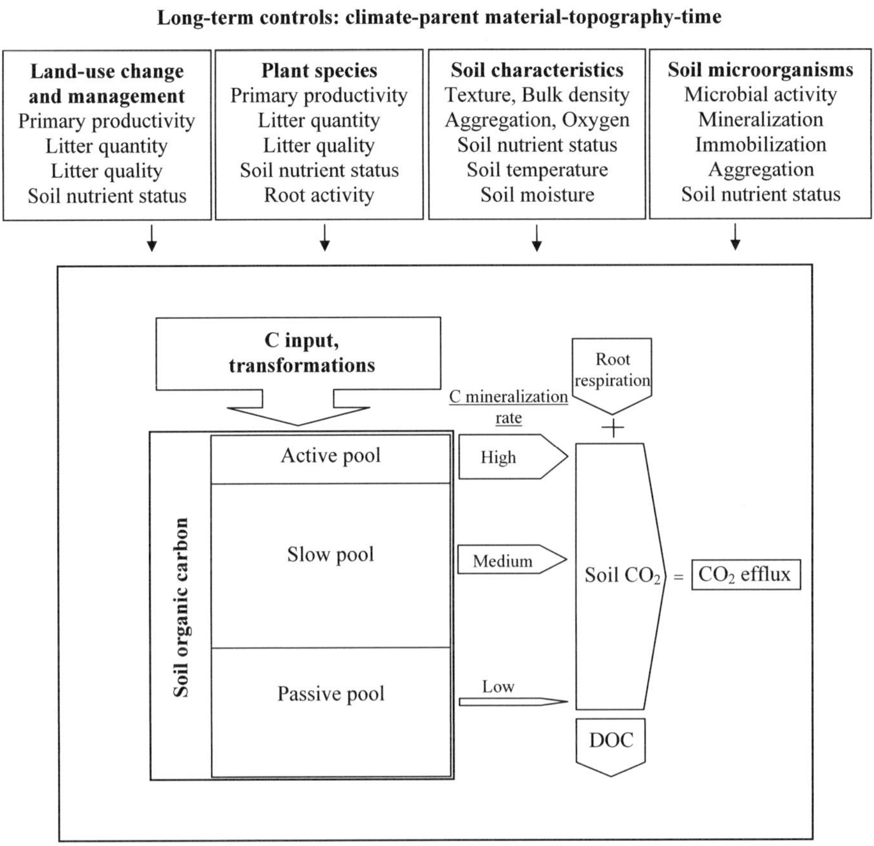

Fig. 1. Simplified model on controls of carbon input, soil organic carbon formation and mineralization

tropics combined with slash removal and/or burning cause significant losses of SOC (Brown and Lugo 1990, Veldkamp 1994). However, pasture establishment in the tropics may maintain or even increase the soil organic carbon content especially when grass species with high percentages of below-ground biomass production are used (Feigl et al. 1995). A recent meta-analysis using data from temperate and tropical regions indicates that soil organic carbon stocks increased by 8% after converting native forests into pastures (Guo and Gifford 2002). In contrast, no clear trend in soil C change following the conversion from forest to pasture was shown in a data compilation for the tropics by Lugo and Brown (1993). Murty et al. (2002) also found no trend in soil C after conversion from forest to pasture in temperate and tropical regions.

In recent years reforestation of degraded and abandoned tropical pastures has been proposed as a measure to mitigate increasing atmospheric CO_2 levels (Montagnini and Porras 1998, Silver et al. 2000). However, the effects of reforestation can result in losses or gains of soil C pools (Rhoades et al. 2000, de Koning et al. 2003, Silver et al. 2004).

Land-use change affects carbon stocks and also soil respiration (also called soil CO_2 efflux) rates (Trumbore et al. 1995). To characterize the carbon exchange in ecosystems, an assessment of the magnitude and dynamics of soil CO_2 efflux is important, considering that soil respiration is a major CO_2 flux in the carbon cycle, second in magnitude to gross canopy photosynthesis (Raich and Schlesinger 1992). The net flux of carbon between the soil and the atmosphere is determined by the rate at which soil organic C is converted to CO_2 by microorganisms and by autotrophic respiration (Figure 1). Soil respiration is influenced mainly by soil temperature and moisture (Orchard and Cook 1983, Howard and Howard 1993) but also by vegetation type (Raich and Tufekcioglu 2000) and substrate availability (Vasconcelos et al. 2004).

The variation in the estimates of soil organic carbon stocks in the tropics is due largely to the limited number of studies used in regional extrapolations (Canadell and Pataki 2002). Thus, soil databases need to be improved to account for different characteristics of soils in tropical regions in order to provide a more accurate estimate of the soil organic carbon pool (Houghton 2003b). Furthermore, with differing trends reported for the conversion from forest to pasture and pasture to plantation, uncertainty remains with respect to changes in soil C after land-use changes. There are still gaps in understanding of the critical processes and properties regulating carbon transfer and storage (Figure 1). A better understanding of the soil carbon cycle and its parameters will ultimately provide better estimates of the global carbon pool and a clearer picture of the impact of human activities, especially on rainforest margins.

Between 1960 and 1990, Asia has lost nearly a third of its tropical forest cover to deforestation while Africa and Latin America each lost about 18 percent (FAO 2001). The Republic of Panama has lost more than two Mill. ha (35%) of its forested area to cropland and agro-pastoral development in the past 50 years (Condit et al. 2001). In the Panama Canal Watershed, where deforestation and development have been associated with increasing rates of erosion and sediment delivery to streams, reforestation efforts are critical to improving soil productivity and water quality (Ibanez et al. 2002). The conversion of degraded pastures to tree plantations is taking place over extensive areas of Panama. In 2003, the total area under plantations was 55,200 ha (ANAM 2004).

The objective of this research is to assess surface carbon stocks, soil CO_2 fluxes and carbon mineralization rates in four different land-use systems in Central Panama. We evaluate C cycling at two study sites with contrasting parent material and soil type with regard to other studies throughout the moist tropics. We compare an undisturbed forest to an adjacent grassland on Barro Colorado Island on soils derived from andesite. Our hypotheses for the BCI study are: (1) surface SOC stocks are lower in the grassland than the forest site and (2) soil respiration and C mineralization are higher under the grassland. We also compare a young native tree plantation to a pasture at Sardinilla on soils developed on limestone. Our hypothesis for the Sardinilla study is that surface SOC stocks, soil respiration and C mineralization are higher under the pasture than the plantation. Across sites we hypothesize that soil C is related to clay content, and that soil respiration is positively correlated with the amount of active pool carbon and soil moisture.

2 Study area and methods

2.1 Description of the study sites

The study was conducted on Barro Colorado Island (BCI; 9°09´N, 79°51´W) and at Sardinilla (9°19' N, 79°38' W), Central Panama. The mean annual temperature of the region is 26°C. Average total yearly rainfall is 2,600 mm, with a distinct dry season from mid-December to April (Leigh et al. 1996).

On BCI we selected an undisturbed forest and an adjacent grassland site to study the impact of forest clearing on carbon stocks and fluxes. Our study sites were located on the north site of the island close to the Miller lighthouse. The soils developed in andesite parent material on a north-facing hill slope of 20 degrees (Yavitt 2000). The vegetation on BCI is classified as Tropical Moist Forest according to Holdrigde's life zone system. The biomass of the forest floor consisted of leaves and woody debris in different stages of decomposition, but no organic horizon was present. At the grassland or so-called clearing site, the forest was cut around 90 years ago to set up a light house for the Panama Canal. During the first half of the 20th century, the clearing was burned to remove vegetation. In recent years, the vegetation has been cut manually several times a year. At the time of sampling, the vegetation was dominated by *Saccharum spontaneum,* a C_4 plant. This site represents a specific case of a tropical grassland established by forest clearing and subjected to frequent burning, but not influenced by grazing. Only mineral soil horizons were found in the grassland.

The second study area was located near the village of Sardinilla, which is approximately 40 km east of BCI. A grazed pasture and a native tree plantation site were selected to assess the effect of afforestation on carbon stocks and fluxes of a young plantation. The soils at Sardinilla are derived from Tertiary limestone and other sedimentary rocks (Wilsey et al. 2002). The original forest vegetation at the Sardinilla site was probably similar to that of BCI. The forest in the area was clear-cut in 1952 and 1953, cropped for two years, and since then has been used as grazed pasture with major species being *Ischaemum indicum* and *Scleria melaleuca* (Potvin et al. 2004). An area 150 m southeast of the grazed pasture was converted to a native tree plantation in July 2001 (Potvin et al. 2004). The site was neither tilled nor burned before seedlings were planted. The species *Luehea seemannii* Triana & Planch, *Cordia alliodora* (Ruiz & Pavon) Oken, *Anacardium excelsum* (Bert.

& Balb. ex Kunth) Skeels, *Hura crepitans* L., *Cedrela odorata* L. and *Tabebuia rosea* (Bertol.) DC were planted in monocultures and mixtures (three and six species plots). The plots have not been fertilized. The grass vegetation is cut manually several times a year and the biomass was left on the site. The stocking density at the time of sampling was ~1100 trees ha^{-1}; in 2003 tree aboveground biomass was ~2.5 Mg ha^{-1} and herbaceous aboveground biomass amounted to ~12 Mg ha^{-1} (Potvin, unpublished data). Ridge and swale topography has important effects on variations in productivity at this site (Potvin, unpublished data).

2.2 Soil sampling and analytical procedures

Each land-use type was represented by one site, with three replicate sampling locations per site. Sampling took place in October 2003. Soil samples were taken at 0-5 cm depth. The bulk density data were collected using the core method. Samples were oven-dried (105 °C for 48 h) and bulk density was estimated as the mass of oven-dry soil divided by the core volume (100 cm^3). Soil organic carbon and nitrogen contents were determined with an elemental analyzer (NCS 2500, CE Elantech, Lakewood, NJ, USA). Total soil carbon stocks were calculated as follows:

$$\mathrm{SOC\,(Mg\,ha^{-1}) = BD\,(Mg\,m^{-3}) \times C\,content\,(g\,C\,kg^{-1}) \times D\,(m)} \qquad (1)$$

where BD is soil bulk density, C is the carbon content estimated by elemental analyzer and D is soil sampling depth. To account for changes in bulk density between forest and clearing we adjusted the soil depth according to the following equation (Veldkamp 1994, Solomon et al. 2002):

$$D_{\mathrm{adjusted}} = \left(\frac{\mathrm{BD_{Forest}}}{\mathrm{BD_{Clearing}}}\right) \times D \qquad (2)$$

Carbon stable isotope technique can be used as a tracer of soil organic carbon turnover in tropical regions, where the main photosynthetic pathway of the plant cover may change, where, for example, C_3 trees with an average $\delta^{13}C$ value of -27‰ grow on soils derived from C_4 vegetation with an average $\delta^{13}C$ value of -11‰ (Smith and Epstein 1971). To measure ^{13}C of soil and plant material, samples were combusted in an elemental analyzer (NCS 2500, CE Elantech, Lakewood, NJ, USA), the evolved CO_2 was then analyzed with an isotope-ratio mass spectrometer (Isoprime, Micromass, Manchester, U.K.). Results are expressed in $\delta^{13}C$ (‰) relative to the V-PDB standard as:

$$\delta^{13}\mathrm{C}(‰) = \frac{(\mathrm{R_{sample}} - \mathrm{R_{standard}})}{R_{\mathrm{standard}}} \times 1000 \qquad (3)$$

where R is the ^{13}C:^{12}C ratio.

The fraction of carbon derived from the current land-use was calculated using a simple mixing model (Balesdent and Mariotti 1996).

$$F_{\text{new}} = \left(\frac{\delta\text{soil}_{\text{current land-use}} - \delta\text{soil}_{\text{former land-use}}}{\delta\text{plant} - \text{residue/litter}_{\text{current land-use}} - \delta\text{soil}_{\text{former land-use}}} \right) \quad (4)$$

where F_{new} is the fraction of new carbon (clearing or plantation derived), $\delta\text{soil}_{\text{current land-use}}$ is the δ^{13}C value of soil carbon under clearing or plantation, $\delta\text{soil}_{\text{former land-use}}$ is the δ^{13}C value of soil carbon under forest or pasture and $\delta\text{plant-residue/litter}_{\text{current land-use}}$ is the δ^{13}C value from the clearing ($\delta^{13}C$ = -16.2 ± 4.2 ‰) or plantation ($\delta^{13}C$ = -29.7 ± 0.5 ‰).

2.3 Measurement of soil CO_2 efflux

We applied the dynamic closed chamber approach for measurement of soil CO_2 efflux (Parkinson 1981). Soil CO_2 efflux was measured using a portable infrared gas analyzer (EGM-4, PP systems, Amesbury, MA, USA) and a dark soil respiration chamber built of PVC (height 150 mm, diameter 100 mm). The soil respiration chamber is equipped with a fan to ensure that the air within the chamber is carefully mixed and to prevent the generation of pressure differences which would affect the evolution of CO_2 from the soil surface. A 10-ml drierite (10 mesh) column combined with a Permapure filter provided dry air to avoid interference in the CO_2 detection from endogenous humidity of the soil air circulating in the closed system. During the measurement the chamber was inserted 1 cm into the soil (grass was cut shortly before the measurement). Air was circulated between the gas analyzer and the respiration chamber. CO_2 concentrations were measured every 5 s and recorded by the internal datalogger. The flux was calculated from the concentration increase over time. Each flux measurement lasted between 3 and 5 minutes, until a good quadratic fit was obtained. Within the forest, clearing and pasture, 15 measurements were carried out at random locations. In the plantation, 50 locations were selected to account for differences in topography. Soil CO_2 efflux was measured for several days during the rainy season (July and October 2003) and at the onset of the dry season in January 2004. A short-term period is sufficient to compare the range of CO_2 efflux of different land-use types during dry and rainy season. However, this dataset would not be sufficient to extrapolate to annual estimates.

Soil temperature (thermocouple K-probe, Extech Instruments Corporation, Waltham, MA, USA) and soil water content (Theta probe ML-X2 with HH2 reader, Delta-T Devices Ltd, Cambridge, UK) were measured adjacent to each flux chamber at 5 cm depth.

2.4 Measuring soil C mineralization rates

By separating the labile SOM pool that is more sensitive than bulk soil organic matter to changes in land-use, management or climate the detection limit for SOM changes is increased. One approach to assess the magnitude and turnover times of the active soil organic carbon pools is by soil incubation measuring the biological mineralization of carbon (Townsend et al. 1997, Paul et al. 2001).

Incubations were performed using ~15 g of field moist soil. Before incubation the soil was sieved and roots were removed. The soil was incubated at 25 °C in sealed canning jars (fitted with inert, butyl septa in the lids). Air samples (headspace samples) from the sealed jar were collected using a syringe. The air sample was then injected in an infrared gas analyzer (LI-820, LI-COR, Linclon, Nebraska, USA) to determine CO_2 concentration. We established a calibration curve with standards of 372, 1000 and 2700 ppm CO_2 and calculated the C mineralization rates as the change in headspace CO_2 concentrations (μg C) per gram soil (dry wt. equivalent) per unit incubation time (day). The concentration was converted to μg C using the universal gas law. After each sampling (at intervals ranging from 1 to 20 days) the jar was flushed with air. When necessary, deionized water was added to maintain constant soil moisture. Carbon mineralization was measured on duplicate subsamples. The active (labile) carbon pool size was determined using a nonlinear regression function (PROC NLIN, Method=Marquardt, SAS, Version 8.2, SAS Institute Inc. Cary, NC, USA) that adjusted for the curvilinear relationship of the C mineralization between sampling points (Paul et al. 2001). We evaluated the active C pool size because it was expected to be the most sensitive to land-use change (Figure 2).

2.5 Statistical analysis

A one-way ANOVA, accompanied by Tukey's HSD post hoc analysis, was used to identify differences in soil organic carbon, stable isotopes, soil CO_2 efflux and carbon mineralization rates between forest and clearing (BCI) and pasture and plantation (Sardinilla). ANOVA was also used to assess seasonal differences in CO_2 efflux, soil moisture and soil temperature. Pearson product-moment correlation was employed to examine relationships between C pool sizes, CO_2 efflux rates and environmental variables. Significant effects were determined at $P < 0.05$. The analysis was carried out using the STATISTICA 7.1 software package (StatSoft Inc., Tulsa, Oklahoma, USA).

3 Results

3.1 Soil carbon, C/N ratios and stable carbon isotopes

On BCI C concentration in the top 5 cm was 38.2 (±4.6) g kg^{-1} in the clearing and 56.0 (±15.2) g kg^{-1} in the forest (Table 1). The surface C stocks amounted to 15.3 (±1.1) and 22.4 (±5.6) Mg C ha^{-1} for clearing and forest, respectively. Due to the high heterogeneity, especially among forest sites, the differences in carbon concentration and stocks between clearing and forest were not significant. C/N ratio in the surface layer was significantly higher under forest (13.0)

Table 1. Soil organic carbon content, C:N ratios, soil organic carbon stocks, $\delta^{13}C$ values and bulk density in the top 5 cm at BCI and Sardinilla, Central Panama. Values are means ± SD of three replicates. Within columns, different lower case letters indicate significant differences between forest and clearing on BCI. Different upper case letters indicate significant differences between pasture and plantation at Sardinilla ($P < 0.05$).

Site	Carbon content (g C kg^{-1})	C:N	Carbon stocks (Mg C ha^{-1})	$\delta^{13}C$ (‰)	Bulk density (Mg m^{-3})
BCI					
Forest	56.0 (15.2) a	13.0 (1.4) a	22.4 (5.6) a	-28.7 (0.4) a	0.85 (0.05) a
Clearing	38.2 (4.6) a	9.9 (0.2) b	15.3 (1.1) a	-21.2 (0.6) b	0.94 (0.08) a
Sardinilla					
Pasture	37.5 (5.9) A	9.4 (0.6) A	15.9 (1.9) A	-15.5 (0.72) A	0.97 (0.08) A
Plantation	68.8 (16.0) B	10.8 (0.5) A	29.4 (6.8) B	-19.5 (0.98) B	0.75 (0.01) B

as compared to clearing (9.9) (Table 1). The soil organic $\delta^{13}C$ values differed significantly between forest and clearing because the forest was dominated by C_3 trees and the clearing was dominated by C_4 grasses (Table 1). Soil in the 90-year-old clearing on BCI was surprisingly depleted (-21.2‰), suggesting a relatively important C_3 component during much of its land-use history. The $\delta^{13}C$ value of the current clearing vegetation ranged between -12‰ and -20‰ (average = -16.2 ± 4.2‰). Using the mixing model (Equation 4) we estimated that 60% of the soil C under the clearing were 'new' (clearing derived) C. The adjacent forest soil had a low $\delta^{13}C$ value (-28.7‰) due to the input of C_3 residue (-29.7‰).

At Sardinilla surface SOC stocks differed significantly between pasture and native tree plantation. We measured 15.9 (±1.9) Mg C ha^{-1} under the pasture and 29.4 (±6.8) Mg C ha^{-1} under the plantation. Pasture soil at Sardinilla was significantly enriched in $\delta^{13}C$ (-15.5‰), reflecting the dominance of C_4 vegetation with an average $\delta^{13}C$ value of -11.8‰. The pasture soil retained 21% rainforest-derived C, with 79% derived from the C_4 vegetation. The nearby plantation, although only 2-3 years old at the time of sampling, had an average soil $\delta^{13}C$ value of -19.5‰, reflecting C_3 inputs from recently planted trees as well as some older, rainforest derived C. The plantation leaf litter had a $\delta^{13}C$ value of -29.7 (± 0.5) ‰. Assuming the plantation soil had a similar amount of rainforest derived C as the pasture (21%), the proportion of plantation derived C in the top 5 cm was 7%.

3.2 Soil CO_2 efflux

On BCI soil CO_2 efflux rates during the wet season were similar in clearing (5.2 µmol CO_2 m^{-2} s^{-1}) and forest (5.7 µmol CO_2 m^{-2} s^{-1}) (Table 2). At the onset of the dry season, soil CO_2 efflux in the forest decreased considerably.

Table 2. Soil CO_2 efflux, soil temperature and soil moisture at BCI and Sardinilla, Central Panama. Standard deviation is given in parentheses (n=15-50). Within columns, different lower case letters indicate significant differences between forest and clearing on BCI. Different upper case letters indicate significant differences between pasture and plantation at Sardinilla. Within rows, differences between seasons are indicated by * ($P < 0.05$). nm = not measured

Sites	Soil CO_2 efflux (μmol CO_2 $m^{-2}s^{-1}$)		Soil temperature (°C)		Soil water content (%)	
	Wet season	Dry season	Wet season	Dry season	Wet season	Dry season
BCI						
Forest	5.7 (2.7) a*	3.6 (1.5)	24.9 (0.2) a	24.7 (0.2)	46.3 (3.3) a	41.9 (3.1)
Clearing	5.2 (1.2) a	nm	27.1 (1.1) b	nm	44.4 (3.8) a	nm
Sardinilla						
Pasture	8.1 (1.8) A	9.3 (3.1) A	26.6 (0.4) A*	24.1 (0.4) A	50.2 (4.4) A*	27.9 (5.6) A
Plantation	5.1 (2.4) B	5.1 (3.8) B	26.1 (1.0) A	25.7 (1.4) B	47.8 (2.4) B*	34.6 (9.3) B

For the forest site we found a significant positive correlation between soil respiration and soil temperature (r=0.3) but no significant relationship with soil moisture.

The soil CO_2 efflux at the Sardinilla pasture site was significantly higher (8.1 µmol CO_2 m^{-2} s^{-1}) as compared to the plantation (5.1 µmol CO_2 m^{-2} s^{-1}) (Table 2). We did not observe any significant seasonal change at the pasture or plantation site. Soil respiration in the plantation tended to be highest at the lower slopes and lowest at the ridge (data not shown). For the pasture and plantation we did not find significant relationships between soil CO_2 efflux and abiotic factors (soil temperature and moisture).

3.3 Carbon mineralization rates

The CO_2 evolution rates from the incubated soil samples followed the same general pattern across all land-use systems (Figure 3). At the beginning of the incubation the amount of CO_2 respired from the soil was high (up to 70 µg C g $soil^{-1}$ d^{-1}). This initial 'active pool phase' lasted for about the first two weeks, then 'slow pool carbon' was mineralized for the remainder of the experiment. We assumed that passive carbon mineralization was not detectable. The fluxes during the 'slow phase' remained relatively constant and were 2-6 µg C g $soil^{-1}$ d^{-1}.

At BCI we found that the absolute amount of mineralized carbon over the 180-day experiment (1.1 mg C g^{-1} $180d^{-1}$) and the active carbon pool (0.31 mg C g^{-1} $180d^{-1}$) did not differ between land-use types, but the relative amounts were higher under the clearing than under the forest (Table 3).

At Sardinilla the absolute amount of mineralized carbon did not differ significantly between pasture (0.97 mg C g^{-1} $180d^{-1}$) and plantation (1.1 mg C g^{-1} $180d^{-1}$). In contrast, the relative amount of mineralized carbon was significantly higher under the pasture (2.6%) as compared to the plantation (1.6%). This finding is consistent with BCI where we also observed a higher proportion of mineralizable carbon under the clearing than under forest. Absolute and relative amount of active soil carbon was significantly higher under the pasture as compared to the plantation.

4 Discussion

4.1 Conversion from forest to pasture

Change in soil carbon

Our comparison revealed that topsoil carbon concentrations and stocks were lower in the clearing than in the undisturbed forest, although the differences were statistically not significant (Table 1). The carbon loss was 31%

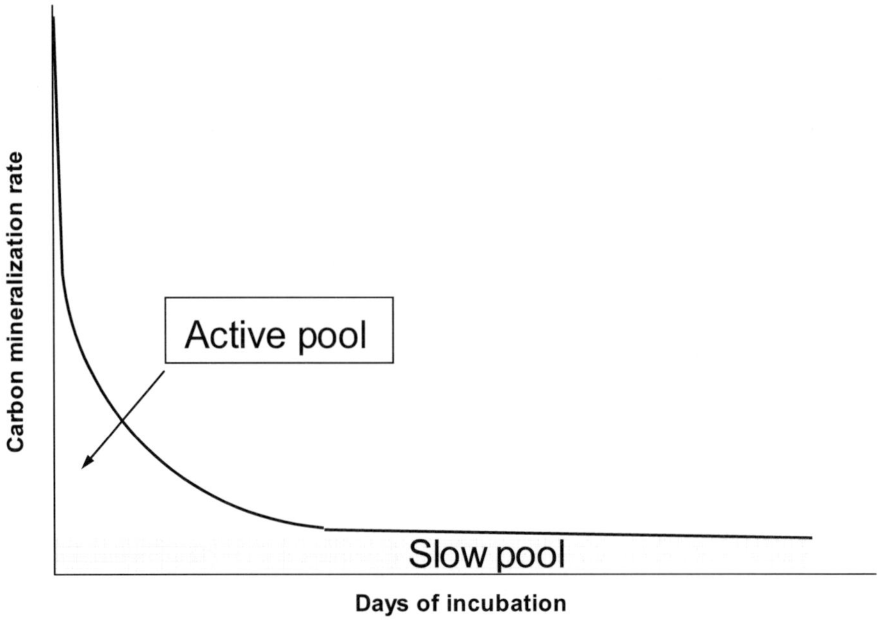

Fig. 2. Contribution of the active and slow pool over the course of a long-term incubation. The carbon mineralized in the initial pulse above the baseline level is identified as the active soil organic carbon pool (after Townsend et al. 1997, Paul et al. 2001).

in 0-5 cm depth. Based on stable isotope techniques we estimated an input rate of new (grass derived) carbon of approximately 0.3 Mg C ha^{-1} year^{-1} (Schwendenmann and Pendall 2005). However, on BCI the loss of old (forest derived) soil carbon exceeded the input.

Our results are consistent with other studies from the Neotropics that have found decreases in carbon and nitrogen content following conversion of tropical forest to pastures. A decrease of 18% (0-10 cm) was estimated for a 36-year old pasture in Costa Rica (Reiners et al. 1994). Desjardins et al. (1994) observed a 5% decrease in carbon stock (0-20 cm) in a 10-year old pasture in Eastern Amazonia. In contrast, Feigl et al. (1995) found an increase in soil carbon (+ 75%) and nitrogen (+ 50%) in the upper 10 cm of a pasture over an 80 year period following forest clearing in Rondonia, Brazil. An increase of C content in surface horizon after conversion to pasture was also reported for other sites in the Amazon region (de Moraes et al. 1996, Koutika et al. 1997).

The variable response in the direction and magnitude of the changes in soil C storage to forest clearing and pasture establishment implies that a number of direct and indirect factors have to be taken into account. For example, when forest is converted into pasture, soil organic carbon transformations are modified due to changes in substrate quality (Feigl et al. 1995), altered mi-

Table 3. The absolute and relative amount of mineralized carbon over the 180-day experiment and the absolute and relative amount of active pool carbon at BCI and Sardinilla, Central Panama. Standard deviation is given in parentheses (n=3). Within columns, different lower case letters indicate significant differences between forest and clearing on BCI. Different upper case letters indicate significant differences between pasture and plantation at Sardinilla ($P < 0.05$).

Sites	Mineralized C (mg C $g^{-1}180d^{-1}$)	Mineralized C (% of total C)	Active pool C (mg C $g^{-1}180d^{-1}$)	Active pool C (% of total C)
BCI				
Forest	1.10 (0.28) a	2.0 a	0.31 (0.09) a	0.6 a
Clearing	0.94 (0.20) a	2.5 b	0.30 (0.07) a	0.8 b
Sardinilla				
Pasture	0.97 (0.18) A	2.6 A	0.45 (0.05) A	1.2 A
Plantation	1.09 (0.38) A	1.6 B	0.30 (0.10) B	0.4 B

crobial community size (Cleveland et al. 2003) and/or changes in soil porosity and water retention (Martinez and Zinck 2004).

Based on conceptual models and studies, initial carbon loss after deforestation is explained by increased mineralization rates, erosion, reduced litter input and changes in litter quality (Figure 1). The long-term effect of forest clearing on the amount of carbon stored in pasture soils and the direction of change (Figure 3) may be related to climate condition and soil type, initial amount of soil carbon, the age of the pasture (Powers and Veldkamp 2005), site fertility, species of grass planted, grazing intensity and pasture management (Trumbore et al. 1995) or other factors that govern the productivity of a site (Lugo and Brown 1993, Fearnside and Barbosa 1998, Guo and Gifford 2002, Murty et al. 2002). The depletion of C in the clearing at BCI is most likely the result of the post harvest treatment (removal of slash) and a reduced input of plant residues (repeated burning of the site in the first half of the 20th century, no fertilizer input).

Change in soil CO_2 efflux and carbon mineralization rate

The soil CO_2 efflux rate reported here for the undisturbed forest (3.6-5.7 µmol CO_2 $m^{-2}s^{-1}$) is consistent with data obtained from various old-growth forest sites on BCI (Kursar 1989) and Gigante Peninsula (ca. 5 km from our study site) (Sayer 2005). In a primary forest, located in the Jambi Province, Central Sumatra, soil respiration during the dry season was 3.6-4.3 µmol CO_2 $m^{-2}s^{-1}$(Ishizuka et al. 2005). Raich and Schlesinger (1992) reported for tropical moist forests a CO_2 efflux of 3.3 (±0.15) µmol CO_2 $m^{-2}s^{-1}$. In general, the comparison of soil respiration data between different studies is difficult due to methodological differences. Soil CO_2 efflux measured by the system which we were using (PP-systems) was shown to be up to 33% higher than fluxes

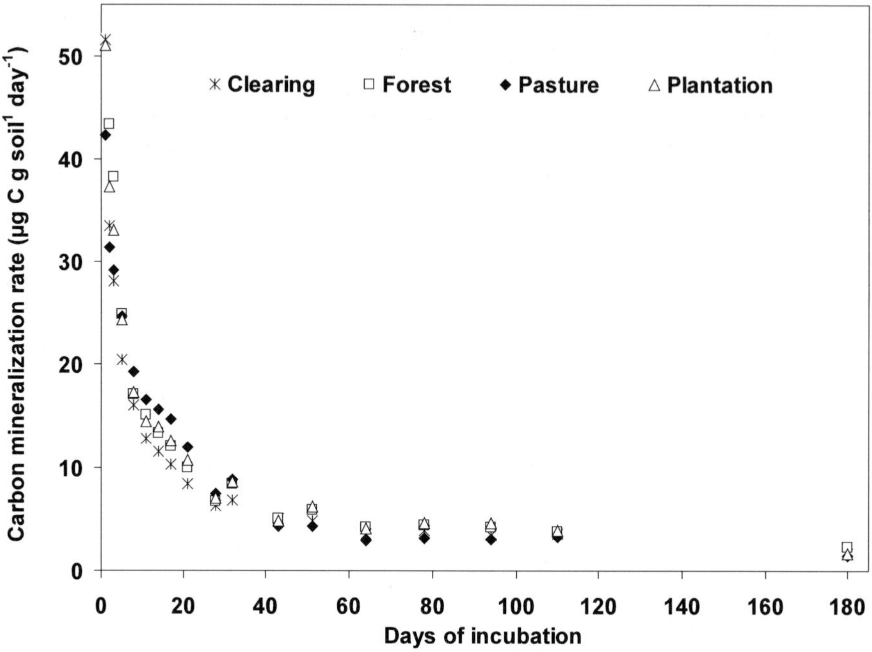

Fig. 3. Carbon mineralization rate over the 180-day incubation across land-use types, Central Panama. The values are mean values of three replicates. Average standard deviation is 4 µg C g soil^{-1} day^{-1} for the initial phase and 0.8 µg C g soil^{-1} day^{-1} for the remaining time.

measured by other closed dynamic systems. The overestimation might be due to the turbulence caused by the fan (Pumpanen et al. 2004).

Our data did not show significant differences in soil CO_2 efflux between forest and clearing. In contrast, a review comparing soil respiration data between forests and grasslands including both tropical and temperature locations revealed that soil respiration rates were greater in grasslands than in forests growing under similar conditions (Raich and Tufekcioglu 2000). Higher CO_2 efflux rates in grasslands might be explained by high belowground biomass production which leads to high root respiration rates (Trumbore et al. 1995). Lower substrate quality and lower net primary production (NPP) might explain why at BCI the respiration at the clearing did not exceed the respiration of the forest.

Around 30% of the variability in soil respiration under the forest could be explained by soil temperature. For the clearing we did not find any relationship between soil respiration and abiotic factors. Measuring soil respiration in an old-growth forest at Gigante Peninsula (ca. 5 km from our study site) over several seasons, Sayer (2005) found that soil water content explained 22% of the variation in soil respiration. In general, soil moisture is a better predictor of soil CO_2 efflux at locations where soil temperatures are high and relatively invariable and the variation of rainfall is high (Schlesinger 1977, Rout & Gupta 1989, Davidson et al. 2000, Schwendenmann et al. 2003). Water might become limiting during the dry season and inhibit root and microbial respiration. In a wet tropical forest, CO_2 flux declined under saturated conditions (Schwendenmann et al. 2003) probably due to a lower diffusion rate and/or reduced microbial activity because of the lack of oxygen.

4.2 Conversion from pasture to plantation

Change in soil carbon

Plantation soil carbon was depleted in $\delta^{13}C$ (-19.5‰) as compared to the pasture site (-15.5‰). This indicates that soil carbon turnover is fairly high as 'new' (tree derived) carbon already replaced some pasture carbon. The fraction of 'new' carbon was around 28%. However, around 20% of the C_3-C found under the plantation is derived from the original rainforest C (Abraham 2004). Still, roughly 10% of C originates from the newly planted trees. This is coherent with results reported earlier for an adjacent teak plantation in which substantial changes in $\delta^{13}C$ signature had been observed five years after planting (Potvin et al. 2004). A possible explanation for this observation is that apparently the rate of incorporation of 'new' carbon in tropical soil is very high. Our results imply that stable isotopes are very sensitive indicators to changes in land-use.

Our carbon inventory showed that the amount of carbon (29.4 Mg C ha^{-1}) stored in the topsoil of the native tree plantation was considerably higher than under pasture (15.9 Mg C ha^{-1}). The difference in surface SOC between the pasture and plantation was 13 Mg C ha^{-1} or equivalent to an unrealistically high carbon gain of 6.5 Mg C ha^{-1} yr^{-1} since the establishment of the plantation, especially considering the size and density of the planted trees at the time of sampling. Based on the isotope approach we would have expected a carbon gain of around 1.5 Mg C ha^{-1} yr^{-1}. This leads to the question whether the SOC pool of the two sampling sites (pasture vs. plantation) were similar prior to land-use change. In this study, the sites were selected according to the chronosequence approach (Yanai et al. 2003). Because the sites are separated in space but studied at the same time, this technique is called a 'space-for-time substitution'. This approach is very common in studying the effects of land-use change as it is almost impossible to perform a real time series as one would have to wait for decades to obtain results. However, a fundamental assumption of this approach is that the initial site conditions (e.g. parent material, soil texture etc) before land-use change are identical for all selected sites. At Sardinilla, pasture and plantation are derived from the same parent material. Differences in carbon between the pasture and plantation are most likely the result of spatial variations due to topographical differences. Abraham (2004) did a carbon inventory of the plantation site in July 2001 before the trees were planted. She observed that carbon concentrations and stocks differed along the slope. She found the highest carbon concentrations (60-70 g kg^{-1}; 0-10 cm) and stocks ($\sim$ 30 Mg ha^{-1}; 0-10 cm) at the ridge where we took our samples. In the swales of the plantation site, Abraham estimated C stocks which were in the range of what we found for the pasture site. Comparing our data with Abraham's data set, carbon concentration and stocks did not increase following the conversion of the site into a tree plantation. On the other hand, eddy covariance measurements at the plantation showed, that the site did not result in a CO_2 source despite the disturbance due to tree planting (Potvin et al. 2004).

Reforestation by plantations on abandoned and degraded agricultural land in the tropics has been proposed as an effective measure to mitigate CO_2 emissions (Brown et al. 1986, Montagnini and Porras 1998). Some results suggest that soil carbon sequestration can occur over time with reforestation of pasture. Silver et al. (2004) reported a net carbon gain of 33 Mg ha^{-1} for a 61-year old plantation in Puerto Rico. Across the tropics a soil carbon accumulation rate of 0.49 Mg ha^{-1} yr^{-1} was observed in secondary forests during the first 100 years after pasture abandonment (Silver et al. 2000). In contrast, some studies found that plantations contribute little to long-term carbon storage. In the Panama Canal Watershed, Kraenzel et al. (2003) suggested that soil C content has changed very little in the 20 years since establishment of teak plantations. Bashkin and Binkley (1998) found no net change in total soil carbon 10-13 years following afforestation with Eucalyptus in Hawaii. A recent meta analysis using data from temperate and tropical locations revealed that losses of soil carbon might occur after reforestation (Guo and Gifford 2002). In a review, Paul et al. (2002) found for temperate locations that coniferous plantations loose soil carbon during the first 5-10 years after conversion form pasture. Changes in soil carbon as a result of tree plantations are small relative to the gains in aboveground biomass (Houghton and Goodale 2004). Factors explaining carbon loss or gain in carbon stocks following pasture to forest conversion are: pasture age (de Koning et al. 2003), time since plantation establishment, primary productivity and availability of limiting nutrients (Lugo et al. 1986) and the pools and turnover time of active and slow carbon pools.

At Sardinilla, soil C stocks in the plantation appear to be influenced by NPP. Higher aboveground biomass stocks were measured at the ridge (Potvin, unpublished data) explaining partly the higher soil C. Productivity in the swales might be reduced due to low soil aeration leading to lower soil C storage.

Change in soil respiration and carbon mineralization rate

Soil CO_2 efflux rates (including microbial decomposition and root respiration) at the pasture site were considerably higher than in the plantation (Table 2). High soil respiration rates from pasture soils might be explained by better quality substrate for microbial respiration (Seto and Yanagiya 1983). Low pasture C:N ratio (9.4), due to the input of N-rich manure and/or N-rich grass residue, implies that the substrate quality under pasture was better as compared to the plantation site. Furthermore, our incubation experiment using root-free samples revealed that pasture soil had a high amount of active carbon, indicating the existence of carbon readily mineralizable by microbes. 'Active' soil organic carbon seemed to be a more sensitive indicator for changes in soil organic carbon following land-use change than the total amount of carbon mineralized. Large soil respiration rates in the pasture might also be explained by high belowground biomass production which leads to high root respiration rates (Trumbore et al. 1995).

Soil respiration in the pasture during the rainy season was similar to measurements of a nearby pasture site in August 1998 (Wilsey et al. 2002). Soil respiration rates in a similar range were also reported by Feigl et al. (1995) and Salimon et al. (2004) for pasture sites in southwestern Amazon. In contrast, CO_2 effluxes from pasture sites in the eastern Amazon were considerably lower (Davidson and Trumbore 1995, Davidson et al. 2000), due to a more pronounced dry season and less fertile soils.

Soil CO_2 efflux within the plantation varied considerably. This is explained by a very heterogeneous soil cover (bare soil, grass cover, tree litter), differences in topography, and species effects. Murphy (2005), who did an intensive soil respiration study at the Sardinilla plantation, measured fluxes of 1.5 µmol CO_2 m^{-2} s^{-1} during the dry season and 7 µmol CO_2 m^{-2} s^{-1} during the wet season. Soil CO_2 efflux from the Sardinilla plantation is higher than values measured at other sites. Li et al. (2005) reported fluxes (annual average) from a Puerto Rican pine plantation and secondary forest of 2.3 μmol CO_2 m^{-2} s^{-1} and 2.6 μmol CO_2 m^{-2} s^{-1}, respectively. In a Malaysian rubber plantation soil respiration was 2.8 μmol CO_2 m^{-2} s^{-1} (Adachi et al. 2005).

5 Cross site comparisons

The relationship between texture and soil organic content in tropical soils has been described by numerous authors (Bird et al. 2003, Desjardins et al. 2004, Powers and Veldkamp 2005). However, we had to reject our hypothesis that differences in soil C stock across sites are explained by differences in clay content. Although clay content varied considerable between BCI (around 20%) and Sardinilla (50-70%) our data set was too small to establish a significant correlation.

Furthermore, we did not find a relationship between soil respiration and soil moisture. This might be partly explained by our short sampling period. An additional complication is that CO_2 is produced over the whole soil rooting depth, but soil temperature and soil moisture were only measured within the top 5 cm. A strong predictor of soil respiration across sites was the absolute (r^2=0.98) and relative amount of active pool C (r^2=0.78). Thus, active soil organic carbon is a sensitive indicator for changes in soil respiration following land use change.

6 Conclusions

Our study showed that surface C stocks varied across sites. The amount of surface SOC stored at BCI and Sardinilla (15-29 Mg C ha^{-1}) is within the range measured for forests, grasslands and plantations throughout the neotropics (Veldkamp 1994, Neill et al. 1996, Rhoades et al. 2000). Although we did not find a significant relationship between carbon and clay content, other studies showed that soil C concentration and stocks can be predicted using the clay content (Feller and Beare 1997, Bird et al. 2003). Topographic variability may have exerted an influence over C stocks at Sardinilla. This was also found for forest and pasture sites form a volcanic landscape in Costa Rica (Powers and Veldkamp 2005). This observation makes clear that underlying site specific factors may play a more important role in determining the magnitude of C than land-use changes. Thus, especially when applying the chronosequence approach, site selection is a crucial factor. Our study revealed that traditional bulk soil C inventory techniques are not sensitive enough to detect short term changes in soil carbon. In contrast, stable isotopes appear to be very sensitive indicators to changes in land-use. Furthermore, our results demonstrate that the active pool C varied with land use and is a good predictor of soil respiration. Besides careful site selection, we recommend the application of carbon stable isotopes and incubation experiments in order to assess land-use related changes in soil C dynamics.

Acknowledgements
This project benefited from field and laboratory assistance from Marco Valdez, Brandy Cline, and Ian Abernethy. We thank Mark Larson for his help with the stable isotope analysis. Partial funding for the research was provided by the Smithsonian Tropical Research Institute, Wyoming NASA Space Grant Consortium, NASA Grant #NGT-40102, Wyoming NASA EPSCoR, NASA Grant #NCC5-578, and an International Travel Grant from the University of Wyoming to E. Pendall.

References

Abraham M (2004) Spatial variation in soil organic carbon and stable carbon isotope signature in a pasture and a primary forest in Central Panamá. MSc Thesis. Department of Geography, McGill University, Montreal, Canada

Adachi M, Bekku YS, Konuma A, Kadir WR, Okuda T, Koizumi H (2005) Required sample size for estimating soil respiration rates in large areas of two tropical forests and of two types of plantation in Malaysia. For Ecol Manage 210: 455-459

ANAM (2004) La reforestación en Panama. Servicio Nacional de Desarrollo y Administracion Forestal de la Autoridad Nacional del Ambiente (ANAM), Panama

Balesdent J, Mariotti A (1996) Measurement of soil organic matter turnover using ^{13}C natural abundance. In: Boutton TW, Yamasaki S (eds) Mass spectrometry of soils. Marcel Dekker, New York, pp 83-112

Bashkin MA, Binkley D (1998) Changes in soil carbon following afforestation in Hawaii. Ecology 79: 828-833

Bird M, Kracht O, Derrien D, Zhou Y (2003) The effect of soil texture and roots on the stable carbon isotope composition of soil organic carbon. Aust J Soil Res 41: 77-94

Brown S, Lugo AE (1982) The storage and production of organic matter in tropical forests and their role in the global carbon cycle. Biotropica 14: 161-187

Brown S, Lugo AE (1990) Effects of forest clearing and succession on the carbon and nitrogen content of soil in Puerto Rica and US Virgin Islands. Plant Soil 35: 53-64

Brown S, Lugo AE, Chapman J (1986) Biomass of tropical tree plantations and its implications for the global carbon budget. Can J Forest Res 16: 390-394

Canadell J, Pataki DE (2002) New advances in carbon cycle research. Trends Ecol Evol 17(4): 156-158

Clark DA (2004) Tropical forests and global warming: slowing it down or speeding it up. Frontiers Ecol Environ 2: 73-80

Cleveland CC, Townsend AR, Schmidt SK, Constance BC (2003) Soil microbial dynamics and biogeochemistry in tropical forests and pastures, southwestern Costa Rica. Ecol Appl 13: 314–326

Condit R, Robinson WD, Ibanez DR, Aguilar S, Sanjur A, Martinez R, Stallard RF, Garcia T, Angehr GR, Petit L, Wright SJ, Robinson TR, Heckadon-Moreno S (2001) The status of the Panama Canal watershed and its biodiversity at the beginning of the 21st century. BioScience 51: 135-144

Davidson EA, Trumbore SE (1995) Gas diffusivity and production of CO_2 in deep soils of the eastern Amazon. Tellus 47B: 550-565

Davidson EA, Verchot LV, Cattanio JH, Ackerman IL, Carvalho JEM (2000) Effects of soil water content on soil respiration in forests and cattle pastures of eastern Amazonia. Biogeochemistry 48: 53–69

de Koning GHJ, Veldkamp E, Lopez-Ulloa M (2003) Quantification of carbon sequestration in soils following pasture to forest conversion in northwestern Ecuador. Global Biogeochem Cycles 17: 1098, doi:10.1029/2003GB002099 de Moraes JFL, Volkoff B, Cerri CC, Bernoux M (1996) Soil properties under Amazon forest and changes due to pasture installation in Rondonia, Brazil. Geoderma 70: 63–81

Desjardins T, Andreux F, Volkoff B, Cerri CC (1994) Organic carbon and ^{13}C contents in soils and soil size-fractions, and their changes due to deforestation and pasture installation in eastern Amazonia, Geoderma 61: 103-118

Desjardins T, Barros E, Sarrazin M, Girardin C, Mariotti A (2004) Effects of forest conversion to pasture on soil carbon and dynamics in Brazilian Amazonia. Agric Ecosyst Environ 103: 365-373

FAO (2001) Global forest resources assessment 2000: main report. FAO Forestry Paper No. 140. Rome (available at http://www.fao.org/forestry/index, 02.02.2006)

Fearnside PM, Barbosa RI (1998) Soil carbon changes from conversion of forest to pasture in Brazilian Amazonia. For Ecol Manage 108: 47-166

Feigl BJ, Melillo J, Cerri CC (1995) Changes in the origin and quality of soil organic matter after pasture introduction in Rondonia (Brazil). Plant Soil 175: 21–29

Feller C, Beare MH (1997) Physical control of soil organic matter dynamics in the tropics, Geoderma 79: 69-116

Guo LB, Gifford RM (2002) Soil carbon stocks and land use change: a meta analysis. Global Change Biol 8: 345-360

Houghton RA (2003a) Revised estimates of the annual net flux of carbon to the atmosphere from changes in land use and land management 1850-2000. Tellus 55B: 378-390

Houghton RA (2003b) Why are estimates of the terrestrial carbon balance so different? Global Change Biol 9: 500-509

Houghton RA, Goodale CL (2004) Effects of land-use change on the carbon balance of terrestrial ecosystems. In: DeFries RS, Asner GP, Houghton RA (eds) Ecosystems and Land Use Change. American Geophysical Union, Washington, DC, pp 85-98

Howard DM, Howard PJA (1993) Relationships between CO_2 evolution, moisture content and temperature for a range of soil types. Soil Biol Biochem 25: 1537-1546

Ibanez DR, Condit R, Angehr G, Aguilar S, Garcia T, Martinez R, Sanjur A, Stallard RF, Wright SJ, Rand AS, Heckadon-Moreno S (2002) An ecosystem report on the Panama Canal: Monitoring the status of the forest communities and the watershed. Environ Monit Assess 80(1): 65-95

Ishizuka S, Iswandi A, Nakajima Y, Yonemura S, Sudo S, Tsuruta H, Murdiyarso D (2005) The variation of greenhouse gas emissions from soils of various land-use/cover types in Jambi province, Indonesia. Nutr Cycl Agroecosys 71:17-32

Jobbagy EG, Jackson RB (2000) The vertical distribution of soil organic carbon and its relation to climate and vegetation. Ecol Appl 10: 423–436

Kraenzel M, Castillo A, Moore T, Potvin C (2003) Carbon storage of harvest-age teak (*Tectona grandis*) plantations, Panama. For Ecol Manage 173: 213-225

Koutika LS, Bartoli F, Andreux F, Cerri CC, Burtin G, Chone Th, Philippy R (1997) Organic matter dynamics and aggregation in soils under rain forest and pastures of increasing age in the eastern Amazon Basin. Geoderma 76: 87–112

Kursar TA (1989) Evaluation of soil respiration and soil CO_2 concentration in a lowland moist forest in Panama. Plant Soil 113: 21-29

Leigh EG Jr, Rand AS, Windsor DM (1996) The ecology of a tropical forest: seasonal rhythms and long-term changes. Second edition. Smithsonian Institution Press, Washington, DC

Li Y, Xu M, Zou XM, Zhang YQ (2005) Comparing soil organic carbon dynamics in plantation and secondary forest in wet tropics in Puerto Rico. Global Change Biol 11: 239-248

Lugo AE, Sanchez MJ, Brown S (1986) Land use and organic carbon content of some subtropical soils. Plant Soil 96: 185–196

Lugo AE, Brown S (1993) Management of tropical soils as sinks or sources for atmospheric carbon. Plant Soil 149: 27-41

Martinez LJ, Zinck JA (2004) Temporal variation of soil compaction and deterioration of soil quality in pasture areas of Colombian Amazonia. Soil Till Res 75: 3-17

Montagnini F, Porras C (1998) Evaluating the role of plantations as carbon sinks: an example of an integrative approach from the humid tropics. Environ Manage 22: 459–470

Murphy M (2005) Biotic and abiotic controls on soil respiration in a biodiversity plantation in the tropics. MSc Thesis. Department of Biology, McGill University, Montreal, Canada.

Murty D, Kirschbaum MUF, McMurtrie RE, McGilvray H (2002) Does conversion of forest to agricultural land change soil carbon and nitrogen? A review of the literature. Global Change Biol 8: 105-123

Neill C, Fry B, Melillo JM, Steudler PA, Moraes JFL, Cerri CC (1996) Forest- and pasture-derived carbon contributions to carbon stocks and microbial respiration of tropical pasture soils. Oecologia 107: 113-119

Orchard VA, Cook FJ (1983) Relationship between soil respiration and soil moisture. Soil Biol Biochem 15: 447-453

Parkinson KJ (1981) An improved method for measuring soil respiration in the field. J Appl Ecol 18: 221-228

Parton WJ, Schimel DS, Cole CV, Ojima D (1987) Analysis of factors controlling soil organic levels of grasslands in the Great Plains. Soil Sci Soc Am J 51: 1173-1179

Paul EA, Morris SJ, Bohm S (2001) The determination of soil C pool sizes and turnover rates: Biophysical fractionation and tracers. In: Lal R, Kimble JM, Follett RF, Stewart BA (eds) Assessment methods for soil carbon. Lewis Publishers, Boca Raton, pp 193-206

Paul KI, Polglase PJ, Nyakuengama JG, Khanna PK (2002) Change in soil carbon following afforestation. For Ecol Manage 168: 241-257

Potvin C, Whidden E, Moore T (2004) A case study of carbon pools under three different land-uses in Panama. Climatic Change 67: 291-307

Powers JS, Veldkamp E (2005) Regional variation in soil carbon and $\delta^{13}C$ in forests and pastures of northeastern Costa Rica. Biogeochemistry 72: 315-336

Pumpanen J, Kolari P, Ilvesniemi H, Minkkinen K,Vesala T, Niinistö S, Lohila A, Larmola T, Morero M, Pihlatie M, Janssens I, Yuste JC, Grünzweig JM, Reth S, Subke JA, Savage K, Kutsch W, Østreng G, Ziegler W, Anthoni P, Lindroth P, Hari P (2004) Comparison of different chamber techniques for measuring soil CO_2 efflux. Agric For Meteorol 123: 159-176

Raich JW, Schlesinger WH (1992) The global carbon dioxide flux in soil respiration and relationship to vegetation and climate. Tellus 44B: 81-99

Raich JW, Tufekcioglu A (2000) Vegetation and soil respiration: Correlations and controls. Biogeochemistry 48: 71-90

Reiners WA, Bouwman AF, Parsons WFJ, Keller M (1994) Tropical rainforest conversion to pasture: changes in vegetation and soil properties. Ecol Appl 4: 363-377

Rhoades CC, Eckert GE, Coleman DC (2000) Soil carbon differences among forest, agriculture, and secondary vegetation in lower montane Ecuador. Ecol Appl 10: 497-505

Rout SK, Gupta SR (1989) Soil respiration in relation to abiotic factors, forest floor litter, root biomass and litter quality in forest ecosystems in Siwaliks in northern India. Acta Oecol/ Oecol Plant 10: 229-244

Salimon CI, Davidson EA, Victoria RL, Melo AWF (2004) CO_2 flux from soil in pastures and forests in southwestern Amazonia. Global Change Biol 10: 833-843

Sayer EJ (2005) Leaf litter manipulation in a tropical forest. PhD thesis. University of Cambridge, UK

Schlesinger WH (1977) Carbon balance in terrestrial detritus. Annu Rev.Ecol Syst 8: 51-81

Schwendenmann L, Veldkamp E, Brenes T, O'Brien JJ, Mackensen J (2003) Spatial and temporal variation in soil CO_2 efflux in an old-growth neotropical rain forest, La Selva, Costa Rica. Biogeochemistry 64: 111-128

Schwendenmann L, Pendall E (2005) Assessing the impact of land-use change on soil organic carbon dynamics in Central Panama using stable carbon isotopes and physical soil fractionation. In: Buzek F, Novák M (eds) Proceedings of the 6th International symposium on applied isotope geochemistry, Czech Geological Survey, Prague, pp 218-219

Seto M, Yanagiya K (1983) Rate of CO_2 evolution from soil in relation to temperature and amount of dissolved organic carbon. Jap J Ecol 33: 199-205

Silver WL, Ostertag R, Lugo AE (2000) The potential for tropical forest restoration and reforestation for carbon accumulation and offset. Restor Ecol 8: 394-407

Silver WL, Kueppers LM, Lugo AE, Ostertag R, Matzek V (2004) Carbon sequestration and plant community dynamics with reforestation of tropical pasture. Ecol Appl 14: 1115-1127

Smith BN, Epstein S (1971) Two categories of $^{13}C/^{12}C$ ratios for higher plants. Plant Physiol. 47: 380-384

Solomon D, Fritsche F, Lehmann J, Tekalign M, Zech W (2002) Soil organic matter dynamics in the subhumid agroecosystems of the Ethiopian highlands: Evidence from natural ^{13}C abundance and particle size-fractionation. Soil Sci Soc Am J 66: 969-978

Townsend AR, Vitousek PM, Des Marais D, Tharpe A (1997) The effects of temperature and soil carbon pool structure on CO_2 and $^{13}CO_2$ fluxes from five Hawaiian soils. Biogeochemistry 38: 1-17

Trumbore SE (1997) Potential responses of soil organic carbon to global environmental change. P Natl A Sci USA 94: 8284-8291

Trumbore SE, Davidson EA, Camargo PB, Nepstad DC, Martinelli LA (1995) Belowground cycling of carbon in forests and pastures of Eastern Amazonia. Global Biogeochem Cycles 9: 515-528

Vasconcelos SS, Zarin DJ, Capanu M, Litell R, Davidson EA, Ishida FY, Santos EB, Araújo MM, Aragão DV, Rangel-Vasconcelos LGT, de Assis Oliveira F, McDowell WH, de Carvalho CJ (2004) Moisture and substrate availability constrain soil trace gas fluxes in an eastern Amazonian regrowth forest. Global Biogeochem Cycles 18: GB2009, doi:10.1029/2003GB 002210

Veldkamp E (1994) Organic carbon turnover in three tropical soils under pasture after deforestation. Soil Sci Soc Am J 58: 175-180

Wilsey B, Parent G, Roulet N, Moore T, Potvin C (2002) Tropical pasture carbon cycling: Relationships between C source/sink strength, aboveground biomass, and grazing. Ecol Lett 5(3): 367-376

Yanai RD, Currie WS, Goodale CL (2003) Soil carbon dynamics following forest harvest: An ecosystem paradigm reviewed. Ecosystems 6: 197-212

Yavitt JB (2000) Nutrient dynamics of soil derived from different parent material on Barro Colorado Island, Panama. Biotropica 32: 198-207

Forest structure as influenced by different types of community forestry in a lower montane rainforest of Central Sulawesi, Indonesia

Johannes Dietz[1,2*], Dirk Hölscher[1], Christoph Leuschner[2], Adam Malik[3], and M. Amran Amir[3]

[1] Tropical Silviculture, Institute of Silviculture, University of Göttingen, Büsgenweg 1, 37077 Göttingen, Germany
[2] Plant Ecology, Albrecht-von-Haller-Institute for Plant Sciences, University of Göttingen, Untere Karspüle 2, 37073 Göttingen, Germany
[3] Institute of Forest Management, Faculty of Agriculture, Bumi Tondo Campus, Tadulako University, Palu 94118, Indonesia

*corresponding author: Johannes Dietz, phone +49 (0)551 399556, fax: +49 (0)551 394019, Email: jdietz@gwdg.de

Summary

Due to high deforestation rates, it is likely that in many regions tropical rainforests will survive in protected areas only. These refuges have to be surrounded by buffer zones where low impact forest management is permitted. In fact, increasing parts of the remaining rainforest areas are encroached by different types of low-intensity forest use. However, until now only few data exist on how different practices affect forest structure. The objectives of this study were to analyze differences in aboveground forest stand structure as caused by different types of forest utilization in the margin zone of a protected rainforest. The study was conducted in a lower montane rainforest region (800 - 1140 m) in the vicinity of the village of Toro in Central Sulawesi, Indonesia. We assessed the structure of twelve forest plots (0.15 ha each) which represented four types of forest use which are widespread in the study region. These types were natural forest without major disturbance, forest extraction of small-diameter timber, forest with selective extraction of large timbers and cacao agroforest plantations under trees remaining from the natural forest. The tree basal area ranged from relatively high values in the natural forests (52.5 m^2 per ha) to 19.4 m^2 per ha in the agroforest, and was paralleled by a decrease in tree height. Stem density peaked in stands where large timbers had been extracted

Tscharntke T, Leuschner C, Zeller M, Guhardja E, Bidin A (eds), The stability of tropical rainforest margins, linking ecological, economic and social constraints of land use and conservation, Springer Berlin 2007, pp 133-148

and abundant tree regeneration had resumed, and it was lowest in the agroforest stands. Canopy openness, as estimated from 30 hemispheric photographs per plot, was lowest in the natural forest after small timber extraction (7 %) and highest in cacao agroforest (15 %). Leaf area index (LAI), as estimated from the same photos, averaged to 6.2 m^2 m^{-2} in the natural forest, 5.3 in the forests with small timber extraction, 5.0 in the forests with large timber extraction, and 5.3 on the agroforest plots. The latter result confirms earlier reports of relatively high leaf areas in the cacao agroforestry system. The relatively high LAI in forest stands after major timber extraction indicates a rapid recovery of leaf area after disturbance. We also analyzed the influence of stand structural parameters on rainfall partitioning into throughfall, stemflow and interception. In all forest stands we observed a close negative correlation between mean throughfall and tree height ($r^2 = 0.63$) indicating a higher interception loss in taller stands. We conclude that local forest management at low to moderate intensities, as it is common in the margin zones of protected forests, has a significant and quantifiable impact on several forest structural parameters which in turn may influence ecosystem functions such as rainfall partitioning in the canopy.

Key words: forest use, leaf area index, rainfall partitioning, rainforest, stand structure, stem density

1 Introduction

Over the past decades, tropical rainforests have been suffering increasing losses by conversion to other land cover types. South East Asia and especially some regions on the Indonesian archipelago including Central Sulawesi were identified as areas with high deforestation rates (Achard et al. 1998, Drigo 2004, FAO 2001). However, more subtle forms of forest disturbance as by selective logging on conservation lands are difficult to identify via satellite and are assumed to be underestimated so far (Asner et al. 2005). Based on recent analyses for five states of the Brazilian Amazon, selective logging doubled previous estimates of the total amount of forest degraded by human activities (Asner et al. 2005).

The high rates of forest conversion together with the rapidly proceeding forest degradation have long since called upon conservation efforts. It has been emphasized that protection requires the interaction with the people that traditionally live in and around the forests (Schweithelm 2004). Contemporary policies advocate buffer zone management by a low impact forest management which, according to Schweithelm (2004), can be developed from the experience of traditional communities that regulate the access to land and forest resources. For the Lore Lindu National Park in Central Sulawesi, such traditional structures and regulations are found in the village of Ngata Toro (Helmi 2005). There, customary law has created a zonation of the forest into different use types which are distinguished by the local population. According to Helmi (2005), the "Wana" type of community forestry is predominantly a natural forest where only occasional rattan extraction takes place. In the "Pangale" small-diameter timbers can be extracted in the understorey, or alternatively, a restricted number of large stems can be selectively logged. Strong human intervention is characteristic for the "Pahawa Pongko" zone, which comprises cacao agroforests under a sparse canopy of large trees that remain from the natural forest (Helmi, 2005).

However, such definitions of land use are subjective and there is a need to quantify such categories by objective means (cf. Drigo 2004). Drigo (2004) suggests the translation from land use practice to land cover types where structural characteristics can be measured and monitored. Following this suggestion, the objective of our study was to identify and quantify the impact of different forest use practices on the stand structure of a lower montane rainforest in the region of the Toro village, Central Sulawesi. In a further study we analyzed the influence of stand structural parameters on rainfall partitioning into throughfall, stemflow and interception in order to relate stand structural characteristics to forest functions with respect to the hydrological cycle.

2 Materials and Methods

2.1 Study Site

This study was carried out in tropical rainforests of the sub-montane to lower montane elevation zone (800 - 1140 m a.s.l.) in the Toro Valley on the western boundary of the Lore Lindu National Park, Central Sulawesi, Indonesia. Twelve sites of 1500 m^2 each were chosen for the study, evenly representing all four investigated forest use types ($n = 3$): natural forest, forest with extraction of small-diameter timbers (hereafter termed 'small timber extraction'), forest with selective extraction of large timbers (hereafter termed 'large timber extraction'), and cacao plantations under trees remaining from the natural forest (agroforest). The four forest use types (natural forest to agroforest) generally occurred in the vicinity of the Village of Toro along a gradient of accessibility with natural forest being most distant and agroforest closest to the settlements. During the investigation period no management was observed on the forested sites while the cacao agroforests received conventional treatment of weeding and pruning. All sites were situated on moderately to severely steep slopes from 17° to 39°. The tree species composition in these forest types was studied by Gradstein et al. (this volume). They found Meliaceae, Lauraceae, and Sapotaceae to be dominant in natural forests, which shifted to Rubiaceae, Fagaceae, and Myristicaceae after both, small and large timber extraction. Moraceae, Myristicaceae, and Melastomataceae dominated in the investigated agroforests. The tree species richness varied little between natural forest stands and after small timber extraction, but was lower in stands after large timber extraction and lowest in the agroforests (Gradstein et al. this volume).

2.2 Stand Inventory

On the forested sites, all trees with a diameter at breast height (dbh) of 10 cm or more were censused for dbh and height on the entire plot according to Kramer and Akça (1995). To account for smaller statured trees all individuals with dbh < 10 cm, but higher than 2 m, were assessed on three randomly located subplots of 5 x 5 m within the plot. In the agroforests, all stems taller than 2 m were censused. Measurement of dbh was done using a metric measuring tape, height measurements were taken with a Vertex III ultrasonic hypsometer (Haglöf, Långsele, Sweden). Stand basal area was corrected for edge and slope effects.

2.3 Hemispherical Photography

Data on canopy structure was derived from hemispherical photographs which were obtained with a Nikon Coolpix 900 digital camera / Nikon FC-E8 fisheye-converter of 2048 x 1536 pixel resolution mounted on a HemiView leveling device (Delta-T, Cambridge, UK). For reduction of horizon effects on the sloped sites, the device was consistently placed at 1.3 m above the ground. Thirty pictures per site and campaign were taken at randomly located points on several occasions during the study period from January 2004 until April 2005. All images were analyzed with the software WinScanopy 2004a (Régent Instruments Inc., Sainte-Foy, Canada) for canopy openness under an opening angle of 30°, and with CanEye V 3.6 (INRA, Avignon, France) for estimates of leaf area index (LAI).

2.4 Statistical Analysis

Data were statistically analyzed using the SAS 8.2 software package (SAS Institute Inc., Cary, NC, US). Multiple comparisons among group means were conducted by analyses of variance (Duncan's test); a significance level of $p < 0.05$ was maintained throughout the analysis.

3 Results and Discussion

3.1 Stem Density

Koop et al. (1995) recognized the density of large trees as a good estimator for human influence on Sumatran rainforests. In our study region, the density of large stems with a diameter at breast height (dbh) larger than 50 cm also

Table 1. Aboveground forest structure in stands influenced by different types of community forestry in the region of Toro, Central Sulawesi. The different forest use types are: natural forest (NF), forest after small timber extraction (STE), forest after large timber extraction (LTE) and agroforest (AF). Significantly different means between forest types ($n = 3$) are indicated by different small letters (analysis of variance, Duncan's test, $p < 0.05$).

Forest use type	Elevation [m asl] plot center	Stem density [n ha^{-1}] height ≥2m	dbh ≥10cm	dbh ≥50cm	Basal area [m^2 ha^{-1}] height ≥2m	dbh ≥10cm	DBH mean [cm] dbh ≥10cm	Top height [m] top 10%	Mid height [m] mean dbh	Leaf area index [m^2 m^{-2}] mean (n=30)	cv (n=30)	Canopy openness [%] mean (n=30)	cv (n=30)
NF1	948	2272	472	103	68.6	62.8	30.3	42.9	25.1	5.7^{ab}	20.7	9.7	53.9
NF4	1042	1806	406	58	50.0	48.1	31.5	47.6	26.7	6.4^{a}	23.9	9.3	52.5
NF5	1130	3455	655	71	51.1	46.6	26.7	35.8	21.5	6.5^{a}	14.9	10.9	58.7
NF mean	1040	2511^{a}	511^{a}	78^{a}	56.6^{a}	52.5^{a}	29.5^{ab}	42.1^{a}	24.4^{a}	6.2^{a}	19.9	10.0^{ab}	55.0
Std Dev		850	129	23	10.4	9.0	2.5	5.9	2.7	0.4	4.6	0.8	3.2
STE2	873	2020	620	61	55.7	43.5	26.8	35.1	20.7	4.3^{de}	19.5	7.7	49.2
STE3	1078	3855	655	91	67.0	62.7	29.5	37.5	22.5	6.3^{a}	18.9	7.0	50.9
STE4	982	2420	620	72	41.4	48.7	23.2	35.3	18.4	5.2^{bc}	20.6	6.4	51.7
STE mean	978	2765^{a}	632^{ab}	75^{a}	54.7^{a}	51.6^{a}	26.5^{ab}	36.0^{ab}	20.5^{ab}	5.3^{a}	19.7	7.0^{a}	50.6
Std Dev		965	20	15	12.8	9.9	3.2	1.3	2.1	1.0	0.9	0.7	1.3
LTE1	974	5495	695	25	41.1	34.9	20.4	33.2	16.8	4.9^{bcd}	30.2	12.9	56.6
LTE3	827	3740	740	64	53.6	50.1	25.9	37.0	21.3	4.9^{cd}	24.4	8.2	48.5
LTE4	959	4052	652	55	34.6	30.2	22.1	27.1	18.7	5.2^{bc}	31.4	9.3	66.7
LTE mean	920	4429^{b}	695^{b}	48^{ab}	43.1^{ab}	38.4^{ab}	22.8^{b}	32.4^{b}	18.9^{b}	5.0^{a}	28.7	10.1^{ab}	57.2
Std Dev		936	44	21	9.6	10.4	2.8	5.0	2.3	0.2	3.7	2.4	9.1
AF2	952	1706	125	6	8.6	8.5	25.6	28.4	15.9	5.7^{ab}	31.6	19.7	85.7
AF3	832	2705	237	25	23.7	23.4	32.7	35.7	20.6	6.3^{a}	26.3	15.9	64.7
AF4	806	2612	247	51	26.5	26.3	38.1	35.0	19.9	4.1^{e}	26.1	9.8	86.7
AF mean	863	2341^{a}	203^{c}	28^{b}	19.6^{b}	19.4^{b}	32.2^{a}	33.0^{b}	18.8^{b}	5.3^{a}	28.0	15.1^{b}	79.0
Std Dev		552	68	23	9.6	9.6	6.3	4.0	2.5	1.1	3.1	5.0	12.4

appeared as good indicator of the intensity of forest use (Figure 1, solid bars). Highest values averaging at 78 and 75 stems per hectare were found in the natural forest and in forest after small timber extraction. Large timber extraction reduces the density to 48 stems per hectare. Agroforests contained an average of only 28 large trees per hectare (ranging between 6 and 51 stems per hectare, Table 1). The very large variability in the density of large-diameter stems in the agroforestry systems may be related to the personal experience and preference of the plot owners. A similar structural diversity in agroforestry systems was previously documented for other tropical regions by Ruf and Schroth (2004). Small stems (< 10 cm in dbh) averaged slightly above 2000 stems per hectare in the natural forest, but reached higher values after large timber extraction (Figure 1, hatched bars). In the gaps created by logging of big trees, abundant regrowth occurred with 3700 stems per hectare on the average. In the agroforest, the class of small stems included most of the planted cacao (*Theobroma cacao*) trees.

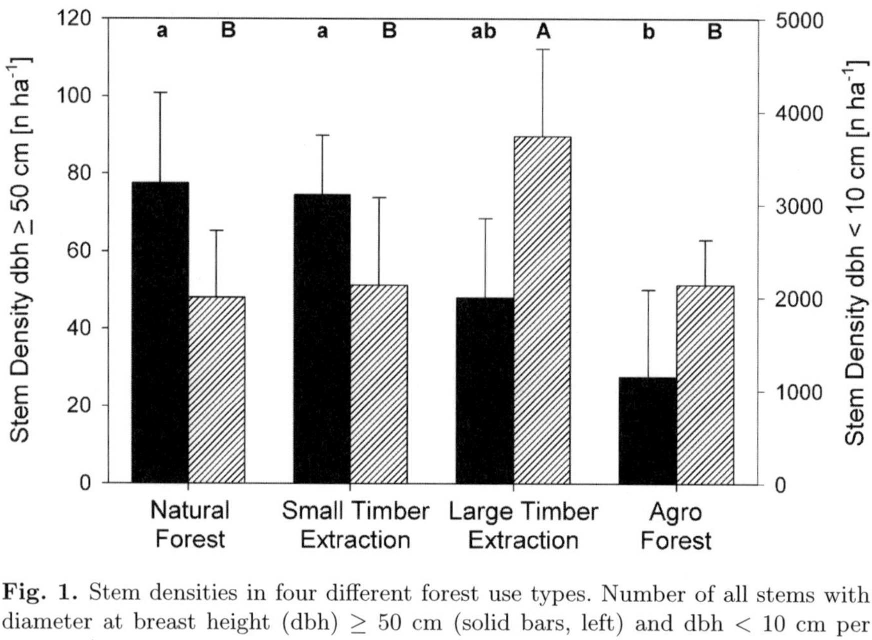

Fig. 1. Stem densities in four different forest use types. Number of all stems with diameter at breast height (dbh) ≥ 50 cm (solid bars, left) and dbh < 10 cm per hectare (hatched bars, right)

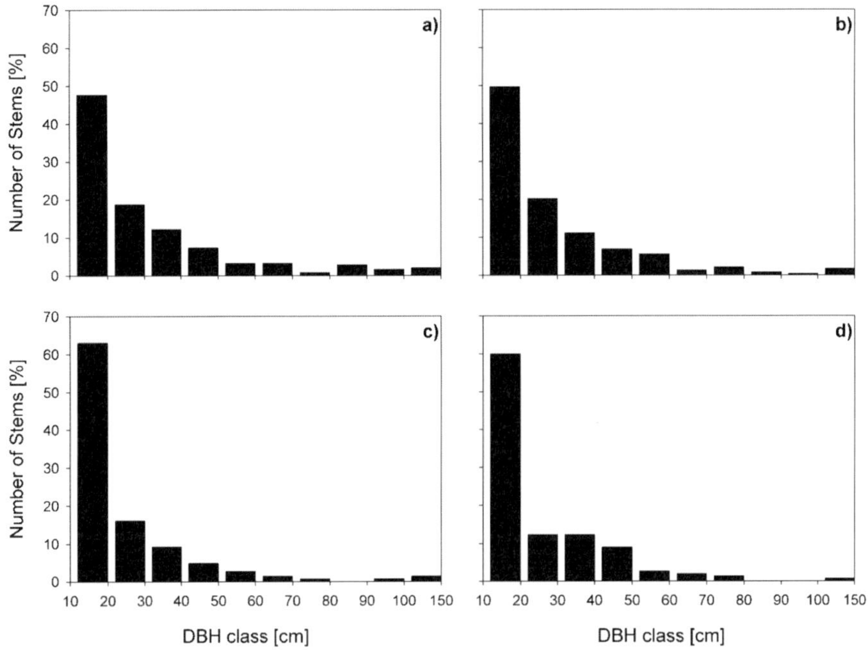

Fig. 2. Diameter distribution of all stems with dbh ≥ 10 cm in natural forest (**a**), after small timber extraction (**b**), after large timber extraction (**c**), and in agroforest (**d**)

Few remaining large trees and a high proportion of small stems were characteristic for both forest stands after large timber extraction and agroforest stands. This contrasted with natural forest stands and stands after small timber extraction. In the latter, the dbh class < 20 cm contained less than 50 % of all stems, (Figures 2a and b), whereas after large timber extraction the proportion of trees in this diameter class reached values above 60 %. In the agroforest, the high proportion of thin stems mostly owed to planted crop trees (Figures 2c and d). Big trees were scarce in these two forest use types.

3.2 Basal Area and Stand Height

The abundance of large-diameter stems is known to be a good estimator of a stand's basal area of stems ≥ 10 cm dbh (McElhinny et al. 2005). Average basal areas of 52.5 and 51.6 m^2 per hectare in the natural forest and in stands after small timber extraction, respectively, seem to be relatively high, especially when the peak value of 62.8 m^2 ha^{-1} is considered. However, a comparison with other studies in natural forests above 600 m asl in the South East Asian region reveals that the values for the natural forests in Toro are indeed

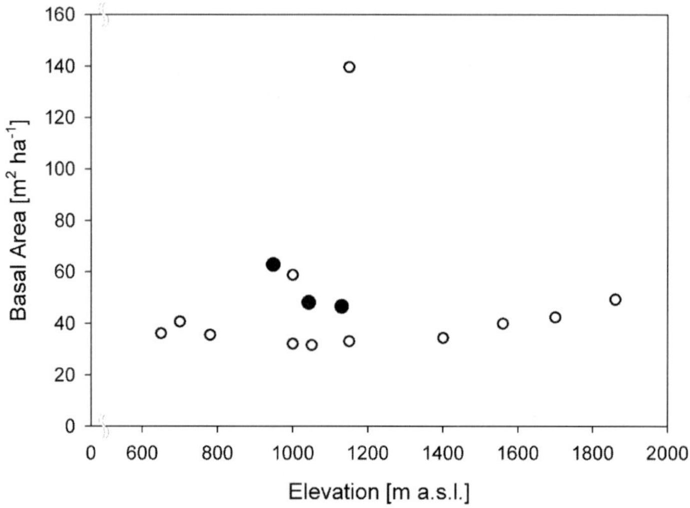

Fig. 3. Basal area of natural forests in the region of Toro, Central Sulawesi (this study, ●), compared to results from other studies in natural forests of South East Asia (○) above 600 m asl. Sources: Aiba and Kitayama 2002, Brodbeck 2004, Dietz et al. unpublished data, Hamann et al. 1999, Kessler et al. 2005, Kitamura et al. 2005, Sri-Gernyuang 2003

slightly above the average, yet they are still within the reported range (Figure 3). In our study region, the average basal area decreased to 38.4 m^2 ha^{-1} in stands after large timber extraction and dropped to 19.4 m^2 ha^{-1} in the agroforest. The decrease in tree basal area along the gradient of use intensity was paralleled by a decrease in tree height. The stand mid height, which is the height of a tree with the arithmetic mean of dbh (Kramer and Akça 1995), decreased from 24.4 m in the natural forest and from 20.5 m in stands after small timber extraction to 18.9 m in stands after large timber extraction and 18.8 m in the agroforest. The apparent drop in tree mean height between natural forest and forest after small timber extraction is statistically not significant. If existent, it may be caused by slight changes in environmental conditions between sites (cf. Takyu et al. 2002).

3.3 Leaf Area Index and Canopy Openness

Leaf area index (LAI) is a basic variable for modeling the energy and matter fluxes in forests; this parameter is also of great importance for vegetation models in the context of global change research (e.g. Asner et al. 2003). A variety of different assessment methods are used in the field (Breda 2003, Jonckheere et al. 2004). A consistent underestimation of LAI by indirect methods such as hemispheric photography was reported by Breda (2003) but their growing potential was acknowledged. On the global scale, Asner et al. (2003) confirmed that indirect optical measurements of LAI were closer to the results from destructive assessments than interpolations derived from litter sampling or estimates obtained from allometric functions. The destructive sampling approach would have been very labor-intensive and would have conflicted with conservation policies in our forest plots, which are located in a national park. Since hemispheric photography is a rapid, sufficiently reliable and non-destructive approach for the assessment of canopy structure (Jonckheere et al. 2004), it was thus chosen in our study.

The LAI of evergreen forests may change seasonally with rainfall volumes and other climatic parameters (Asner et al. 2003, Breda 2003). Therefore, we analyzed most stands for their leaf area at three different times. However, in none of the stands, a distinct seasonality in LAI was observed (see Figure 4a for the LAI of a natural forest). This may be due to the largely non-seasonal rainfall distribution in the study region. The following results are based on photos taken during a measurement campaign in April 2005.

The LAI as derived from 30 photos within a plot varied only moderately in space within a plot. In the natural forest, the coefficient of variation (CV) ranged between 15 and 24 % which is similar to results found in natural forests of the Brazilian Amazon using a different, yet quite comparable method (Licor LAI 2000; Aragão et al. 2005). In our study region, the CV values in forest stands after large timber extraction and the agroforest stands ranged from 24 to 32 %. This corresponds with CVs measured in Amazonian secondary forests (Aragão et al. 2005). The average LAI of the natural forest stands in Sulawesi ranged from 5.7 to 6.5 m^2 m^{-2}. In their global review Asner et al. (2003) reported a mean ($\pm$ SD) LAI of 4.9 (2.0) for tropical evergreen broadleaf forests. Aragão et al. (2005) found values between 3.6 and 6.6 for natural forest in Amazonia, Trichon et al. (1998) reported values between 3.9 and 6.1 from rainforests in Sumatra, and similarly Kumagai et al. (2004) found values between 4.8 and 6.8 in Borneo. Roberts et al. (2004) obtained LAI values ranging from 4.1 to 8.0 for tropical lowland rainforests, with a tendency for higher values in Asia. Thus, the values obtained in Sulawesi are well within the range of other studies and they may also support slightly higher values in Asian tropical forests (Roberts et al. 2004). The average LAI of a given forest use type was 6.2 m^2 m^{-2} in the natural forest, 5.3 in forests with small timber extraction, 5.0 in forests with large timber extraction, and 5.3 in the agroforest. The variation between plots of the same use type was largest in the agroforestry systems, where values between 4.1 and 6.3 were observed (Figure 4b). The large variability between the agroforest sites may be explained with the already mentioned densities of large and small stems that differ greatly under permanent human influence (Ruf and Schroth 2004). Our relatively high average LAI values for agroforest confirm studies claiming that plantations in general (Asner et al. 2003), and cacao in Central Sulawesi in particular (Falk 2004), can maintain a comparably large leaf area. The relatively high LAI val-

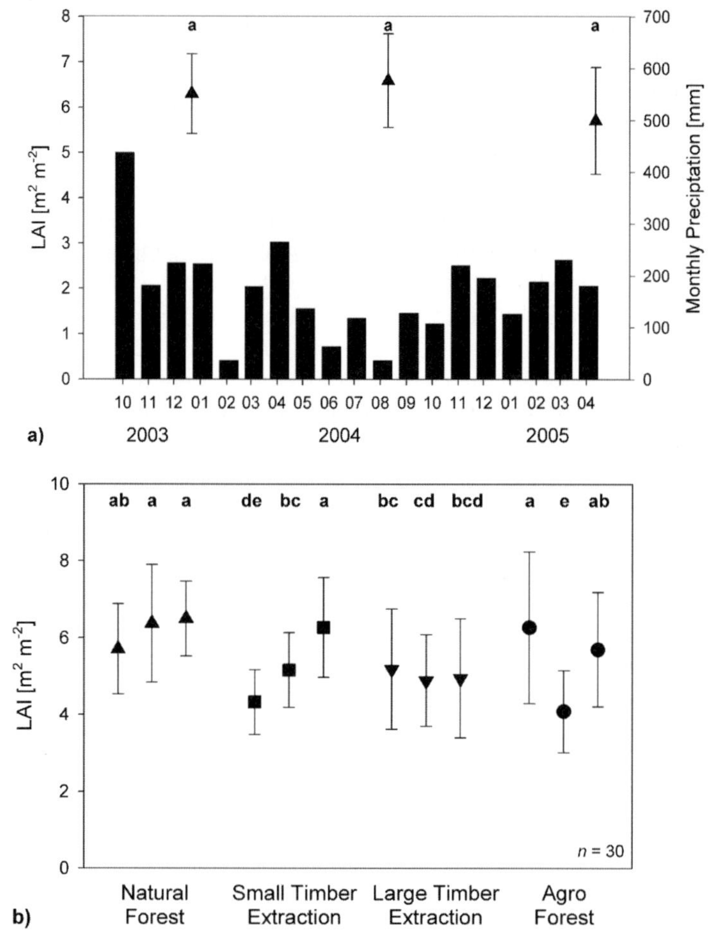

Fig. 4. Leaf area index (LAI) estimates from a natural forest on three different dates compared to the monthly rainfall volumes (**a**), and LAI values with standard deviation for all twelve plots grouped by forest use types (**b**); (n = 30 photos per stand, analysis of variance, Duncan's test $p < 0.05$)

ues in managed stands of Sulawesi are comparable to results from Amazonia, where only little differences in LAI between natural forests and selectively logged sites were observed (Aragão et al. 2005). The high proportion of regrowth after large timber extraction may contribute substantially to such high LAI values in Sulawesi.

The average canopy openness, as determined from photographs taken at 1.3 m above the ground, was between 10 and 11 % per site in the natural forest; it was considerably higher in cacao agroforest (between 16 and 20 %;

Table 1). The values obtained in the forest stands after small and large timber extraction did not differ significantly from those in the natural forest. Trichton et al. (1998) found canopy openness values below 10 % in mature Sumatran rainforests but they report much larger values for gap phases. Tobón Marin et al. (2000) found canopy openness values from 9 to 17 % in natural rainforests of western Amazonia. In the agroforest systems of Sulawesi, many large trees have been removed arbitrarily and to different extents by the plantation owners who subsequently plant crop trees in different densities. Thus, in the agroforest, we found the highest average canopy openness which was associated with a relatively high spatial variation in this parameter in a given stand and between different stands (Table 1).

4 Outlook – the relationship between stand structure and ecosystem functions

The aboveground structure of tropical forests has been shown to influence many ecosystem properties that control energy and matter fluxes, including radiation transmission, the temperature regime, the atmospheric saturation deficit (Leigh 1975, Montgomery 2004, Parton et al. 1996, Trichon et al. 1998), soil temperature, fine root biomass and turnover (Hertel et al. this volume), as well as the diversity of the soil fauna (Migge-Kleian et al. this volume) and of butterflies (Fermon et al. 2005). Forest structure is also considered to be an important parameter in the hydrological cycle of forests (e.g. Bigelow 2001, Bruijnzeel et al. 1993, Chappell et al. 2001, Hall 2003, Hölscher et al. 2004).

In our study plots in Central Sulawesi, we monitored throughfall with 30 funnel collectors each on all 12 plots, stemflow with 10 collectors per plot, and gross rainfall with rain gauges adjacent to each plot. The throughfall median was 70 % of incident precipitation in natural forest, 79 % in forest with small timber extraction, and 80 % in forest with large timber extraction and 81 %in the agroforest. Stemflow was estimated to be less than 1 % in all studied forest use types. Among the stand structural parameters that correlate with rainfall partitioning we found the best correlation between throughfall and mean tree height (the arithmetic mean height of all stems $>$ 10 cm dbh). Considering all 12 plots from all four forest use types, the coefficient of determination (r^2) was 0.63 (Figure 5). We conclude that tree height is an easy-to-measure parameter that allows for an estimation of the throughfall volume along the disturbance gradient from natural forest to agroforest. In other studies, LAI has often been used for modeling the partitioning of rainfall into throughfall, stemflow and interception (Gash 1979, Gash et al. 1995), and has been described as a good parameter to estimate throughfall volumes (Hall 2003, Holwerda et al. in press). In contrast, LAI alone was not significantly correlated to throughfall (or rainfall interception) in our study. A possible explanation is that the total LAI and the associated canopy water storage differed only slightly between the studied stands while the vertical distribution of the foliage varied

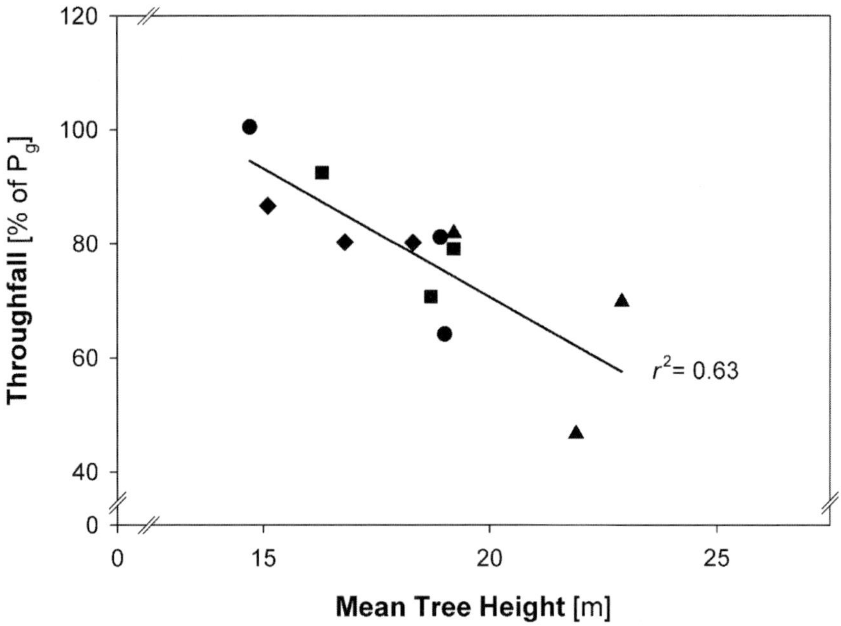

Fig. 5. Relationship between mean canopy height (arithmetic mean of all trees $\geq$ 10 cm dbh) and throughfall expressed in percent of gross rainfall (P_g) for 15 days of rainfall per plot ($r^2 = 0.63$, $p < 0.01$). Considered in the regression are three natural forest plots (▲), three forests after small timber extraction (■), three forests after large timber extraction (◆), and three agroforest sites (●).

greatly. Taller trees presumably increase the aerodynamic roughness length of the stand, resulting in a more effective turbulent energy exchange with the atmosphere. As a consequence, the canopy dries up more often between rainfall events in tall stands, while it remains wet for longer periods in stands with lower-statured trees and with a more clumped distribution of the leaf area.

5 Conclusions

We conclude that the different types of community forestry in Central Sulawesi changed stand structural characteristics considerably. We found strong effects on stem density, basal area and stand height, whereas the leaf area index seemed to be less affected. The changes in forest structure were associated with differences in the forest hydrological cycle as the volume of net precipitation increased with decreasing stand height.

Acknowledgements
This study was conducted in the framework of the joint Indonesian-German research project "Stability of Tropical Rainforest Margins, Indonesia" (STORMA) funded by the German Science Foundation (SFB 552). We thank the village of Toro for their friendly help and cooperation and André Twele for substantial support in processing the hemispherical photographs.

References

Aiba SI, Kitayama K (2002) Effects of the 1997-98 El Niño drought on rain forests of Mount Kinabalu, Borneo. J Trop Ecol 18: 215-230

Aragão LEOC, Shimabukuro YE, Espírito Santo FDB, Williams M (2005) Landscape pattern and spatial variability of leaf area index in Eastern Amazonia. For Ecol Manage 211: 240-256

Achard F, Eva H, Glinni A, Mayaoux P, Richards T, Stibig HJ (1998) Identification of deforestation hot spot areas in the humid tropics. TREES Publication Series B4, European Commission, Luxembourg

Asner GP, Knapp DE, Broadbent EN, Oliveira PJC, Keller M, Silva JN (2005) Selective logging in the Brazilian Amazon. Science 310: 480-482

Asner GP, Scurlock JMO, Hicke JA (2003) Global synthesis of leaf area index observations: implications for ecological and remote sensing studies. Global Ecol Biogeogr 12: 191-205

Aylward B (2004) Land use, hydrological function and economic valuation. In: Bonnell M, Bruijnzeel LA (eds) Forests, water and people in the humid tropics. Cambridge University Press, Cambridge, pp 99-120

Bigelow S (2001) Evapotranspiration modeled from stands of three broad-leaved tropical trees in Costa Rica. Hydrol Proc 15: 2779-2796

Breda NJJ (2003) Ground-based measurements of leaf area index: a review of methods, instruments and current controversies. J Exp Bot 54: 2403-2417

Brodbeck F (2004) Structure and Processes in Traditional Forest Gardens of Central Sulawesi, Indonesia. Cuvillier Verlag, Göttingen

Bruijnzeel LA, Waterloo MJ, Proctor J, Kuiters AT, Kotterink B (1993) Hydrological observations in montane rainforests on Gunung Silam, Sabah; Malaysia, with special reference to the 'Massenerhebung' effect. J Hydrol 81: 145-167

Chappell NA, Bidin K, Tych W (2001) Modelling rainfall and canopy controls on net-precipitation beneath selectively-logged tropical forest. Plant Ecol 153: 215-229

Drigo R (2004) Trends and patterns of tropical land use change. In: Bonnell M, Bruijnzeel LA (eds) Forests, water and people in the humid tropics. Cambridge University Press, Cambridge, pp 9-39

Falk U (2004) Turbulent fluxes of CO_2, H_2O and energy in the atmospheric boundary layer above tropical vegetation investigated by Eddy covariance measurements. PhD thesis, University of Göttingen

FAO (2001) Global forest resources assessment 2000. Main Report. FAO Forestry Paper 140

Fermon H, Waltert M, Vane-Wright RI, Mühlenberg M (2005) Forest use and vertical stratification in fruit-feeding butterflies of Sulawesi, Indonesia: impacts for conservation. Biodiv Conserv 14: 333-350

Gash JHC (1979) An analytical model of rainfall interception of forests. Quart J Roy Met Soc 105: 43-55

Gash JHC, Lloyd CR, Lachaud G (1995) Estimating sparse forest rainfall interception with an analytical model. J Hydrol 170: 79-86

Gradstein SR, Kessler M, Pitopang R (this volume) Tree species diversity relative to human land uses in tropical rain forest margins in Central Sulawesi. In: Tscharntke T, Leuschner C, Guhardja E, Zeller M, Bidin A (eds) The stability of tropical rainforest margins, linking ecological, economic and social constraints of land use and conservation. Springer, Berlin Heidelberg New York

Hall RL (2003) Interception loss as a function of rainfall and forest types: stochastic modelling for tropical canopies revisited. J Hydrol 280: 1-12

Hamman A, Barbon EB, Curio E, Madulid DA (1999) A botanical inventory of a submontane tropical rainforest on Negros Island, Philippines. Biodiv Conserv 8: 1017-1031

Hart MJ (1928) Stamtal en dunning. Dissertation, University of Wageningen

Helmi I (2005) Local people are well-placed to develop zonation plans in Indonesia's Lore Lindu National Park. Tropical Forest Update 15: 31-32

Hertel D, Leuschner C, Harteveld M, Wiens M (this volume) Fine root mass, distribution and activity in disturbed primary forests and secondary forests of the moist tropics. In: Tscharntke T, Leuschner C, Guhardja E, Zeller M, Bidin A (eds) The stability of tropical rainforest margins, linking ecological, economic and social constraints of land use and conservation. Springer, Berlin Heidelberg New York

Hölscher D, Mackensen J, Roberts JM (2004) Forest recovery in the humid tropics: changes in vegetation structure, nutrient pools and the hydrological cycle. In: Bonnell M, Bruijnzeel LA (eds) Forests, water and people in the humid tropics. Cambridge University Press, Cambridge, pp 598-621

Holwerda F, Scatena FN, Bruijnzeel LA (2006) Throughfall in a Puerto Rican lower montane rain forest: a comparison of sampling strategies. J Hydrol, in press

Jonckheere I, Fleck S, Nackaerts K, Muys B, Coppin P, Weiss M, Baret F (2004) Review of methods for in situ leaf area index determination Part I. Theories, sensors and hemispherical photography. Agri Forest Meteor 121: 19-35

Kessler M, Keßler PJA, Gradstein SR, Bach K, Schmull M, Pitopang R (2005) Tree diversity in primary forests and different land use systems in Central Sulawesi, Indonesia. Biodiv Conserv 14: 547-560

Kitamura S, Suzuki S, Yumoto T, Chuailua P, Plongmai K, Poonswad P, Noma N, Maruhashi T, Suckasam C (2005) A botanical inventory of a tropical seasonal forest in Khao Yai National Park, Thailand: implications for fruit–frugivore interactions. Biodiv Conserv 14: 1241-1262

Koop H, Rijksen HD, Wind J (1995) Tools to diagnose forest integrity: an appraisal method substantiated by Silvi-Star assessment of diversity and forest structure. In: Boyle TJB, Boontawee B (eds) Measuring and Monitoring Biodiversity in Tropical and Temperate Forests. CIFOR, Bogor, Indonesia, pp 309-333

Kramer H, Akça A (2002) Leitfaden zur Waldmeßlehre. Sauerländer Verlag, Frankfurt

Kumagai T, Katul GG, Saitoh TM, Sato Y, Manfroi OJ, Morooka T, Ichie T, Kuraji K, Suzuki M, Porporato A (2004) Water cycling in a Bornean tropical rain forest under current and projected precipitation scenarios. Water Resour Res 40: W01104

Leigh EG (1975) Structure and Climate in Tropical Rain Forest. Ann Rev Ecol Syst 6: 67-86

McElhinny C, Gibbons P, Brack C, Bauhus J (2005) Forest and woodland stand structural complexity: Its definition and measurement. For Ecol Manage 218: 1-24

Migge-Kleian S, Woltmann L, Schaefer M (this volume) Impact of forest disturbance and land use change on soil invertebrate fauna in tropical rainforest margins. In: Tscharntke T, Leuschner C, Guhardja E, Zeller M, Bidin A (eds) The stability of tropical rainforest margins, linking ecological, economic and social constraints of land use and conservation. Springer, Berlin Heidelberg New York

Montgomery RA (2004) Effects of understorey foliage on patterns of light attenuation near the forest floor. Biotropica 36: 33-39

Parton WJ, Haxeltine A, Thornton P, Anne R, Hartman M (1996) Ecosystem sensitivity to land-surface models and leaf area index. Global and Planet Change 13: 89-98

Roberts JM, Gash JHC, Tani M, Bruijnzeel LA (2004) Controls on evaporation in lowland tropical rainforest. In Bonnell M, Bruijnzeel LA (eds) Forests, water and people in the humid tropics. Cambridge University Press, Cambridge, pp 287-313

Ruf F, Schroth G (2004) Chocolate forests and monocultures: A historical review of cocoa growing and its conflicting role in tropical deforestation and forest conservation. In: Schroth G, da Fonseca GAB, Harvey CA, Gascon C, Vasconcelos HL, Izac AMN (eds) Agroforestry and biodiversity conservation in tropical landscapes. Island Press, Washington, pp. 107-134

Schweithelm J (2004) Forest people and changing tropical forestland use in tropical Asia. In Bonnell M, Bruijnzeel LA (eds) Forests, water and people in the humid tropics. Cambridge University Press, Cambridge, pp 66-74

Sri-Gernyuang K, Kanzaki M, Mizuno T, Noguchi H, Teejuntuk S, Sungpalee C, Hara M, Yamakura T, Sahunalu P, Dhanmanonda P, Bunyavejchewin S (2003) Habitat differentiation of Lauraceae species in a tropical lower montane forest in northern Thailand. Ecol Res 18: 1-14

Takyu M, Aiba SI, Kitayama K (2002) Effects of topography on tropical lower montane forests under different geological conditions on Mount Kinabalu, Borneo. Plant Ecol 159: 35-49

Tobón Marin C, Bouten W, Sevink J (2000) Gross rainfall and its partitioning into throughfall, stemflow and evaporation of intercepted water in four forest ecosystems in western Amazonia. J Hydrol 237: 40-57

Trichton V, Walter JMN, Laumonier Y (1998) Identifying spatial patterns in the tropical rain forest structure using hemispherical photographs. Plant Ecol 137: 227-244

Willott SJ, Lim DC, Compton SG, Sutton SL (2000) Effects of selective logging on the butterflies of a Bornean rainforest. Conserv Biol 14: 1055-1065

Impact of forest disturbance and land use change on soil and litter arthropod assemblages in tropical rainforest margins

Sonja Migge-Kleian[1*], Lars Woltmann[1], Iswandi Anas[2], Wenke Schulz[1], Andrea Steingrebe[1], and Matthias Schaefer[1]

[1] Ecology Group, Institute of Zoology, University of Göttingen, Berliner Strasse 28, 37073 Göttingen, Germany
[2] Laboratory of Soil Biology, Soil Department, Faculty of Agriculture, Institut Pertanian Bogor (IPB), Darmaga Campus, Bogor 16680, Indonesia

*corresponding author: Sonja Migge-Kleian, phone: +49 (0)551 395557, fax: +49 (0)551 395448, Email: smigge@gwdg.de

Summary

Land use intensification at rainforest margins increases the pressure on the primary rainforest ecosystem, known for its high biodiversity and important functional role in global climate and carbon balances. Studying this system at different levels of disturbance increases our knowledge about its stability and mechanisms of diversity changes. While most studies investigating the effects of disturbance on animal populations focus on canopy communities, the present study reviews the driving factors for soil and litter arthropod communities. In addition, results from a study along a land use gradient in Central Sulawesi (Indonesia) will be discussed.

Within natural forests, litter composition, patchiness and especially litter abundance seem to be key factors driving litter arthropod diversity and densities, by creating microhabitats for many species and individuals. With increasing levels of disturbance in tropical forests, and with forest conversion into agroforestry and agricultural systems, soil/litter temperature and moisture levels become extremely important, leading to reduced densities and species richness. Changes of soil pH, a reduction of the litter habitat and increased anthropogenic disturbance via site management practices add to degradation. In the present study, litter arthropod communities were compared between three levels of forest use (none, small and large timber extraction) and two types of cacao plantations (under natural shade trees and under a polyculture of planted shade trees). The negative effects of increasing gap fraction

Tscharntke T, Leuschner C, Zeller M, Guhardja E, Bidin A (eds), The stability of tropical rainforest margins, linking ecological, economic and social constraints of land use and conservation, Springer Berlin 2007, pp 149-165

and thus increasing temperatures accompanied by decreasing moisture levels and litter depth could be confirmed. Abundances of Collembola, Oribatida and Symphyla in litter and soil and the number of ant species in the litter declined with increasing land use intensity. In contrast, spider density and ant activity increased when natural forest sites were compared with the two types of cacao plantations. This can be explained by the shifting community structure of these two taxa, indicating that few, but abundant species of open habitats, such as Lycosidae, were replacing forest species. Surprisingly, the decline in microarthropod abundances along the land use gradient was not linearly correlated with decreasing moisture and litter depth or increasing temperature or litter quality (C-to-N ratio). Rather low microarthropod densities were already measured in little disturbed natural forest sites with small timber extraction. A similar pattern was observed for fungal abundances and root production, indicating that even small levels of disturbance influence the litter and soil system in tropical forests. On the one hand, changing microclimate, limited litter space, reduced availability of microbial resources and increased disturbance by plot maintenance inhibit higher densities of litter and soil animal populations. On the other hand, the simplified habitat structure favours populations of open habitats, especially predatory groups adapted to higher temperatures and/or lower soil moisture. It has to be investigated how a reduction of decomposers and an increase in predators could influence pest control and thus yields in these agroforestry systems.

Keywords: ants, cacao plantations, disturbance, forest conversion, litter arthropods, litter habitat, microarthropods, moisture, rainforest, shade trees, soil fauna, spiders, temperature

1 Introduction

Tropical forests are well known for their high invertebrate diversity (Whitmore 1998, Chazdon and Whitmore 2002, Primack and Corlett 2005), being linked directly to the high tree diversity (Wagner 2000, Hill and Hill 2001). Intensification of land use practices and forest conversion into secondary forests, agroforestry systems or agricultural fields often result in strong declines of the natural vegetation and its associated fauna (Franklin et al. 2001) while intermediate levels of disturbance can lead to increased invertebrate diversity (Sheil and Burslem 2003, Armbrecht et al. 2005). Comparisons between natural forest habitats and disturbed systems help us to gain insight into mechanisms leading to diversity changes by studying the changes in habitat characteristics. Geographical distance between those sites often poses problems when interpreting the data, thus the transition zones where natural forest habitats connect directly to agroforestry systems are highly valuable for comparisons in geographical proximity. Most studies focus on herbivores and canopy dwelling arthropods (Basset 2001, Basset et al. 2003) thereby ignoring the litter and soil inhabiting fauna. Canopy and litter arthropod communities often differ quite strongly in abundance and diversity patterns (Paoletti ct al. 1991, Wunderle 1992, Brühl et al. 1998). Although soil fauna communities play an important role in litter decomposition and nutrient cycling of tropical forests and plantations (e.g. Heneghan et al. 1999, Vasconcelos and Laurance 2005), the current knowledge of the soil fauna in the tropics is still "miserable" (Noti et al. 2003, Martius et al. 2004, Giller et al. 2005).

This chapter briefly reviews the current knowledge of the factors driving soil and litter arthropod communities in tropical forests and then focuses on recent results from a study along a land use gradient at the forest margin.

2 Arthropod communities in soil and litter of tropical rain forests

Arthropod assemblages in the litter and soil of undisturbed tropical rainforests are rather homogenous when compared to the canopy communities. However, when litter patches are compared, a strong heterogeneity can be observed, for instance in ant communities (Kaspari 1996). Data from natural forests, but also from plantations, indicate that litter types varying in structure and nutrient content influence diversity and abundance of the litter fauna (Tian et al. 1993, Richardson et al. 2005). Thus, high tree diversity in undisturbed forests can create patchy litter habitats on a relatively small scale (Kaspari 1996), including woody litter which has been shown to harbour different arthropod communities depending on its decomposition stage (Kelly and Samways 2003). Other studies found that while local heterogeneity of litter had no influence on ant species richness, the abundance of litter was more important (Carvalho and Vasconcelos 1999, McGlynn et al. 2002, Campos et al. 2003), indicating that an increased litter layer also offers larger habitat space and food resources.

Abiotic parameters such as litter/soil temperature and moisture become important factors driving arthropod communities in disturbed tropical forests. In addition to isolation effects, forest fragments often harbour less diverse or less abundant communities of litter arthropods due to changing microclimate conditions at the edges (Carvalho and Vasconcelos 1999). These changes also occur when natural forest is compared to secondary forest (Hsieh et al. 2003), woodlands (Noti et al. 2003) or over logged forests (Brühl 2001) and is even more pronounced when natural forest is converted into agroforestry systems (Franklin et al. 2001) or tree plantations (Tsukamoto and Sabang 2005). Only few studies showed no or even positive effects of forest conversion into plantations on arthropod species richness or composition (Belshaw and Bolton 1993, Armbrecht et al. 2005). However, these studies were investigations on ants, indicating that despite observed dramatic declines in some areas (Brühl 2001) this group also comprises many species being able to survive at or even profit from intermediate levels of disturbance and increased temperatures, as long as the systems are diverse and not extremely simplified (Andersen 2000, Watt et al. 2002, Armbrecht et al. 2005).

The effect of temperature often cannot be separated from changing moisture regimes, since an increased gap fraction and elevated temperatures also allow for higher evaporation (Dietz et al. 2006) and drying of the upper litter layer. In the tropics, litter inhabiting arthropods are known to react negatively to decreasing moisture levels (Gonzalez and Seastedt 2000, Brühl 2001, Reddy 1995) especially due to land use change (Hodda et al. 1997, Martius et al. 2001) while seasonal variation within natural forest only plays a minor role (Leigh 1999).

Studies in temperate forests have investigated the influence of soil pH on the arthropod community in litter and soil (Wolters and Schaefer 1994, Carcamo et al. 1998), showing that sharp acidification gradients can lead to strongly declining abundances of ground spiders and carabid beetles (Carcamo et al. 1998). In contrast, microarthropods tend to be sensitive to changing pH, but generally can withstand low values (Hagvar and Amundsen 1981). Studies in tropical forests are rare and often other factors, such as moisture levels or litter quality, are correlated with pH differences influencing litter arthropod abundances (Plowman 1981). Generally, base rich soils are favourable habitats for earthworm (Fragoso and Lavelle 1992, Edwards and Bohlen 1996, Tian 1998). It can be hypothesised that elevated levels of pH would negatively influence litter and soil arthropods by enhancing earthworm activity and thus increasing the levels of disturbance and reducing the litter habitat. Studies from temperate zones show a strongly negative impact of earthworm activity on microarthropod communities (Migge-Kleian et al. 2006). However, studies for tropical systems are still missing.

Generally, species richness declines with increasing levels of forest disturbance in tropical systems (e.g. Lawton et al. 1998, Brown and Gurewitch 2004, Schulze et al. 2004,), but these disturbances were measured on a landscape scale. Anthropogenic disturbances by plantation management and plot maintenance could also affect local species richness and abundance patterns, especially of the litter dwelling arthropods when litter or herb layers are removed frequently.

3 soil fauna in habitats of the transition zone: rainforest margins

Rain forest margins are especially vulnerable to disturbance since they are easily accessed by the local human population (Laurance et al. 2002). Additionally, re-colonisation of disturbed habitats by forest invertebrate species can only occur from one side. Within the German research project "Stability of Rainforest Margins" the litter and soil arthropod communities were investigated along a land use gradient also representing a disturbance gradient at the edge of the Lore Lindu National Park, Central Sulawesi (Indonesia).

The study sites were located around the village of Toro at approximately 800 m a.s.l.). The aim of the study was to investigate the driving factors being responsible for changes in the arthropod community and to evaluate the different management practices from an ecological point of view. The land use gradient consisted of three types of forest use (none, small and large timber extraction named A, B and C respectively) and two types of cacao plantations (under natural shade trees and under planted shade trees named D and E respectively) with four replications each. Four soil and litter samples were taken from each replicate of these land use types,. Microarthropods were extracted at two sampling times (May and August 2004) using a high-gradient heat extractor (Macfadyen 1961), while spiders and ants from three litter samples per replicate were extracted in May and June 2005 using Winkler-bags (Besuchet et al. 1987). In addition, ant activity was measured using eight fish and honey baits per replicate which were exposed for 20 minutes on leaves of the litter layer in the early afternoon. Ants were identified to morphospecies level (Pik et al. 1999) while for oribatid mites, collembolans, symphylans and spiders individual numbers were counted. All data were analysed using analyses of variance (ANOVA). During the course of the year, litter depth, moisture and C-to-N ratio, temperatures at 10 cm above the litter, soil pH and, in June 2004, the number of fungal propagules in the soil were also measured at each site. Since there is only little seasonal variation of these parameters at the sites, only mean values will be discussed.

3.1 Litter and soil properties

Comparable to many other studies on forest conversion, the mean annual temperatures in this study were similar in all forest sites and increased along the disturbance gradient being positively correlated to canopy gap fraction (Figure 3.1 a, Swift et al. 1979, Brühl 2001, Martius et al. 2004). The canopy cover decreased from 90% in natural forest sites to 75% in plantations (Leuschner et al. 2006), leading to decreasing moisture levels along the land use gradient (Figure 3.1 b). The depth of the forest litter also decreased from natural forest to cacao plantations. Fast decomposition rates due to high metabolic rates of microorganisms limit the accumulation of litter in the tropics (Swift et al. 1979). Metabolic activity of microorganisms was not measured in this study, however, parallel to increasing temperatures the quality of the litter changed from forest sites with a higher C-to-N ratio to plantations under planted shade trees with the lowest C-to-N ratio, thereby favouring microbial decomposition and nutrient cycling (Figure 3.1 d). Thus, higher metabolic rates can be expected. In the present study, soils varied between replicates of land use types consisting mainly of cambisols and several ferrasols, lixisols and regosols (Häring et al. 2005), but generally soil pH was high for tropical soils (Kauffman et al. 1998) and slightly increased with forest disturbance (Figure 3.1 e). Only plantations under planted shade trees had generally lower soil pH than plantations under natural forest cover.

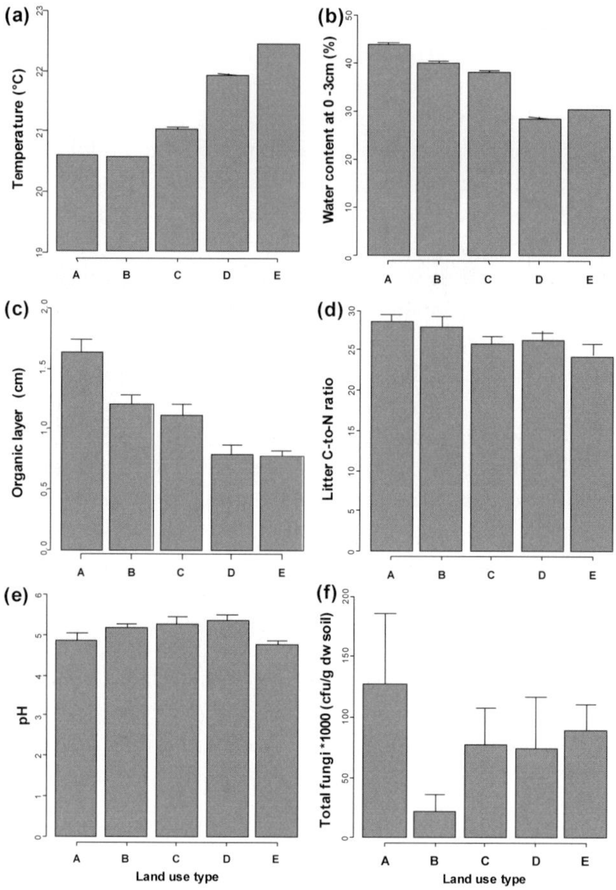

Fig. 1. Mean annual temperatures (a), moisture levels (b), litter depth (c), C-to-N ratio (d), pH (e) and number of fungal propagules (f) at rainforest margins in Central Sulawesi investigating three forest use intensities (A: none, B: small timber extraction, C: large timber extraction) and two cacao plantations (D: under natural forest shade trees, E: under polyculture of planted shade trees). Means and standard error are given.

3.2 Animal communities

The mean density of spiders in the litter did not differ significantly between the natural forest sites and plantations ($F_{[2, 25]}=1.2$, $p=0.3$, Figure 2). The observed increase in spider densities in cacao plantations under planted shade trees was correlated with a change in community structure: Lycosidae increased in density and dominance with increasing levels of disturbance (Schulz et al. 2005). These spiders prefer habitats with higher temperatures and open patches for hunting (Andersen 1995) and thus are even favoured by decreasing

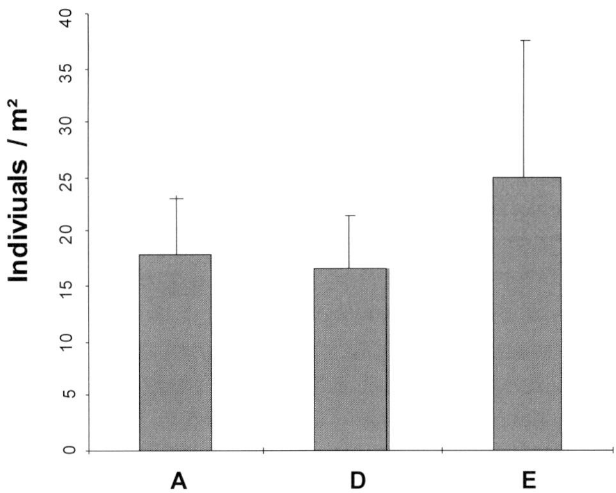

Fig. 2. Mean densities of spiders at different land use types (A: natural forest, D: cacao plantation under natural trees, E: cacao plantation under polyculture of planted shade trees). Means and standard error are given.

litter layers. In comparison, other spider groups need the litter layer for their web structures, as "hunting" area or for shelter (Bess et al. 2002, Pearce et al. 2004). Lycosid spiders are known to withstand different levels of disturbance (Hsieh et al. 2003) and carrying their egg sacs with them is a trait adapted to unstable habitats (Uetz 1976). Detailed data on species or guild composition in disturbed and undisturbed tropical forests are rare.

Ants and termites are the most abundant macroarthropods in tropical litter and soil (Tsukamoto and Sabang 2005). However, termite densities in the Toro region were very low. Generally, the density and diversity of termites on Sulawesi is very low compared to other tropical regions and mostly wood-feeding termite species occur (Gathorne et al. 2000). Thus only ants will be discussed in more detail. When the ant community was compared between natural forest and the two types of cacao plantations, different patterns could be observed regarding litter dwelling ants and the activity of ants on the litter surface. Mean ant density tended to be highest in cacao plantations under natural forest ($F_{[2,\ 25]}=2.58$, $p=0.09$, Figure 3 a), indicating that the diverse litter habitats created by cacao and forest trees in combination with increasing temperatures are favourable for ants. However, in cacao plantations under planted shade trees resources for ants seem to be limited, since densities and species richness in the litter were lowest. This is in agreement with other studies where litter depth seemed to be a factor limiting ant density and diversity (Carvalho and Vasconcelos 1999, McGlynn et al. 2002, Campos et al. 2003). The ants seemed to be forced to hunt in larger areas which is indicated by the high number of morphospecies caught at baits and the high-

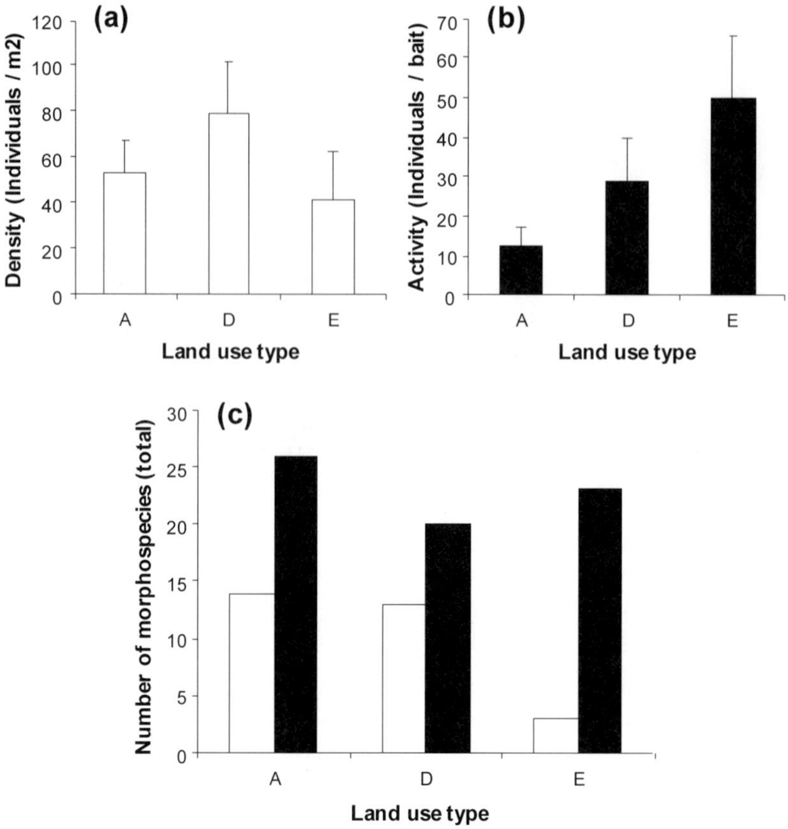

Fig. 3. Mean densities (a), activity (b) and total number of morphospecies (c) of ants on different land use types (A: natural forest, D: cacao plantation under natural trees, E: cacao plantation under polyculture of planted shade trees). White bars show data from litter samples, black bars show data from bait samples. Means and standard error are given.

est levels of activity in these plantations compared to the other plantations or natural forest ($F_{[2,60]}$=4.36, p=0.02, Figure 3 b). In addition, morphospecies composition was different between natural forest and cacao plantations under planted shade trees (Steingrebe et al. 2005) which has also been shown for Columbian coffee cultures (Armbrecht et al. 2005). While increasing temperature along this land use gradient seem to favour ant activity, comparisons between non-shaded and shaded plots in the plantations having low and high ant activity, respectively, indicate that in non-shaded areas temperature becomes a limiting factor (Steingrebe et al. 2005). Ants are common predators in the soil and litter food web and act as pest control agents (Vandermeer et al. 2002). Consequently, a successive elimination of shade trees during the growth of cacao and coffee plantations can also influence crop production and pest management by leading to strongly simplified ant communities (Armbrecht et al. 2005).

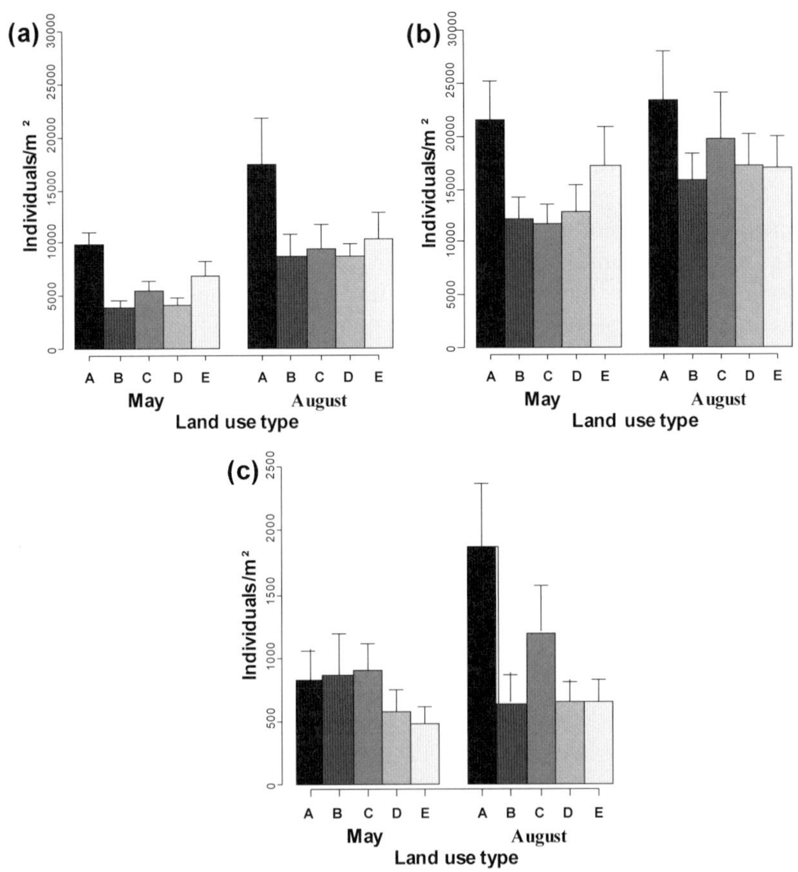

Fig. 4. Mean densities of Collembola (a), Oribatida (b) and Symphyla (c) in May and August at three forest use intensities (A: none, B: small timber extraction, C: large timber extraction) and two cacao plantations (D: under natural forest shade trees, E: under polyculture of planted shade trees). Means and standard error are given.

Mean density of microarthropods in the natural forest was within the range typically known from undisturbed rainforests (Franklin et al. 2001) which is comparable to densities in mull forests of temperate zones where the litter layer is reduced by earthworm activity. Due to generally fast decomposing organic material in tropical forests, habitat for these, mainly litter and humus dwelling organisms, is restricted. In the present study, densities of microarthropods (Collembola, Oribatida, Symphyla) were nearly always highest in the natural forest were litter depth and moisture content were also highest and mean annual temperatures were lowest (Figure 4). This difference was significant for collembolans in May ($F_{[4,71]}=4.71$, $p=0.002$) and symphylans in August ($F_{[4,72]}=2.58$, $p=0.04$), while the pattern was visible for Oribatida but not significant due to high variability within sites. The results confirm the hypothesis already stated for ants, that litter as a habitat plays a major role for microarthropod communities as well. However, densities did not differ significantly between the other forest sites or plantations, thus not following the gradient pattern of the other environmental variables presented in Figure 3.1. This differing pattern was consistent for all groups and both sampling times irrespective of seasonal differences in densities (exception: Symphyla in May 2004). Most surprisingly, densities of microarthropods declined strongly from undisturbed natural forests to natural forests with small timber extraction, indicating that even low and hardly visible levels of forest disturbance can have harmful effects on microarthropod populations. Mechanical disturbance has been shown to be detrimental for collembolans and oribatid mites (Maraun et al. 2003, Migge-Kleian et al. 2006), but fine roots and fungi also seem to be affected, since the number of fungal propagules and fine root production decreased in a similar pattern (Figure 3.1 f, M. Harteveld unpubl. data, Leuschner et al. 2006). This indicates that small decomposer organisms and fine root production can be used as indicators of forest disturbance and soil health (Hertel et al. this volume, Ponge et al. 2003, Saito 2003). Further analyses of the data will reveal if single species of collembolans or oribatid mites are more sensitive to disturbance than others.

4 Conclusions

We observed a clear gradient in the sequence natural forest - disturbed forest - cacao plantation under natural shade tree - cacao plantation under planted shade tree. The abiotic environment changed from lower to higher temperatures and from higher to lower moisture levels with increasing land use intensity. Litter and soil habitat characteristics were modified by human disturbance: in principal, litter depth and C-to-N ratios of litter decreased. All these habitat variables should exert a major influence on litter and soil communities. Until now, we found the following patterns with increasing disturbance: a decrease of microbial biomass and of microarthropod populations; in contrast, the density (and partly diversity) of predatory groups among spiders and ants were favoured by human impact.

Some controlling factors for community change are evident and are in accordance with the findings of other authors (Bloemer et al. 1997, Hodda et al. 1997, Franklin et al. 2001, Hanagardh and Brändle 2001, Martius 2001). On the one hand, an extreme microclimate, limited litter space, increased disturbance by plot maintenance and decreased microbial resources prevent

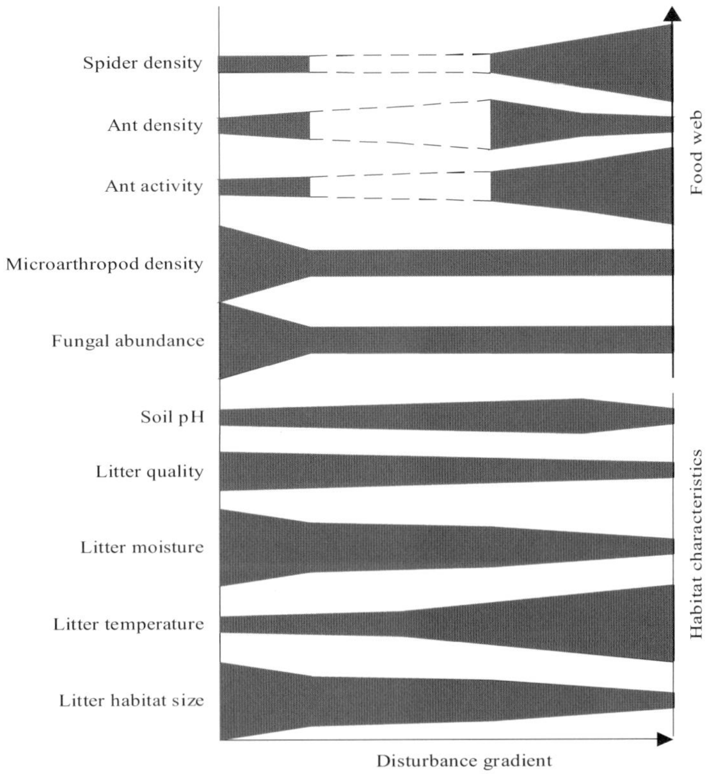

Fig. 5. Summary of environmental factors and disturbance influencing litter arthropod communities along a land use gradient from natural forest to cacao plantation under planted shade trees at the rainforest margin in Central Sulawesi.

litter and soil animal populations from attaining higher densities; on the other hand the simplified habitat structure favour populations of open habitats, especially predatory groups adapted to higher temperature and/or lower soil moisture (Figure 4).

The decisive question is how decomposition processes are influenced by changes in the detritus food web (Heneghan et al. 1999). Preliminary data from experiments with litter bags in the study area demonstrate that mass loss of litter is lower in more open spaces, that is in the direction of increasing disturbance (Woltmann unpubl. data). However, studies in other tropical forests did not find a tight correlation between litter decay and soil fauna (Barajas-Guzman and Alvarez-Sanchez 2003). Vasconcelos and Laurance (2005) studied different forest types (forest fragments and second growth forests) in the Amazonian landscape: exclusion of soil fauna decreased decomposition rates, but there was no evidence that difference in soil fauna community between sites altered litter decomposition.

A further aspect of the biota in the forest margins is the possible effect of increasing predator density in plantations on herbivorous insects. It is far from easy to quantify predator-prey interactions in the upper litter layer in relation to herbivores in the upper strata.

Thus, it has to be investigated how a reduction of decomposers and an increase in predators could influence pest control and thus yields in these agroforestry systems.

References

Andersen AN (1995) A classification of Australian ant communities, based on functional-groups which parallel plant life-forms in relation to stress and disturbance. J Biogeogr 22: 15-29

Andersen AN (2000) Global ecology of rainforest ants: functional groups in relation to environmental stress and disturbance. In: Agosti D, Majer JD, Alonso LE, Schultz TR (eds) Ants: standard methods for measuring and monitoring biodiversity. Biological Diversity Handbook Series Smithonian Institution Press, Washington, pp 25-34

Armbrecht I, Rivera L, Perfecto I (2005) Reduced diversity and complexity in the leaf-litter ant assemblage of Colombian coffee plantations. Conserv Biol 19: 897-907

Barajas-Guzman G, Alvarez-Sanchez J (2003) The relationships between litter fauna and rates of litter decomposition in a tropical rain forest. Appl Soil Ecol 24: 91-100

Basset Y (2001) Invertebrates in the canopy of tropical rain forests - how much do we really know? Plant Ecol. 153: 87-107

Basset Y, Novotny V, Miller SE, Kitching RL (2003) Arthropods of tropical forests: spatio-temporal dynamics and resource use in the canopy. Cambridge University Press, New York

Belshaw R, Bolton B (1993) The effect of forest disturbance on the leaf litter ant fauna in Ghana. Biodiv Conserv 2: 656-666

Bess EC, Parmenter RR, Mccoy S, Molles MC (2002) Responses of a riparian forest-floor arthropod community to wildfire in the middle Rio Grande Valley, New Mexico. Environ Entomol 31: 774-784

Besuchet C, Burckhardt D, Loebl I (1987) The "Winkler/Moczarski" eclector as an efficient extractor for fungus and litter Coleoptera. Coleopt Bull 41: 392-394

Bloemers GF, Hodda M, Lambshead PJD, Lawton JH, Wanless FR (1997) The effects of forest disturbance on diversity of tropical soil nematodes. Oecologia 111: 575-582

Brown KA, Gurevitch J (2004) Long-term impacts of logging on forest diversity in Madagascar. Proc Natl Acad Sci USA 101: 6045-6049

Brühl CA, Gunsalam G, Linsenmair KE (1998) Stratification of ants (Hymenoptera, Formicidae) in a primary rain forest in Sabah, Borneo. J Trop Ecol 14: 285-297

Brühl CA (2001) Leaf litter ant communities in tropical lowland rain forests in Sabah, Malaysia: effects of forest disturbance and fragmentation . Julius-Maximilians-Universität Würzburg, Würzburg

Campos RBF, Schoereder JH, Sperber CF (2003) Local determinants of species richness in litter ant communities (Hymenoptera : Formicidae). Sociobiology 41: 357-367

Carcamo HA, Parkinson D, Volney JWA (1998) Effects of sulphur contamination on macroinvertebrates in Canadian pine forests. Appl Soil Ecol 9: 459-464

Carvalho KS, Vasconcelos HL (1999) Forest fragmentation in Central Amazonia and its effects on litter-dwelling ants. Biol Conserv 91: 151-157

Chazdon RL, Whitmore TC (2002) Foundations of tropical forest biology. University of Chicago Press, Chicago

Dietz J, Hölscher D, Leuschner C, Hendrayanto (2006) Rainfall partitioning in differently managed maontane forest stands of Sulawesi, Indonesia. For Ecol Manage. in press.

Edwards CA, Bohlen PJ (1996) Biology and ecology of earthworms, 3rd edition. Chapman and Hall, London

Fragoso C, Lavelle P (1992) Earthworm communities of tropical rain-forests. Soil Biol Biochem 24: 1397-1408

Franklin EN, Morais JW, dos Santos EMR (2001) Density and biomass of Acari and Collembola in primary forest, secondary regrowth and polycultures in central Amazonia. Andrias 15: 141-153

Giller KE, Bignell D, Lavelle P, Swift M, Barrios E, Moreira F, Van Norrdwijk M, Barios I, Karanja N, Huising J (2005) Soil biodiversity in rapidly changing tropical landscapes: scaling down and scaling up. In: Bargett RD, Usher MB, Hopkins DW (eds) Biological diversity and function in soils. Cambrigde University Press, Cambridge, pp 295-318

Gonzalez G, Seastedt TR (2000) Comparison of the abundance and composition of litter fauna in tropical and subalpine forests. Pedobiologia 44: 545-555

Heneghan L, Coleman DC, Zou X, Crossley DA, Haines BL (1999) Soil microarthropod contributions to decomposition dynamics: tropical-temperate comparisons of a single substrate. Ecology 80: 1873-1882

Hertel D, Leuschner C, Harteveld M, Wiens M (this volume) Fine root mass, distribution and activity in disturbed primary forests and secondary forests of the moist tropics. In: Tscharntke T, Leuschner C, Guhardja E, Zeller M, Bidin A (eds) The stability of tropical rainforest margins, linking ecological, economic and social constraints of land use and conservation. Springer, Berlin Heidelberg New York

Hill JL, Hill RA (2001) Why are tropical rain forests so species rich? Classifying, reviewing and evaluating theories. Prog Phys Geography 25: 326-354

Hodda M, Bloemers GF, Lawton JH, Lambshead PJD (1997) The effects of clearing and subsequent land-use on abundance and biomass of soil nematodes in tropical forest. Pedobiologia 41: 279-294

Hsieh Y-L, Lin Y-S, Tso I-M (2003) Ground spider diversity in the Kenting uplifted coral reef forest, Taiwan: a comparison between habitats receiving various disturbances. Biodiv Conserv 12: 2173-2194

Hagvar S, Amundsen T (1981) Effects of liming and artificial acid rain on the mite (Acari) fauna in coniferous forest. Oikos 37: 7-20.

Jocque R, Samu F, Bird T (2005) Density of spiders (Araneae : Ctenidae) in Ivory Coast rainforests. J Zool 266: 105-110

Kauffman S, Sombroek W, Mantel S (1998) Soils of rainforests. Characterization and major constraints of dominant forest soils in the humid tropics. In: Schulte A, Ruhiyt D (eds) Soil of tropical forest ecosystems. Characteristics, ecology and management. Springer, Berlin, pp 9-20

Kelly JA, Samways MJ (2003) Diversity and conservation of forest-floor arthropods on a small Seychelles island. Biodiv Conserv 12: 1793-1813

Laurance WF, Albernaz AKM, Schroth G, Fearnside PM, Bergen S, Venticinque EM, Da Costa C (2002) Predictors of deforestation in the Brazilian Amazon. J Biogeogr 29: 737-748

Lawton JH, Bignell DE, Bolton B, Bloemers GF, Eggleton P, Hammond PM, Hodda M, Holt RD, Larsen TB, Mawdsley NA, Stork NE, Srivastava DS, Watt AD (1998) Biodiversity inventories, indicator taxa and effects of habitat modification in tropical forest. Nature 391: 72-76

Leigh EGJ (1999) Tropical forest ecology: a view from Barro Colorado Island. Oxford University Press, Oxford

Leuschner C, Wiens M, Harteveld M, Hertel D, Tjitrosemito S (2006) Patterns of fine root mass and distribution along a disturbance gradient in a tropical montane forest, Central Sulawesi. Plant Soil (in press)

Macfadyen A (1961) Improved funnel-type extractors for soil arthropods. J Animal Ecol 21: 87-118

Maraun M, Salamon JA, Schneider K, Schaefer M, Scheu S (2003) Oribatid mite and collembolan diversity, density and community structure in a moder beech forest (Fagus sylvatica): effects of mechanical perturbations. Soil Biol Biochem 35: 1387-1394

Martius C, Höfer H, Garcia MVB, Römbke J, Forster B, Hanagarth W (2004) Microclimate in agroforestry systems in Central Amazonia: does canopy closure matter to soil organisms? Agrofor Sys 60: 291-304

Martius C, Römbke J, Verhaagh M, Höfer H, Beck L (2001) Termiten, Regenwürmer und Ameisen - prägende Elemente der Bodenfauna tropischer Regenwälder. Andrias 15: 15-27

Martius C, Tiessen H, Vlek PLG (2001) The management of organic matter in tropical soils: what are the priorities? Nutrient Cycling Agroecosystems 61: 1-6

McGlynn TP, Hoover JR, Jasper GS, Kelly MS, Polis AM, Spangler CM, Watson BJ (2002) Resource heterogeneity affects demography of the Costa Rican ant *Aphaenogaster araneoides*. J Trop Ecol 18: 231-244

Migge-Kleian S, McLean M, Macrz J, Hcncghan L (2006) The influence of invasive earthworms on indigenous fauna in ecosystems previously uninhabited by earthworms. Biological Invasions (in press)

Noti MI, Andre HM, Ducarme X, Lebrun P (2003) Diversity of soil oribatid mites (Acari : Oribatida) from high Katanga (Democratic Republic of Congo): a multiscale and multifactor approach. Biodiv Conserv 12: 767-785

Olson DM (1994) The distribution of leaf-litter invertebrates along a neotropical altitudinal gradient. J Trop Ecol 10: 129-150

Paoletti MG, Taylor RAJ, Stinner BR, Stinner DH, Benzing DH (1991) Diversity of soil fauna in the canopy and forest floor of a Venezuelan cloud forest. J Trop Ecol 7: 373-383

Pearce JL, Venier LA, Eccles G, Pedlar J, Mckenney D (2004) Influence of habitat and microhabitat on epigeal spider (Araneae) assemblages in four stand types. Biodiv Conserv 13: 1305-1334

Pik AJ, Oliver I, Beattie AJ (1999) Taxonomic sufficiency in ecological studies of terrestrial invertebrates. Aust J Ecol 24: 555-562

Plowman KP (1981) Inter-relation between environmental factors and cryptostigmata and mesostigmata (Acari) in the litter and soil of two Australian subtropical forests. J Animal Ecol 50: 533-542

Ponge JF, Gillet S, Dubs F, Fedoroff E, Haese L , Sousa JP, Lavelle P (2003) Collembolan communities as bioindicators of land use intensification. Soil Biol Biochem 35: 813-826

Primack R, Corlett R (2005) Tropical rain forests: an ecological and biogeographical comparison. Blackwell, Malden

Reddy MV (1995) Litter arthropods. In: Reddy MV (ed) Soil organisms and litter decomposition in the tropics. Westview Press, Boulder, pp 113-140.

Richardson BA, Richardson MJ, Soto-Adames FN (2005) Separating the effects of forest type and elevation on the diversity of litter invertebrate communities in a humid tropical forest in Puerto Rico. J Animal Ecol 74: 926-936

Saito M (2004) Can soil biodiversity be used for an indicator of soil health? Case studies in Japan. In: Francaviglia R (ed) Agricultural Impacts on Soil Erosion and Soil Biodiversity: Developing Indicators for Policy Analysis. Proceedings from an OECD Expert Meeting. Istituto Sperimentale per la Nutrizione delle Piante, Rome, pp 1-9

Schulz W, Woltmann L, Migge S, Schaefer M (2005) Influence of habitat structure on the predatory macroarthropod community along a land use gradient (cacao plantation to natural forest) in Central Sulawesi. In : Stietenroth D, Lorenz W, Tarigan S, Malik A (eds) The Stability of Tropical Rainforests margins: Linking ecological, economic and social constraints of land use and conservation. Universitätsverlag, Göttingen, pp 193

Schulze CH, Waltert M, Kessler PJA, Pitopang R, Shahabuddin, Veddeler D, Mühlenberg M, Gradstein SR, Leuschner C, Steffan-Dewenter I, Tscharntke T (2004) Biodiversity indicator groups of tropical land-use systems: comparing plants, birds, and insects. Ecol Appl 14: 1321-1333

Sheil D, Burslem DFRP (2003) Disturbing hypotheses in tropical forests. Trends Ecol Evol 18: 18-26

Steingrebe A, Woltmann L, Platner C, Migge S (2005) Influence of microclimate changes on ant communities in litter habitats along a tropical land use gradient from cacao plantations to natural forest . In: Stietenroth D, Lorenz W, Tarigan S, Malik A (eds) The Stability of Tropical Rainforests margins: Linking ecological, economic and social constraints of land use and conservation. Universitätsverlag, Göttingen, pp 202

Swift MJ, Heal OW, Anderson JM (1979) Decomposition in terrestrial ecosystems. Blackwell Scientific Publications, Oxford

Taylor AR, Schroter D, Pflug A, Wolters V (2004) Response of different decomposer communities to the manipulation of moisture availability: Potential effects of changing precipitation patterns. Global Change Biol 10: 1313-1324

Tian G (1998) Effect of soil degradation on leaf decomposition and nutrient release under humid tropical conditions. Soil Sci 163: 897-906

Tian G, Brussaard L, Kang BT (1993) Biological effects of plant residues with contrasting chemical-compositions under humid tropical conditions - effects on soil fauna. Soil Biol Biochem 25: 731-737

Tsukamoto J, Sabang J (2005) Soil macro-fauna in an *Acacia mangium* plantation in comparison to that in a primary mixed dipterocarp forest in the lowlands of Sarawak, Malaysia. Pedobiologia 49: 69-80

Uetz GW (1976) Gradient analysis of spider communities in a streamside forest. Oecologia 22: 373-385

Vandermeer J, Perfecto I, Nunez GI, Phillpott S, Ballinas AG (2002) Ants (*Azteca* sp.) as potential biological control agents in shade coffee production in Chiapas, Mexico. Agrofor Sys 56: 271-276

Vasconcelos HL, Laurance WF (2005) Influence of habitat, litter type, and soil invertebrates on leaf-litter decomposition in a fragmented Amazonian landscape. Oecologia 144: 456-462

Wagner T (2000) Influence of forest type and tree species on canopy-dwelling beetles in Budongo Forest, Uganda. Biotropica 32: 502-514

Watt AD, Stork NE, Bolton B (2002) The diversity and abundance of ants in relation to forest disturbance and plantation establishment in Southern Cameroon. J Animal Ecol 39: 18-30

Whitmore TC (1998) An introduction to tropical rain forests. 2nd. ed. Oxford University Press, Oxford

Wolters V, Schaefer M (1994) Effects of acid deposition on soil organisms and decomposition processes. In: Hüttermann A, Godbold DL (eds) Effects of acid rain on forest processes. Wiley, New York, pp 83-127

Wunderle I (1992) Arboricolous and edaphic Oribatei (Acari) in the lowland rain-forest of Panguana, Peru. Amazoniana 12: 119-142

From ecological to political buffer zone: ethnic politics and forest encroachment in Upland Central Sulawesi

Melani Abdulkadir-Sunito* and M.T.Felix Sitorus

Department of Communication and Community Development, Faculty of Human Ecology, Bogor Agricultural University, Indonesia

*corresponding author: Melani Abdulkadir-Sunito, phone: +62 251 420252, Email: sunito@indo.net.id

Summary

This paper discusses inter-ethnic relations and patterns of land-use and landownership, and how ethnicity is defined, used to grant preference, and to allocate socio-economic resources, by analysing two forest-margin villages in Sulawesi.

The heavy influx of migrants has altered the proportion of local people to migrants in Sintuwu and Watumaeta, villages located on the eastern border of Lore Lindu National Park. In 2001, there were over one-third households of Bugis – the largest group of ethnic migrants. Social distance, inability to adapt and unwillingness to integrate (Charras 1993, Human Rights Watch 1998) were indicated by house-clustering of the ethnic groups.

As most agricultural land inside the village has been sold to migrants, mainly to Bugis, the locals have been transformed from landlords to landless, while the migrants became landlords. In a land-based economy, land scarcity leads to declining socio-economic security. Social destabilization occurs as ethnic frictions result from this shift in resource control.

To gain back their economic power and socio-economic security, locals adopted a strategy of land expansion into ecological buffer zone areas. By outsiders, particularly the State, this is viewed as an encroachment into the national park forest, while to locals this is an attempt to re-secure their economic base. Locals justified their actions by disputing the park boundaries. Subsequently their "illegal" actions were accepted and formalized when they received usufruct rights in the form of SKPL (Letter of Land Utilization) issued by the village head and KKM (Community Conservation Agreement),

Tscharntke T, Leuschner C, Zeller M, Guhardja E, Bidin A (eds), The stability of tropical rainforest margins, linking ecological, economic and social constraints of land use and conservation, Springer Berlin 2007, pp 167-180

as was made by the Village Council with Lore Lindu National Park administrators.

Land expansion is also an attempt to re-stabilize relationship between locals and migrants in the face of ethnic conflict. The economic buffer zone has become a socio-political buffer that prevents local tensions from becoming open conflicts. In the absence of ethnic politics at national level, this local innovation in ethnic politics is a means to neutralize destabilization of inter-ethnic relations. Such innovation, however, cannot go on endlessly, for once formalized the locals' new economic space will most likely be taken over by the migrants, leading to a vicious circle of forest margin destabilization.

The implication of this is the need for a land-use modeling that accounts for social constraints of inter-ethnic relations and land-use systems. Encroachment is not only perceived as an economic action, but also as a political action, as forest has not only economic value, but is also an arena of ethno-political action. Disregarding this reality can trigger, and at times has triggered, ethnic conflicts.

Keywords: ethnic politics, ethnic relations, forest encroachment, Central Sulawesi

1 Introduction

If one defines forest buffer zone as the interface between a natural forest ecosystem and man-made (as in agriculture or village) ecosystems [1], then forest margins of a park are a good example. The role of such buffer zones, as described by Ruff and Schroth (2004), is in conserving "a degree of biodiversity", which is a specifically ecological role. The buffer zone is also an extractive area, where people harvest rattan, timbers, and other forest products and which provides spatial buffering between the villages and the forest (Naughton-Treves and Salafsky 2004). This view entails recognition of the fact that people live inside and around the forest, and their livelihoods depend upon it. These people have lived there for generations, while some other migrants only recently arrived in search of land and livelihood (Li 2002). This fact indicates that forest buffer zones also have socio-economic roles.

The causes and mechanisms behind deforestation have been comprehensively discussed by Wunder (2004), who identifies three main approaches to discuss the phenomenon: the impoverishment, the neo-classical, and the po-

[1] This definition is an attempt to combine the definition of agro forestry buffer zone (Naughton-Treves and Salafsky 2004) as "any land use system combining trees with agricultural crops that lies adjacent to a park or reserve" and description of forest margins (Wunder 2004) as "the spatial transition zone between tropical forests and converted land uses".

litical ecology[2]. This paper highlights the interplay of aspects within those approaches by elaboration on local's – migrants' social relations, land-use and access to land resources, and its impact on forest ecosystems in the margin area.

The Sintuwu and Watumaeta communities are two forest-margin villages on the eastern border of Lore Lindu National Park (LLNP). At the time of research, social instability in these villages was imminent. This was due to high in-migration that significantly – and rapidly – changed the composition of locals to migrants. Changes in land-use created unequal distribution of access to land resources in the villages proximity in and around to Poso, where ethno-religious conflict has intermittently flared since 1998 (Abdulkadir-Sunito 2004, Sitorus 2004).

In the pre- and post-1997 economic crisis, Indonesia was rife with communal conflicts. Analysts (i.e. Human Rights Watch 1998, Sudagung 2001, Van Dijk 2001) noted ethnicity as a component in the more persistent and bloody ones, and pointed the causes to cultural differences between communities, inability of migrants to adjust to local culture, and competition for politico-socio-economy resources. Furthermore, Tirtosudarmo (2001) traced the roots of conflict to the lack of policies on ethnicity and ethnic relations, while Van Dijk (2001) noted the (unanticipated) consequences of the greater intra-Indonesian mobility. Tirtosudarmo attributed the roots of conflict to the perception of the New Order regime, that national development and national integration necessitated the delineation of pluralism and ethnic heterogeneity. They are perceived as sources of ethnic conflict, and thus not to be discussed[3]. As a result, people do not learn to manage differences, including ethnic differences. At a local level, ethnic friction is often resolved with local-level innovation. If this fails and the friction escalates into conflicts, the state often solves them by way of military intervention or enforces harmony by deliberation (*musyawarah*). The intra-Indonesian mobility (Van Dijk 2001) – as in transmigration[4] and migrant workers to large-estate or timber-company – not only place different ethnics in one area, but often done by way of diminishing

[2] Wunder (2004) distinguished three main approaches: the impoverishment, the neo-classical, and the political ecology – each identify differently the main drivers, agents, and mechanisms behind forest loss. The impoverishment approach points to a combination of poverty and growing number of population. The neo-classical approach points to the ill-defined forest property rights that encourage smallholders and large investors to open up the forest and claim land-rights afterwards. The political ecology approach points to externally driven processes with capitalist entrepreneurs demand additional land, disrupting local land-use systems, and pushing smallholder farmers into the forest.

[3] The Indonesian policy on ethnicity is to prohibit discussion on SARA (*Suku, Agama, Ras, Antar- golongan* or Ethnic, Religion, Race, Inter-Group).

[4] Transmigration as a formal policy, is an economic and demographic policy, with integration as its objective. It did, however, have strong influence on the ethnic composition at the local level (see Elmhurst 1999)

access to resources of the indigenous communities. Stewart (1998) added, that old motives for conflict, rather than new motives for conflict, may erupt due to a weakening of the state authority and its ability to suppress conflicts. This view is important, since it acknowledges that conflicts exist in any situation and not just a by-product of some transitional or crisis era.

Esman (as quoted in Tirtosudarmo 2001) wrote, "Ethnicity as politics is meaningful only in a relational framework". As a collective identity, ethnicity is shaped by self-definition and by constraints imposed by outsiders. Ethnic solidarity arises when competitive outsiders must be confronted, by creating distinctions between 'we' and 'they'. Sociologically, the 'we' and 'they' refer to concepts of in-group and out-group (Sumner as quoted in Light and Keller 1982)[5], which implies a sense of common identity and rights. The formation of in-group-ness is a process of sharing similar history, ethnicity, religion, and/or occupation. Yet, some will remain as 'they' because the ethno-religion-occupational differences remain significant. Writing on ethnic conflict, Stavenhagen (as quoted by Stewart 1998) concluded, "....when regional and social disparities in the distribution of economic resources also reflect differences between identified ethnic groups, then conflict over social and economic issues readily turns into ethnic conflict". This, as we shall discuss, is the situation in Sintuwu and Watumaeta. Yet, how does a 'we' vs 'they' distinction develop? How is it used to gain position and socio-economic security? To what extent, then, does ethnicity have impact on the distribution of politico-socio-economic resources?

This study was carried out as part of a larger research project on agrarian formation, land-use and migration within the STORMA project. The study areas of Sintuwu and Watumaeta are villages directly bordering Lore Lindu National Park, so changes in the park's forest buffer-zone can be observed. The villages also experienced, or are experiencing, a high influx of migration, and are undergoing a rapid move towards a more capitalistic economy. It was, indeed, the similar findings in the two villages that brought the authors together in the first place. The study itself combines qualitative and quantitative approaches. The quantitative methods comprise a land-holding survey, a population census, and a migration survey. The qualitative methods are transecting walks, individual as well as group interviews, and direct observation. Fieldwork was carried out intermittently from 2001 – 2004.

2 Sintuwu and Watumaeta: a continuum

As one ascends from Palu to the eastern border of Lore Lindu National Park, Palolo valley – where Sintuwu is located – is reached after one-hour of travel.

---

[5] "In-groups breed a consciousness of kind ... (and) are typified by ... religious and ethnic groups. The members ...develop their sense of we-ness as a result of sharing common experiences" (Light & Keller 1982:192)

At the altitude of 500 to 700 meters, and with Gumbasa River at its northern border, Sintuwu accommodates both wet and dry types of agriculture. The community expanded from 5 Kaili households in 1961 to 1,205 people of 328 households in 2001. About 50 km southeast of Palolo, on a higher altitude in the Napu Valley, Watumaeta is located. As settlement, Watumaeta exists since the early 1930s, and was formally established as *desa* in 1950. In 2001, the population of Watumaeta was 1,125 people, a three-fold increase from 1985's 365 inhabitants.

Table 1. Some geographic and population characteristics of Sintuwu and Watumaeta.

	Sintuwu	Watumaeta
Distance from Palu	54 km	105 km
Altitude	500-700 m asl	1100 m asl
Road construction to Palu	1970s	1982
migration (r = annual population rate)	Increasing since 1960s, peak in 1970s (Kaili); 1980s (Bugis from South Sulawesi); late 1980s (Kulawi)	1st wave 1983-87 (r = 14.2%) – Bugis by way of Palu 2nd wave 1995-1998 ($r_{1997\text{-}2000}$ = 17.4%) – Bugis from South Sulawesi 3rd wave 2000-02 – Sunda/ Java ex transmigrants, Palu, Poso refugees
cocoa introduction	mid 1980s	late 1990s

Sources: Syahyuti (2002) and Sitorus (2004) for Sintuwu; Abdulkadir-Sunito (2004) for Watumaeta

The road connecting Palu and Napu (distance 105 km) was constructed more than a decade after the Palu – Palolo connection (distance 54 km) was build (see Table 1). Although historically Napu has closer connections to Poso (some 80 km to the East), the road connecting these villages had not been constructed until 1985. This Palu – Palolo - Napu – Poso road has become a highway for migrants from South Sulawesi, and later for migrants from Poso, into the highlands of Central Sulawesi.

The correlation of road construction, in-migration, and introduction of cocoa is apparent in Sintuwu, but is not the case in Watumaeta. The first wave of in-migrants to Watumaeta followed the construction of the Palu – Napu road. Introduction of cocoa, however, did not occur until late 1990s, following the second wave of migration. The reason is the agricultural agency's (*Dinas Pertanian*) recommendation to plant coffee, which was considered suitable for high altitude areas, instead of cocoa, which was considered suitable only for lowland cultivation. Several early migrants recalled how their family cut their cocoa trees and replaced them with coffee. As a result, newcomer migrants to Watumaeta, who defied such advice and were emboldened by the success of cocoa planting in other areas, were in a better position to take advantage of the late 1990s cocoa boom.

Sintuwu was established by a process of multi-ethnic immigration, with the Kaili, who were the pioneer migrants, declaring themselves to be 'indigenous' – and at the same time label the Bugis as 'newcomers'[6]. Following the pioneers, Kaili immigrants, mainly from Sigi-Biromaru, came in the 1960s, their numbers peaked in the 1970s. The first Bugis immigrants came in 1968 as saw-mill company laborers. Pulled-in by the success of earlier migrants, Bugis immigrations from South Sulawesi peaked in the 1980s. Following the Bugis were the Kulawi migrants, whose numbers peaked in the late 1980s.

Watumaeta's pioneer settlers are the Napu or Pekurehua ethnicies, who came from the neighboring villages of Wuasa and Sedoa. The first migrants, who came in a trickle in the 1970s from the Lindu area, were Napu returnees who had migrated to Lindu in the 1950s. The Bugis migrants of the early 1980s were those who had worked in Palolo earlier on, or had relatives in Palu. The migrants arriving between 1995 and 1998 were mostly Bugis from South Sulawesi. The most recent wave of migration came between 2000-2002, a mix of Sunda/Java former transmigrants to Kebun Kopi (Central Sulawesi), the educated migrants from Palu, and conflict-related refugees from Poso areas.

Table 2. Autochton and allochton parts of the population in Sintuwu and Watumaeta in 1985 and 2001.

	1985		2001	
	Sintuwu	Watumaeta	Sintuwu	Watumaeta
% autochton	n.a.	60	58	37
% allochton	n.a.	40	42	63
% Bugis to total population	n.a.	18.5	29	60

Sources : Abdulkadir-Sunito (2004) for Watumaeta ; Sitorus (2004) for Sintuwu

Between 1982 and 1983, the population of Watumaeta grew at an annual rate of 9.6 percent (natural increase was around 2.1 percent), and at an annual rate of 14.2 percent from 1983 to 1987. Migration significantly altered the composition of locals (autochthon) to migrants (allochthon) (see Table 2). In 1985, two-thirds of the population of Watumaeta were locals. By 2001 – a mere 16 years later – the situation was reversed: two-thirds of the populations were migrants. In terms of religion, the population also changed from a christian to a muslim majority. The situation in Sintuwu is similar, with 60 percent locals

[6] This paper uses interchangeably the terms indigenous-newcomers, local-migrants, autochthon-allochthon to indicate the distinctions of in-group – out-group, and (in a more competitive situation) 'we' – 'they'

in 2001. In both villages, Bugis became the largest group of ethnic migrants. 69 percent of the allochton population in Sintuwu and 95 percent of that in Watumaeta are Bugis.

Charras (1993) noted that interaction of locals with migrants is often less integrated in communities with two sizable ethnic groups. Such is the situation in Sintuwu and Watumaeta, one indication being the house-clustering by ethnicity. In Sintuwu, each of the ethnic groups has established exclusive settlements, either on the same hamlet (as the Kaili in Hamlet-3) or in a settlement cluster within one hamlet (Bugis, Kaili, Kulawi in Hamlet-1 and Bugis, Kaili in Hamlet-2). In Watumaeta, the local Napu clustered at the village center (Hamlet-1). Hamlet-2 is the face-off area, in which the Napu settle closer to the village center, the Bugis further away. Hamlet 3, the area farthest from the village center and down the road to Poso, is exclusively populated by the Bugis[7]. The area was largely developed by the incoming of Bugis migrants.

This house-clustering indicates a tendency to exclusive social life. Different religions of Napu and Bugis with their own rituals and gatherings further emphasize the exclusive social life. The Bugis are even clustered in units according to their place of origin in South Sulawesi, rendering social solidarity to each sub-ethnic group of settlement cluster only.

The spatial and social distances are further emphasized by the ethno-religious conflict in Poso. As one of the destination areas for Christian refugees from Poso, Watumaeta set up a refugee settlement in the area of Hamlet-1 in 2002. This policy differed from other villages in the Napu area, which preferred to perceive refugees as temporary residents, or, if they expressed their intention to become permanent residents, to integrate them with the locals. This policy is viewed as the local's attempt to rebalance the composition of locals versus migrants. Since the refugees have the same religion and regional history [8] as the locals, they are perceived as part of the 'we'. Although most refugees are petty commodity cocoa farmers much like the Bugis, lack of capital and access to land puts them closer to the locals. Similar attempts were made in Sintuwu with the inclusion of later-come migrants, the Kulawi ethnic, who have the same history as the indigenous Kaili.

3 Land transfer in a land-based economy and its effects

The introduction of cocoa to Sintuwu in the 1980s shifted the dominant mode of production from the subsistence to the petty commodity or the capitalist

[7] Interestingly, Watumaeta's Hamlet-3 is also bordering Wuasa's separate Bugis settlement of Hamlet-4, implying that the Bugis of the two villages are connected closer to each other than to their respective villages.

[8] In the past, the coastal areas of Poso were a trade destination for the highland Napu people.

one[9]. Earlier, in the 1960s and 1970s, the main subsistence production was of rice in wet land areas and corn in dry land areas. The crops were produced mainly for household consumption. In the petty commodity sector, the villagers cultivated – extensively, and with minimum care – perennial cash crops, mainly coffee.

Unlike coffee, cocoa needs more intensive care and capital. Consequently, as cultivation of the new "golden crop" extended all around the village in the early 1990s, Sintuwu experienced a drastic change in composition of land utilization. Between 1992 and 1998, nearly 72 percent of wet-rice fields in Sintuwu were converted to cocoa plantation, the main reason being the good price of cocoa. Wet-rice fields shrunk from 270 hectares in 1992 to a mere 75 hectares in 1998. Conversion of dry land and house gardens – the other subsistence agricultural areas – was also observed. The Bugis did most of the land conversion, and they purchased land mainly from the Kaili. Another way of expansion, albeit illegal, was by encroaching into forest areas. By mid 2001, the villagers of Sintuwu, mainly Kaili, had cultivated nearly 100 ha of cocoa inside LLNP.

In Watumaeta, petty commodity and capitalist sectors are represented by cocoa, cultivated mainly by the Bugis, and highland vegetables (i.e. onion green, cauliflower, potatoes, chinese cabbage, carrot, tomatoes) and mandarin orange fruit cultivated mainly by the Sunda/Java transmigrants. The Bugis also cultivate vegetables as a cash crop, and as temporary crop before the main crop (primarily cocoa) begins to produce. Traditionally, the locals owned the wet-rice fields, located in flat lands between *kampong* dwelling and the steep forest-hill. Corn is the subsistence dry land crop. Coffee, as well as kidney beans and nutmeg, are cultivated for the market. Coffee became more commercialized with the introduction of 'Kartika', a dwarf type Arabica coffee with a high, but relatively short, productive period. In 2002,the locals cautiously started to plant onion-green [10], which they prefer because of its short harvest time, the minimum care needed, and the possibility to plant it in home-gardens – which is the only bit of agricultural plot that is left to many locals.

These different modes of production between locals and migrants form the 'we' – 'they' distinction. Locals generally carry on a subsistence production,

[9] Assuming three distinguishable modes of production (Kahn 1974), namely subsistence, petty commodity, and capitalist.

[10] One reason behind this change was a suggestion made by members of *Forum Kerukunan Umat Beragama* (Inter-Religion Harmony Forum; formed in 2001 in every refugee-destination village at the Napu valley after some frictions due to misunderstanding occurred), who realized, that the income gap between locals and migrants was one crucial aspect that lead to inter-ethnic conflicts in different parts of Central- and South Sulawesi. They concluded that Watumaeta people needed to reduce this gap, and planting onion green was thought to give the locals the cash they needed.

while Bugis carry on agriculture in a more capitalistic way. Some Bugis also became traders of agricultural products, which further accumulates surplus.

The mid 1990s Bugis migrants in Watumaeta opened ex production forest (HPH) to the hills and develop wet-rice fields at the flat land of Hamlet-3. While the 1980s Bugis migrants generally got their land from Village administration's allotment[11], the mid 1990s onward migrants had to buy it, either from locals or other fellow migrants. Similar data were obtained from Wuasa (Sunito forthcoming), where by 1997 land transactions were not only made between land-hungry immigrants and land-rich local people, but to a substantial part between migrants and between locals. The intra-migrant transaction is exemplified by the Bugis migrants from Palopo, who came to Watumaeta as a group in 1997. Its leader bought a large plot of agricultural land, mainly through contact with village administrators, and alloted plots to the group-members. Over the years, the group-members paid for their plots with money and labor at the leader's plots. Locals also sell land because of economic pressures and to gain quick cash – i.e. for weddings or funeral celebrations and, more recently, to pay back credits for motorcycles.

By 2003, even migrants felt the shortage of land. As one migrant simply put it, "locals have no more land to sell, Bugis will not resell what they own". The response to such a situation appears to be segmented by ethnicity, which also reflects access to, and ownership of, capital. Having accumulated capital from cocoa, the Bugis try to open plots in other villages (i.e. Alitupu), or work in different land-tenure arrangements (*pinjam-pakai* – land borrowing, *bapajak* – land leasing, *bagi-hasil* – share cropping). Without capital, locals generally encroach upon forests.

From landlords to landless

A 2001 survey in Sintuwu showed, that Kaili households on average owned 1,9 ha of agricultural land, with 0.69 ha of dry land and 0.12 ha of rice fields. Bugis households had an average of 2,32 ha with 0.95 ha dry land – the Bugis own no rice fields since all have been converted into cocoa field. Assuming that land ownership is the base of social stratification, the land transfer downgraded Kaili households in Sintuwu and Napu households in Watumaeta from the status of 'land-owners' to 'landless'. At the same time, Bugis households were upgraded from the status of 'landless' to 'land-owners'. Land transfer, thus, shifted economic power from locals to migrants.

Losing land in a land-based economy, where non-farming activities are still undeveloped, means losing ones' basis of socio-economic security. The cocoa boom not only widened the economic gap between locals and migrants, but

[11] This is in part a strategy to increase the population of a village. The more populated a village was, the higher was the government's development aid. In the 1980s, villages were also supposed to be merged, if the population was less than a specific number – this has, however, never happened. A similar situation has been observed in Sintuwu.

also between early-come migrants and the later-come migrants. The loss of the socio-economic security base was felt earlier by the locals, when the LLNP border was formalized. Many local households, which traditionally made use of the forest areas for planting corn in its flat areas, harvesting coffee, and gathering dammar resin[12] or other forest products on the steeper slopes suffered, when these activities were forbidden. This traditional relation to the forest then became the reason for contesting the park border.

The situation where locals are losing access to resources to migrants, by way of land transfer or by other means, is now common in many areas of Indonesia. This has led to social tensions between locals and migrants, who, in both Sintuwu and Watumaeta happen to be of two major ethnic groups of two different religions. In Watumaeta the tensions are expressed by hurling stones into Bugis houses, by ruining ripe vegetables, by ethnic stereotyping, or, worse, by excessive suspicion towards any activities of the other ethnic group (see Abdulkadir-Sunito 2004). Poso's ethno-religious conflicts heightened this tension. In Sintuwu, social jealousy was rife as the Kaili wanted to cultivate cocoa and reap its profit, like the Bugis, but were faced with the problem of land scarcity inside the village area[13].

The 'we' – 'they' distinction has been solidified as economic power shifted from locals to migrants. Bugis migrants as 'they' became apparent, not only because of their more capitalistic mode of production, but also of their expansive nature toward land resources. Economically, Bugis became the "competitive outsiders" (Esman as quoted in Tirtosudarmo 2001) and a threat.

...And from landless back to land-owners

In an effort to search for new bases outside the village area, locals realized that forest area is their best alternative. This activity, however, brought them into land dispute with the LLNP office and the State. On the one hand, LLNP claimed that the Kaili in Sintuwu and the Napu in Watumaeta have transgressed park borders, on the other hand, the Kaili and Napu dispute the park borderline. In Sintuwu, the Kaili accept the borderline that was established by the colonial administration, 2 km further inside the forest area than the border that was drawn later (in 1982). The Kaili also pointed out to the fact that they had cleared and cultivated this part of the forest prior to 1982. In Watumaeta, the Napu claimed that the flatland area of LLNP along

[12] The dammar resin lost its value in the 1980s (Sunito forthcoming)

[13] Sunito (forthcoming) observed in Wuasa, that negative rhetorics against immigrants are still restricted by indigenous domination of the administrative system, appreciation of the agricultural know-how and work ethics of migrants, and fear of uncontrolled conflicts as flaring up in Poso. This situation is dissimilar to Sintuwu and Watumaeta, not only because of Wuasa's different composition of local - migrants, proportion of Bugis as the largest group of ethnic migrants, and land concentration, but also because Wuasa, being the capital of subdistrict Lore Utara, is able to provide more non-farm works

the Western border of the village is their *powanuangaa*, or inherited forest (*hutan waris*)[14].

The formalization of this claim fell into the responsibility of the village head, who was assisted by other village elites. Although his economic power, like other fellow locals, is limited, the village head still has a political power. And as Stewart (1998) pointed out, "political power is an important instrument of economic power, setting the rules and determining allocation of.... economic resources". In the past, under the collective regime of land ownership, the village head – or the leader of a pioneer settlement – had authority over the distribution of land to ensure the land-based livelihood of his/her people. Land as a collective resource belonged to the community; therefore settlers needed approval when clearing forests. Though one might clear as much forest area as one could, the leader (chief of pioneer settlement, or village head) had the discretionary power to grant cultivation rights. This distributive authority was no longer valid in the 1980s (Sitorus 2002) due to privatization of land ownership.

The village head's way to re-enact his land distributive authority, hence exercising his political power, was by issuing SKPT (*Surat Keterangan Pengolahän Tanah* or Letter of Land Utilization)[15]. The letter is not officially accepted by the supra-village government institutions, especially by BPN (*Badan Pertanahan Nasional* or National Agency for Land Affairs). The argument put forth by the Village Head was, that such letter is not a document of land-ownership but of land-use, and that it aims at preventing borderline disputes between cultivators. The "illegal" activity of cultivating forest land was thus formalized.

To further formalize the activity, the conflicting claims over the LLNP border were put forth as a way to reclaim village territory. Forest margins as buffer zones, in this sense, become politically contested areas. This action was timely, as similar claims were made by other forest-margin villages, while the Dongi-dongi case of large-scale forest encroachment was in the news. In Watumaeta, under the pressure of *Lembaga Adat* (Customary Council) and BPD (*Badan Perwakilan Desa* or Council of Village Representatives), a team was formed to deal with the distributive aspects among the landless locals. Forest-land distribution is seen as a solution to problems of social inequality and socio-economic insecurity. In effect, it only formalizes existing encroachment and more forest clearing.

For the Bugis, this recent acquisition of forest-land by the distributive authority of the village head is not a preferable way to acquire land. Legality

---

[14] Powanuangaa is the opened forest-land which is not immediately cultivated, but reserved for inheritance and expansion. The folklore explained that *powanuangaa* is the ancestors' dwellings, as indicated by remnants of stone mortars, stones for housing-foundation (*umpak*), and coffee trees. With coffee trees, such claim is without foundation, since only young trees are found in the *powanuangaa* forest

[15] Each applicant is charged between Rp.50,000 and Rp.100,000 for each hectare of forest area demanded for cultivation

of land holding is important for the Bugis to assure continuation of farming in their new residence. They also regard the clearing of forest area as time consuming, labor intensive, and costly. Purchasing is more efficient than grabbing. However, there are cases of Bugis owning land inside LLNP. Especially among the more recent Bugis migrants, who came when the scarcity of land was already felt, SKPT seems to be accepted as a legal document. Another reason is that by 2004 forest encroachment was so widespread, that everybody seemed to be convinced that their newly acquired plot would somehow be legalized.

In the case of Watumaeta, former Poso refugees took to clearing up forest since 2002. They argued that they did it because they had no capital to buy land, having been stripped of all resources while fleeing conflict areas, and that without land to work on they would always depend on government aid. Remarkably, the activities of these refugees – who are among the more recent migrants – are endorsed by locals, perhaps because the same religion puts them as the 'we' people.

4 Implications

A forest margin area as a buffer zone not only fulfills an ecological role, i.e. to preserve biodiversity of natural forest ecosystems as they interact with ecosystems modified by humans, but also plays a political role. A buffer zone is a contested area of different actors (subsistence locals, commercial migrants, local government and the State), whose decisions may affect forest use or conversion, and power balances among ethnic groups. As a result, a forest becomes an arena of ethno-political action.

In Sintuwu and Watumaeta, the 'we' – 'they' distinction between ethnic groups is formed by social-spatial distances as indicated by house-clustering, different religions, and ethnic proportion. It is then solidified as economic power shifts by way of more capitalistic mode of production and expansive nature to land resources. All these put the Bugis migrants as the distinct 'they'. In a land-based economy, land scarcity leads to declining socio-economic security. Social destabilization occurs as ethnic frictions result from this shift in resource control.

In the absence of ethnic politics at national level[16], people who live at the forest margins – and face serious threat of ethnic conflict due to rapid changes of access to socio-economic resources – have to innovate ethnic politics at a local level. Re-securing the economic base of the 'we' by exploiting their political power is done by distributing forest-margin land – an action perceived by the State as illegal encroachment. This innovation is a means to neutralize destabilization of inter-ethnic relation. Yet, social (re)stabilization is reached at the expense of forest margins (ecological) destabilization.

[16] Tirtosudarmo 2001 – see discussion at the beginning of the paper

Can such innovation continue? The cases of Sintuwu and Watumaeta reveal that it cannot go on endlessly. Once formalized, the 'we' opened up new economic space (land "distribution") will most likely be taken over by the 'they' (land purchasing), leading to a vicious circle of forest margin destabilization.

The main implication of this is the need to recognize forest encroachment as an economic- as well as a political action. Its impact on forest margins is environmental, as well as political. Therefore, any land-use modeling should take into account the social constraints of inter-ethnic relations and land-use systems. Disregarding this variable in the modeling is ignoring a reality that can trigger, and at times has triggered, ethnic conflicts.

Acknowledgements
The authors wish to thank Chip Fay, Senior Policy Analyst at the World Agroforestry Center, based in Bogor, for his valuable inputs and language-editing while revising this paper.

References

Abdulkadir-Sunito M (2004) *Orang Kampung* and *Pendatang*: Analysis of Demographic Structure and Migration in Two Forest-Margin Villages, Central Sulawesi. In: Gerold G, Fremerey M and Guhardja E (eds) Land Use, Nature Conservation and the Stability of Rainforest Margins in Southeast Asia. Springer, Berlin

Charras M, Pain M (Eds) (1993) Spontaneous Settlements in Indonesia: Agricultural Pioneers in Southern Sumatra. Orstrom-CNRS/Departemen Transmigrasi

Van Dijk K (2001) A Country in Despair: Indonesia Between 1997-2000. KITLV Press, Jakarta

Elmhurst R (1999) Space, Identity Politics and Resources Control in Indonesia's Transmigration Programme. Political Geography 18: 813-835

Human Rights Watch (1998) Indonesia: Communal Violence in West Kalimantan. Report 9(10), Dec.1997 (http://www.hrw.org, 9/19/02)

Li TM (2002) Keterpinggiran, Kekuasaan, dan Produksi: Analisis terhadap Transformasi Daerah Pedalaman (*Marginality, Power, and Production: Analysis toward Indonesian Upland Transformation*). In: Li TM (ed) Proses Transformasi Daerah Pedalaman di Indonesia (*Transforming the Indonesian Uplands*). Obor, Jakarta

Light Jr D, Keller S (1982, 3rd edition) Sociology. Alfred A.Knopf, New York

Naughton-Treves L, Salafsky N (2004) Wildlife Conservation in Agroforestry Buffer Zones: Opportunities and Conflict. In: Schroth G, da Fonseca GAB, Harvey CA (eds) Agroforestry and Biodiversity Conservation in Tropical Landscapes. Island Press, Washington-Covelo-London

Ruff F, Schroth G (2004) Chocolate Forests and Monocultures: A Historical Review of Cocoa Growing and Its Conflicting Role in Tropical Deforestation and Forest Conservation. In: Schroth G, da Fonseca GAB, Harvey CA, Gascon C, Vasconcelos HL and Izac AN (eds) Agroforestry and Biodiversity Conservation in Tropical Landscapes. Island Press, Washington-Covelo-London

Schroth G, da Fonseca AB, Harvey CA, Vasconcelos HL, Gascon C, Izac A-MN (2004) Introduction: The Role of Agroforestry in Biodiversity Conservation in Tropical Landscapes. In: Schroth G, da Fonseca GAB, Harvey CA, Gascon C, Vasconcelos HL and Izac AN (eds) Agroforestry and Biodiversity Conservation in Tropical Landscapes. Island Press, Washington-Covelo-London

Sitorus MTF (2004) *Revolusi Cokelat*: Social Formation, Agrarian Structure, and Forest Margins in Upland Sulawesi, Indonesia. In: Gerold G, Fremerey M and Guhardja E (eds) Land Use, Nature Conservation and the Stability of Rainforest Margins in Southeast Asia. Springer, Berlin

Stewart F (1998) The Root Causes of Conflict: Some Conclusions. QEH Working Paper Series no.16. Oxford University, London

Sudagung HS (2001) Mengurai Pertikaian Etnis: Migrasi Swakarsa Etnis Madura ke Kalimantan Barat (*Analyzing Ethnic Conflict: Voluntary Migration of Madurese Ethnic to West Kalimantan*). ISAI-Yayasan Adikarya Ikapi-The Ford Foundation

Sunito S (forthcoming) Continuation and Discontinuation of Local Institution in Community Based Natural Resource Management. Dissertation – University of Kassel.

Syahyuti (2002) Pembentukan Struktur Agraria pada Masyarakat Pinggiran Hutan: Studi Kasus di Desa Sintuwu dan Desa Berdikari, Kecamatan Palolo, Kabupaten Donggala, Sulawesi Tengah (*The Formation of Agrarian Structure in Forest-Margin Communities*). Master Thesis – Bogor Agricultural University (IPB)

Tirtosudarmo R (2001) Pengantar: Heterogenitas Etnik dan Konflik Komunal di Kalimantan Barat (*Introduction: Ethnic Heterogeneity and Communal Conflict in West Kalimantan*). In: Sudagung HS Mengurai Pertikaian Etnis: Migrasi Swakarsa Etnis Madura ke Kalimantan Barat. ISAI-Yayasan Adikarya Ikapi-The Ford Foundation

Assessing economic preferences for biological diversity and ecosystem services at the Central Sulawesi rainforest margin – a choice experiment approach

Jan Barkmann[1,4*], Klaus Glenk[1,4], Handian Handi[2], Leti Sundawati[2], Jan-Patrick Witte[3,4], and Rainer Marggraf[4]

[1] Junior Scientist Group "Economic Valuation of Ecosystem Functions and Services", Department of Agricultural Economics and Rural Development, Georg-August-Universität Göttingen, Platz der Göttinger Sieben 7, D-37073 Göttingen, Germany
[2] Laboratory for Social Forestry, Faculty of Forestry, Institute Pertanian Bogor, Bogor, Indonesia
[3] Landscape Ecology, Institute of Geography, Georg-August-Universität Göttingen, Goldschmidtstr. 5, D-37077 Göttingen, Germany
[4] Environmental and Resource Economics, Department of Agricultural Economics and Rural Development, Georg-August-Universität Göttingen, Platz der Göttinger Sieben 7, D-37073 Göttingen, Germany

*corresponding author: Jan Barkmann, Email: jbarkma@gwdg.de

Summary

Recognition of the importance of economically sound conservation strategies is one of the foundations of the Convention on Biological Diversity. Because of their exceptional contribution to global biological diversity, the conservation of the Central Sulawesi (Indonesia) rainforests is a particularly important case for a successful application of such strategies. One of the obstacles to the design and implementation of economically sound conservation strategies is the lack of knowledge on the economic value of non-market benefits generated by tropical forest ecosystems and the agricultural land use systems that replace them.

In this paper, we identify and quantify economically relevant preferences towards biological diversity at the rainforest margin that operate at the regional level around Lore Lindu National Park in Central Sulawesi. For economic valuation, choice experiments (CE) were used. We focused on indirect use benefits generated from different local and regional ecosystem functions.

Tscharntke T, Leuschner C, Zeller M, Guhardja E, Bidin A (eds), The stability of tropical rainforest margins, linking ecological, economic and social constraints of land use and conservation, Springer Berlin 2007, pp 181-208

An ecosystem service approach was applied to avoid problems of respondent unfamiliarity with ecosystem functioning. One CE study covers preferences for the conservation of the endemic dwarf buffalo anoa, the preponderance of shading trees in cacao plantations, and the availability of water and of rattan. It was conducted in 12 villages across the entire project region (n=249). A second CE was conducted in three selected villages in the Kulawi subdistrict investigating preferences for protection against soil erosion, flooding and – to our knowledge, for the first time ever – for protection against *uncertain* future environmental threats (preferences for "ecosystem health"; n=585).

The results indicate substantial economic preferences for all biodiversity and ecosystem goods and services studied. Preference variations can be explained by attitudes of and perceptions on the human-environment relation. In particular, (i) several attitudinal variables from Protection Motivation Theory explain preference heterogeneity; (ii) preferences for ecosystem health depend on stated respect for *Katawua* – a normative principle on harmony between humans, supreme beings, and the environment. For an optimistic resource and biodiversity protection scenario, total aggregated mean WTP in the project region is 1.6 billion IDR/yr$_{i}$ $\sim$136,000 €/yr). This local demand for the studied environmental non-market services does not suffice to compensate financial benefits from rainforest conversion or agroforestry intensification that accrue to individual households.

Keywords: stated preferences, economic valuation, ecosystem services, ecosystem health, water availability, rattan, anoa, cacao shading

1 Introduction

The recognition of the importance of economically sound conservation strategies is one of the foundations of the Convention on Biological Diversity (CBD). The Central Sulawesi (Indonesia) rainforests are part of the global Wallacea biodiversity hotspot (Myers et al. 2000), an important centre of species endemism (Waltert et al. 2004, Whitten et al. 1987). In particular, the Lore Lindu National Park (LLNP) harbors many of Sulawesi's endemic mammals and birds. Central Sulawesi and the area around the LLNP are an object of intensive international biodiversity conservation efforts, e.g., by the Indonesian branch of The Nature Conservancy (TNC). Because of their exceptional contribution to global biological diversity, the conservation of the Central Sulawesi rainforests is a particularly important case for a successful application of conservation strategies in line with the requirements of the CBD, and its ecosystem approaches.

One of the most severe obstacles to the design and implementation of economically sound conservation strategies is the lack of knowledge on the economic value of benefits generated by tropical forest ecosystems in comparison to the agricultural land use systems that replace them (Balmford et al.

2002). With regard to economic non-market values of Indonesian rainforests, only very few studies are available (e.g., van Beukering et al. 2003). One study commissioned by TNC on the value of the water resources of LLNP (ESG International 2001) does not meet economic valuation criteria as substitution effects are ignored (cf. Balmford et al. 2002 on criteria). Also, very few stated preference studies on the valuation of ecosystem services or of biological diversity have been carried out in rural Indonesia (one example: Pattanayak & Kramer 2001).

The application of stated preference techniques for economic valuation is a matter of dispute. With regard to the fundamental suitability of stated preference methods to provide information for societal decision-making, we have to refer to specialized contributions (e.g., Bateman & Willis 1999, Sudgen 2005). A particular challenge has been posed, however, with regard to the applicability of stated preference techniques to functional ecosystem values, such as the provisioning of water, or flooding and erosion control (Nunes & Bergh 2001). The critics argue that survey respondents are usually unfamiliar with the scientific description of the ecosystem functions on which the functional ecosystem values rest. Consequently, respondents are unable to make informed preference statements that can meaningfully inform decision-making. As a response to this critique, SFB-STORMA project A5 used a specifically developed *ecosystem service approach* to environmental valuation that addresses the critique that non-expert respondents lack sufficient insight into the scientific background of ecosystem functions (Barkmann et al. *subm.*).

Following a methods section (2), we present the main results for two choice experiments, one focusing on a representative coverage of the project area ("12-villages study"; 3.1), and one focusing on additional functional ecosystem benefits in two selected villages ("Toro/Salua-study", 3.2). In two specific analyses, we test if risk-related individual attitudes, and fundamental value orientations affect preferences (3.2). For both choice experiments, exemplary policy scenarios are assessed and discussed in section 4.

2 Methods

2.1 Study Area

The studies took place in the humid tropics in the area around Lore Lindu National Park (LLNP) in Central Sulawesi, Indonesia (1° 03'- 1° 57'S, 119° 57'- 120° 21' E; Fig. 1). LLNP covers some 2,200 km^2 of mountainous rainforest. An UNESCO Man & Biosphere reserve since 1978, the national park was officially established in 1993, and its permanent border fixed in 1999 (Maertens 2004). A large number of endemic species can be found in and around LLNP (Waltert et al. 2004) including the mammals anoa (*Bubalus spp.*), babirussa (*Babyrousa babyrussa*), and many endemic bird species. In the project area of 7,220 km^2, ~130,000 citizens – mainly smallhold farmers – live in about 120 villages.

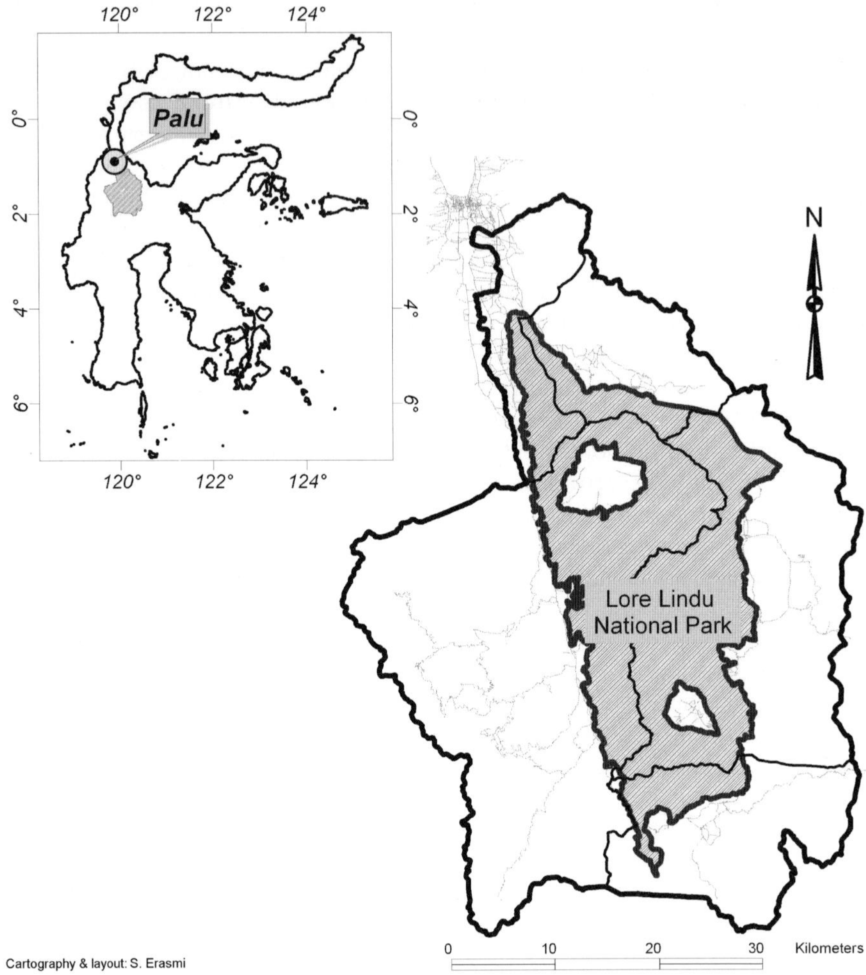

Fig. 1. Location of the project area, and the focus villages of Toro, Salua and Bolapapu.

2.2 The Choice Experiment Technique

The development of instructions for a standardized design of stated preference studies has been a research priority for nearly two decades. The findings of an expert panel convened by the United States National Oceanic and Atmospheric Administration (NOAA) in 1992 represent an important milestone (Arrow et al. 1993). Importantly, the NOAA panel recommended using dichotomous preference elicitation questions. In dichotomous question formats, respondents are not directly asked to state their maximum willingness-to-pay for some environmental improvement. Instead, they agree to or disagree to pay a certain sum for a mix of environmental improvements presented to them. This requirement is fulfilled by the choice experiment (CE) method.

CEs are based on Lancaster's attribute-based consumer theory (Lancaster 1966). Lancaster stressed that consumers do not have preferences for goods per se, but for the characteristics of the goods. In line with this idea, choice experiments 'decompose' the preferences for a complex good into a small number of components, each of which refers to one of the characteristics (*attributes*) of the good. All attributes are quantified by a small number of *levels*. For example, a village development and conservation program could be characterized by different impact levels on farmer access to extension services, on wild animal habitat, and on erosion protection.

After introducing the topic of the survey, attributes are explained to respondents. At the central stage of the choice experiment, respondents are asked to compare three cards on which different versions of the good are represented. If the good is a village development and conservation program, each of the versions represents a *scenario* of a potential future development characterized by a specific mix of attribute levels. Two of the scenarios are generated according to experimental design rules. If these two scenarios are not particularly labeled (e.g., "cacao intensification", or "conservation of forest gardens"), they form 'generic' scenarios. The third scenario implies that no program is implemented. To enable economic valuation, each scenario includes a monetary attribute. If scenarios tend to represent overall improvements of the situation, the monetary attribute is designed as an additional cost to the respondent to be paid when the program is implemented.The status quo scenario carries no additional cost. Respondents are asked to select the scenario that they like to be implemented subject to personal budget constraints.

Several detailed manuals on the application, design and analysis of choice experiments have recently been published (Louviere et al. 2001, Bateman et al. 2002, Hensher et al. 2005). For an overview of empirical valuation studies using the CE method, see Bennett & Blamey (2001).

2.3 Attribute construction

Particular care is needed to ensure that valuation scenarios are well understood by respondents (Bateman et al. 2002:119ff). With regard to functional ecosystem values, one of the most relevant objections against the application of stated preference methods is the unfamiliarity of respondents with the importance of changes in ecosystem functioning. According to Nunes & Bergh (2001), lay respondents will usually lack sufficient insight into "ecosystem life support functions and processes". Consequently, the average non-expert respondent is not able to make meaningful statements how much s/he is willing to pay for a change in the level of provisioning of an ecosystem function. Thus, Nunes & Bergh (2001 pp. 208f, 214) point out that the application of stated preference methods to the valuation of ecosystem functions is "less appropriate". The critique is well taken with regard to the valuation of ecosystem *functions* - understood as a scientific term that refers to functional aspects in the interrelated configuration of ecosystem structures, states and processes.

As a response to this challenge, we applied an *ecosystem service approach* to the valuation of functional ecosystem phenomena that translates ecosystem functions into benefits that directly relate to the life-world of Central Sulawesi stakeholders and respondents. We describe attributes and scenarios in terms of direct impacts of different village development and conservation programs on respondent lives. For example, scenario impacts on hydrological ecosystem functions are not described in hydrological terms. Instead, our hydrological attribute refers to a locally relevant ecosystem service, namely, changes in water availability during the dry season for irrigation of wet rice. The econometric advantages of this approach are briefly touched upon in section 2.6..

For applying the ecosystem service approach, we conducted a qualitative investigation of the pre-theoretic patterns of the perception of nature and the human-nature-relation (Barkmann et al. 2005*b*). Extensive qualitative pre-studies were carried out during field visits in 2003 and 2004. Data from the first STORMA phase did already indicate high awareness of several natural resource-related issues (Birner & Mappatoba 2002). Focus group interviews and 26 semi-structured individual interviews were conducted in several villages in the Kulawi and Palolo subdistricts. We found high awareness of functional (indirect) benefits provided by forested lands including the provisioning of rattan and water, erosion and flood protection (Fig. 2).

For the **12-village study,** four environmental attributes were selected: availability of rattan and water, preponderance of shading trees in cacao plantations, and population size of the endemic dwarf buffalo 'anoa'. All attributes except shading in local cacao plantations can be classified as *demand-driven* as they showed up strongly in the focus groups/interviews (Blamey et al. 1997:7). Cacao shading was included as a *policy-driven* attribute: While previous research had indicated the importance of shade tree cover for biological diversity (e.g., Perfecto et al. 1996), a trend for the removal of shading trees in cacao plantations had become apparent in the project region. We decided to complement the ecological studies on a gradient of cacao agroforestry intensification (see Tscharntke et al. this volume) by an investigation of local farmer preferences for the preponderance of cacao shading at the village level.

Attribute levels were selected that mirror the range of realistically expectable changes/improvements by a village development and conservation program. Because of heterogeneous environmental conditions, the status quo situation differed strongly between villages in respect to most attributes. Thus, we asked respondents to compare the current situation with attribute levels, and to identify the attribute level that most closely described the local situation for the water, rattan and the cacao shading attributes (self-explicated status quo). The cost attribute was split-sampled: (i) a rise in house land tax *versus* a donation to a village funds; (ii) a monthly *versus* a yearly payment scheme. A third split sample was used to assess ordering effects. Results of the

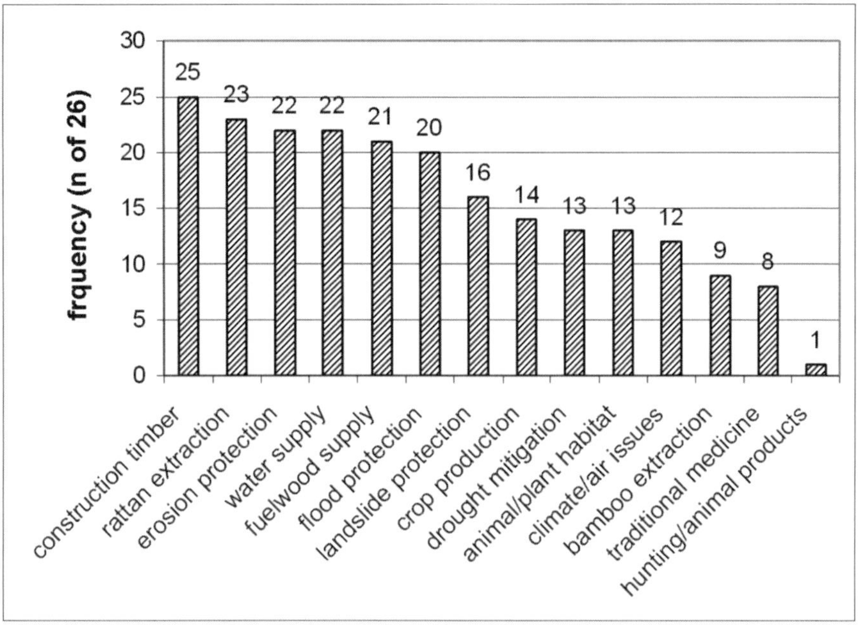

Fig. 2. Perceived forest ecosystem benefits: prestudy results of 26 in-depth interviews conducted in the Kulawi and Palolo Districts in 2003.

split-sample analyses will not be reported here. For a detailed specification of attribute levels, see Glenk et al. (2006*a*).

For the **Toro/Salua study**, a focus group meeting on unknown future threats to the human-nature relation was carried out in September 2004 in Toro. Protection against such threats is a commandment of sustainable development in the face of a fundamentally uncertain future (Barkmann & Windhorst 2000; "*land health*" attribute). Materially, protection against unknown threats to the human-nature relation can only be expected by protecting and strengthening the most essential factors on which ecosystem self-organisation depends (Kutsch et al. 2001, Barkmann et al. 2001). In line with the ecological insurance hypothesis (Yachi & Loreau 1999, McCann 2000), one of these factors is the availability of biological information. Three additional factors are: the availability of (i) energy, and (ii) of water and nutrients, as well as (iii) the current degree of ecosystem self-organisation. Economically, ecological insurance provides insurance benefits - including insurance benefits in the face of unknown threats to the human-nature relation (Barkmann & Marggraf 2004). Preferences for this conception of land health had never been assessed with stated preference methods.

The focus group yielded encouraging results with respect to the applicability of a land health attribute: During the focus group presentation, one of the participants spontaneously identified the suggestion to protect ecosystem self-organisation in order to protect the long-term flow of ecosystem services with the traditional Katawua-Principle of the Kulawi people. All other local participants agreed with this interpretation. Katawua is a normative principle of harmony between humans, nature and supreme being(s).

For quantifying the demand for land health, a specific CE design approach was applied. In the frame of the instrument, we first explained two well-known environmental threats – flooding and erosion – to respondents. Using the health metaphor, we gave a simplified explanation of the self-organisation capacity of "the land", which constitutes "land health". Respondents were informed that land health can protect against *unknown* environmental threats the same way a healthy body provides some protection against the impact of *unknown* diseases. Next, we included exactly the examples given for well-known environmental dangers – i.e., flooding protection and erosion protection – *as attributes in the CE alongside the land health attribute.* This way, respondents can make trade-offs between additional protection against two well-known environmental dangers and additional protection against unknown environmental dangers. As a result, we reduce the potential problem that the well-known benefits of the examples carry over to the target preferences for land health. For details of the specification of the attribute levels, see Witte (2006).

Before application, CE instruments were repeatedly pre-tested in both studies to improve the verbal and visual representation of attributes and attribute levels. We heavily relied on visualizations crafted by a local artist (e.g., see Fig. *3a/b*).

2.4 Attitudinal analysis

Attitudinal variables are included in stated preference studies (i) to test for expectations-based constrict validity of the utility coefficients on which WTP estimates rest (Bateman et al. 2002 p.340; see section 2.6.), and (ii) to explain preference variations. Below, we present two sets of analyses with attitudinal variables. The first set focuses on the performance of risk-related attitudinal constructs based on the data of the 12-village study. The second set addresses the influence of approval of the Katawua principle on preferences for land health based on the Toro/Salua study.

Does risk perception explain preference variations for ecosystem services and biodiversity?

Standard procedures in stated preference research can be improved by a theory-based treatment of WTP as a behavioral intention in line with social psychological theories. One such theory, Protection Motivation Theory focuses on risk perceptions and coping perceptions of respondents (PMT; Rogers & Prentice-Dunn 1997). PMT supplies a suitable foundation for the construction of attitudinal variables because demand for many ecosystem services is

Fig. 3. a) Artist sketch on local availability and extraction of the forest vine rattan (*Calamus spp.*) as an ecosystem service; b) artist sketch on degree of shading in cacao agroforestry representing 65% canopy cover by shading trees.

motivated by risk and uncertainty considerations. PMT variables were successfully used to explain WTP for biodiversity protection previously (Menzel 2003, Cerda et al. 2006).

Within the 12-village household study, 38 PMT items on each of the attributes and, additionally, on forest-related risks and coping perceptions were sampled. Among these, we operationalised an efficacy scale via items on the perception of trust in successful government action that respondents ascribe to their peers. Factor analysis singled out a group of four items that tends to emphasize the local context of government action.

We had set up an a priori hypothesis that risk perception and coping variables from PMT explain a substantial share of preference variation. We test the hypotheses by calculating interaction terms of PMT variables with attributes (see section 2.6.1), and comparison of model fit between models that either do (Tab. 1, model d) or do not include PMT variables (Log Likelihood ratio-test). Testing of interactions was restricted to matching attributes; PMT items on water availability, for example, were interacted only with the water attribute. As changes affecting 'the forest' may be causally prior to the availability of rattan, water and the conservation status of anoa populations, we interacted forest-related PMT items with these attributes. The four-item scale on trust in government is interpreted as an indicator for self- and response-efficacy: Do I think that the money that I am willing to pay for any village development and conservation program will actually precipitate the promised effects, which requires some effective government action?

Table 1 contains two additional models. Model (b) features a quadratic form of the cacao shading attribute, and an interaction term between cacao shading and relative poverty (Zeller et al. 2006). Model (c) includes interaction terms of the rattan, water, and cacao attributes with dummy variables that indicate if a household is actively involved in rattan extraction, in wet rice or cacao farming. By comparing the log likelihood values of models (a) to (c) with the PMT model (d), the relative importance of the predictors for explaining choices can be estimated.

Do traditional beliefs in the harmony of humans, nature and supreme beings explain preference variations for land health?

For the Toro/Salua-study, several attitudinal items were included. One of the attitudinal questions addresses practical respect for the Katawua Principle. The Katawua-item asked for the importance of the principle in daily decision-making (5-point Likert scale answer format: 1: "not at all" important to 5: "very much" important). We had hypothesized that practical respect for the Katawua principle would correlate strongly with preferences for land health. This hypothesis is tested by calculating an interaction variable of the score of the Katawua item with land health attribute levels.

2.5 Sampling frame and administration of the studies

The **12-village household study** took place from December 2004 to March 2005 in 12 villages that constitute a stratified sample of the about 120 villages of the project area (n=301). An extensive pilot study took place in September 2004 (n=96). The 12-village household sample is also used for a panel data study of household economics and land use (Zeller et al. 2002). The survey was administered face-to-face by 6 well-trained local enumerators, who are all B.Sc. graduates from the provincial university in Palu (Universitas Tadulako). To minimize interviewer effects, enumerators were randomly assigned to households. The results reported here are based on a subsample of 249 respondents that did (i) complete the choice experiment successfully, and (ii) for whom also the full set of land use and attitudinal data could be collected in collaboration with the household panel data study. Stratification weighing factors are not applied in the following analyses.

The **Toro/Salua**-study was conducted from July to September 2005 (Toro: n=280; Salua: n=205). Preparatory pretests and the pilot study took place in May and June 2005 in Bolapapu (n=100). Random samples were drawn from household lists and randomly assigned to enumerators. For Toro and Salua, roughly every second household was interviewed. All three studies were conducted face-to-face by a team of intensively trained enumerators. The main study took place from June-September 2005. Because the choice experiment instrument was only minimally adjusted after the pilot study, we also present Bolapapu results.

2.6 Estimation of marginal and aggregate WTP values

Econometric background

The econometric analysis of CEs is based on Lancaster's 'characteristics' theory of goods (Lancaster 1966), and on random utility theory (RUT, McFadden 1973). We can express an individual's utility function U as a function of the provision with market goods represented by income Y, and the levels of provision with different non-market – in our case mainly environmental – goods Z_1 to Z_q. At the present situation (status quo: SQ, superscript 0), we obtain

$$U(\mathrm{SQ}) = U^0(Y^0; Z_1^0; \ldots; Z_q^0). \tag{1}$$

Only a subset of the universe of utility generating environmental goods, say Z_1 to Z_m (vector $\mathbf{Z}_m$), will be impacted by a village development and conservation program (scenario 1, S^1). Because the program scenarios offered during the CE include a tax increase, Y also changes. Focusing on the affected environmental goods, utility changes to

$$U(\mathrm{S}^1) = U^1(Y^1; Z_1^1; \ldots; Z_m^1) = U^1(Y^1; \mathbf{Z}_m^1). \tag{2}$$

Each choice scenario of the CE represents a different version of the village development and conservation program. Thus, each scenario implies a vector of changes versus the present situation, and forms a complex, multi-attribute good that can be treated in accordance with the characteristics theory of goods.

Because the full specification of the utility function is usually not known, U is broken down into a deterministic, observable part V, and an unobservable error part, e. This RUT specification results in

$$U = V(Y, \mathbf{Z}_m) + e(Y, \mathbf{Z}_m) = V_Y + \sum_m V_m + e(Y, \mathbf{Z}_m). \tag{3}$$

In practice, only those utility components are observable with respect to a specific CE implementation that are included as CE attributes (Z_1 to Z_g; vector $\mathbf{Z}$). This methodological restriction can result in systematically biased coefficient estimates because respondents may (i) *know* that the village development program results in additional changes among attributes relevant to them but not covered by the CE attributes (Z_h to Z_m), or (ii) they erroneously *assume* that the provision of some relevant environmental goods will change although they are not affected by the program (Z_n to Z_q). The ecosystem service approach used in this study can be viewed as an effort to reduce these problems. It describes the impact of the village development and conservation program directly in terms of the ecosystem services that respondents obtain, e.g., as changes in the availability of irrigation water during the dry season to focus respondent attention on the intended impacts on Z_1 to Z_g. If, instead, watershed afforestation had been offered to improve water availability, respondents are more likely to associate additional, unintended impacts on Z_h to Z_q with the scenarios offered. For example, watershed afforestation may have been associated also with improved availability of fuel wood, or with (non-existing) effects on rare medical herbs.

For the econometric estimation of the influence of the CE attributes on respondent choices, assumptions on the functional relation of Y and $\mathbf{Z}$ on the one hand, and the corresponding V terms on the other must be made. We indicate this assumption by replacing the observable utility components V by the product of a coefficient b and a function f of the attributes:

$$U = b_Y * f(Y) + \sum_g b_g * f(Z_g) + e(Y, \mathbf{Z}). \tag{4}$$

Individual choices do not only depend on scenario attribute levels but also on socio-economic and/or attitudinal parameters of the respondents. To include such influences, interaction terms of the respective variables with attributes are calculated. Interaction terms can improve the predictive capacity of the model, and are used to test hypotheses on socio-economic and attitudinal influences on preferences. For example, if a deterministic utility component V_1 is hypothesized to be a linear function of Z_1 itself plus an interaction term of Z_1 with an individually varying attitudinal variable A, V_1 can be expressed as

$$V_1(Z_1, A) = b_A * Z_1 * A + b_1 * Z_1 \tag{5}$$

with b_A: utility coefficient of the interaction term.

For the empirical estimation of the utility coefficients, respondents are asked to choose between three different village development and conservation scenarios. Respondents are assumed to compare the utility s/he obtains from the three scenarios concurrently presented, and then select the scenario with the highest utility. In the CE scenarios, Y does regularly not represent respondent income directly, but a change in income, for example, brought about by a tax rise. Defining the "do nothing" scenario as the present situation (status quo, cf. Eq. (1))[1], let us consider the choice between a scenario S^1 and the status quo. The status quo is assigned the deterministic utility V^0 (resulting from $\mathbf{Z}^0$ and 0 income change) while S^1 results in V^1 based on $\mathbf{Z}^1$, and Y^1 income change. The probability that a respondent chooses S^1 over the status quo, i.e. $P(U^1 > U^0)$, is

$$P[(V^1 + e^1) > (V^0 + e^0)] = P[(V^1 - V^0) > (e^0 - e^1)]. \tag{6}$$

Replacing the V terms (see Eq. (4)), we obtain after re-arrangement with $Y^0 = 0$:

$$P[(V^1 - V^0) > (e^0 - e^1)] = P[(b_y * \{f(Y^1) - f(0)\} + \sum_g b_g * \{f(Z_g^1) - f(Z_g^0)\}) > (e^0 - e^1)]. \tag{7}$$

By recording the decisions on systematically constructed sets of scenarios and a suitable specification of the distribution of the error term, maximum likelihood estimation procedures can be used to estimate the coefficients of the utility function parameters (vector $\mathbf{b}$)[2]. If interaction terms were included (see Eq. (5)), $\mathbf{b}$ also includes the coefficients of the interaction terms.

---

[1] Because of an apparently deteriorating trend in the provision with a number of ecosystem services, we initially considered defining the "do nothing" scenario in terms of its – negative – future consequences. The predictive capacity to generate sufficiently reliable prognostic values for the attributes of interest was not available, however.

[2] For applicability of the most simple Gumbel error distribution that results in a multi-nomial logit model, restrictive statistical conditions must hold (independence from irrelevant alternatives/"IIA" condition; Bateman et al. 2002 p.280; see next subsection). Preliminary data analyses had indicated a risk to violate this condition. Because Nested Logit (NL) analysis relaxes the IIA requirement, we used this estimation procedure. For a detailed description of the estimation of NL models, we have to refer to more specialised literature, e.g., to Hensher et al. (2005 pp.518ff).

The estimation of statistically significant attribute coefficients allows for the calculation of welfare measures. For attributes with a linear relation between V and Z, the maximum willingness-to-pay for a 1 unit change of scenario attribute Z ("marginal" WTP) equals the ratio of the respective coefficient b_g and the coefficient of the monetary attribute b_y:

$$mWTP(Z_g) = -\frac{b_g}{b_y}. \tag{8}$$

In practice, an alternative specific constant, ASC, is additionally included in the maximum likelihood estimation procedures. Using two generic options and a status quo option, the ASC can be coded as a dummy variable with ASC=1 for the generic scenarios and 0 for the status quo scenario ("non-status quo ASC"). The ASC picks up systematic choice variations between the status quo scenario and the offered alternatives that cannot be explained by the attributes.

The welfare changes that result from a move from the status quo situation SQ to a policy scenario S^1 is calculated as a *compensating variation* measure ($V^1 - V^0$; Marggraf & Streb 1997). Assuming an interaction term according to Eq. (5) with $\overline{A}$ = mean of A, the resulting maximum willingness-to-pay can be expressed as

$$WTP(\mathrm{SQ} \rightarrow \mathrm{S}^1) = \\ -\frac{b_{ASC} + b_A * \overline{A} * \{f(Z_1^1) - f(Z_1^0)\} + \sum_g b_g * \{f(Z_g^1) - f(Z_g^0)\}}{b_y * f(Y^1)} \tag{9}$$

for an 'average' household.

Model estimation

For the Central Sulewesi CEs, linear functions are assumed for the relation between attributes and observable utility for most attributes. One exception is the cacao shading attribute in the 12-village study. It appears plausible to include a quadratic shading term reflecting a unimodal utility curve with an optimal shading level. For specific analysis, we calculate interaction variables with socio-economic and attitudinal variables. All models include a non-status quo alternative specific constant (ASC).

Because Nested Logit (NL) analysis relies on less restrictive assumptions, we applied this maximum likelihood estimation procedure (also see footnote 2). For an eligible NL tree structure, the inclusive value was set to 1.0 for the degenerated branch, and all models were initiated with starting values obtained from a non-nested NL model (Hensher et al. 2005 pp.530ff). All scale parameters were normalized at the lowest level ("RU1"). Estimation of NL models was carried with NLOGIT 3.0.

For the analysis of policy scenarios, we enter status quo and scenario values into in linear form of Eq. (9) (i.e., $f(Z) = Z, f(Y) = Y$). To calculate the

WTP contribution of the quadratic cacao attribute in the 12-village study we use $f(\%\text{ cacao shading}) = (\%\text{ cacao shading})^2$. Resulting values are added up, and multiplied by the number of respondent households in the project area to generate a first approximation of aggregate WTP in the project region. Protest responses are regarded as zero WTP statements.

Normative problems of aggregated WTP estimates

The direct utilization of aggregate WTP values for decision advice carries a controversial normative loading. Even if based on a truly representative data set, accounting for all market and non-market benefits, and calculated with utmost technical precision, the aggregated WTP value does not provide a decision criterion that meets the requirements of Pareto-optimality (Hanley & Spash 1993). A Pareto-improvement is achieved if at least one individual is better off while no-one is made worse off. No matter how big the aggregate advantage for all others, if just one individual suffers a deterioration of her/his economic status, a Pareto-improvement is not achieved. In fact, differences in aggregated WTP are designed to meet only the weaker Kaldor-Hicks criterion of *potential* Pareto-optimality (Jakobsson & Dragun 1996 p.42). Potential Pareto-optimality means that those "winning" individuals who will be better off after the implementation of some project *could* compensate all losing individuals, and still be better off themselves. If this compensation is actually paid, however, is ignored. This has obvious distributional implications – which are highly problematic to ignore when individuals are affected who are poor in absolute terms.[3]

Because poverty is evident in the Central Sulawesi project area (van Rheenen et al. 2004), we tested for the influence of relative poverty (Zeller et al. 2006) on preferences. Although there are – statistically non-significant – hints pointing at interesting non-linear interactions of relative poverty with preferences for rattan, water, and anoa (Glenk et al. 2006*b*), only a linear interaction of relative poverty with cacao shading could be clearly documented (Tab. 1, model b). These results indicate that wealth differences are not likely to grossly bias aggregate WTP estimates for the non-market benefits of the investigated ecosystem services. Further potential distributional problems that may be inherent in the actual implementation of any of the investigated scenarios – for example on market goods – have, yet, to be analyzed however.

[3] Additionally, the international discussion on public choice alternatives to standard welfare theory points at the importance of the initial distribution of property rights with respect to natural resources and biodiversity – even challenging the normative foundation of Pareto optimality. A more detailed discussion of the limitations and advantages of standard welfare economic theory is beyond the scope of this paper, however. A biodiversity focused discussion is provided by Jakobsson & Dragun (1996 pp. 35-49).

3 Results

3.1 Quantification of preferences for ecosystem services: the 12-villages study

All calculated models are overall highly significant (Tab. 1; $P(\chi^2) < 0.0001$). The base model (a) displays mostly the expected signs: disutility for longer distance to the next rattan site, disutility of worse water availability, positive utility for bigger anoa populations, disutility of higher costs. However, shade in cacao plots is valued *negatively* on average. A more detailed look at the cacao attribute (model b) reveals that the negative influence can be decomposed into one component representing an inverted U-shaped utility function (sign of the quadratic cacao term), and a positive linear term. Optimum shading conditions are perceived towards the lower end of the shading gradient (Fig. 4). A linear correlation of relative poverty can be found for the cacao shading attribute ($p < 0.05$; Tab. 2, model b) resulting in differing optimal shading levels for the different poverty groups (Fig. 4). Interaction terms with relative poverty did not become significant for any other attribute.

We base non-market valuation on a simple, economically optimistic ad hoc policy scenario (Tab. 2), and on model (b) parameter estimates. The scenario assumes that the next rattan sites are 10 km closer to the villages than currently, water shortage is reduced by one month, cacao shading is reduced by 20%, and anoa numbers increase by 150. Estimated total aggregated WTP for the scenario changes is about 1,565 million IDR/yr (~136,000 €/yr). Contributions to this total value range from about 98 million IDR/yr for the shading reduction to about 873 million IDR for improved water availability.

3.2 Quantification of preferences for ecosystem services: the Toro/Salua study

In the four models of the Toro/Salua study (Tab. 3), most ecosystem service attributes and the cost attribute are significant at least at the 0.1 % level, and display the expected sign. The only exception is "land health" in Salua ($p = 0.265$). The interaction of land health with Katawua is also for the Salua data highly significant, however ($p = 0.005$). The resulting mean WTP/household/year is calculated for an exemplary scenario: flooding damage -25%; soil loss by erosion: -25%, and implementation of additional measures to keep "land health" constant for "many years".

While considerable variability exists among villages, the stated economic importance of protection against unknown environmental dangers is very high on average: Total WTP for land health is regularly higher than WTP for flooding or erosion protection, and accounts for clearly more than 1/3 of the combined WTP for the investigated environmental improvements.

Table 1. Valuation of biodiversity-related economic goods and services in Central Sulawesi (NL results; n=249, STORMA 12-villages household sample)

Variable	(a) linear attributes only	(b) linear + cocoa^2 + cocoa * poverty index	(c) linear + interaction with use dummies	(d) linear+ interaction with PMT variables
Rattan availability	-0.0392***	-0.0468***	-0.0276**	-0.0645*
Water for irrigation of wet rice	-0.9303***	-0.9725***	-0.5289***	-0.0586
Cacao Shade (linear)	-0.0104***	0.0194**	-0.0016	-0.0121***
Anoa Population Size	0.0012***	0.0013***	0.0013***	-0.0008
Cost (Tax rise/village fund donation)	-0.0251***	-0.0248***	-0.0263***	-0.0121
ASC (non-status quo choice)	0.3471*	0.5918***	0.2661	0.3117
Cacao Shade (quadratic)		-0.00032***		
Cacao*relative poverty index		-0.0058*		
Rattan*rattan collector			-0.1039***	
Water*wet rice involvement			-0.7194***	
Cacao*cacao farmer			-0.0118**	
Rattan*Rattan Value				0.0279**
Rattan*Probability Rattan Loss				-0.0278**
Water*Water Value				0.2233***
Water*Severity of Forest Loss				-0.1044*
Cacao*Cacao Value				0.0045*
Cacao*Severity of Cacao Problem				0.0058**
Cacao*Probability Cacao Problem				-0.0061***
Anoa*Severity Anoa Reduction				0.0016**
Cost*Response Efficacy Government				0.01196***
Log-likelihood	-742.3	-729.9	-706.3	-695.1
P(Chi^2); DF	<0.0001; 7	<0.0001; 9	<0.0001; 10	<0.0001; 16
IV of degenerated branch	0.9135	0.8389	0.9226	0.8974
Adj. ϱ^2 (Pseudo-R^2) §	0.261	0.273	0.297	0.305

***: significant at $p \leq 0.001$; **: significant at $p \leq 0.01$; *:significant at $p \leq 0.05$; cost coefficients for 1,000 IDR/yr; §Pseudo-R^2 (constants only) values between 0.2 and 0.4 correspond to R^2 values of 0.7 to 0.9 value in OLS regression.

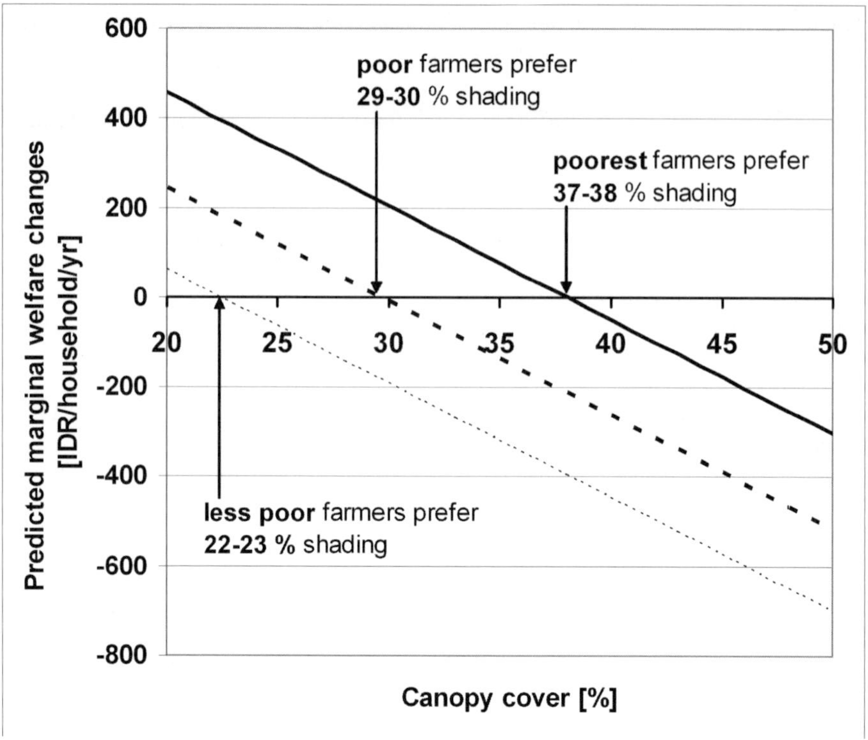

Fig. 4. First derivative of inverted U-shape preference function for shading in cacao agroforestry: relative position of the shading optima of poverty terciles according to relative poverty index (Zeller et al. 2006).

3.3 Attitudinal hypotheses

Influences of risk perception

From the 38 PMT items sampled, 9 interaction terms could be generated that are significant at least at the 10%-level (1 four-item-variable, 5 two-items variables, 4 single-item-variables). In total, model (d) (Tab. 1) includes 18 of the initially sampled 38 items. Compared to the linear base model (a), inclusion of PMT variables improves pseudo-R^2 substantially from 0.261 by about 0.04 to a very reasonable 0.305 in model (d). A Log Likelihood ratio test confirms a significant improvement of model fit ($P<0.0001$). Model fit is also improved in comparison with model (b), and with model (c) that includes (rough) production-oriented predictors of respondent preferences ($P=0.002$).

In general, value, severity and probability variables become significant. Additionally, the efficacy scale shows a clear and considerable influence on the disutility of program cost. Only one of the interaction variables with the forest-related PMT items influenced choices (water*severity of forest loss).

Table 2. Marginal non-market WTP and exemplary policy scenario calculation based on Tab. 1 model (b) statistics.

	Rattan	Water	Cacao	Anoa
Marginal Unit	harvesting location 1 km closer	1 month less with water scarcity	1 % less shading in plantations of village	1 individual more
Marginal mean WTP/household [IDR/yr]	1,884	39,145	408$	52
Scenario Change	-10 km	-1 month	-20 %§	+150
Mean scenario WTP/household	18,837	39,145	4,384#	7,796
Aggregated scenario WTP(scenario)/project area (Mio. IDR/yr)	420.1	872.9	97.8	173.9

Aggregation based on 23,307*(301-13)/301 = 22,300 households in the project region that did not file protest responses; 1 € ~ 11,500 IDR (winter/spring 2005); $value calculated for household with mean relative poverty index; §for respondents perceiving 5 % shading, scenario change = -5%; #because of non-linear preferences for cacao shading, mean scenario WTP ≠ marginal mean WTP*scenario change.

For water, anoa and cost, the interaction terms are so good predictors, that the corresponding linear terms of the attributes do not become significant. In other words, the entire influence of the attribute on choices can be explained by the interaction term alone.

Influence of stated respect for Katawua on land health

Residents of the three sampled villages of the Toro/Salua study report, on average, a practical respect score for the Katawua principle of 4.03 (Katawua is "much" important). Mean values for the single villages range from 3.83 to 4.27. The interaction of land health and practical respect for the Katawua principle is a most highly significant predictor of choice for the entire sample of the Toro/Salua-study (Tab. 3, model a) as well as for the models for the single villages of Bolapapu (b), Toro (c), and Salua (d) ($p<0.001$ in all cases; not all data shown). Thus, traditional beliefs in the harmony of humans, nature and supreme being(s) do explain preference variations for land health.

For the entire sample as well as for the villages of Bolapapu and Toro, total mean annual household WTP for land health is composed of a negative constant (-47,000 to -76,000 IDR), and a positive contribution of the interaction term with the Katawua item. Calculated for Katawua mean values, the contribution of the interaction term is highest for Toro (~152,000 IDR) and lowest for Bolapapu (~55,000 IDR) with an average of ~118,000 IDR for the entire sample. Based on these values, residents with a self-reported

Table 3. Mean WTP/household and exemplary scenario calculation (NL analysis; scenario: flooding damage -25%; soil loss by erosion: -25%; implementation of additional measures to keep "land health" constant for "many years"

Attributes	coefficients	Mean scenario WTP [IDR/household/yr][§]			
	(a)	(a)	(b)	(c)	(d)
	ALL	ALL	Bolapapu	Toro	Salua
Cost (tax increase)	-0.00027***				
Flood protection	0.0550***	60,112	75,691	55,385	62,989
Soil Erosion Protection	0.0530***	57,922	62,404	54,679	65,321
(i) Ecosystem Risk Protection ("land health")	-1.0808***	-47,246	-76,044	-66,379	23,259
(ii) interaction land health*Katawua[#]	0.6663***	117,531	121,134	152,460	54,719
Sum "land health" WTP (i) + WTP (ii)		70,285	45,090	86,081	77,979
Non-Status Quo (ASC)	-0.1510	-6,734	24,426	7,083	-49,959
Total Scenario WTP		181,716	207,611	203,228	156,329
Katawua mean score	4.04		4.27	4.16	3.86
Katawua "break even" score	1.62		2.68	1.81	n.a.[$]
IV of degererated brach	0.5895		0.9374	0.5352	0.4106
N; Adj. ϱ^2 (Pseudo-R^2)	585; 0.131		100; 0.177	280; 0.145	205; 0.106

***: significant at p≤0.001; for models (b), (c) and (d), p≤0.01 for all coefficients with the following exceptions: model (b): Non-Status Quo (p=0.037), model (c): Non-Status Quo (P=0.440), model (d) land health (p=0.265); [§]for details of the calculation of WTP values, see text; [#]WTP values are calculated for an 'average' household based on Katawua mean scores of the respective villages; [$]not applicable as value for (i) land health is positive.

practical respect for Katawua below 1.62 in the entire sample (between "not" and "hardly" important) to 2.68 in Bolapapu (between "hardly" and "a bit" important) are not expected to have a positive WTP for securing land health. All other residents are expected to be willing to pay for improved protection of land health in the sense of protecting the self-organising capacity of local ecosystems.

4 Discussion

4.1 Applicability of the choice experiment methodology

For the 12-village study as well as the Toro/Salua study, a choice experiment approach to the quantification of environmental preferences could be applied. The studies investigated preferences for ecosystem services such as the provision of water or protection against erosion and flooding risks. These are ecosystem services that rely on regulation functions at the ecosystem level. This is a non-trivial result because the applicability of stated preference methods to the economic assessment of functional – particularly hydrological - ecosystem values had been questioned (e.g., Nunes & Bergh 2001).

This positive assessment is supported by the results of the attitudinal analyses. Risk and coping perceptions from PMT explain a substantial share of the surveyed preferences in the 12-villages-study as hypothesized a priori. Also the significance of the (rough) indicators of economic activity (Tab. 1, model c) on preferences suggests consistency in respondent choices with the necessities of household farming activities. Similarly, preferences for land health depend on stated practical respect for the Katawua Principle in daily life as expected. Failure to document clear evidence for a correlation of preferences with individual attitudes would have falsified the expectations-based construct validity assumption that preferences are an expression of more fundamental attitudes and value orientations.

Conceptually, we attribute the successful application to an explicit ecosystem service approach (Barkmann et al. *subm.*) to the design of the CEs. The strategy to translate the impact of changes in ecosystem functions into the everyday language of benefits that people can directly experience has worked sufficiently well. The ecosystem service approach necessitated an extended design process for the CE instrument that included qualitative pre-studies (individual in-depth interviews, focus groups), repeated extensive pre-testing including the drafting and optimization of visual aids (Fig. 3a/b), and a quantitative pilot study. In addition, the socio-demographic and ecological heterogeneity of the study region was accounted for by the utilization of a self-explicated status-quo alternative for the 12-village study. As a result, the vast majority of respondents was able to complete the choice tasks, and to make meaningful choices.

4.2 Interpretation of results from the 12-villages study

Substantial economic preferences were detected for all biodiversity and ecosystem goods and services. Based on the analysis of an ad hoc policy scenario, improved water availability by one month is perceived as 5 to 9 times as important as a reduction of shading in cacao plantations by 20%, or a higher population of the endemic dwarf buffalo anoa (+150 animals). Preferences for rattan stands 10 km closer to the villages are about half as strong as preferences for improved water availability. Aggregated WTP by the households in the project region for the implementation of this economically optimistic scenario is about 1.6 billion IDR/yr (~136,000 €/yr).

Mean WTP for one month of reduced water scarcity (~39,000 IDR/household/yr) can be compared to a WTP estimation for drought mitigation services by watershed protection on Flores/Indonesia calculated in a contingent valuation study (Pattanayak & Kramer 2001). This study reports a WTP of 2-3 $ for comparatively dry Flores Island. The value that is slightly higher than our estimate for the wetter Central Sulawesi project area. Although operationalisation differs somewhat, our results strengthen the conclusion by Pattanayak & Kramer (2001) that rural farming households in Indonesia have substantial economic preferences for protecting water resources at the ecosystem level.

For the interpretation of the presented aggregate WTP values, one has to keep in mind that the figures only represent preferences for selected non-market ecosystem services; they are not based on a comprehensive assessment of all cost and benefits associated with the implementation of the scenario changes. Consequently, our focus on local, income restricted WTP results in much lower benefit values as, e.g., calculated in a simulation study by van Beukering et al. (2003) for Leuser National Park on Sumatra that includes a much broader range of market costs and benefits.

Preferences for cacao shading show an inverted U-shape pattern with the highest optimum shading values for the poorest relative poverty group. With a few 1,000 IDR/household/year (Tab. 2), absolute welfare differences brought about by changes in shading in local cacao plots are small compared with the average gross margin differences per hectare between more shaded (usually less intensively managed) and less shaded (more intensively managed) plots. The removal of shading trees is one component of an intensification syndrome, however, that also includes increasing application of fertiliser and pesticides. For example, the average gross margins for the less intensive, more shaded, and ecologically more diverse "D" agroforestry plots in Toro are much lower than gross margins of the less shaded "E" plots (~3.8 versus ~6.3 Mio. IDR/ha; Schwarze, *pers. comm.*). Respondents have between 0.6 ha (mean poorest third) and 1.2 ha (mean least poor third) cacao plots. Thus, the above per hectare differences are about three orders of magnitude bigger than differences brought about by shading effects alone. With >13,300 ha of closed agroforestry in the project area in 2002 (Erasmi et al. this volume), an intensification equivalent to the transition from "D" to "E" plots on less 5 % of the agroforestry area would already exhaust total aggregate WTP for the presented environmental improvements of the ad hoc scenario.

4.3 Interpretation of results from the Toro/Salua study

In the Toro/Salua study, substantial WTP for the improvement of erosion and flooding protection as well as for the protection against as yet unknown threats to the human-nature relation (i.e., secured "land health") was found. Overall WTP of an average household for a 25% reduction in flooding and erosion risks and for safeguarding the current level of "land health" with ~182,000 IDR/yr amounts to about 6.6% of household cash income (Witte 2006). This is about 2.6 times the scenario WTP estimated for in the 12-villages-study. Because of limited geographical coverage and non-random selection of the studied villages, the Toro/Salua WTP values cannot be extrapolated to the entire project region, however.

The interaction of land health and importance of the Katawua-Principle in daily decision-making is most highly significant in all models (Tab. 3). Respondents who report only low levels of respect for Katawua do not have a positive WTP for ecosystem health. For the entire sample, the "break even" score for positive WTP values (1.63) is more than two Likert scale points lower than the sample mean (4.03).

The Toro/Salua-study represents the first empirical evidence for WTP for securing "land health". Securing land health safeguards the self-organisation capacity of ecological systems and its four fundamental constituents (see section 2.3). The benefits from land health are insurance benefits in the face of uncertain risks to the human-nature relationship. The only comparable study also detected substantial evidence for preferences for an insurance value in face of uncertain risks to the human-nature relationship (Cerda et al. 2006, Barkmann et al. 2005*a*). However, this study focused on only one instead on all four constituents of the capacity of ecological systems to self-organise.

4.4 Summary and Caveats

In sum, we showed the applicability of a CE approach to the economic valuation of functional ecosystem services in a developing country setting. Resulting WTP values indicate that, on average, inhabitants of the project region are interested in additional resource protection efforts even if they have to pay for these efforts. The construct validity of the WTP results is strengthened by several meaningful correlations of individual preferences and attitudinal as well as economic activity-oriented variables.

For a full utilization of the results presented here for decision-making in conservation and development planning, three factors must be given. First, we used *ad hoc* scenarios for WTP calculations. These scenarios are only insufficiently based on scientific research on the question which actual management intervention would result in which actual non-market effects.

Second, the aggregate WTP values do not account for distributional effects. At the theoretical level, this is a consequence of the fact that the aggregation of individual WTP values relies on the – welfare economically accepted, but normatively controversial - Kaldor-Hicks principle of *potential* Pareto optimality. For the 12-villages study, we could document, for example, that the utility of shading in cacao plots differs between respondents with different relative poverty. A program that would try to bring about the shading level favored by the average population could hurt the poorest respondents who prefer more shade than relatively better off respondents. For the other attributes, no clear statistical evidence was obtained for poverty influences. Still, the actual implementation of any contemplated village development and conservation options informed by our preference data may result in unacceptable distributional effects – for example on rattan collecting households (cf. Tab. 1, model c). Such undesirable effects need to be analyzed and minimized.

Erasmi S, Kappas M, Twele A, Ardiansyah M (this volume) From global to regional scale: remote sensing-based concepts and methods for mapping land-cover and land-cover change in tropical regions. In: Tscharntke T, Leuschner C, Zeller M, GuhardjaE, Bidin A (eds) The stability of tropical rainforest margins, linking ecological, economic and social constraints of land use and conservation. Springer, Berlin Heidelberg New York

ESG International Inc. (2001) The Value of Water Resources in Lore Lindu National Park, Central Sulawesi, Indonesia. Report prepared for The Nature Conservancy Indonesia Program, Jakarta (Indonesia)

Glenk K, Barkmann J, Marggraf R (2006*a*) Unveiling regional preferences for biological diversity in Central Sulawesi: a choice experiment approach. STORMA-Discussion paper series no 16, Göttingen and Bogor

Glenk K, Barkmann J, Schwarze S, Zeller M, Marggraf R (2006*b*) Differential influence of relative poverty on preferences for ecosystem services: Evidence from rural Indonesia. Paper presented at the 26th Conference of the International Association for Agricultural Economists (IAAE) "Contribution to Critical Political Issues", Brisbane (Australia) Conference on Environment and Development, August 12-18, 2006

Hanley N, Spash CL (1993) Cost-Benefit Analysis and the Environment. Edward Elgar, Aldershot (UK)

Hensher DA, Rose JM, Greene WH (2005) Applied choice analysis: a primer. Cambridge University Press, Cambridge (UK)

Jakobsson KM, Dragun AK (1996) Contingent Valuation and Endangered Species – Methodological Issues and Applications. Edward Elgar, Cheltenham (UK)

Kutsch WL, Steinborn W, Herbst M, Baumann R, Barkmann J, Kappen L (2001) Environmental Indication: A Field-test of an Ecosystem Approach to Quantify Biological Self-organization. ECOSYSTEMS 4: 49-66

Lancaster K (1966) A New Approach to Consumer Theory. Journal of Political Economy 74: 132-157

Louviere J, Hensher D, Swait J (2001) Stated Choice Methods – Analysis and Application. Cambridge University Press, Cambridge (UK)

Maertens M (2004) Economic Modelling of Agricultural Land-Use Patterns in Forest Frontier Areas. Ph.D. Thesis, Institute of Rural Development, Georg-August-Universität Göttingen

Marggraf R, Streb S (1997) Ökonomische Bewertung der natürlichen Umwelt: Theorie, politische Bedeutung, ethische Diskussion. Spektrum, Heidelberg

McCann KS (2000) The diversity-stability debate. Nature 405: 228-232

McFadden, D (1973) Conditional Logit Analysis of Qualitative Choice Behavior. In: Zarembka P (ed) Frontiers in Econometrics. Academic Press, New York, pp 105-42

Menzel S (2003) Der Beitrag der Protection Motivation Theory für die Zahlungsbereitschaftsäußerungen zur Erhaltung biologischer Vielfalt. Umweltpsychologie 7: 92-112

Myers N, Mittermeier RA, Mittermeier CG, Fonseca GABd, Kent J (2000) Biodiversity hotspots for conservation priorities. Nature 403: 853-858

Nunes P, Bergh Jvd (2001) Economic valuation of biodiversity: sense or non-sense? Ecological Economics 39: 203-22

Pattanayak K, Kramer R (2001) Pricing ecological services: Willingness to pay for drought mitigation from watershed protection in eastern Indonesia. Water Resources Research 37: 771-778

Perfecto I, Rice RA & Greenberg R (1996) Shade coffee: A disappearing refuge for biodiversity. Bioscience 46: 598-608

Rogers RW, Prentice-Dunn S (1997) Protection motivation theory. In: Gochman DS (ed) Handbook of Health Behavior Research. Plenum, New York, pp 113-132

Sugden R, (2005) Anomalies and Stated Preference Techniques: A Framework for a Discussion of Coping Strategies. Environmental and Resource Economics 32: 1-12

Tscharntke T, Leuschner C, Zeller M, Guhardja E, Bidin A (this volume) The stability of tropical rainforest margins, linking ecological, economic and social constraints of land use and conservation – an introduction. In: Tscharntke T, Leuschner C, Zeller M, Guhardja E, Bidin A (eds) The stability of tropical rainforest margins, linking ecological, economic and social constraints of land use and conservation. Springer, Berlin Heidelberg New York

van Beukering PJH, Cesar HSJ, Janssen MA (2003) Economic valuation of the Leuser National Park on Sumatra, Indonesia. Ecological Economics 44: 43-62

van Rheenen T, Elbel C, Schwarze S, Nuryartono N, Zeller M, Sanim B (2004) Encroachments on Primary Forest: Are They Really Driven by Despair? In: Gerold G, Fremerey M, Guhardja E (eds) Land Use, Nature Conservation and the Stability of Rainforest Margins in Southeast Asia. Springer, Berlin, pp 199-213

Waltert M, Mardiastuti A, Mühlenberg M (2004) Effects of land use on bird species richness in Sulawesi, Indonesia. Conservation Biology 18: 1339-46

Whitten AJ, Henderson GS, Mustafa M (1987) The Ecology of Sulawesi. Gadjah Mada University Press, Yogyakarta (Indonesia)

Witte J-P (2006) Determination of willingness to pay for "Ecosystem Health" in terms of prevention of unknown environmental dangers in Central Sulawesi (Indonesia) – A Choice Experiment Approach. Treffpunkt Biologische Vielfalt 6: 155-161

Yachi S, Loreau M (1999) Biodiversity and ecosystem productivity in a fluctuating environment: The insurance hypothesis. Proceedings of the National Academy of Sciences of the United States of America 96: 1463-1468

Zeller M, Schwarze S, Rheenen Tv (2002) Statistical Sampling Frame and Methods Used for the Selection of Villages and Households in the Scope of the Research Programme on Stability of Rainforest Margins in Indonesia (STORMA). STORMA Discussion Paper Series no 1, Göttingen and Bogor

Zeller M, Sharma M, Henry CJ, Lapenu C (2006) An Operational Method for Assessing the Poverty Outreach Performance of Development Policies and Projects: Results of Case Studies in Africa, Asia, and Latin America. World Development 34: 446-464

Forest Products and Household Incomes: Evidence from Rural Households Living in the Rainforest Margins of Central Sulawesi

Stefan Schwarze[1*], Björn Schippers[2], Robert Weber[2], Heiko Faust[2], Adhitya Wardhono[1], Manfred Zeller[3], and Werner Kreisel[2]

[1] Institute of Rural Development, Georg-August-University, Waldweg 26, D-37073 Göttingen, Germany
[2] Department of Cultural and Social Geography, Institute of Geography, Georg-August University, Goldschmidtstr. 5, D-37077 Göttingen, Germany
[3] Institute of Agricultural Economics and Social Sciences in the Tropics and Subtropics, University of Hohenheim (490), D-70593 Stuttgart, Germany

*corresponding author: Stefan Schwarze, phone: +49-(0)551 393906, Email: s.schwarze@agr.uni-goettingen.de

Summary

Tropical forests play a key role in the world carbon cycle and in maintaining biodiversity, but agricultural activities as well as the extraction of forest products are threatening these functions. Empirical evidence from developing countries suggests that forest products play an important role as a source of income for rural households, particularly for the rural poor. There is, however, still a lack of quantitative studies on the link between poverty and forest products. The research presented in this chapter seeks to fill the gap in general knowledge on the link between poverty, livelihood systems, and extraction of forest products. Considering as an example the vicinity of the Lore-Lindu National Park (LLNP) in Central Sulawesi/Indonesia, this chapter analyses the importance of forest products, especially for the rural poor, and identifies underlying factors which drive households into the forest. Moreover, the paper investigates similarities and differences in the use of forest products in the village of Toro, where an agreement with the national park authority on the use of forest areas exists, and in the research area at large, where such agreements did not exist.

In the vicinity of the LLNP, 76% of the households collect forest products, with firewood being the most important product. The sale of forest products contributes only 7% to the total household income of all households, with 17%

Tscharntke T, Leuschner C, Zeller M, Guhardja E, Bidin A (eds), The stability of tropical rainforest margins, linking ecological, economic and social constraints of land use and conservation, Springer Berlin 2007, pp 209-224

of the households participating in this activity. Almost three-quarters of the income from forest products originates from the sale of rattan. Differentiating forest product income by wealth groups shows the importance of forest products, especially rattan, as a source of income for the poorest households. 21% of the total household income of the poorest households originates from the selling of forest products and 30% of these households reported to have income from forest products. Participation in the sale of forest products is influenced by the wealth of the household, the area of land owned, education, ethnicity, and access to road infrastructure.

Based on participatory mapping, the area of the village of Toro is divided into six sections for which different use options are defined. In the so-called *pangale*, a 20-25 years old secondary forest, forest products can be collected for home consumption, but not for sale. Our empirical results reflect these regulations. 4 years after implementation of the village agreement, the share of households collecting forest products is significantly higher in Toro than in the research area. But, the share of households which sell their products as well as the mean sales value is much lower in Toro implying lower levels of extraction. This result is strengthened by the econometric analysis, which found that the village agreement has a strong negative influence on the likelihood of selling forest products beyond other factors.

Keywords: forest products, household income, poverty, village agreements, Central Sulawesi, econometric modelling

1 Introduction

Natural tropical forests play a key role in the world carbon cycle and in maintaining biodiversity, but agricultural activities as well as the extraction of forest products are threatening these functions. In this book, for example, the impact of human interventions on soil invertebrate fauna is investigated by Migge-Kleian et al. and on amphibian communities by Fiedler et al. Empirical evidence from developing countries suggests that forest products play an important role as a source of income for rural households. Besides timber, also non-timber forest products provide income to a large number of households in many countries (Neumann and Hirsch, 2000). In a meta-study, Vedeld et al. (2004) estimate that forest income contributes 22% to the total household income of rural households in developing countries. They conclude that even "if encumbered with substantial uncertainties and variations, the figures suggest that forest environmental incomes contribute significantly to the economic production of goods and services and to welfare levels in these societies" (Vedeld et al. 2004 p. 60). They also show that this share is much higher for poorer households than for the better-off households. There is, however, still a lack of quantitative studies on the link between poverty and forest products (Sunderlin et al. 2005). The research presented in this chapter seeks to fill the gap in general knowledge of the role of forest products in rural livelihoods. While there have been many studies of forest products and livelihoods, the link between poverty, livelihood systems, and extraction of forest products is still poorly understood.

Considering as an example the vicinity of the Lore-Lindu National Park (LLNP) in Central Sulawesi/Indonesia, this chapter analyses the importance of forest products, especially for the rural poor, and identifies the underlying factors that drive households into the forest. Moreover, the paper investigates similarities and differences in the use of forest products in the village of Toro, where an agreement with the national park authority on the use of forest areas exists, and in the research area at large, where such agreements did not exist.

Specifically, the following research questions will be addressed: (1) Which forest products are gathered and sold? (2) How important are forest products as a source of income? (3) Do poor households differ in this respect from better-off households? (4) Which underlying factors influence participation in the sale of forest products? (5) What are the similarities and differences between the village of Toro and the research area concerning the use of forest products?

This research seeks to identify the underlying causes that drive rural household to collect forest products, and to detect so-called win-win outcomes. These are situations in which livelihood improvements are matched by gains in environmental protection. In this context, we investigate the influence of village agreements on the use of forest products which might be a possibility to conserve forests together with local communities.

Sources of data are the 2004 census of Project A1 in the village of Toro and the 2001 household survey of Project A4. During the census all 521 households living in the village of Toro have been interviewed. The Project A4 household survey interviewed 291 randomly selected households living in 12 villages in the vicinity of the Lore-Lindu National Park. Details on the sampling frame and on the selection of villages and households are described in Zeller et al. (2002). During the census and the survey the same questions concerning the use of forest products were used in the questionnaires.

2 Conceptual framework

The analysis is based on the livelihood approach, which emphasises the role of the household's resources as determinants of activities, and highlights the link between assets, activities, and incomes. Ellis (2000, p.10) defines a livelihood as consisting of "[...] the assets (natural, physical, human, financial, and social capital), the activities, and the access to these (mediated by institutions and social relations) that together determine the living gained by the individual or household". Although livelihood and income are not synonymous, they are nevertheless inseparably connected; income "at a given point in time is the most direct and measurable outcome of the livelihood process" (ditto).

Figure 2 illustrates the conceptual framework used in this work, which builds on the features of the livelihood approach.

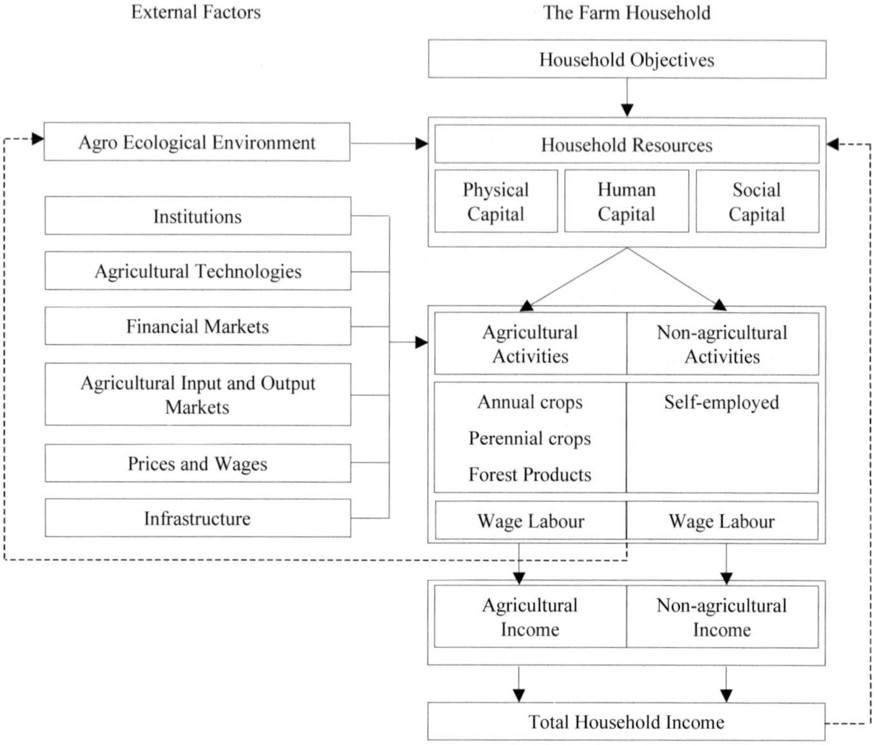

Fig. 1. Conceptual framework

According to its objectives, the household allocates its resources to activities, such as the collection of forest products, subject to factors which are external to the household (Figure 2). The collection of forest products generates outcome which will meet the objectives and which has an effect on the stock of resources available to the household as well as on the availability of forest products in the future. External determinants of the decision to collect forest products are the agro-ecological and socio-economic environment.

3 Hypothesis

An integral part of this study is the analysis of the link between poverty and the collection of forest products. As in Vedeld et al. (2004), we expect that the share of households collecting forest products as well as the share in total household income is higher for the poorest households than for the better-off households.

As outlined in the conceptual framework, the household's capital endowment plays an important role in activity choice. We include the mayor physical assets needed for crop and livestock production into the econometric analysis, and expect that they have a negative impact on the selling of forest products, because they compete for family labour. Education is expected to have a negative influence because better-educated people have access to a wider range of income possibilities. For the research area, Schwarze et al. (2005) showed that better-off households have better access to non-agricultural income activities than poorer households.

Among the external factors conditioning household decisions, we include the ethnicity of the head of the household, the access to credit, and the access to road infrastructure into the econometric analysis. The influence of ethnicity highly depends on context. Non-indigenous groups may have no access to forest products in some area, whereas in others they may be restricted from other activities and may therefore be pushed into forest activities. In the research area, non-indigenous ethnic groups are particularly involved in perennial crop production and non-agricultural activities (Schwarze 2004), and do not need to get involved in the collection of forest products. As Vedeld et al. (2004) point out, an important role of forest income in rural livelihoods is to overcome unexpected reductions in income or cash needs. Access to credit can play the same role and, thus, we expect that access to credit has a negative influence on the selling of forest products. The influence of access to road infrastructure cannot be hypothesised a priori. Forest products are often associated with remote areas lacking road access, but road infrastructure is important for the marketing of forest products. It is nevertheless difficult to distinguish the effect of the distance to roads from other spatially fixed effects. To control for these effects, we include location dummies, which are equivalent to the four sub districts in the research area.

4 Measurement of income

Income measures both cash and in-kind contributions, where the cash-earning component consists of crop, livestock, and forest product sales, wages, rents etc. The consumption of own-farm products (crops and livestock) is taken into account in the in-kind contributions component. All the goods produced are valued at market producer prices regardless of their use. So all own-farm products are valued at the same price as if they were being sold. Cash outgoings incurred by the household in its production are subtracted from the gross-contribution to the material standard valued at market prices (Ellis, 2000).

In collaboration with village authorities and researchers working on forest gardens, we compiled a list of 30 forest products which can be collected in the forest including hunting of animals. Firewood, which is mainly deadwood, is not traded locally and, thus, no market prices exist. A valuation using surrogate prices for alternative fuels, like for example kerosene, is not appropriate, because in large parts of the research area such alternatives are not available. A valuation might have been possible based on opportunity costs of time, which is, however, very data demanding. Information on who is collecting firewood, the time needed, and a measure of the value of time is required, but very intricate to collect. Due to these problems in valuing firewood, we decided to focus our analysis on the sale of forest products, which is much more relevant with respect to the LLNP. The main forest product sold is rattan, which is collected inside the park threatening its integrity. Rattan collectors are often the first people encroaching an area paving the way for others. All the other forest products, like firewood, are mainly collected at agro forestry plots or at the forest margin, but not further inside the national park. Income from forest products refers to the value of products sold in the last twelve months.

5 Measurement of poverty

As a measure of wealth we applied the poverty assessment tool as described in Zeller et al. (2005), which employs principal component analysis to select and eventually aggregate various indicators of poverty into a (0,1) normally distributed poverty index. It increases with wealth and measures the medium term relative welfare of households. Unlike the commonly used absolute measures of poverty such as a monetary poverty line, this method takes also into account other dimensions of poverty, such as education, food consumption and the condition of the dwelling. Details of this method are reported in Henry et al. (2001). The poverty index was estimated for each of the sample households (Abu Shaban 2001). It is computed from three asset-related indicators, four dwelling indicators, and two consumption indicators (Table 1). The poverty groups are the terciles of the poverty index: poorest (poverty group 1), poor (poverty group 2), and less-poor households (poverty group 3).

6 Econometric issues

Participation in the sale of forest products is measured by a binary variable, which is zero if the household does not sell any forest products. The binary variable takes on the value one if the household sells forest products. We are interested in how the vector of explanatory variables influences the possibility that the binary dependent variable takes on the value 1. The binary response Probit model is estimated by Maximum Likelihood Estimation (MLE) using the computer package Stata 8.2. Similar models have been widely used in the

Table 1. Indicators used for calculating the poverty index.

Asset-based indicators
- Total value of electronic appliances
- Value of transport assets
- Number of televisions owned
Dwelling-related indicators
- Access to electricity
- Type of wall
- Type of roof
- Type of floor
Consumption indicators
- per capita expenditures on clothes and footwear
- share of income spent on food out of a hypothetical increase in income of IDR 20,000 per week

Source: Abu Shaban (2001)

literature in similar settings by, for example, Corral and Reardon (2001) and Lanjouw et al. (2001).

A Probit model is estimated for the research area to identify factors influencing the households' decision to sell forest products. In addition, a second Probit model has been estimated to evaluate the influence of the village agreement on the sale of forest products beyond other factors. Therefore a random sample of 291 households from the village census in Toro has been drawn and merged with the data from the research area at large. The regression model includes variables from the first Probit model as well as a dummy variable indicating whether the household lives in a village with conservation agreement or not.

7 Collection and sale of forest products

In the vicinity of the LLNP, 76% of the households collect forest products. Apart from firewood, which is gathered by 72% of the households, they collect wood for construction, rattan, bamboo and the juice of the sugar palm. Except for rattan all of the products are collected at agro forestry plots or at the forest margin. Rattan collectors have to leave their villages for days or even weeks because rattan is only found farther inside the forest. They drag the rattan to rivers and streams and float it downstream to roadside access points, from where it is brought to the district capital of Palu on trucks. None of the households reported to gather or hunt animal products. Firewood and bamboo is collected for home consumption only, whereas rattan is mainly collected for selling. 17% of the households collect rattan and 12% also reported to sell it (Figure 7).

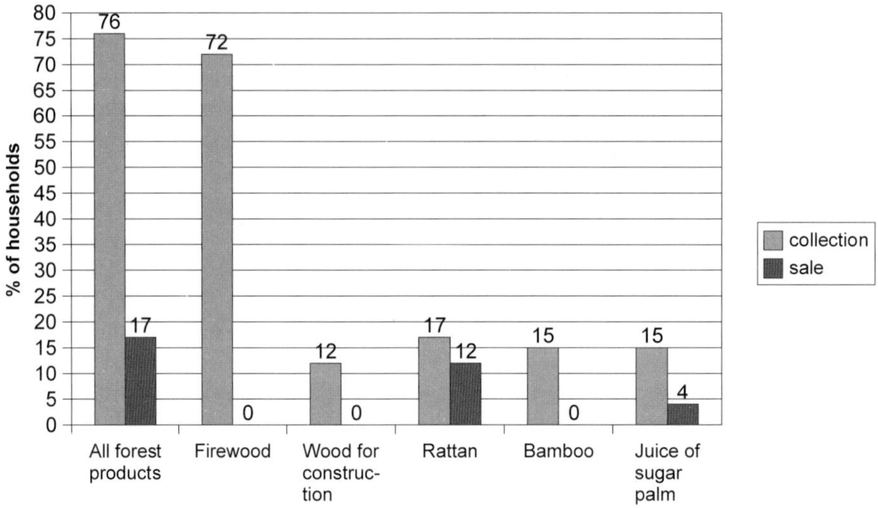

Fig. 2. Collection and sale of forest products. Source: STORMA project A4 household survey; Number of observations = 290

Overall, 17% of all households sell forest products, which accounts for 7% of the total household income (Table 2). Differentiating the forest income by product reveals the importance of rattan. Almost three-quarters of the income from forest products originates from the sale of rattan. The selling of the juice of the sugar palm accounts for 20% of the income from forest products and the remainder from selling wood for construction (Figure 3).

The analysis so far referred to all households, but how does this change if we look at different wealth groups? Table 2 shows the importance of forest products as a source of income for the poorest households. 21% of the total household income of the poorest households originates from the selling of forest products and 30% of these households reported to have income from forest products. The share of households selling forest products and the contribution to the total household income decreases with increasing wealth. Among the less-poor households only 4% collect forest products contributing to just 1% of their total household income.

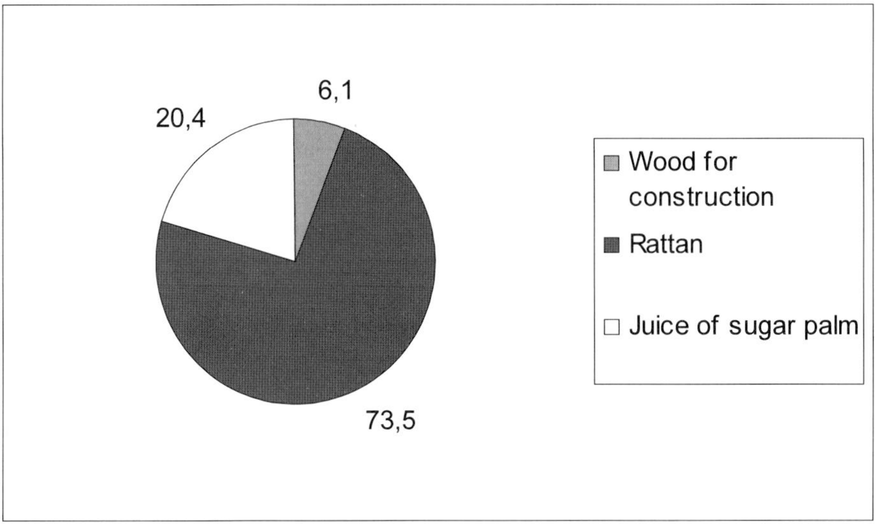

Fig. 3. Income from forest by product. Source: STORMA project A4 household survey; Number of observations = 290

Table 2. Income gained from forest products by poverty group.

	All households	Poverty-group 1 (very poor)	Poverty-group 2 (poor)	Poverty-group 3 (less poor)
Households selling forest products (%)	17	30	14	4
Share in total household income (%)	7	21	7	1

Source: STORMA project A4 household survey
Number of observations=291

8 Results of the econometric analysis

Table 3 presents a description and summary statistics of the regressors used in the econometric analysis of participation in the selling of forest products.

Table 4 presents the estimated coefficients as well as their standard errors of the Probit model, with coefficients with a significance level greater than 90% in bold. Participation in the sale of forest products is significantly influenced by the poverty index, the area of land owned, the education level, ethnicity, and the access to road infrastructure. As expected, the poverty index has a negative influence on the probability of selling forest products. This means that with increasing wealth it is less likely that households are involved in the selling of forest products. An additional hectare of irrigated land owned decreases the likelihood of participation in the sale of forest products by 12%,

their products is much lower in Toro. Whereas 17% of all households in the research area sell forest products, this share is only 11% in Toro. Moreover, the mean income gained from the selling of forest products is significantly lower in Toro than in the research area (Table 5).

Table 5. Income and participation in the research area and in the village of Toro

	Research area	Toro	z-value
Households collecting forest products (%)	76	93	-6.796***
Households selling forest products (%)	17	11	-3.168***
Mean income gained from selling conditional on selling (1000 IDR)	2384	489	-5.284***
Mean income gained from selling (1000 IDR)	309	51	-2.466**

** Mann-Whitney test significant at the 5% level of error probability.
*** Mann-Whitney test significant at the 1% level of error probability.
Source: Census of Project A1 and Project A4 household survey

From the descriptive analysis we can conclude that there is a difference in the sale of forest products between the research area at large and the village of Toro. We can, however, not deduct that the village agreement has caused the difference in behaviour. There might be other influencing factors. Because of a lack of data from before the village agreement, we conduct a regression analysis which controls for all other variables influencing the use of forest products. Both datasets have been merged and Table 6 presents a description and summary statistics of the regressors used in the econometric analysis of the influence of village agreements on participation in the selling of forest products. The regressors used are the same except for access to credit and access to road infrastructure because no data was available from the village of Toro.

Table 7 presents the estimated coefficients as well as their standard errors of the Probit model with coefficients with a significance level greater than 90% in bold. Participation in the sale of forest products is significantly influenced by the poverty index, the area of irrigated land owned, and ethnicity. All of these variables show the same signs as in chapter 9 referring to the research area at large. The dummy variable measuring the influence of the village agreement is also significant. The village agreement decreases the likelihood of participation in the sale of forest products by 11.5%.

10 Conclusions

Forest products, which are mainly collected inside the LLNP, are particularly important for the poorest third of the households. They contribute to 21% of

Table 6. Descriptions and summary statistics of the variables used in the econometric analysis of the influence of village agreements on participation in the selling of forest products

Variable	Mean	Std. dev.	Min.	Max.
Irrigated area owned (ha)	0.32	0.54	0	4.52
Rainfed area owned (ha)	1.34	1.67	0	10.50
Poverty index	-0.06	1.01	-1.84	3.48
Livestock units owned	0.58	0.91	0	6.36
Number of adult household members	3.42	1.63	1	8
Highest level of schooling of the adult household members				
not finished primary school	0.05	0.21	0	1
finished secondary school	0.29	0.45	0	1
finished tertiary school	0.27	0.44	0	1
Ethnicity of head of household (1=non-indigenous)	0.19	0.39	0	1.00
Access to credit (1=yes)	0.89	0.32	0	1.00
Walking distance house - road (hours)	1.43	3.62	0	13.00
Sub district dummies				
Sub district dummy for Palolo	0.22	0.42	0	1.00
Sub district dummy for Sigi-Biromaru	0.36	0.48	0	1.00
Sub district dummy for Kulawi	0.23	0.43	0	1.00

Source: STORMA project A4 household survey
Number of observations=582

their total household income. So any improvements in law enforcement concerning the collection of forest products within the national park will hit the poorest households hardest. Moreover, they are already the most constrained in terms of income generation as well as the most vulnerable concerning malnutrition and hunger. Our econometric analysis reveals possibilities to reduce the collection of forest products and, at the same time, improve the livelihood of the poorest households. Improving the access to primary education for the poor reduces the likelihood that they sell forest products and also improves their possibilities to find other sources of income. Another policy option is to foster the construction and improvement of irrigated rice fields, which are then cultivated by the poorest households. The labour demanding cultivation of irrigated rice reduces the time left to go into the forest. The harvested rice can improve the nutrition status of the family or can be sold on the market to gain additional income.

The distance to the tarmac road also has a negative influence on the likelihood that households sell forest products. This result strongly suggests to rethink the proposed road extension plans. In the political discussion, there are still plans to build new roads inside the national park, linking, for example the sub district of Lore Utara with Lore Selatan (ANZDEC, 1997). Alternative routes, which are further away from the national park but more costly (ANZDEC, 1997), are better suited to protect the integrity of the LLNP.

Table 7. Probit results for the influence of village agreements on participation in the selling of forest products

Regressors	Coeff.	std.error
Irrigated area owned (ha)	**-12.17**	**0.039**
Rainfed area owned (ha)	**1.60**	**0.007**
Poverty index	**-5.58**	**0.017**
Livestock units owned	-1.54	0.017
Number of adult household members	-0.19	0.009
Highest level of schooling of the adult household members		
not finished primary school[a]	3.27	0.064
Finished secondary school[a]	**-5.59**	**0.027**
Finished tertiary school[a]	-3.39	0.032
Ethnicity of head of household (1=non-indigenous)	**-7.37**	**0.023**
Village agreement (1=yes)?	**-11.52**	**0.026**
Sub district dummy for Palolo[b]	-4.31	0.040
Sub district dummy for Sigi-Biromaru[b]	**-8.36**	**0.028**
Sub district dummy for Kulawi[b]	2.02	0.042
Pseudo R^2	0.20	
% correctly predicted	85	

Source: Census of Project A1 and Project A4 household survey
Number of observations=582
Notes: The coefficients in the table are the percentage change in the probability for an infinitesimal change in each independent, continuous variable and the discrete change in the probability for dummy variables. Coefficients with a significance level greater than 90% are in bold.
[a] In contrast to finished primary school.
[b] In contrast to the sub district of Lore Utara.

The analysis also suggests that community agreements have an impact on the use of forest products. According to the agreement, the collection of forest products is only allowed for home consumption, not for sale. Our empirical results reflect these regulations. In the village with a community agreement the share of households collecting forest products is higher than in the research area, but less households sell the products. Hence, the use of forest products is mainly for home consumption and less for commercial use implying lower levels of extraction. This result is strengthened by the econometric analysis, which found that the village agreement has a strong negative influence on the likelihood of selling forest products beyond other factors.This outcome is even more striking when we consider that the village is located right close to the forest and rattan can still be found in the surroundings of the village. Our results suggest that community conservation agreements are a promising way to protect the integrity of the park. The outcome of such regulations, however, always depends on the support of the local leaders as well as on the acceptance of the population.

References

Abu Shaban A (2001) Rural poverty and poverty outreach of Social Safety Net programs in Central Sulawesi – Indonesia. MSc-Thesis, Institute of Rural Development, Georg-August University, Göttingen

ANZDEC (1997) Report on the Central Sulawesi Integrated Area Development and Conservation Project. Central Sulawesi Integrated Area Development and Conservation Project, Palu

Corral L, Reardon, T (2001) Rural nonfarm incomes in Nicaragua. World Development 29: 427-442

Ellis, F (2000) Rural livelihoods and diversity in developing countries. Oxford University Press, Oxford

Henry, C, Sharma, M, Lapenu, C, Zeller M (2001) Assessing the relative poverty of microfinance clients: A CGAP operational tool. The Consultative Group to Assist the Poor (CGAP), The World Bank, Washington D. C.

Keil, A (2004) The socio-economic impact of ENSO-related droughts on farm households in Central Sulawesi, Indonesia. Shaker, Aachen.

Lanjouw, P, Quizon, J, Sparrow, R (2001) Non-agricultural earnings in peri-urban areas of Tanzania: Evidence from household survey data', Food Policy 26: 385-403

Neumann, RP, Hirsch, E (2000) Commercialisation of non-timber forest products: Review and analysis of research. Center for International Forestry Research, Bogor, FAO, Rome

Schwarze, S, Zeller, M (2005) Income diversification of rural households in Central Sulawesi, Indonesia. Quarterly Journal of International Agriculture 44: 61-73

Schwarze, S (2004) Determinants of income generating activities of rural households: a quantitative study in the vicinity of the Lore Lindu National Park in Central Sulawesi/Indonesia. Doctoral thesis, Institute of Rural Development, Georg-August University, Goettingen http://webdoc.sub.gwdg.de/diss/2004/schwarze/schwarze.pdf

Sunderlin, WD, Angelsen, A, Belcher, B, Burgers, P, Nasi, R, Santoso, L, Wunder, S (2005) Livelihoods, forests, and conservation: An overview. World Development 33: 1383-1402

Vedeld, P, Angelsen, A, Sjaastad, E, Berg, GK (2004) Counting on the environment: Forest incomes and the rural poor. Environment Department Papers No. 98. The World Bank, Washington DC

Zeller, M, Sharma, M, Henry, C, Lepenu, C (2006) An operational method for assessing the poverty outreach of development policies and projects:

Results of case studies in Africa, Asia, and Latin America. World Development 34: 446-464.

Zeller, M, Schwarze, S, van Rheenen, T (2002) Statistical sampling frame and methods used for the selection of villages and households in the scope of the research programme on Stability of Rainforest Margins in Indonesia (STORMA). STORMA Discussion Paper Series No. 1, Georg-August University, University of Kassel, Institut Pertanian Bogor, Universitas Tadulako, Bogor.

Part II

Sustainable management of agroforestry systems

Shaded coffee and the stability of rainforest margins in northern Latin America

Ivette Perfecto[1*], Inge Armbrecht[2], Stacy M. Philpott[3], Lorena Soto-Pinto[4], and Thomas V. Dietsch[5]

[1] School of Natural Resources and Environment, Dana Building, University of Michigan, 440 Church St. Ann Arbor, MI 48109, USA
[2] Departamento de Biología, Universidad de Valle, Apartado Aéreo 25360, Cali, Colombia
[3] Smithsonian Migratory Bird Center, National Zoological Park, 3001 Connecticut Avenue NW, Washington, DC 20008, USA
[4] El Colegio de la Frontera Sur, Unidad San Cristóbal, Carretera Panamericana y Periférico Sur, Apartado Postal 63. C.P. 29290, San Cristóbal de las Casas, Chiapas, México
[5] Institute of the Environment, Center for Tropical Research, University of California, Los Angeles, La Kretz Hall, Suite 300, Box 951496, Los Angeles, California 90095-1496, USA

*corresponding author: Ivette Perfecto, Email: perfecto@umich.edu

Summary

Most native forests in Latin America are highly fragmented. In the mid elevation areas of Northern Latin America, the agricultural matrix is frequently composed of coffee. In this region, coffee has been traditionally cultivated under the diverse canopy of shade trees, representing a high quality matrix that can contribute to the social and ecological stability of the region. This agroforestry system has been proven to be important for biodiversity conservation. Studies over the last fifteen years have shown that shaded coffee plantations maintain a high diversity of vertebrates, invertebrates and plants. These organisms play an important role in the functioning of coffee agroecosystems. Shaded coffee plantations promote a high abundance and diversity of natural enemies that help to regulate herbivores, weeds and diseases. Shaded plantations also harbor a higher diversity of native pollinators which have been shown to contribute to higher coffee yields. Likewise, the diverse shade-tree component contributes to soil fertility and soil conservation and has been shown to contribute significantly to carbon sequestration. As a matrix, coffee agroforests also contribute to the conservation of biodiversity within

Tscharntke T, Leuschner C, Zeller M, Guhardja E, Bidin A (eds), The stability of tropical rainforest margins, linking ecological, economic and social constraints of land use and conservation, Springer Berlin 2007, pp 227-263

forest fragments by promoting migration among fragments and facilitating a metapopulation structure. Three "sustainable" coffee certification programs have been developed to help farmers cope with the vagaries of the market: organic, fair-trade and biodiversity-friendly (or shade-grown). Although certified coffees still represent a small niche market, they have the potential to promote conservation and benefit the livelihoods of small producers. Especially under conditions of low international coffee prices, as those experienced in the first years of this century, these certification programs have contributed to the ecological and socio-economic stability of the coffee growing regions of northern Latin America.

Keywords: coffee agroecosystems, Latin America, certification programs, function of biodiversity, intensification

1 Introduction: Coffee agroecosystems and the stability of forest margins: a general framework

Most of the forest in the Neotropics is already highly fragmented. The predominant landscape in these regions consists of small to medium size forest fragments surrounded by a sea of managed systems, the agroecological matrix. The ecological and socio-economic stability of such landscapes depends on how this matrix is managed. A mosaic of diverse cropping and land use systems that are managed for both ecological and socio-economic goals of the local population would prevent the further erosion of the natural areas and would maintain biodiversity at the landscape level. In this chapter we review the role of the coffee agroecosystem in mid elevation regions of Northern Latin America in maintaining both the ecological and socio-economic stability of the region.

Before we examine the case of coffee in particular it is important to establish a more general framework for the need to integrate the agroecological matrix in conservation strategies. There are 5 reasons for doing so: 1) most tropical habitats in the region are already highly fragmented, 2) extinction rates are high even in large forest fragments, 3) the matrix itself is sometimes an important repository of biodiversity, 4) the matrix provides migration pathways from fragment to fragment, and 5) agriculture is not a permanent activity (Vandermeer et al. in press). With this framework we argue that conservation needs to be refocused, away from preservation areas and towards the matrix in which fragments of native habitat are situated. Vandermeer and colleagues (in press) argue that this is especially so in the light of what we now know about extinction patterns in fragments, metapopulation dynamics, biodiversity patterns in agroecosystems, movement patterns of various organisms, and postagricultural succession. Yet a focus on the matrix means taking part in the debates on the nature of development within the agricultural context, and, more specifically, modifying the paradigm that there is an intrinsic conflict between managed areas and conservation with one that distinguishes between different types of agricultural development and recognizes the critical role of the agricultural matrix in the conservation of biodiversity. The rest of this chapter analyzes these ideas within the context of the coffee landscapes in Northern Latin America. We first present a profile of the coffee agroecosystems in northern Latin America (section 2), then examine the ecological stability of coffee by reviewing the literature of biodiversity in coffee farms (section 3), the function of that biodiversity as it relates to coffee production and sustainability (section 4), and the role of shade coffee as a high quality matrix (section 5). Finally, we discuss how diverse coffee farms can contribute to the social stability of the region (section 6).

2 Coffee in northern Latin America

Coffee agroecosystems are particularly important both ecologically and economically in northern Latin America (Mexico, Central America, the Caribbean, and Colombia). Coffee does not cover as much land area as other agricultural activities – land in coffee production is roughly 3.6 million ha (FAO 2002). However, the ecological importance of coffee is a consequence of where it is produced, rather than how much land is under production. Coffee production greatly overlaps with the world's biodiversity hotspots (Hardner and Rice 2002) and may overlap with key forest habitats containing a large number of endemics (Moguel and Toledo 1999). Generally, coffee is grown on mid-elevation mountain ranges that in northern Latin America have been largely deforested. In El Salvador, for example, more than 90% of original forest cover has been lost, but shade grown coffee (representing 92% of production; Rice and Ward 1996) now accounts for roughly 80% of remaining forested area (Panayotou et al. 1997).

Economically, coffee is also exceedingly important for Latin America. Coffee makes up a large percentage of total agricultural export revenue in countries like Mexico and Peru (Nolasco 1985, Greenberg and Rice, 2000, Calo and Wise 2005). Throughout Latin America, there are large numbers of people producing coffee (Calo and Wise 2005) and several million depend on coffee income (Nolasco 1985). Approximately 95% of Mexican coffee producers (65.5% of total production area) are small producers, most of whom still manage coffee under traditional multi-species canopies and manage less than 5 ha each (Rice and Ward 1996).

Coffee was traditionally grown under a diverse, dense shade canopy, but recent intensification includes reducing shade tree density and diversity and agrochemical use (Figure 1; Moguel and Toledo 1999, Mas and Dietsch 2003). Much of the ecological importance of coffee directly relates to the role of traditional or shaded coffee in providing a high quality agricultural matrix and related ecosystem services not provided by intensive coffee systems. Many studies have measured biodiversity loss across the coffee intensification gra-

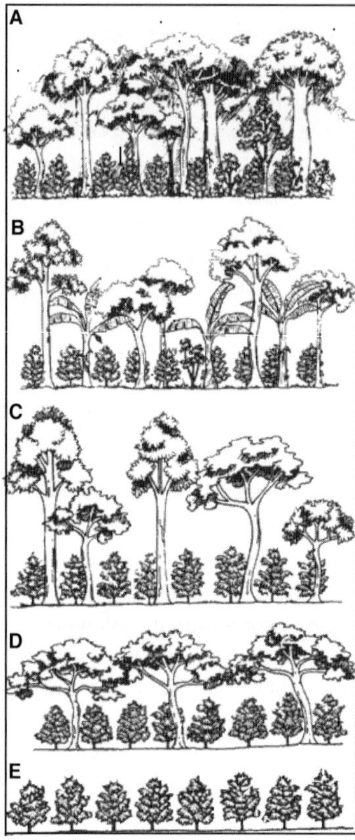

MANAGEMENT SYSTEM	%SHADE* COVER	SHADE TREE* RICHNESS
RUSTIC	71-100	> 50
TRADITIONAL POLYCULTURE	41-70	21-50
COMMERCIAL POLYCULTURE	31-40	6-20
SHADED MONOCULTURE	10-30	1-5
UNSHADED (SUN) MONOCULTURE	0	0

Modiefied from: Moguel and Toledo 1999; Rain Forest Alliance.

* Figures for percent shade and tree species richness are approximates based on studies cited by Moguel and Toledo, 1999 and our own research (Perfecto et al. 2003).

Fig. 1. Diagram of the different coffee management systems with shade cover and shade tree richness.

Figure 1.1 Diagram of the different coffee management systems with shade cover and shade tree richness.

dient. Rustic or traditional coffee farms with a high density and diversity of shade trees and high percentages of canopy cover conserve a large number and proportion of forest species (Perfecto et al. 1996, Greenberg et al. 1997, for example) and reduction of different aspects of shade negatively affect species richness (reviewed in Perfecto et al. 1996, Moguel and Toledo 1999, Perfecto and Armbrecht 2003, Donald 2004). Patterns of biodiversity loss strongly depend on the particular taxa studied (Daily et al. 2001, Perfecto et al. 2003), and matrix species composition may differ from native habitats (Rappole et al. 2003), yet most studies corroborate that a high density and diversity of shade trees in coffee plantations help preserve forest species. The conservation benefits of shade coffee are further enhanced by the proximity and connectedness of intact natural habitats (Ricketts 2001, Vandermeer and Carvajal 2001, Perfecto and Vandermeer 2002, Steffan-Dewenter 2002) and may allow for increased migration out of forest patches for resource use (Perfecto and Vandermeer 2002, Armbrecht and Perfecto 2003). Finally, shaded coffee farms are also prized for their contribution to various ecosystem services – functions that also can be negatively affected by intensification.

Coffee intensification, at microeconomic scales, is associated with increased yields and revenues, but also with increased costs for labor, fertilizers, and other on-farm products needed to carry out some of the functions that the missing biodiversity can no longer provide. In a process largely backed by INMECAFE (the National Coffee Federation of Mexico) coffee farmers throughout Mexico transformed their traditional coffee farms to shade monocultures, with drastic effects on biodiversity, and temporary increases in yields – ultimately drastically transforming the landscape (Nestel 1995, Rice 1997). More recently, in Vietnam and Indonesia, national initiatives and a drive for high profits have resulted in the development of large scale intensive coffee production (O'Brien and Kinnaird 2003). One study comparing the productivity and profitability in coffee farms in Costa Rica found that conventional farms produced on average 22% higher yields than did organically-managed farms, leading to overall higher profits (Lyngbæk et al. 2001). Yet, increased chemical inputs into coffee agroecosystems carry well documented costs for biodiversity and habitats bordering intensive agricultural systems (Perfecto et al. 1996). Furthermore, intensive techniques with high associated direct costs may make farmers more financially vulnerable to lower prices.

The economic importance of coffee to Latin Americans has been further demonstrated by recent international price crashes in coffee markets (Calo and Wise 2005). Between 1999 and 2002, coffee prices, determined by the commodities market of the NY stock exchange reached lows of $0.42 /lb (FAO 2002) – the lowest prices in 100 years based in real terms (Perfecto et al. 2005, Calo and Wise 2005). This recent crash, stemming largely from overproduction due to intensification efforts in Latin America and more production in Asia, has resulted in widespread environmental and social disasters (Gresser and Tickell 2002). Reportedly over 300,000 coffee growers in Mexico have abandoned their farms (LaFranchi 2001), and many other farmers have intensified their production further hoping to increase yields, or worse, have converted coffee farms to pastures or illegal crops (Perfecto and Armbrecht 2003). Although some argue that biodiverse farms may buffer against such price swings, those small farmers growing in a manner that protects biodiversity may not have sufficient capital to wait until prices swing back up. Low, but consistent fluctuations in coffee prices do affect smallholders (the majority of producers) more deeply. Thus finding techniques to provide economic and social stability in coffee growing regions will benefit ecological stability as well. If these techniques are not supported, it will enormously impact conservation efforts of any kind in the region.

3 Ecological stability: the coffee agroecosystem as a reservoir of biodiversity

3.1 Vertebrates

Bird use of coffee agroecosystems was an early focus of coffee biodiversity research. Perfecto and colleagues (1996) and others have summarized this early bird work that compared coffee management with high and low levels of shade (Dietsch 2005). Consistently high levels of bird diversity in heavily shaded coffee agroecosystems, comparable in many cases to tropical forests, were a primary motivation behind the movement to certify shade-grown coffee (R. Greenberg, personal communication). Based on their research, the Smithsonian Migratory Bird Center coined the term Bird-friendly^{TM} to name their shade-grown coffee certification program. Their criteria were shown to certify farms with significantly higher diversity of forest-associated birds and butterflies (Mas and Dietsch 2004). Recent work has focused on evaluating bird diversity across a fuller spectrum of management practices to identify the shape of biodiversity loss curves (Dietsch 2003).

While earlier research consistently showed a loss of diversity and abundance for all birds, these new studies are showing a more complex pattern of loss, with some components of the avifauna more sensitive to management intensity than others (Komar and Dominguez 2002, Dietsch 2003, Tejeda-Cruz and Sutherland 2004). In particular, forest-associated resident birds are sensitive to the introduction of even low-intensity coffee management (Dietsch 2003, Tejeda-Cruz and Sutherland 2004). However, a sizeable component of the bird diversity is coffee-associated and only begins to drop at higher levels of intensification (Dietsch 2003). There is also a seasonal component to bird diversity changes with the (northern winter) dry season presenting a significant bottleneck for forest-associated resident birds (Dietsch 2003). During this period, migratory birds are abundant and resource availability may be a serious constraint and may affect resident resource use (Jedlicka et al. 2006).

In farms with relatively low shade tree diversity, species of the genus *Inga* may act as a keystone resource to birds, but as the shade is diversified, other resources become more important. When *Inga* are in flower, they provide resources that help many bird species, especially migrants, persist through the dry winter months (Calvo and Blake 1998, Johnson 2000). The *Inga* flowers produce high quantities of nectar, attracting many insects as well as nectivorous and insectivorous birds (Johnson 2000). However, in a Chiapas farm augmented with native shade tree species, birds made disproportionately greater use of the additional resources provided by the increased tree diversity than in the farm dominated by *Inga* trees (Dietsch 2003). In El Salvador, Komar and Dominguez (2002) found that both structural and floristic habitat components contributed to resident bird diversity and abundance, with thresholds of 44% canopy cover and 15 tree species per 0.5 hectare important for the conservation of species sensitive to perturbation. In addition, shade-coffee farms with epiphytes maintained higher abundance and diversity of the inhabitant bird fauna than farms without epiphytes (Cruz-Angon and Greenberg 2005).

Small mammals including rodents and bats show similar diversity patterns as birds. For example, small mammal diversity was lowest in technified coffee when compared to forest and an organic shaded farm (Witt 2001). Gallinas and colleagues (1996) found that the diversity of medium-sized mammals was related to the vegetative structure of the coffee shade canopy. They recommended the maintenance of high tree diversity to provide food resources and protection for the mammal community. In Guatemala, bat diversity was higher in coffee with diverse shade and may be limited by roost site availability in more intensive coffee management systems (Valle and Calvo 2002). However, in Veracruz, Mexico, Estrada and Coates-Estrada (2001) attributed high bat diversity in both shaded and unshaded coffee systems to the high mobility of bats allowing these mammals to use habitats ephemerally even if roost sites are unavailable in technified areas. In Costa Rica, Daily and colleagues (2003) reported that small forest fragments contiguous with coffeee plantations did not differ from more extensive forests in species richness of non-flying mammals and where richer than other agricultural habitats, demonstrating that the quality of the coffee matrix affects diversity within the forest fragments. Though large mammals are probably limited by hunting pressure, particularly near rural communities, species without hunting pressure can persist and even maintain territories in shaded coffee farms. For example howler monkeys in Nicaragua successfully maintained territories in shade coffee making use of the tree diversity for leaf forage (McCann et al. 2003, Williams-Guillen 2003).

Coffee plantations have been found to be less effective in maintaining the diversity of reptiles and amphibians (Komar and Dominguez 2002, Pineda et al. 2005). It has been suggested that a more open canopy structure may create warmer, drier conditions that adversely affect some species within these taxa. In addition, amphibians are considered to be sensitive to pesticide and herbicide use. In particular, atrazine and glyphosate, regularly used herbicides in coffee, are toxic to amphibians (Hayes 2004, Relyea 2005). Another major hazard for reptiles, especially snakes, in agricultural landscapes is that workers usually kill them regardless of whether they are venomous or not (Dietsch, personal observation).

3.2 Invertebrates

Most studies show that traditional shaded coffee farms harbor a high diversity of invertebrates, and that species richness declines along the intensifiction gradient (for reviews see: Perfecto et al. 1996, Perfecto and Armbrecht 2003, Donald 2004, Somarriba et al. 2004). In Latin America, special attention has been devoted to ants. Of 21 studies covering soil, leaf-litter, arboreal and army ants, 18 show a significant decline in ant species richness with intensification (Table 1). Some studies show comparable values between adjacent forest remnants and coffee agroforests with respect to ant species richness (Perfecto and

Table 1. Studies examining the effect of coffee intensification on ant diversity and interactions

Country	Reference	group/theme	Effect
Colombia	Armbrecht et al. 2005	leaf litter	yes
Colombia	Armbrecht et al. 2004	leaf litter	yes
Colombia	Sadeghian 2000	soil	yes
Colombia	Sossa & Fernández 2000	leaf litter	yes
Colombia	Garcia & Botero 2005	Ponerinae	yes
Costa Rica	Benítez & Perfecto 1990	soil	yes
Costa Rica	Perfecto & Snelling 1995	soil	yes
		coffee plants	no
Costa Rica	Perfecto & Vandermeer 1994	soil	yes
Costa Rica	Barbera et al. 2004	soil	yes
Costa Rica	Perfecto & Vandermeer 1996	competition interactions	yes
Costa Rica	Perfecto et al. 1997	arboreal	yes
Mexico	Philpott et al. 2006	arboreal	yes
Mexico	Armbrecht & Perfecto 2003	soil and leaf litter	yes
Mexico	Perfecto et al. 2003	soil	yes
Mexico	Perfecto & Vandermeer 2002	soil	yes
Mexico	Lachaud & Garcia-Ballinas 1999	Ponerinae/Cerapachinae	yes
Mexico	Ibarra-Nuñez et al. 1995	coffee plants	yes
Mexico	Nestel & Dickschen 1990	foraging dynamics	yes
Mexico	Ramos-Suárez et al. 2002	soil	no
Panama	Roberts et al. 2000	army ants	yes
Puerto Rico	Torres 1984	soil	no

Vandermeer 2002, Ramos-Suárez et al. 2002, Perfecto et al. 2003). Identified mechanisms responsible for the reduction of ant species richness include loss of nesting sites (Philpott and Foster 2005), reduction in the leaf-litter complexity (Armbrecht et al. 2005), microclimatic changes (Perfecto and Vandermeer 1996), changes in ant competitive hierarchies (Perfecto 1994), and an enigmatic preference for diversity (Armbrecht et al. 2004).

Coffee agroforests have been found to maintain high diversity of other athropods such as beetles (Moron and López-Méndez 1985, Nestel et al. 1993, Perfecto et al. 1997, Estrada et al. 1998, Molina 2000, Pineda et al. 2005), butterflies (Botero and Baker 2002, Mas and Dietsch 2003, 2004, Valencia 2004, Krantz 2005), homopterans (Rojas et al. 2001, Franco et al. 2003), spiders (Ibarra-Núñez and García-Ballinas 1998), and non-formicid hymenopterans (Hanson 1991, Monro and Gauld 2002, Klein et al. 2002, Tylianakis et al. 2004). Although most studies do show a significant decline in arthropod species richness with the intensification of coffee, a few studies failed to show differences (for example, Ricketts et al. 2001, Klein et al 2002, Ramos-Suárez et al. 2002). The failure to detect a significant effect due to intensification could be due to landscape level features (Tscharntke et al. 2005), or could suggest that some organisms may even benefit from the more open habitats related to agricultural production (Klein et al. 2002). Several studies have found that closeness to the forest positively influenced arthropod richness within the coffee farms (Ricketts et al. 2001, Perfecto and Vandermeer 2002, Klein et al. 2003c, Armbrecht and Perfecto 2003, Horner-Devine et al. 2003, Ricketts 2004, DeMarco and Monteiro Coelho 2004, Krants 2005). Furthermore, different arthropod taxa or different guilds within a taxon show varying patterns of richness loss along the intensification gradient (Perfecto et al. 2003, Schulze et al. 2004, Pineda et al. 2005, Rivera and Armbrecht 2005, Armbrecht et al. in press) possibly due to different mechanisms such as spatial scale, landscape features, dispersion of the organism, or degree of diet specialization.

3.3 Plants

Ironically, although most of the biodiversity benefits of shade coffee are attributed to their high floristic diversity, which provides habitat to a variety of other organisms, very few studies have focused on plant diversity *per se*. Some researchers have compared woody-plant diversity among different types of coffee plantations and natural forest and have found similar species richness and structure (Reynoso 2004, Bandeira et al. 2005, Cruz-Lara et al. 2004). However, not all shaded coffee is the same. For example, rustic plantations are more like natural forest than traditional polycultures, which are dominated by one or two genera (Figure 1, Table 2). In Chiapas, Mexico, at least 40% of the species recorded in coffee plantations corresponded to the natural surrounding vegetation: tropical forest, pine-oak forest and cloud forest; the last of which is itself in a threatened status (Soto-Pinto et al. 2001). 224 plant species and 53 families, among trees, shrubs and palms were recorded from coffee shade in several studies in Chiapas. 97% of the total were native, most belonging to the Fabaceae (16.4%), Asteraceae (7.1%), Lauraceae (7.5%), Euphorbiaceae (4.5%), Tiliaceae (3.6%) and Moraceae (3.6%). Floristic composition and management intensity make a difference in terms of the potential for biodiversity conservation. The more complex the structure, given by tree height and diameter and canopy cover, the greater the potential for diversity conservation. Rustic coffee farms seem to present the best characteristics for conservation, followed by traditional polycultures (Figure 1, Table 2). Bandeira and colleagues (2005) studied 100 coffee stands in Oaxaca, Mexico. They stated that many of these coffee stands harbor as many as 34 species of wild trees. The floristic structure of rustic coffee farms was highly variable, given by a combination of factors such as human management, original stand cover and the asynchrony in development stage of different farms, promoting a large beta-diversity at the landscape level. Similar results have been reported by Méndez (in press), and Monro (2002) reported 200 tree species for coffee farms in El Salvador. Thus, although a single plantation may have a limited potential to preserve native tree species, it is the whole ensemble of floristically heterogeneous farms which renders this agroforestry system valuable

Table 2. Comparison of vegetation structure variables, woody-species richness and soil attributes between traditional polyculture coffee plantations, rustic coffee plantations and natural forest.

Variables	Traditional Polyculture Coffee Plantation	Rustic Coffee Plantations	Natural Forest (Tropical forest)
Number of vegetation strata	2b (Jitotol, Sepultura)	3-4a (Jitotol, Sepultura)	More than 4 (Sepultura)
Vegetation cover (%)	49.7c (Sepultura)	80.7b (El Triunfo)	93.1a(El Triunfo)
		87.5 (Lacandona)	94.7 (Lacandona)
		68.4b(Sepultura)	91.2a (Sepultura)
	77a (Jitotol)	71b (pine/oak Jitotol)	
Tree height (m)	11a (Chilon & El Triunfo)	11a (Chilón)	12a (El Triunfo)
Tree diameter (cm)	27a (Jitotol)	24a (Jitotol)	
		18.7a (El Triunfo)	21.1a (El Triunfo)
Litter thickness (cm)	3.2a (Jitotol)	2.2b (Jitotol)	N.D.
Total aboveground biomass (ton/ha)	42b (Jitotol)	93.8a (Jitotol)	
Tree & shrub density	176 (Jitotol)	336b (El Trinfo)	887a (El Triunfo)
(individuals/ha)	282b (Jitotol)	316a (Jitotol)	
		463 (Chilón)	
		457 (Jitotol)	
Trees & shrub species	33/2700m^2 (Jitotol)	46/2700m^{2}b (Jitotol)	
richness (species/m^2)		21/1310m^2 (Lacandona)	39/1310m^2 (Lacandona)
		10/1000m^{2}b (El Triunfo)	30/1000m^{2}a (El Triunfo)
		18/1000m^2 (Chilón)	
		45/1000m^2 Oaxaca	35/1000m^2 Oaxaca
Organic matter	5.2a (Chilón)	5.3 a(Chilón)	6.06a (El Triunfo)
Phosphorous	5.5a (Chilón)	4b (Chilón)	3.2b (El Triunfo)
Ph	5.6 (Chilón)	5.7 (Chilón)	6 (El Triunfo)
Nitrogen	0.33a (Chilón)	0.33 a(Chilón)	0.21b (El Triunfo)
CIC	22.3a (Chilón)	35.4b (Chilón)	21.4a (El Triunfo)
Shannon diversity index	1.9307b (El Triunfo)	N.D.	3.0433b (El Triunfo)

A T-test was used to test for differences between two habitat types; a one-way ANOVA followed by Duncan's Multiple Range test was used to test for differences between three habitat types. Different letters between columns for the same locality mean significant difference at $p<0.05$.
Source: Soto-Pinto et al. 2000, Ramos-Suárez et al. 2002, Peeters et al. 2003, Cruz-Lara et al. 2004, Reynoso 2004, Bandeira et al. 2005.

for plant diversity conservation, particularly in a region where native forest vegetation has almost disappeared.

Studies on vascular epiphytes in coffee farms have reported a high species richness (Solís-Montero et al. 2005, Hietz 2005). Hietz (2005) reported 89 and 104 vascular epiphyte species in coffee and natural forest, respectively. This author stated that farms with small trees and sparse shade hosted fewer epiphytes than those with large trees, and pointed out the value of traditional polycultures for epiphyte diversity. However, Hietz (2005) also suggested that traditional polycultures may not be suitable for all epiphytes. In Veracruz, Mexico, Solís-Montero and colleagues (2005) reported high population densities (800 plants/ha) of three orchid species, pointing out that coffee farms may not replace the original conditions of a forest, but it is possible that orchids may survive and reproduce in farms that provide appropriate microclimate conditions for the orchids and their pollinators.

Finally, it is important to remember that agriculture is not a permanent activity (Kaimowitz 1996, Aide and Grau 2004) and that the regeneration of secondary and eventually mature forest is affected by the type of agriculture that is practice (Ferguson et al. 2003). In this respect, agroforestry systems and traditional shaded farms have been shown to play an important role in the regeneration of tropical forests in Nicaragua (Griffith 2000), El Salvador (Hecht et al. 2002) and Puerto Rico (Brasch 1987, Nir 1988).

4 Ecological stability: the function of biodiversity in the coffee agroecosystem

For more than a decade, studies addressing agricultural intensification in general (Tscharntke et al. 2005) and the intensification of coffee plantations in particular (Perfecto and Armbrecht 2003) have noted the loss of important ecosystem services and ecosystem functions as a consequence of reduced biodiversity. However, with a few exceptions, the role of biodiversity in maintaining ecosystem services and function have not been rigorously examined. Shaded coffee farms conserve plants, arthropods and vertebrates that have been shown to carry out important ecological functions such as nitrogen fixation, carbon sequestration, pest regulation, pollination and seed dispersal (Witt 2001, Klein et al. 2002, Callo-Concha et al. 2004, Perfecto et al. 2004, Ricketts et al. 2004). In the following section we summarize these studies and reflect on their influence with regard to the stability of tropical forest margins.

4.1 Pest regulation

Pest regulation depends both on bottom-up and top-down ecological forces (e.g. Dyer and Letourneau 1999). In the first case, agroforestry systems should rely mostly on a strategy of increased resilience - and other internal control mechanisms through enhancing diversity, but this does not automatically guarantee pest control (Rao et al. 2000). Top-down regulation in diverse traditional shaded coffee farms relies partly on the "natural enemies hypothesis" (Root 1973, Andow 1991) in which more resources provided by the complex vegetation structure leads to a higher abundance and richness of natural enemies of herbivores. Some ecological functions from natural enemies include parasitization, predation or deterrence of herbivores, all of which form part of complex food webs in traditional polyculture shaded coffee plantations (Philpott et al. 2004) and have proven to be an insurance preventing pest outbreaks (Perfecto et al. 2004). Langellotto and Denno's meta-analysis (2004) showed that increasing habitat complexity leads to increases of invertebrate natural enemy abundances. This is consistent with studies in complex shaded coffee farms (Vandermeer et al. 2003), where trait mediated effects by parasitic influence, competitive interactions among arboreal ant predators (Philpott 2005), or predation by birds is affected by dominant ant species in trees (Philpott et al. 2005).

The most important coffee pest in LatinAmerica, the coffee berry borer, *Hypothenemus hampei* (Coleoptera: Scolitidae), was found to be predated by ants at a significantly higher rate in shaded coffee than in sun coffee plantations of Colombia (Gallego and Armbrecht in press). Vélez and colleagues (2003) also found that *Gnamptogenys* pos. *sulcata*, an ant abundant in forests and shaded coffee but very rare or completely absent in sun coffee, is a very effective predator on adult berry borers. Other results from Costa Rica tested predation by three ant species (two of them abundant in open habitats) without any significant decrease in the berry borer in field trials (Varón 2002). Both, the intensification of coffee management and dry season negatively affected important ant forager predators in Mexico (Philpott et al. 2006a) and Colombia (Armbrecht 2003).

Another way to examine the potential effect of diversity on pest regulation is the insurance hypothesis (Yachi and Loreau 1999, Loreau et al. 2003), the idea that diversity protects functional properties of communities in the face of environmental perturbation. Perfecto and colleagues (2004) provided a test of the insurance hypothesis by examining insect predation by birds in coffee farms with different levels of floristic diversity. They simulated an outbreak of lepidopteran larvae on coffee plants and noted the larval disappearance rates inside and outside bird exclosures in two farms with distinct levels of shade. Significant differences were found associated with the exclosure treatment, indicating that birds can potentially prevent pest outbreaks. Interestingly, the effect was significant only for the farm with a high floristic diversity, providing partial evidence in support of the insurance hypothesis. Similarly, Soto-Pinto and colleagues (2002) found that the higher the number of vegetation strata, the lesser the incidence of coffee leaf rust (*Hemileia vastatrix*), suggesting that vegetation structure in coffee farms conditions the ecological relationships related to this important disease. These authors also noted that the higher the shade species richness the lesser the weed cover, suggesting that species richness promotes a larger number of habitats for herbivores and other weed-controlling organisms.

Given these results, increasing diversity in agroecosystems has been proposed as a prophylactic strategy for pest control. Yet almost a third of reported studies in a recent review do not support the hypothesis that there is greater pest control in diversified agroforestry systems (Schroth et al. 2000). Some of the possible mechanisms mentioned include introducing plant species that harbor pests, changes in microclimate which benefit pest and diseases, and the physical protection of mammal and bird pests offered by trees (Schroth et al. 2000, Muriel and Vélez 2004). Therefore, this strategy should be promoted only in those cases where its effectiveness has been firmly established (Vandermeer and Perfecto 2000).

4.2 Pollination

Another ecosystem service affected by declines in arthropod diversity and abundance with intensification is coffee pollination. Although much discussion in the pollination biology literature debates the relative role of pollinator biomass versus diversity (Balvanera et al. 2001), several recent studies indicate that both are important for coffee pollination. Arabica coffee is a self-compatible species that does not require outcross pollination, but may nevertheless benefit from pollinator presence. High numbers of visits of honeybees (*Apis mellifera*) correlate to increased coffee fruit set and fruit weight (Raw and Free 1977, Manrique and Thimann 2002, Roubik 2002). This brings up an interesting conservation dilemma since *A. mellifera* is an introduced, and partly invasive, species in Latin America. There is an ongoing debate as to how dangerous this bee is for native (competitively inferior) bee species. So any advantages that can only be achieved at the cost of having this non-native species in the system are debatable. On the other hand, a study in coffee agroforests in Indonesia demostrated an increase in fruit set of *Coffea arabica* with the number of flower visiting bee species (Klein et al. 2003b). Furthermore, this and other studies have shown the importance of a diverse suite of pollinators, including both social and solitary bees, for pollination (Klein et al. 2003a) and pollen deposition (Ricketts 2004). Two recent studies in Brazil (DeMarco and Monteiro Coelho 2004) and Costa Rica (Ricketts et al. 2004) calculated that for coffee plants located near forest fragments, native bees increased yields by more than 14% and 20% respectively. Ricketts estimated that this represents a total dollar value of $62,000 for the farm studied in Costa Rica (Ricketts et al. 2004). This represents substantial benefits to farmers and highlights the importance of maintaining forest fragments within coffee landscapes, even if small. Finally, another study has documented that under high shade management, where activity of pollinators (including ants) is greater, the presence of ants, or some interaction between ants and flying pollinators affords higher fruit weights than under low shade conditions

(Philpott et al. 2006b). Although some researchers have found no effects of pollinators on coffee pollination (Nogueira-Neto et al. 1959), more recent evidence clearly indicates that in fact both diversity and abundance of bees, and potentially other pollinators, do increase yields, weights, and quality of coffee.

4.3 Vegetation structure, diversity and ecological functions

Some studies have emphasized the role of trees as "host" and food source for other organisms. Carlo and colleagues (2004) recorded bird species foraging on fruit trees in coffee plantations and natural forests. In their study, fruits comprised more than 50% of the diets for four focal bird species. They also reported differences in the number of foraging records for focal bird species between the commercial polyculture and forest habitats but few differences between rustic coffee and forests. In their study, the genera *Cecropia, Miconia, Schefflera, Phoradendron*, and *Guarea* were reported as the most important, providing birds with a fairly constant fruit supply. Other species such as *Brosimum alicastrum* have also been shown to be important food resources for frugivorous birds, deer and wild boar. As mentioned above *Inga* flowers offer an important resource for insectivorous and nectivorous birds in the winter months (Calvo and Blake 1998, Johnson 2000).

Other studies have shown the ecological function of the tree species most commonly planted with coffee. For instance *Ochroma pyramidale*, *Alchornea latifolia*, *Simarouba glauca* and *Theobroma cacao* are useful for soil conservation, erosion control and restoration of degraded soils (Vázquez-Yanez et al. 1999). Most legume species increase ecosystem–level nitrogen availability through nitrogen fixation, while other species, like *Trema micrantha* have been shown to have abundant mycorrhizal associations thus enhancing nutrient uptake (Dalla Rosa 1993).

Comparing coffee farms dominated by *Inga* species with rustic coffee Peeters and colleagues (2003) reported a significantly higher tree biomass in rustic coffee (Table 2). Similarly, Soto-Pinto (2001) and Soto-Pinto and colleagues (2000) found that soil cation exchange capacity, soil K and soil Ca were higher in rustic farms as compared to *Inga* dominated systems and that in rustic farms shade-species richness was positively correlated with soil Ca and soil cation exchange capacity. Overall, these studies suggest that maintaining diversity and managing shade cover (to around 45%) might be a good way of maintaining yields and balancing the relationships between soil nutrients, nutrient uptake, yields and floristic diversity.

4.4 Carbon sequestration

Recently the role of coffee agroforests in carbon sequestration has been recognized (De Jong et al. 1995, 1997). Coffee agroforests accumulate a high amount of aboveground biomass. Results from research in Chiapas (Peeters et al. 2003) showed 42 and 138 tonnes ha^{-1}of aboveground biomass contained in shade vegetation in polyculture coffee and rustic coffee systems respectively (coffee shrubs, soils, litter and roots not included). Callo-Concha and colleagues (2004) estimated 195.6 ton $ha^{-1}$for coffee with shade, similar to the amount found in homegardens. These data reveal that rustic coffee systems accumulate a significant amount of carbon, equivalent to 1/3 of the amount sequestered by primary forest (307 ton ha^{-1}, Rice and Greenberg 2000; 465.8 ton ha^{-1},Callo-Concha et al. 2004). Additionally, soils contain 6.8% of organic matter in these agroecosystems (Romero-Alvarado et al. 2002, Soto Pinto et al. 2001). Callo-Concha and colleagues (2004) reported that soil retained in coffee with shade, homegardens, silvopastoral systems and pastures represents more than 50% of total carbon. Agroforestry coffee systems may increase carbon sequestration by incorporating more trees, increasing density from 229 trees ha^{-1} to 393 trees ha^{-1}, which is the average density in rustic coffee plantations. According to Soto-Pinto and colleagues (2000) this increase in tree density does not affect yields as long as cover is managed at around 45%.

4.5 Floristic diversity and farmers perceptions of their function

An important source of knowledge about the effect of species diversity on ecosystem function comes from farmers who frequently manage floristic diversity for specific purposes related to the productivity, stability or sustainability of the agroecosystem (Méndez in press). In Chiapas, Mexico, farmers classified trees based primarily on tree morphology and functional attributes such as nurse characteristics, micro-climate modification, leaf litter production and decomposition rate and pest, disease and weed control. Pioneers, such as *Heliocarpus* aff. *popayensis*, *Callicarpa acuminata*, *Lippia myriocepahala*, *Liabum glabrum*, *Vernonia* spp. and *Croton* spp., were retained by farmers because of their nurse features (Soto-Pinto et al. in press). Farmers also mentioned that trees increase organic matter and therefore fertilise the soil, which is consistent with other research on farmer knowledge about tree-crop interactions (Thapa et al. 1995, Grossman 2003, Albertin and Nair 2004, Joshi et al. 2004). The role of trees in either encouraging or controlling pests, diseases and weeds was another feature perceived by farmers in Chiapas, as has also been reported from Nicaragua (Staver et al. 2001). According to farmer's knowledge in Chiapas, some trees "control weeds", while others "call for spiders and ants". In El Salvador and Nicaragua, coffee farmers that maintain tree diversity in their farms see themselves as contributing to environmental services, such as soil and water conservation, as well as providing habitat for birds and other animals (Méndez et al. 2002, Méndez and Bacon 2005).

5 Ecological stability: shade coffee as a high quality matrix

As previously discussed, biodiversity conservation efforts benefit from the habitat provided by shade coffee management. However, habitat quality is only one variable in determining the biodiversity found at a particular location. Landscape context and configuration are also important (Perfecto and Vandermeer 2002, Armbrecht and Perfecto 2003, Tscharntke and Brandl 2004). The diversification of habitat structure seems to encourage use of agricultural habitats by a wide range of bird species (Mellink 1991). However, this habitat use is further enhanced by the proximity and connectedness of intact natural habitats. Estrada and colleagues (1997) studied anthropogenic landscapes in Los Tuxtlas, Mexico and found higher avian diversity in arboreal agricultural habitats (coffee, cacao, and other tree crops) than non-arboreal (corn, hot peppers, and bananas) with isolating distance and disturbance regime as important variables affecting species richness. Agricultural islands and live fencerows were important elements that reduce physical and biotic isolation among remaining forest fragments. Comparing Brazilian cacao agricultural landscapes with high and low forest cover, Faria and colleagues (2006) found that cacao agroforests (*cabrucas*) contained a bird diversity similar to that in nearby forests, but that in the heavily deforested landscape context the proportion of forest-associated bird species was lower. Similarly, Petit and colleagues (1999) suggest that proximity of extensive forest and presence of riparian vegetation may enhance plantation habitat value for many forest-associated bird species in Panama. While the landscape context determines which species may use an agricultural habitat, habitat features determine the permeability of agricultural landscapes to forest associated biodiversity.

Permeability refers to how "hard" or "soft" a habitat edge or boundary is for organisms moving from one habitat to another (Stamps et al. 1987). The change in habitat at a boundary affects the movement behavior of individual species differently depending on variables such as available foraging habitat and perceived predation risk. In general, there is an assumption that the more similar the vegetation in an adjacent habitat, the "softer" the edge – in other words, the more willing an individual will be to cross the boundary. In Mexico, Witt (2001) found that small mammals were more willing to forage for and to transport seeds across the boundary between a forest and coffee grown under diverse shade than the boundary between the same forest and coffee grown under a simplified and heavily managed shade monoculture. Different taxa can have differential permeability into the same habitat. For example, native and "Africanized" bees traveled different distances from forest fragments containing their hives into an intensive Costa Rican coffee plantation. There were high levels of native bee diversity visiting coffee flowers only near forest edges ($\tilde{5}0$ m) while "Africanized" honeybees were able to enhance pollination at intermediate distances up to 800 m (Ricketts 2004).

Willingness to move between patches and to use agriculture may vary depending on the type of agriculture at the edge (i.e., the sharpness or fuzziness of the edge) with greater use by forest-associated birds at lower contrast edges (Kirk et al. 1996). Certification efforts to raise the quality of the agricultural matrix (i.e., increased density and diversity of shade trees) may help in decreasing the sharpness of the edge. Furthermore, at the landscape level, as more farmers adopt certification methods, increases in matrix habitat quality may show threshold effects by improving overall connectivity and reducing fragmentation. Forest-associated species may then be less constrained by dispersal abilities and distance from source populations in reserves.

Another characteristic of boundaries affected by the sharpness of the habitat change across the boundary is referred to as edge effect. The environmental characteristics found at distinct habitat boundaries, for example increased ambient light, exposure to wind, and lower humidity at the interface between sun coffee and primary forest, can dramatically increase gap formation and alter the structure and diversity of vegetation found at the edge. In tropical forests, this edge effect can penetrate 200 m into tropical forests, effectively reducing the acreage of primary forest habitat (Laurance 1991, Kapos et al. 1997). The use of less intensive agricultural practices, such as shade coffee, along habitat boundaries and in buffer zones can ameliorate the environmental characteristics that produce this edge effect. Reducing edge effect effectively increases the size of the remaining forest habitat.

Shade coffee management may produce high quality habitat for buffer zones around tropical forest reserves and biological corridors between reserves (Dietsch 2005). This intervening habitat between intact forest patches is often referred to as the matrix and is part of the generalized landscape mosaic, composed of farms and other land-uses with varying degrees of habitat quality. This is particularly true in agricultural landscapes where each farmer makes management decisions more or less independently. Similar to the permeability of boundaries, the composition and configuration of this agricultural patchwork determines how readily and far forest-associated biodiversity will move into the mosaic.

Observed species loss with increased intensification raises questions about the mechanisms that underlie the loss of forest-associated biodiversity as coffee agriculture is intensified. Habitat selection theory suggests that abundance for many species is linked to factors associated with vegetative structure (i.e., appropriate nest-site, food, and other resource availability, Cody 1985). Nonetheless, caution should be used when using census data to demonstrate conservation benefits (Van Horne 1983). Source-sink dynamics, a special case of metapopulation theory, suggest that in lower quality habitat patches, populations can be maintained by the influx of individuals from source populations (Hanski 1999, Robinson et al. 1995). Though this may mask species loss near habitat fragments, proximity or connectedness to higher quality habitat can facilitate movement between suitable patches (Noss 1983, Noss and Harris 1986). Theoretically, this increased connectivity is the basis for focusing some conservation attention on the matrix. In metapopulation models, a higher quality matrix that facilitates movement between populations generally buffers against extinctions but in some circumstances the outcome is less predictable (Vandermeer and Carvajal 2001).

As a matrix between forest patches, coffee agroecosystems can also provide a range of habitat benefits for forest-associated biodiversity (from Dietsch 2005):

- *Improved Connectivity* – Forage and cover allow for movement of species through the matrix that would otherwise be isolated.
- *Temporary Resources* – Individuals leave their preferred habitat for short-term forays to make use of available resources (daily or seasonal migration).
- *Temporary Refuge* – Allows some reproduction (perhaps very low and the benefit is reproductive experience) for individuals which would prefer to breed in higher quality habitats but are excluded by members of their own species or prevented from reaching those areas (by dispersal barriers). When able, adults (or the next generation) move to higher quality territories and younger individuals replace them.
- *Permanent Residents* – Individuals breed successfully and may hold high quality territories. Adults remain on the territories; young disperse because the habitat is saturated (i.e., when there are no available territories).

In general, specific data are needed to assess whether coffee agroecosystems represent viable habitats for forest biodiversity or transitory habitats providing short-term resources or dispersal corridors.

It is important to note that not all taxa are constrained by the quality of the matrix, particularly groups with high mobility. In Costa Rica, Rickets and colleagues (2001) found that moth diversity formed "halos" of relatively high species richness and abundance extending over 1 km around forest patches irrespective of the land management. Yet other studies have shown distinct losses of moth diversity and species composition on far smaller spatial scales around natural tropical forest remnants (Beck et al. 2002, Fiedler et al. this volume). In open agricultural habitats of Costa Rica, Daily and colleagues (2001) found no correlation between distance from large fragments and bird species richness or abundance, though a significant portion of the avian diversity were restricted to forest fragments. This may be due to a significant proportion of the avifauna that seem to readily use and even prefer the dense shrub habitat produced by coffee management (Dietsch 2003, Lindell et al. 2004).

While matrix habitat is generally viewed as less desirable and only of temporary value for conservation, this is more the case for mobile organisms. For some less mobile taxonomic groups, less intensively managed coffee agroecosystems may be utilized as suitable habitat or may require several generations of residence time within the matrix before gene flow or population connectivity can occur between habitat fragments. For many ant taxa, dispersal only occurs during the winged reproductive period, after mating the queen removes her wings and builds a nest. Consequently successful dispersal across managed landscapes may require several colonies within intervening habitat of adequate quality to ensure success (Perfecto and Vandermeer 2002). Epiphytes and other plants face similar obstacles. Similarly, in more degraded landscapes, shade coffee farms can act as a refuge for biodiversity, providing forest-like habitat where none is available. For example, shade coffee may have provided refuge habitat for birds and orchids in Puerto Rico when much of the island was deforested (Brash 1987, Nir 1988).

In coffee landscapes, farm or patch size (i.e., grain or resolution) of the landscape mosaic is another important consideration affecting matrix quality (Forman 1995). Coffee farm size varies greatly by region. In Chiapas, Mexico, producers holding 5 ha or less account for 91% of the coffee growers in the state and control around 61% of the coffee land area (Rice 1997). Nonetheless, large farms are common and generally 300 ha in size with some family holdings over 1000 ha managed contiguously (Dietsch, personal observation). Though some large farms are managed to maintain high levels of diverse shade, the management is still relatively uniform across the entire farm. In contrast, small farmer landscapes form a patchwork of differing management systems with hedgerows or boundary vegetation to demarcate farms. In at least one study, heterogeneous small farmer landscapes have higher levels of bird diversity than homogenous large farmer landscapes (Dietsch, unpublished data). This may be due to the added habitat value from boundary vegetation or a greater accumulation of shade tree diversity as individual small farmers make different management choices for the shade canopy.

At the landscape level, there are many confounding effects that must be considered when comparing farms or management systems. More research is needed to identify and quantify conservation value from landscape features in coffee areas. In particular, care must be taken to distinguish conservation benefits that derive from vegetation management practices from those that result from the landscape context of the farm (Mas and Dietsch 2004). In addition, the role of habitat heterogeneity within agricultural landscapes needs more attention. Both the landscape context and quality of the matrix contribute to the conervation value of coffee agriculture. With most mid-altitude forests in northern Latin America already converted to coffee, understanding the role of small reserves and forest fragments in contributing to landscape diversity will be an important part of establishing a conservation baseline (Schelhas and Greenberg 1996). This baseline can be connected with studies aiming to understand how sensitive taxa and species select and use habitat. In particular, knowledge of survival and reproductive rates under different management practices will be necessary to understand the population ecology of species in managed landscapes. Apparently poor habitat may provide some population benefits (Foppen et al. 2000, Murphy 2001). Thus, metapopulation dynamics should be evaluated at the landscape level and combine fecundity and mortality estimates to predict patch persistence. Though shade-grown coffee may benefit species that prefer closed canopy forests, there may be conservation trade-offs with those that prefer more open habitats. More research is needed to identify how management activities affect each species and identify characteristics associated with those adversely affected. This will involve an assessment of species benefits (resources) and hazards (predators and disease) within management systems and the surrounding landscape.

6 Socio-economic stability

6.1 Coffee certification programs

One area that may contribute to ecological and socio-economic stability for coffee-growing regions is sustainable coffee certification. Generally, sustainable certification programs attempt to promote ecologically sustainable use of traditional coffee agroecosystems and promote social justice for coffee farmers. The established programs currently fall into three distinct categories: organic, fair-trade, and biodiversity friendly (shade-grown).

Organic certification generally addresses the adverse effects of agrochemical use with the broad goal to certify intercropping and alternative agrochemical-free practices that maintain soil fertility and control pests (Vandermeer 1995). All organic certification programs must comply with international standards maintained by the International Federation of Organic Agriculture Movements (IFOAM). The history of organic certification for coffee farms originated with Finca Irlanda in Chiapas, Mexico. Today, Mexico remains one of the largest suppliers of organic coffee worldwide. Higher prices and good international consumer recognition facilitate the spread of organically certified coffees, though certification costs are still a barrier for many farmers (Dudley et al. 1997, Gobbi 2000). National certification initiatives in producer countries (i.e. Certimex in Mexico) may, however, help to reduce costs at the producer level (Giovannucci 2003).

Fair-trade certification focuses on small coffee farmers. Under fair-trade programs, farmers belonging to smallholder cooperatives are guaranteed a minimum price (US$1.26/lb and US$1.41/lb for organic coffee or at least US$0.05/lb above market prices in 2004), receive financing aid from buyers, and are expected to use some added income towards social goals or improving cooperative infrastructure. Until 2004, farmers did not pay fees to be certified, but now cooperatives pay between roughly US$2500-3500 initially and US$650 per year to renew their fair-trade status. Additionally, they are charged about US$0.02/kg of fair-trade coffee exported (http://www.fairtrade.net). The fair-trade movement traces its history to the creation in 1988 of Max Havelaar in the Netherlands. Since then, fair-trade expanded into other European countries, the US and Canada. All fair-trade certification falls under the Fair Trade Labeling Organizations International (FLO) based in Germany. FLO coordinates certification and standards, but national initiatives are responsible for marketing fair-trade in country. Although fair-trade does not require organic certification, FLO encourages farmers to use integrated crop management techniques and a large proportion of cooperatives with fair-trade certification have organic certification. In some regions, such as Chiapas, Mexico, farmer organizations report that FLO inspectors require organic certification before allowing fair-trade certification (S. Philpott, personal observation, 2004).

Biodiversity friendly coffee certification (known also as eco-friendly, shade-grown, or Bird-friendly™) is the newest and least regulated of the three certifications. At the international level, shade-grown coffee is certified by two programs, the Smithsonian Migratory Bird Center Bird-Friendly™ and Rainforest Alliance Certified (formerly Eco-OK). Bird-Friendly™ criteria are generally acknowledged as more stringent, and more difficult for farmers to meet (Philpott and Dietsch 2003, Mas and Dietsch 2004). Biodiversity friendly certification is based on the concept that complex vegetative structure and diversity protect associated biodiversity and thus certification criteria relate to characteristics such as shade cover, tree diversity, density, and height, species composition and distribution, and maintenance of epiphytes. An important distinction between the two programs is that Bird-Friendly™ requires that farms also be organically certified. Rainforest Alliance Certified, although promoting soil conservation and integrated pest management techniques, does not require organic certification. Costs for biodiversity friendly certification are used to pay travel and per-diem expenses for inspectors. Both SMBC and Rainforest Alliance try to minimize costs by using local certification agencies.

When the international price of coffee is below susbsistance level, many small farmers are forced out of the market or decide to sell their land or plant other crops that may not be as environmentally friendly as shade coffee. During this last coffee crisis many coffee areas in Colombia were transformed into cattle pasture or were bought by large producers who intensified the coffee in order to increase their production. Given this, certification programs that offer a premium price may contribute to the ecological and social stability of coffee growing regions.

6.2 Biodiversity, shade, and yield

The basic complaint that farmers state regarding biodiversity friendly certification is based on the conventional wisdom that as vegetation cover and complexity increases, yield and thus profits will decline. A growing number of studies have investigated the relationships between shade and yield finding either that 1) yields increase with shade removal, 2) yields do not differ in moderately shaded and sun plantations, or 3) maximum yields are found from approximately 35-65% shade cover, and are lower with either lower or higher cover (Soto-Pinto et al. 2000, Staver et al. 2001). Yet more indirect tests of this relationship show that high levels of shade is associated with increased bee diversity and increased fruit set (i.e. yield) (Klein et al. 2003a,b) and that increased shade levels reduce leaf rust infection and weed growth (Soto-Pinto et al. 2002). These increases in ecosystem services in turn directly relate to increased yields and/or profits. Thus some scientific and economic evidence indicates that biodiverse farms may be financially preferable due to 1) alternative products harvested, 2) price premiums for organic and / or biodiversity friendly certified coffee, and 3) complex (and not necessarily negative) relationships between shade, yield, and net incomes.

Shaded coffee farms offer additional income to farmers from timber and non-timber products made available from the shade trees (Somarriba et al. 2004). In Peru, shade tree products may account for about 30% of revenues – especially fruits and firewood rather than timber (R. Rice, unpublished data 2002). Escalante and colleagues (1987) found that fruits from the shade canopy accounted for 55-60% of income, and timber for 3%. In Costa Rica, fruits sales accounted for 5-11% of income from coffee growing areas (Lagemann and Heuveldop 1983). Having available products from the shade tree canopy reduces vulnerability to market fluctuations and household dependence on outside products while increasing local commerce. Especially when coffee prices are low, the alternative income can protect farmers from financial ruin.

The coffee certification programs mentioned above are currently providing some compensation to farmers for the trade-off between protecting biodiversity and yield. Gobbi (2000) found that making the transition from intensive monocultures or traditional polycultures to biodiversity friendly production and certification is economically viable because of price premiums made for certified coffee. Another perspective regarding price premiums for biodiversity friendly production is presented by Perfecto and colleagues (2005). By superimposing data from studies investigating both the relationship between shade cover and biodiversity and shade cover and yield, they investigated theoretically how changes in species richness of two taxa (ants and butterflies) related to coffee yields, and in turn to premiums needed to compensate for potential yield losses with increased biodiversity. For those taxa relatively less sensitive to shade (and thus yield) reduction (such as ants), only moderate price premiums are needed to compensate for yield losses. In contrast, for forest specialists or those taxa highly sensitive to habitat changes (such as butterflies), increasing shade levels needed to protect a critical number of species may result in lowers yields making larger price premiums necessary to compensate farmers.

But how high a premium are consumers willing to pay? According to Giovannuci (2003), in 2002, premiums paid to farmers were: organic \$0.15 - \$0.30/lb, fair-trade \$0.66/lb, and shade-grown \$0.10-\$0-.60/lb. The majority of coffee industry participants support price premiums, and consumers are somewhat willing to pay higher prices for sustainable coffees. This willingness is reflected in the high growth rates for this sector of the coffee market (26.5% annually in 2001-2002). Although consumption may be relatively low (1-2% of total market), sustainable coffees affect many coffee farm households and these price premiums offer substantial economic incentives for participants (Giovannucci 2003, Calo and Wise 2005).

6.3 The dilemma of a stable price for an internationally traded commodity

Price fluctuations in the coffee market may have widespread ecological effects. Low coffee prices in the early 1990's, following the disintegration of the International Coffee Agreement, led to landscape transformations (Nestel 1995). Many smallholders became more indebted and lost their lands (Renard 1992). Many other farmers transformed their coffee crops to more environmentally destructive crops as a result of low prices (Nestel 1995). The recent and more drastic price crashes in the late 1990's are having similar effects. Blackman and colleagues (2003) in a land-use survey of coffee-growing areas of Oaxaca, Mexico found that proximity to urban centers and receiving higher coffee prices (via membership in coffee cooperatives) reduced the probability of land clearing. They suggest that the collapse of coffee prices has increased deforestation but non-timber agroforestry crops, when offering price advantages, may help in forest conservation. Persistent low prices, below production costs have motivated farmers to abandon their crops. This may lead to short-term conservation benefits as forests regenerate in abandoned coffee areas. But in the long-term, farmers are more likely to convert to more intensive land uses. Furthermore, workers on large farms will be laid off or smallholders will necessarily turn to other crops thus contributing to more forest clearing (Vandermeer and Perfecto 2005).

Whatever the motive, fluctuating prices of coffee lead to widespread economic and ecological instability in regions where coffee is grown. At the global level, failure to address fluctuating coffee prices will incur additional conservation costs. With the selling price below production costs, farmers can no longer afford to harvest coffee, and are forced to seek other options to feed their families leading to greater deforestation. While shade-grown coffee premiums paid to farmers may encourage shade coffee in forests (Rappole et al. 2003, O'Brien and Kinnard 2003), encouraging less intensive agricultural practices and re-educating agricultural extension agents in biodiversity techniques may instead stabilize local economies reducing pressure on primary forests. Although farmers may hesitate to adopt shade-grown coffee on an individual basis, creating price incentives for shade, organic and fair trade coffee may help break this cycle and is precisely why making sure premiums are sufficient is an important consideration. A recent study has shown that organic and fair trade coffee reduce small farmers' livelihood vulnerability (Bacon 2005). While some farmers and researchers might be doubtful about potential benefits of biodiversity friendly certification (Rappole et al. 2003, O'Brien and Kinnard 2003), with proper price premiums and income alternatives, gains for individual farmers due to biodiversity friendly certification can have similar effects as organic and fair trade certification (Philpott and Dietsch 2003, Dietsch et al. 2004).

7 Conclusions

Shaded coffee as a reservoir of biodiversity and as a high quality matrix can provide ecological and socioeconomic stability to rainforest areas. At least a decade of research on the potential of coffee agroforests to conserve biodiversity has demonstrated that this management system has a high conservation value for biodiversity. More recent research, though, highlights the complexities of refocusing conservation in the matrix. While shade coffee has been shown to maintain high species richness for most taxa examined, not all taxa respond equally to varying levels of disturbance associated with coffee management. Furthermore, different components of a taxon can exhibit different levels of susceptibility to disturbance. Some forest specialists birds, for example, are eliminated even in highly shaded coffee plantations.

This reality highlights the need to have a landscape approach to biodiversity conservation, where the forest fragments and the mosaic of agroecosystems are seen as interdependent. At the landscape level shaded coffee farms can contribute to overall biodiversity in two ways. First, the coffee agroforests themselves serve as a habitat for many organisms and can maintain high levels of biodiversity. Second, the coffee agroforests represent a high quality matrix that facilitates migration between forest fragments allowing for the maintenance of diversity within a metapopulation structure. Therefore, even if the most rustic coffee plantations do not provide a permanent high quality habitat for certain forest specialists, it can serve as a permeable matrix that facilitates interfragment movement and allows the maintenance of a metapopulation structure.

The diversity contained both within the coffee agroforests as well as in the forest fragments adjacent to coffee farms have been shown to provide ecological services and maintain ecosystem functions of importance for the productivity and sustainability of the coffee agroecosystem, as well as more general environmental services. Functions and services such as pest control, pollination, enhanced soil fertility and carbon sequestration have been shown to be be diminished with the intensification of the agroecosytem. Small farmers in Latin America rely on these ecosystem services provided by biodiversity more than large resource-rich farmers who may substitute some biological functions with agrochemicals. Therefore, the maintenance of biodiversity and its associated functions is important for the ecological stability of the coffee growing regions of Latin America where most of the farmers own/manage less than 5 hectares. Likewise, coffee certification programs that encourage environmental protection and biodiversity conservation and provide a premium price can improve farmer's livelihoods increasing socioeconomic stability within the coffee growing regions of Latin America.

References

Aide TM, Grau HR (2004) Globalization, migration and Latin American ecosystems. Science 305: 1915-1916

Albertini A, Nair PKR (2004) Farmers' perpective on the role of shade trees in coffee production systems: an assesment from the Nicoya peninsula, Costa Rica. Hum Ecol 32: 443-463

Andow DA (1991) Vegetational diversity and arthropod population response. Ann Rev of Entomol 36: 561-586

Armbrecht I (2003) Diversity and function of leaf-litter ants in Colombian coffee agroecosystems. PhD Dissertation. University of Michigan, 242 pp.

Armbrecht I, Gallego MC, Rivera L, Chacón P (in press) Efecto de la intensificación del cultivo de café sobre Poneromorfas de Risaralda. In: Jiménez E, Lozano F, Fernández F (eds) Poneromorfas de Colombia. Instituto Alexander von Humboldt (Colombian Ministry of Environment) and Instituto de Ciencias Naturales, Universidad Nacional de Colombia, Bogotá Colombia.

Armbrecht I, Perfecto I (2003) Litter-twig dwelling ant species richness and predation potential within a forest fragment and neighbouring coffee plantations of contrasting habitat quality in Mexico. Agric Ecosyst Environ 97: 107-115

Armbrecht I, Perfecto I, Vandermeer J (2004) Enigmatic biodiversity correlations: ant diversity responds to diverse resources. Science 304: 284-286

Armbrecht I, Rivera L, Perfecto I (2005) Reduced diversity and complexity in the leaf litter ant assemblage of Colombian coffee plantations. Conserv Biol 19: 897-907

Bacon C (2005) Confronting the coffee crisis: can fair trade, organic, and specialty coffees reduce small-farme vulnerability in northern Nicaragua? World Dev 33: 497-511

Balvanera P, Daily GC, Ehrlich PR, Ricketts TH, Bailey SA, Kark S, Kremen C, Pereira H (2001) Conserving biodiversity and ecosystem services. Science 291: 2047-2047

Bandeira FP, Martorell C, Meave J, Caballero J (2005) The role of rustic coffee plantations in the conservationof wild tree diversity in the Chinantec region of Mexico. Biodiversity Conserv 14: 1225-1240

Barbera N, Hilje L, Hanson P, Longino JT, Carballo M, de Melo E (2004) Diversidad de especies de hormigas en un gradiente de cafetales orgánicos y convencionales. Manejo Integrado de Plagas y Agroecología (Costa Rica) 72: 60-71

Beck J, Schulze CH, Linsenmair KE, Fiedler K (2002) From forest to farmland: diversity of geometrid moth along two habitat gradients on Borneo. J Trop Ecol 18: 33-51

Benitez J, Perfecto I (1990) Patrón de distribución de hormigas en diferentes sistemas de café. Agroecología Neotropical 1: 11-15.

Blackman A, Albers H, Ávalos-Sartorio B, Crooks B (2003) Land cover in a managed forest ecosystem: Mexican shade coffee. Discussion Paper 03–60 Resources for the Future, Washington, DC.

Botero JE, Baker PS (2002) Coffee and biodiversity: a producer-country perspective. Coffee Futures, Colombia.

Brash AR (1987) The history of avian extinction and forest conversion on Puerto Rico. Biol Conserv 39: 97-111

Callo-Concha D, Krishnamurthy K, Alegre J (2004) Carbon sequestration by Amazonian agroforestry systems. Revista Chapingo. Serie Ciencias Forestales y del Ambiente, 8: 101-106

Calo M, Wise TA (2005) Revaluing peasant coffee productioin: organic and fair trade markets in Mexico. Global Development and Environment Institute, Tuft University

Calvo L, Blake J (1998) Bird diversity and abundance on two different shade coffee plantations in Guatemala. Birds Conserv Int 8: 297-308

Carlo TA, Collazo JA, Groom M J (2004) Influences of fruit diversity and abundance on bird use of two shaded coffee plantations. Biotropica 36: 602-614

Cody ML (1985) Habitat Selection in Birds. Physiological Ecology. Academic Press, San Diego, California.

Cruz-Angon A, Greenberg R (2005) Are epiphytes important for birds in coffee plantations? An experimental assessment. J Appl Ecol 42: 150-159

Cruz-Lara L, Lorenzo C, Soto L, Naranjo E, Ramírez-Marcial N (2004). Diversidad de mamíferos en cafetales y selva mediana de Las Cañadas de la Selva Lacandona, Chiapas, México. Acta Zool Mex 20: 63-81

Daily GC, Caballos G, Pacheco J, Suzan G, Sánchez-Azofeifa A (2003) Countryside biogeography of neotropical mammals: conservation opportunities in agricultural landscapes of Costa Rica. Conserv Biol 17: 1814-1826

Daily GC, Ehrlich PR, Sánchez-Azofiefa GA (2001) Countryside biogeography: Use of human-dominated habitats by the avifauna of southern Costa Rica. Ecol Appl 11: 1-13

Dalla Rosa KR (1993) Nitrogen fixing trees as atoll soil builders. Agroforestry for the Pacific Technologies. No. 4. Winrock International. USA.

De Marco P Jr, Monteiro Coelho F (2004) Services performed by ecosystems: forest remnants influence agricultural cultures' pollination and production. Biodiversity Conserv 13: 1245-1255

Dietsch TV (2003) Conservation and ecology of birds in coffee agroecosystems of Chiapas, Mexico. PhD Dissertation. University of Michigan, School of Natural Resources and Environment, Ann Arbor, MI.

Dietsch TV (2005) Eco-labeling in Latin America: Providing a scientific foundation for consumer confidence in market-based conservation strategies. In: Environmental issues in Latin America and the Caribbean, Romero A, West S (eds), Springer, Berlin, pp 175-203

Dietsch TV, Philpott SM, Rice RA, Greenberg R, Bichier P (2004) Conservation policy in coffee landscapes. Science 303: 625

Donald PF (2004) Biodiversity impacts of some agricultural commodity production systems. Conserv Biol 18: 17-37

Dudley N, Elliott C, Stolton S (1997) A framework for environmental labeling. Environment (Wash. DC) 39: 42-45

Dyer LA, Letourneau DK (1999) Relative strengths of top-down and bottom up forces in a tropical forest community. Oecologia 119: 265-274

Escalante E, Aguilar, A Lugo R (1987) Identificación, evaluación y distribución espacial de especies utilizadas como sombra en sistemas tradicionales de café (*Coffea arabica*) en dos zonas del estado de Trujillo, Venezuela. Venezuela Forestal 3: 50-62

Estrada A, Coates-Estrada R (2001) Species composition and reproductive phenology of bats in a tropical landscape at Los Tuxtlas, Mexico. J Trop Ecol 17: 627-646.

Estrada A, Coates-Estrada R, Dadda AA, Cammarano P (1998) Dung and carrion beetles in tropical rain forest fragments and agricultural habitats at Los Tuxlas, Mexico. J Trop Ecol 14: 577-593

Estrada A, Coates-Estrada R, Meritt Jr. DA (1997) Anthropogenic landscape changes and avian diversity at Los Tuxtlas, Mexico. Biodiversity Conserv 6: 19-43

FAO (Foor and Agriculture Organization) (2002). Atricultural production statistics. FAO, Rome. Available from http://apps.fao.org/ (accessed March 2002). FAO, Rome.

Faria D, Laps RR, Baumgarten J, Cetra M (2006) Bat and bird assemblages from forests and shade cacao plantations in two contrasting landscapes in the Atlantic Forest of southern Bahia, Brazil. Biodiversity Conserv 15: 587-612

Ferguson BG, Vandermeer JH, Morales H, Griffith DM (2003) Post-agricultural succession in El Petén, Guatemala. Conserv Biol 17: 818-828

Fiedler K, Hilt N, Brehem G, Schulze CH (this volume) Moth at tropical forest margins – how mega-diverse insect assemblages respond to forest disturbance and recovery. In Tscharntke T, Leuschner C, Zeller M, Guhardja E, Bidin A (eds) The stability of tropical rainforest margins, linking ecological, economic and social constrains. Springer, Berlin Heidelberg New York

Foppen RPB, Chardon JP, Liefveld W (2000) Understanding the role of sink patches in source-sink metapopulations: Reed Warbler in an agricultural landscape. Conserv Biol 14: 1881-1892

Forman RTT (1995) Land mosaics: The ecology of landscapes and regions. Cambridge University Press, Cambridge

Franco A, Cárdenas R, Montoya EC, Zenner de Polanía I (2003) Hormigas asociadas con insectos chupadores en la parte aérea del cafeto. Revista Colombiana de Entomología 29: 95-105

Gallego MC, Armbrecht I (in press) Depredación por hormigas sobre la broca del café en cafetales cultivados bajo dos niveles de sombra en Colombia. Revista Manejo Integrado de Plagas, Costa Rica.

Gallinas SE, Mandujano S, González-Romero A (1996) Conservation of mammalian biodiversity in coffee plantations of Central Veracruz, Mexico. Agroforestry Systems 33: 13-27

García R, Botero J (2005) Hormigas cazadoras (Formicidae: Ponerinae) en paisajes cafeteros de Colombia. Simposio Insectos Sociales en Paisajes Rurales y Agroecosistemas. *En* Libro de Resúmenes V Coloquio de la Unión Internacional para el Estudio de los Insectos Sociales. Universidad del Valle, Cali, Septiembre 7-9, 2005

Giovannucci D (2003) The state of sustainable coffee: A study of twelve major markets. World Bank, Washington DC

Gobbi JA (2000) Is biodiversity-friendly coffee financially viable? An analysis of five different coffee production systems in western El Salvador. Ecol Econ 33: 267-281

Greenberg R, Bichier P, Sterling J (1997) Bird populations in rustic and planted shade coffee plantations of eastern Chiapas, Mexico. Biotropica 29: 501-514

Greenberg R, Rice R (2000) Shade-grown coffee and biodiversity in Peru. Smithsonian Migratory Bird Center, Washington, DC.

Gresser C, Tickell S (2002) Mugged: Poverty in your cup. Oxfam International, Washington, D.C.

Griffith DM (2000) Agroforestry: a refuge for tropical biodiversity after fire. Conserv Biol 14: 325-326

Grossman JM (2003) Exploring farmer knowledge of soil processes in organic coffee systems of Chiapas, Mexico. Geoderma 111: 267-387

Hanski I (1999) Metapopulation Ecology. Oxford University Press, New York

Hanson P (1991) Los parasitoides asociados al cafeto en Costa Rica. Manejo Integrado de Plagas (Costa Rica) 20-21: 8-10

Hardner J, Rice R (2002) Rethinking green consumerism. Sci Am 286: 88-95

Hayes TB (2004) There is no denying this: Defusing the confusion about atrazine. BioScience 54: 1138-1149

Hecht S, Kandel S, Rosa H (2002) Globalization, forest resurgence and environmental politics in El Salvador. San Salvador, PRISMA.

Hietz P (2005) Conservation of vascular epiphyte diversity in Mexican coffee plantations. Conserv Biol 19: 391-399

Horner-Devine MC, Daily GC, Ehrlich PR, Boggs CL (2003) Countryside biogeography of tropical butterflies. Conserv Biol 17: 168-177

Ibarra-Núñez G García-Ballinas JA (1998) Diversidad de tres familias de arañas tejedoras (Araneae:Araneidae, Tetragnathidae, Theridiidae) en cafetales del Soconusco, Chiapas, México. Folia Entomológica Mexicana 102: 11-20

Ibarra-Núñez, G, García JA Moreno MA (1995) Diferencias entre un cafetal orgánico y uno convencional en cuanto a diversidad y abundancia de dos grupos de insectos. Memorias Primera Conferencia Internacional IFOAM sobre Café Orgánico. Universidad Autónoma Chapingo. México, pp 115-129

Jedlicka J, Greenberg G, Perfecto I, Philpott S, Dietch TV (2006) Seasonal foraging niche shift of tropical avian residents: resource competition at work? J Trop Ecol 22: 1-11

Johnson M D (2000) Effects of shade-tree species and crop structure on the winter arthropod and bird communities in a Jamaican shade coffee plantation. Biotropica 32: 133-145

Jong De BH, Montoya-Gómez G, Nelson K, Soto-Pinto ML, Taylor J, Tipper R (1995) Community forest management and carbon sequestration: a feasibility study from Chiapas, Mexico. Interciencia 20: 409-416

Jong De BH, Soto-Pinto ML, Montoya-Gómez G, Nelson K, Taylor J, Tipper R (1997) Forestry and agroforestry land-use systems for carbon mitigation: an analysis in Chiapas, Mexico. In: Adger WN, Pettenella D, Whitby M (eds), Climate-change mitigation and European land-use policies. CAB International, pp 269-246

Joshi L, Shrestha PK, Moss C, Sinclair FL (2004) Locally derived knowledge of soil fertility and its emerging role in integrated natural resource management. In: van Noordwijk M, Cadisch G, Ong C (eds.). Below-ground interactions in tropical agroecosystems: concepts and models with multiple plant components. CAB International, Wallingford, pp 17-39

Kaimowitz D (1996) Livestock and deforestation in Central America in the 1980s and 1990s: A policy perspective. CIFOR Special Publication, Bogor

Kapos V, Wandelli E, Camargo JL, Ganade G (1997) Edge-related changes in environment and plant responses due to forest fragmentation in central Amazonia. In: Laurance WF, Bierregaard Jr. RO (eds), Tropical forest remnants: Ecology, management, and conservation of fragmented communities, The University of Chicago Press, Chicago, Illinois, pp 33-44

Kirk DA, Evenden MD, Mineau P (1996) Past and current attempts to evaluate the role of birds as predators of insect pests in temperate agriculture. In: Nolan Jr., Ketterson ED (eds), Current Ornithology Vol. 13, V. Plenum Press, New York, New York, pp 175-269

Klein AM, Steffan-Dewenter I, Buchori D, Tscharntke T (2002) Effects of land-use intensity in tropical agroforestry systems on coffee flower-visiting and trap-nesting bees and wasps. Conserv Biol 16: 1003-1014

Klein AM, Steffan-Dewenter I, Tscharntke T (2003a) Bee pollination and fruit set of *Coffea arabica* and *C. canephora* (Rubiaceae). Am J Bot 90: 153-157

Klein AM, Steffan-Dewenter I, Tscharntke T (2003b) Fruit set of highland coffee increases with the diversity of pollinating bees. Proc R Soc Lond, B 270: 955-961

Klein AM, Steffan-Dewenter I, Tscharntke T (2003c) Pollination of *Coffea canephora* in relation to local and regional agroforestry management. J Appl Ecol 40: 837-845

Komar 0, Dominguez JP (2002) Efectos del estrato de sombra sobre poblaciones de anfibios, reptiles y aves en plantaciones de café de El Salvador: Implicaciones para programas de certificación. Actas del Simposio de Café y Biodiversidad, Revista Protección Vegetal, Año 12, No. 2: 61-62. Proceedings from Biodiversity and Coffee Symposium at Sociedad Mesoamericana para la Biología y la Conservación Fifth Annual Meeting in San Salvador, El Salvador, October 2001.

Krantz S (2005) Butterfly diversity in a coffee agroecosystem mosaic in Chiapas, Mexico. Masters Thesis, University of Michigan, Ann Arbor, MI

Lachaud JP, García-Ballinas JA (1999) Biodiversidad de la myrmeco-fauna (Hymenoptera:Formicidae) en los agroecosistemas de café y cacao en Chiapas, México: Enfoque sobre las subfamilias Ponerinae y Cerapachinae. In: Memorias del XXXIV Congreso Nacional de Entomología, Sociedad Mexicana de Entomología, Aguascalientes, México, pp 79-83

LaFranchi H (2001) Economic upheaval over coffee-rising production and falling demand bankrupt farms from Indonesia to Mexico. Christian Science Monitor, August 15, 2001 (http://www.jubilee2000uk.org/finance/economic_unheaval_coffee.htm)

Lagemann J, Heuveldop J (1983) Characterization and evaluation of agroforestry systems: the case of Acosta-Puriscal, Costa Rica. Agroforestry Systems 1: 101-115.

Langellotto GA, Denno RF (2004) Responses of invertebrate natural enemies to complex-structured habitats: a meta-analytical synthesis. Oecologia 139: 1-10

Laurance WF (1991) Edge effects in tropical forest fragments: Application of a model for the design of nature reserves. Biol Conserv 57: 205-219

Lindell CA, Chomentowski WH, Zook JR (2004) Characteristics of bird species using forest and agricultural land covers in southern Costa Rica. Biodiversity Conserv 13: 2419-2441

Loreau M, Mouquet N, González A (2003) Biodiversity as spatial insurance in heterogeneous landscapes. Proc Natl Acad Sci USA 100: 12765-12770

Lyngbaek AE, Muschler RG, Sinclair FL (2001) Productivity and profitability of multistrata organic versus conventional coffee farms in Costa Rica. Agroforestry Systems 53: 205-213

Manrique AJ, Thimann RE (2002) Coffee (*Coffea arabica*) pollination with Africanized honeybees in Venezuela. Interciencia 27: 414-416

Mas AH, Dietsch TV (2003) An index of management intensity for coffee agroecosystems to evaluate butterfly species richness. Ecol Appl 13: 1491-1501

Mas AH, Dietsch TV (2004) Linking shade coffee certification programs to biodiversity conservation: Butterflies and birds in Chiapas, Mexico. Ecol Appl 14: 642-654

McCann C, Williams-Guillén K, Koontz FW, Roque Espinoza AA, Martínez Sánchez JC, Koontz C (2003) Shade coffee plantations as wildlife refuge for mantled howler monkeys (*Alouatta palliata*) in Nicaragua. In: Marsh LK (ed.) Primates in fragment, Kluwer Academic Press, New York, pp 321-341

Mellink E (1991) Bird communities associated with three traditional agroecosystems in the San Luis Potosi Plateau, Mexico. Agric Ecosyst Environ 36: 37-50

Méndez VE (in press) Farmer livelihoods and biodiversity conservation in a coffee landscape of El Salvador. In: Bacon C, Méndez VE, Gliessman SR, Goodman D, Fox JA (eds) Confronting the coffee crisis: Sustaining livelihoods and ecosystems in Mexico and Central America, Food, Energy and Environment Series, MIT Press, Canbridge, MA

Méndez VE Bacon C (2005) Medios de vida y conservación de la biodiversidad arbórea: Las experiencias de las cooperativas cafetaleras en El Salvador y Nicaragua, LEISAR Revista de Agroecología, abril 2005: 27-30

Méndez VE, Herrador D, Dimas L, Escalante M, Diaz O, Garcia M (2002) Café con sombra y pagos por servicios ambientales: Riesgos y oportunidades para impulsar mecanismos con pequeños agricultores de El Salvador, PRISMA, Fundación Ford, San Salvador, El Salvador.

Moguel P, Toledo VM (1999) Biodiversity conservation in traditional coffee systems of Mexico. Conserv Biol 13: 11-21

Molina J (2000) Diversidad de escarabajos coprófagos (Scarabaeidae: Scarabaeinae) en matrices de la zona cafetera (Quindío-Colombia). In: Federación Nacional de Cafeteros de Colombia (eds) Memorias Foro Internacional Café y Biodiversidad. Agosto 10-12, Chinchiná, Colombia. p 29

Monro A, Gauld I (2002) Apoyando el inventario y manejo de la diversidad biológica en cafetales de sombra salvadoreños. In: Monro A, Peña MC (eds.) Actas del Simposio Café y Biodiversidad. Revista Protección Vegetal, 12(2). San Salvador, pp 39-50

Moron MA, López-Méndez JA (1985) Analysis of the necrophilic fauna of a coffee plantation in Soconusco Chiapas Mexico. Folia Entomológica Mexicana 63: 47-60

Muriel SB, Vélez LD (2004) Evaluando la diversidad de plantas en los agroecosistemas como estrategia para el control de plagas. Manejo Integrado de Plagas y Agroecología (Costa Rica) 71: 13-20

MurphyMT, (2001) Source-sink dynamics of a declining Eastern Kingbird population and the value of sink habitats. Conserv Biol 15: 737-748

Nestel D (1995) Coffee and Mexico: International market, agricultural landscape, and ecology. Ecological Economics 15: 165-179

Nestel D, Dickschen F (1990) Foraging kinetics of ground ant communities in different Mexican coffee agroecosystems. Oecologia 84: 58-63

Nestel D, Dickschen F, Altieri MA (1993) Diversity patterns of soil macrocoleoptera in Mexican shaded and unshaded coffee agroecosystems: An indication of habitat perturbation. Biodiversity Conserv 2: 70-78

Nir MA (1988) The survivors: Orchids on a Puerto Rican coffee finca. Am Orchid Soc Bul 57: 989-995

Nogueira-Neto P, Carvalho A, Antunes H (1959) Efcito da exclusa do dos insetos polinizadores na produciao do cafe´ Bourbon. Bragantia 18: 441-468

Nolasco M (1985) Café y sociedad en México. Centro de Ecodesarrollo, México, DF.

Noss RF (1983) A regional landscape approach to maintain diversity. BioScience 33: 700-706

Noss RF, Harris LD (1986) Nodes, networks, and MUMs: preserving diversity at all scales Environ Manage 10: 299-309

O'Brien TG, Kinnaird MF (2003) Caffeine and conservation. Science 300: 587-587

Panayotou T, Restrepo C, Faris R (1997) Green book: From peace to sustainable development: A call to action. FUSADES/CODES, El Salvador

Peeters LYK, Soto-Pinto L, Perales H, Montoya G, Ishiki M (2003) Coffee production, timber and firewood in Southern México. Agric Ecosyst Environ 95: 481-493

Perfecto I (1994) Foraging behavior as a determinant of asymmetric competitive interactions between two ant species in a tropical agroecosystem. Oecologia 98: 184-192

Perfecto I, Armbrecht I (2003) The coffee agroecosystem in the Neotropics: Combining ecological and economic goals. In: Vandermeer J (ed), Tropical agroecosystems, CRC Press, Boca Raton, pp 159-194.

Perfecto I, Mas A, Dietsch TV, Vandermeer J (2003) Conservation of biodiversity in coffee agroecosystems: A tri-taxa comparison in southern Mexico. Biodiversity Conserv 12: 1239-1252

Perfecto I, Rice RA, Greenberg R, VanderVoort ME (1996) Shade coffee: A disappearing refuge for biodiversity. Bioscience 46: 598-608

Perfecto I, Snelling R (1995) Biodiversity and tropical ecosystem transformation: ant diversity in the coffee agroecosystem in Costa Rica. Ecol Appl 5: 1084-1097

Perfecto I, Vandermeer J (2002) Quality of agroecological matrix in a tropical montane landscape: Ants in coffee plantations in southern Mexico. Conserv Biol 16: 174-182

Perfecto I, Vandermeer J, Lopez-Bautista G, Ibarra-Núñez G, Greenberg R, Bichier P, Langridge S (2004) Greater predation in shaded coffee farms: The role of resident neotropical birds. Ecol 85: 2677-2681

Perfecto I, Vandermeer J, Mas AH, Soto-Pinto L (2005) Biodiversity, yield, and shade coffee certification. Ecol Econ 54: 435-446

Perfecto I, Vandermeer JH (1996) Microclimatic changes and the indirect loss of ant diversity in a tropical agroecosystem. Oecologia 108: 577-582

Perfecto I, Vandermeer V, Hanson P, Cartin V (1997) Arthropod diversity loss and the transformation of a tropical agroecosystem. Biodiversity Conserv 6: 935-945

Petit LJ, Petit DR, Christian DG, Powell HDW (1999) Bird communities of natural and modified habitats in Panama. Ecography 22: 292-304

Philpott SM (2005) Trait-mediated effects of parasitic phorid flies (Diptera: Phoridae) on ant (Hymenoptera: Formicidae) competition and resource access in coffee agroecosystems. Environ Entomol 34: 1089-1094

Philpott SM, Dietsch TV (2003) Coffee and conservation: a global context and the value of farmer involvement. Conserv Biol 17: 1844-1846

Philpott SM, Foster PF (2005) Nest-site limitation in coffee agroecosystems: artificial nests promote maintenance of arboreal ant diversity. Ecol Appl 15: 1478-1485

Philpott SM, Greenberg R, Bichier P (2005) The influence of ants on the foraging behavior of birds in an agroforest. Biotropica 37: 468-471

Philpott SM, Greenberg R, Bichier P, Perfecto I (2004) Impacts of major predators on tropical agroforest arthropods: comparison within and across taxa. Oecologia 140: 140-149

Philpott SM, Perfecto I, Vandemeer J (2006a) Effects of management intensity and season on arboreal ant diversity and abundance in coffee agroecosystems. Biodiversity Conserv 15: 139-155

Philpott SM, Uno S, Maldonado J (2006b) The importance of ants and high-shade management to coffee pollination and fruit weight in Chiapas, Mexico. Biodiversity Conserv 15: 487-501

Pineda E, Moreno C, Escobar F, Halffter G (2005) Frog, bat and dung beetle diversity in the cloud forest and coffee agroecosystems of Veracruz, Mexico. Conserv Biol 19: 400-410

Ramos-Suárez MP, Morales H, Ruiz-Montoya L, Soto-Pinto L, Rojas-Fernández P (2002) ¿Se mantiene la diversidad de hormigas con el cambio de bosque mesófilo a cafetales? Revista Protección Vegetal 12 (2), The Natural History Museum (London) & Universidad de El Salvador.

Rao MR, Singh MP, Day R (2000) Insect pest problems in tropical agroforestry systems: Contributory factors and strategies for management. Agroforestry Systems 50: 243-277

Rappole JH, King DI, Vega Rivera JH (2003) Coffee and conservation. Conserv Biol 17: 334-336

Raw A, Free JB (1977) Pollination of coffee (*Coffea arabica*) by honeybees. Trop Agric 54: 365-370

Relyea RA (2005) The lethal impact of Roundup on aquatic and terrestrial amphibians. Ecol Appl 15: 1118–1124

Renard M (1992) El cafe en el tratado de libre comercio. Cuadernos Agrarios 2: 129-135

Reynoso SR (2004) Estructura, composición florística y diversidad del bosque y cafetales de la Reserva de la Biosfera El Triunfo, Chiapas, México. BSc. Thesis. Universidad de Ciencias y Artes de Chiapas. Tuxtla Gutierrez, Chiapas. 94 p.

Rice R, Ward J (1996) Coffee, conservation, and commerce in the Western Hemisphere (White Paper 2). Smithsonian Migratory Bird Center, Washington, DC.

Rice RA (1997) The land use patterns and the history of coffee in eastern Chiapas, Mexico. Agriculture and Human Values 14: 127–143

Rice RA, Greenberg R (2000) Cacao cultivation and the conservation of biological diversity. Ambio 29: 167-173.

Ricketts T (2004) Tropical forest fragments enhance pollinator activity in nearby coffee crops. Conserv Biol 18: 1262-1271

Ricketts TH (2001) The matrix matters: Effective isolation in fragmented landscapes. Am Nat 158: 87-99

Ricketts TH, Daily GC, Ehrlich PR, Fay JP (2001) Countryside biogeography of moths in a fragmented landscape: biodiversity in native and agricultural habitats. Conserv Biol 15: 378-388

Ricketts TH, Daily GC, Ehrlich PR, Michener CD (2004) Economic value of tropical forest to coffee production. Proc Natl Acad Sci USA 101: 12579-12582

Rivera L, Armbrecht I (2005) Diversidad de tres gremios de hormigas en cafetales de sombra, de sol y bosques de Risaralda. Revista Colombiana de Entomología 31: 89-96

Roberts DL, Cooper RJ, Petit LJ (2000) Use of premontante moist forest and shade coffee agroecosystems by army ants in western Panama. Conserv Biol 14: 192-199

Robinson SK, Thompson FR, Donovan TM, Whitehead DR, Faaborg J (1995) Regional forest fragmentation and the nesting success of migratory birds. Science 267: 1987-1990

Rojas L, Godoy C, Hanson P, Kleinn C, Hilje L (2001) A survey of homopteran species (Auchenorrhyncha) form coffee shrubs and poro and laurel trees in shaded coffee plantations, in Turrialba, Costa Rica. Rev Biol Trop 49: 1057-1065

Romero-Alvarado Y, Soto-Pinto L, García-Barrios L, Barrera-Gaytán JF (2002) Coffee yields and soil nutrients under the shade of Inga sp. vs. multiple species in Chiapas, Mexico. Agroforestry Systems 54: 215-224

Root RB (1973) Organization of plant-arthropod associations in simple and diverse habitats: the fauna of collards (Brassica oleracea). Ecol Monogr 43: 95-124

Roubik DW (2002) Tropical agriculture - the value of bees to the coffee harvest. Nature 417: 708

Sadeghian S (2000) Diversidad de mesoorganismos del suelo en la zona cafetera. In: Federación Nacional de Cafeteros de Colombia (eds.) Memorias Foro Internacional Café y Biodiversidad. Agosto 10-12, Chinchiná, Colombia, pp 20

Schelhas J, Greenberg R (1996) Introduction: The value of forest patches. In: Schelhas J, Greenberg R (eds.) Forest patches in tropical landscapes, Island Press, Washington DC.

Schroth G, Krauss U, Gasparotto L, Duarte-Aguilar JA, Vohland K (2000) Pests and diseases in agroforestry systems of the humid tropics. Agroforestry Systems 50: 199-241

Schulze CH, Waltert M, Kessler PJA, Pitopang R, Shahabuddin, Veddeler D, Mühlenberg M, Gradstein SR, Leuschner C, Steffan-Dewenter I, Tscharntke T (2004) Biodiversity indicator groups of tropical land-use systems comparing plants, birds, and insects. Ecol Appl 14: 1321- 1333

Solís-Montero L, Flores-Palacios A, Cruz-Angón A (2005) Shade-coffee plantations as refuges for tropical wild orchids in Central Veracruz, México. Conserv Biol 19: 908-916

Somarriba E, Harvey CA, Samper M, Anthony F, González J, Staver C, Rice RA (2004) Biodiversity conservation in Neotropical coffee (*Coffea* arabica) plantations. *In* Schroth, G, da Fonseca GA, Harvey C, Gascon C, Vasoncelos HL Izac AMN (eds) Agroforestry and biodiversity conservation in tropical landscapes. Island Press, Washington, D.C.pp 1-14

Sossa J, Fernandez F (2000) Himenopteros de la franja cafetera del departamento del Quindio. In: Numa C, Romero LP (eds) Biodiversidad y sistemas de producción cafetera en el departamento del Quindio. Instituto Alexander von Humboldt. Dic. 2000. Bogota, Colombia, pp 168-180

Soto-Pinto L (2001) Estudio Agroecológico del cafetal con sombra en comunidades indígenas de Chiapas. PhD Thesis. Universidad Nacional Autónoma de México. México D.F. 171 pp

Soto-Pinto L, Perfecto I, Caballero-Nieto J (2002) Shade over coffee: Its effects on berry borer, leaf rust and spontaneous herbs in Chiapas, Mexico. Agroforestry Systems 55: 37-45

Soto-Pinto L, Perfecto I, Castillo-Hernandez J, Caballero-Nieto J (2000) Shade effect on coffee production at the northern Tzeltal zone of the state of Chiapas, Mexico. Agric Ecosyst Environ 80: 61-69

Soto-Pinto L, Romero-Alvarado Y, Caballero-Nieto J, Segura-Warnholtz G (2001) Woody plant diversity and structure of shade-grown-coffee plantations in Northern Chiapas, México. Rev Biol Trop 49: 977-987

Soto-Pinto L, Villalvazo-López V, Jiménez-Ferrer G, Ramírez-Marcial N, Montoya G, Sinclair FL (in press) The role of local knowledge in determining shade composition of multistrata coffee systems in Chiapas, México. Biodiversity Conserv

Stamps JA, Buechner M, Krishnan VV (1987) The effects of edge permeability and habitat geometry on emigration from patches of habitat. Am Nat 129: 533-552

Staver C, Guharay F, Monterroso D, Muschler RG (2001) Designing pest-suppressive multistrata perennial crop systems: shade-grown coffee in Central America. Agroforestry Systems 53: 151-170

Steffan-Dewenter I (2002) Landscape context affects trap-nesting bees, wasps, and their natural enemies. Ecol Entomol 27: 631-637

Tejeda-Cruz C, Sutherland WJ (2004) Bird responses to shade coffee production. Anim Conserv 7: 169-179

Thapa B, Sinclair FL, Walker DH (1995) Incorporation of indigenous knowledge and perspectives in agroforestry development. Part two: Case-study on the impact of explicit representation of farmers' knowledge. Agroforestry Systems 30: 249 - 261

Torres J (1984) Diversity and distribution of ant communities in Puerto Rico. Biotropica 16: 296-303

Tscharntke T, Brandl R (2004) Plant-insect interactions in fragmented landscapes. Annu Rev Entomol 49: 405-430

Tscharntke T, Klein AM, Kruess A, Steffan-Dewenter I, Thies C (2005) Landscape perspectives on agricultural intensification and biodiversity – ecosystem service management. Ecol Let 8: 857-874

Tylianakis J, Veddeler D, Lozada T, Lopez RM, Benítez P, Klein AM, de Koning GHJ, Olschewski R, Veldkamp E, Navarrete H, Onore G, Tscharntke T (2004) Biodiversity of land-use systems in costal Ecuador and bioindication using trap-nesting bees, wasps, and their natural enemies. Lyonia 6: 7-15

Valencia CA (2004) Las mariposas diurnas como indicadores biológicas en el cultivo del café. Trabajo de grado para optar el título de agronomo. Universidad de Caldas, Manizalez. Colombia.

Valle L, Calvo L (2002) Diversidad y abundancia de quirópteros en plantaciones de café bajo sombra en Palajunoh, Quetzaltenango, Guatemala. Rev Protec Veg 12 (2): 53-60

Van Horne B (1983) Density as a misleading indicator of habitat quality. J Wildl Manage 47: 893-901

Vandermeer J (1995) The ecological basis of alternative agriculture. Annu Rev Ecol Syst 26: 201-224

Vandermeer J, Caraval R (2001) Metapopulation dynamics and the quality of the matrix. Am Nat 158: 211-220

Vandermeer J, Perfecto I (2000) La biodiversidad y el control de plagas en sistemas agroforestales. Manejo Integrado de Plagas (Costa Rica) 55: 1-5

Vandermeer J, Perfecto I (2005) Breakfast of biodiversity: The political ecology of rainforest destruction. 2nd Edition, Food First Books, Oakland, 256 pp

Vandermeer J, Perfecto I, Ibarra-Núñez G, Philpott SM, García Ballinas JA (2003) Ants (*Azteca* sp.) as potential biological control agents in shade coffee production in Chiapas, Mexico. Agroforestry Systems 56: 271-276

Vandermeer J, Perfecto I, Philpott SM, Chappell JM (in press) Reenfocando la conservación en el paisaje: La importancia de la matriz. In: Saenz J, Harvey C (eds), Evaluación y conservación de la biodiversidad en paisajes fragmentados en Mesoamerica. Editorial de la Universidad Nacional Autonoma de Costa Rica.

Varón EH (2002) Distribución espacio-temporal de hormigas con potencial como depredadoras de *Hypothenemus hampei* e *Hypsipyla grandella,* en sistemas agroforestales de café, en Costa Rica. Master Thesis. Catie, Turrialba, Costa Rica, December 2002, 97 p.

Vázquez-Yanes C, Batis Muñoz AI, Alcocer Silva MI, Gual Díaz M, Sánchez Dirzo C (1999) Árboles y arbustos potencialmente valiosos para la restauración ecológica y la reforestación.Reporte técnico del proyecto J084. CONABIO - Instituto de Ecología, UNAM.

Vélez M, Bustillo AE, Posada F (2003) Depredación de *Hypothenemus hampei* por *Solenopsis geminata* y *Gnamptogenys* sp. (Hymenoptera:Formicidae).

In: Sociedad Colombiana de Entomología (eds) Libro de Resumenes XXX Congreso. Cali, Colombia, pp 26

Williams-Guillén K (2003) The behavioral ecology of mantled howling monkeys (*Alouatta palliata*) living in a Nicaraguan shade coffee plantation. Ph.D. thesis, New York University, New York.

Witt E (2001)Seed dispersal by small terrestrial mammals in shaded coffee farms in Chiapas, Mexico. Masters Thesis, University of Michigan, Ann Arbor, MI.

Yachi S, Loreau, M (1999) Biodiversity and ecosystem productivity in a fluctuating environment: The insurance hypothesis. Proc Natl Acad Sci USA 96:1463-1468

Economic evaluation of ecosystem services as a basis for stabilizing rainforest margins? The example of pollination services and pest management in coffee landscapes

Roland Olschewski[1*], Teja Tscharntke[2], Pablo C. Benítez[3], Stefan Schwarze[4], and Alexandra-Maria Klein[2]

[1] Institute of Forest Economics, University of Göttingen, Büsgenweg 5, D-37077 Göttingen, Germany
[2] Agroecology, University of Göttingen, Waldweg 26, D-37073 Göttingen, Germany
[3] Department of Economics, University of Victoria, PO Box 1700 STN CSC, Victoria BC V8W 2Y2, Canada
[4] Institute of Rural Development, University of Göttingen, Waldweg 26, D-37077 Göttingen, Germany

*corresponding author: Roland Olschewski, phone: +49 (0)551 39 8714, fax: +49 (0)551 39 3420, Email: rolsche@uni-forst.gwdg.de

Summary

The determination and evaluation of ecosystem services provides crucial information for a comprehensive strategy to preserve near natural habitats such as rainforest margins. We show results of an economic evaluation of coffee bee pollination services and pest control in two distinct tropical regions: a low human-impact area in Indonesia with continuous near natural forests neighboring agroforestry and a high-impact landscape in Ecuador with almost no forest fragments left. We evaluate bee pollination services comparing forest destruction scenarios, where coffee yields depend on forests providing nesting sites and foraging habitats for bees and present three novel approaches: first, we show how net coffee revenues depend on pollination services of adjacent forests considering berry weight in addition to fruit set, thereby providing a more comprehensive evaluation. Second, we combine our findings on pollination with an assessment of pest management affecting coffee production. Third, we determine net welfare effects of land-use changes including the fact that former forestland is normally used for alternative crops. In both regions, crop revenues exceed coffee pollination values, generating incentives to convert forest margins even if owners would be compensated for pollination services.

Tscharntke T, Leuschner C, Zeller M, Guhardja E, Bidin A (eds), The stability of tropical rainforest margins, linking ecological, economic and social constraints of land use and conservation, Springer Berlin 2007, pp 265-278

The promotion of certified "biodiversity-friendly" coffee is a feasible option to maintain shade-coffee systems for conservation purposes. This is of special importance in high-impact areas where only small forest fragments remain. We conclude that a comprehensive economic analysis is necessary to adequately evaluate rainforest preservation for the enhancement of ecosystem services within a mosaic of competing land-use systems.

Keywords: bees, coffee-berry-borer, ecosystem services, pest management, certified shade coffee , cost-benefit analysis

1 Introduction

Tropical landscapes often consist of a mosaic of different land-use systems. The importance of such a landscape composition to ecosystem services has recently been shown for coffee landscapes, where pollinator activity and coffee fruit set increased in agroforestry sites adjacent to forest fragments (Ricketts 2004, Klein et al. 2003a,b). The preservation of rainforest margins and related ecosystem services such as pollination and pest control can only be addressed adequately if the local (habitat) management is considered in the context of the surrounding landscape matrix (Daily et al. 1997, Kremen et al. 2004, Kremen 2005). Within such a mosaic, biodiversity conservation through land-use systems on private land is becoming a pressing environmental policy issue (IISD 2005). The economic evaluation of these services can provide information for a comprehensive strategy to preserve natural source habitats of plants and animals invading agricultural systems (Tscharntke et al. 2005). Although highland coffee, *Coffea arabica L.*, is known to be self-compatible, recent studies have shown that bee pollination increases fruit set and berry weight, whereas the occurrence of pea-berries is reduced (Manrique and Thimann 2002, Roubik 2002, De Marco and Coelho 2003, Gillison et al. 2004, Klein et al. 2003a, Ricketts 2004). Many bee species depend on natural habitats such as forest fragments (Heard and Exley 1994, Ghazoul et al.1998, Cunningham 2000, Aizen and Feinsinger 2003, Chacoff and Aizen 2006 , e.g., it was shown that coffee fruit set (De Marco and Coelho 2003, Klein et al. 2003a, Ricketts 2004, Ricketts et al. 2004) and yield of other crop species like atemoya, macadamia, longan, and canola increases when natural rainforest patches are adjacent to these agricultural systems (Blanche and Cunningham 2005, Blanche et al. 2006, Morandin and Winston 2006). In consequence, natural forest ecosystems play an important role for bee pollination services. The determination of their economic value is one aim of the present article.

In the case, where natural habitats are destroyed and decreasing pollination services lead to lower revenues, coffee growers might be forced to reduce their costs by diminishing their management activities. This behavior bears the danger of provoking additional negative economic effects because the occurrence of pests and diseases might be encouraged. The coffee berry borer (CBB), *Hypothenemus hampei* Ferrari (Coleoptera: Scolytidae), is the most important insect pest affecting coffee world wide (Le Pelley 1968). CBB-infestation rates vary considerably between five and 100 percent depending on coffee regions and harvesting time. Higher rainfalls and highly shaded sites positively influence the pest insect (Vijayan et al. 1999, Wegbe et al. 2003). Fecundity of CBB is reported to be higher in overripe berries than in unripe and ripe berries (Mathieu et al. 1999). This means that management and harvesting methods play an important role when trying to combat CBB infestation. We determine infestation rates and evaluate the economic impact of an improved pest management.

In this article, we present three novel approaches: first, we analyze the impact of reduced pollination services on net revenues due to forest destruction adjacent to coffee agroforestry taking berry weight in combination with fruit set into account. Second, we determine net welfare effects of destroying rainforest margins including the fact that former forestland is normally used for alternative crops. Third, we combine our findings on pollination with an evaluation of coffee management regimes. Furthermore, we compare the results of our study areas in Sulawesi (Indonesia), characterized by a low human impact and the neighborhood of a continuous near-natural rainforest, and southern Manabí (Ecuador) as a high-impact area with a highly fragmented landscape, to show to which extent pollination values depend on the surrounding landscape matrix. Finally, we show possible consequences of shade-coffee certification for conservation purposes.

2 Methods

The ecological data were collected in agroforestry systems of the low human-impacted area in Central Sulawesi (Indonesia) at the south-eastern border of the Lore-Lindu National Park. The sites were selected in a way that forest distance was not correlated with shade density. In each agroforestry system flower-visiting bee species were counted and fruit set after open pollination was determined on four distinct branches of different coffee plants (for detailed information see Klein et al. 2003a). Additionally, we randomly collected 50 ripe and 50 overripe berries per site. In the laboratory, all ripe berries were freshly weighted and mean values per site were calculated.

Furthermore, we dissected the ripe berries to determine CBB-infestation rates. The proportion of overripe berries on coffee plants was estimated based on the counting of all green, ripe, and overripe berries on coffee plants in two different 10 m^2 plots per site. Additionally, we estimated the proportion of overripe berries lying on the ground in the same two 10 m^2 plots per site. Furthermore, to ensure the explanatory power of the sampling size, we collected all ripe and overripe berries of five coffee plants in apparently not infested sites, and in fact no berry borer infestation could be found.

For our economic analysis we assumed that coffee growers strive for a maximization of their net revenues, defined as the difference between revenues and costs. A decrease of pollination services caused by ongoing destruction of adjacent forest margins has impact on profits in two ways: first, reduced yields lead to lower gross revenues if the market price remains constant, and second, variable harvest and transportation costs decrease. Data on Indonesian yields, prices, and production costs were collected through household surveys (for details on the sampling procedure see Zeller et al. 2002).

In order to compare the economic impact of our ecological findings in different landscapes, we choose southern Manabí in Ecuador as a second study area, which is characterized by a high degree of landscape fragmentation with a small number of remaining forest fragments. Our economic analysis is based on published findings in both Latin America (Le Pelley 1968, Ricketts 2004) and Indonesia (Klein et al. 2003a) that the ecological mechanisms for coffee pollination services and CBB infestation are similar in both regions. In Ecuador we determined net revenues through an economic survey of coffee farmers combined with time series analysis of our study region (SICA 2003). In addition, we conducted an economic analysis of the local coffee markets in both regions to predict possible price changes. Based on our ecological results concerning average fruit set and berry weight, we used scenarios to determine the regional net welfare effect when former forest land is converted to alternative crops. Furthermore, we showed the impact of increasing pest infestation rates, in the case where coffee systems are extensively managed and determined the conditions under which certified shade-coffee production would be economically attractive to small-scale coffee growers.

3 Results

3.1 Pollination services

Our economic survey in Sulawesi shows that an extensively managed coffee system adjacent to forests generates an average annual yield of 269 kg of cleaned and dried coffee beans per hectare. Based on a regression analysis this quantity was assumed to represent a 100-percent yield reflected by a combination of maximum fruit set (85,22%) and maximum berry weight (1.65g/berry) (for further details see Olschewski et al. 2006). Net revenues sum up to USD 100 per ha, when considering an average price of USD 0.48 per kg. We estimate the economic value of pollination services taking into account that coffee fruit set (Klein et al. 2003a) as well as berry weight are negatively correlated with forest distance. This implies a stronger reduction of pollination services as fruit set is reduced and the resulting fruits have a lower weight. At a maximum distance of 1500m, the combined effect results in a 45% reduction of yields per hectare, whereas considering fruit set alone would give a yield decrease of 29%, only. Net revenues diminish by 47% from

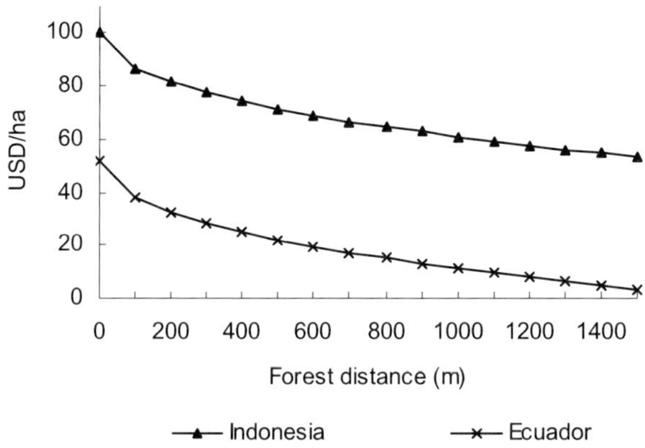

Fig. 1. Net coffee revenues in relation to forest distance in Indonesia and Ecuador. Calculated as revenues minus costs combining fruit set and berry weight impact (compare Olschewski et al. 2006)

USD 100 to 53 per ha (see Figure 1). Additionally, we calculated the marginal net revenue decrease when forest distance increases step by step by 100m and found that the main effect on yields (about 55% of the overall reduction) takes place within a distance of 400 meters from adjacent forests (Olschewski et al. 2006).

According to Ricketts (2004), similar effects of forest distance occur also in other tropical coffee regions such as Latin America. This motivated us to compare the economic results of our Indonesian low-impact study area with a highly fragmented region in Ecuador. Here, coffee is usually not cleaned but sold as dried berries without further processing. This results in an average coffee yield of about 850 kg per hectare. Taking the Ecuadorian long-term average price of USD 0.2 per kg of dried coffee berries into account (SICA 2003) net revenues sum up to USD 53 per hectare. At a maximum distance of 1500m, yields decline by 45%, whereas net revenues are reduced by 93% from USD 52 to USD 3 per ha, i.e., by far more than expected when looking at the yield reduction, only. Note that as yields decline, variable harvesting and transportation costs are reduced, too, but due to a more intensive management (with a higher percentage of fixed costs for cleaning, pruning and shadow control) compared to Sulawesi net revenues decline at a higher rate.

As a result, increasing distance to forests leads to a substantially diminished attractiveness of coffee production. Additional calculations based on the even lower average coffee prices over the last five years resulted in negative net revenues per ha, forcing many coffee growers to reduce management intensity or to switch to other land-use systems like maize or rice.

3.2 Economic impact of deforestation on pollination services

Based on our findings concerning fruit set and berry weight, we calculated the economic impact on coffee sites with increasing distance to a 100 ha circular forest area. For this purpose, the coffee area around the forest is divided into four 100m wide concentric circles (adding up to 192 ha), where net coffee revenues depend on forest distance (Olschewski et al. 2006). We took this as a reference scenario for comparison with the effects of forest destruction, and assumed in a first step that a 100m wide area of forest margins is destroyed, i.e., the forest area diminishes from 100 to 68 ha. In Sulawesi, average net revenues of the adjacent coffee area are reduced by 7% from USD 86 to 80 per ha due to increasing distance to the remaining forest. In Manabí, net revenues decline from USD 38 to 31 per ha, which reflects a reduction of about 17%. Finally, we considered the extreme situation that the forest area of 100 ha is completely destroyed. In this case we assumed that coffee sites receive only minor pollination services from other forest patches in about 1000m distance. Complete forest destruction reduces average net revenues by 28% to USD 61 per ha in Sulawesi, whereas in Manabí, the decline is about 70% to USD 11 per ha (Olschewski et al. 2006).

In case that the forest is not destroyed, the value of pollination services can be calculated by assigning the avoided coffee net revenue loss to the forest area conserved. In contrast to the substantial differences in country specific relative changes determined above, the absolute value of pollination services is similar in both countries. Deforestation of the first 100m (32 ha) would reduce total net revenues in the adjacent coffee area from USD 16,347 to 15,214 in Sulawesi and from USD 7,005 to 5,825 in Manabí. In both cases, the pollination service value is about USD 35 per ha of forest. Converting the last 22 ha of forest would lead to a pollination services loss of USD 47 per ha of forest in Indonesia and USD 49 in Ecuador (Olschewski et al. 2006). To determine the impact of coffee yield and price volatility on our results, we conducted a sensitivity analysis and found that a price variation of ±10% causes a pollination value change of about ±15% in Ecuador and ±12% in Indonesia. Coffee yield changes lead to similar effects.

3.3 Net welfare effect of deforestation including alternative crops

When determining the value of forest patches for coffee production we assumed that an existing forest area is partly or entirely destroyed. The resulting net revenue reductions were assigned to the deforested area and the respective value per hectare reflects the positive economic effect of pollination services when maintaining the forest. This procedure neglects that the former forest land will generally be used for crop production or pasture and, thus, income will be generated, which would be forgone if the forest was maintained.

Consequently, the net income reduction of farmers in the coffee area (192 ha) has to be compared with the net income increase generated on the deforested land (32, 58, 78 or 100 ha, respectively). For this purpose, we conducted cost-benefit analyses of alternative land uses like maize, rice and pasture. In both countries, we found that the best alternative land-use systems exceed the value of pollination services lost by forest destruction. In Sulawesi, net revenues of maize (USD 160 per ha) are substantially higher than the average pollination service value of USD 47 per ha forest land. The same holds for Manabí as far as maize (USD 108 per ha) is considered (Benítez et al. 2006). In contrast, alternatives like rice or pasture, which only achieve levels between USD 57 and 53 per ha, are closer to the average value of USD 49 per ha generated by pollination services of forests.

Adding up the reduced net coffee revenues and the additional net crop revenues after deforestation leads to a positive net welfare effect in both study regions. In Sulawesi, net welfare in the overall area increases by 69% from USD 16,347 to 27,697, in Ecuador by 84% from USD 7,005 to USD 12,911 (Olschewski et al. 2006).

3.4 CBB infestation and harvest management

A more extensive management of coffee systems is a rational strategy of coffee growers faced with declining revenues: the cost reduction leads to a new profit maximum on a lower level. But this isolated vision seems to be short-sighted and might generate unexpected results, if further important aspects are not taken into account.

Our results show that the percentage of infested coffee berries correlates with the percentage of overripe berries hanging on as well as lying under coffee plants. CBB-infestation increases with increasing proportion of overripe berries on the coffee sites. The proportion of infested berries in 50 randomly sampled ripe berries significantly increases with the proportion of overripe berries on coffee plants sampled in 10 m^2 (Figure 2A). In consequence, an extensive management bears the danger of increasing infestation rates, whereas selective harvesting of ripe berries and collecting overripe berries fallen off the plant reduces infestation rates and, thus, generates higher yields per ha.

Additionally, we examined a possible impact of adjacent forests on the infestation rates. As shown in Figure 2B, we did not find a correlation between the proportion of infested berries and forest distance, which means that we can exclude a contrary effect between pollination services and CBB infestation with respect to forest distance.

4 Discussion

Our results show that the value of pollination services might be underestimated when referring to fruit set only and excluding the additional effect on berry weight. In contrast, studies might overestimate this value when calculations are based on yields or gross revenues instead of net revenues. Furthermore, the net welfare effect of land-use changes has to be taken into account,

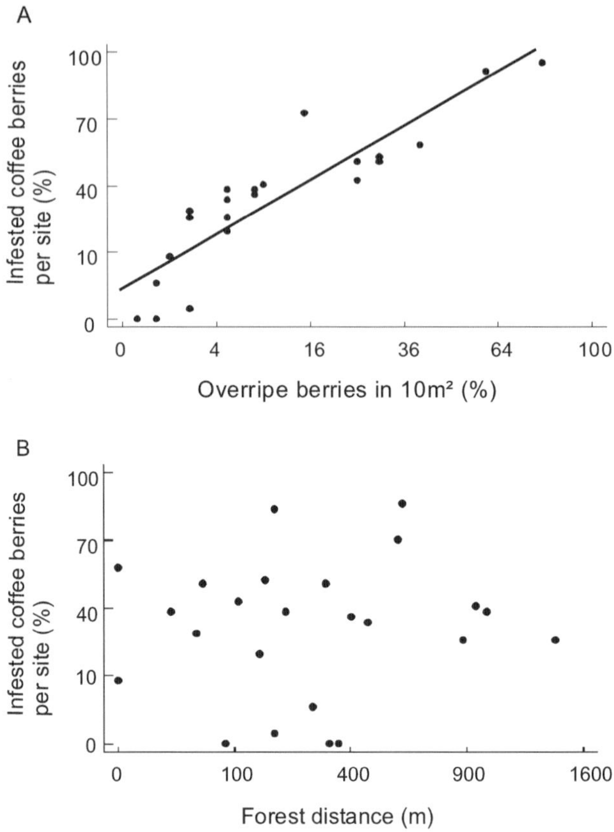

Fig. 2. Mean percentage of infested berries in 50 randomly sampled ripe berries in relation to (A) Estimated proportion of overripe berries on coffee plants calculated in 10 m^2 per agroforestry system: $\mathrm{asin}(\sqrt{y}/100) = 0.28 + 2.55 \log(x+1)$, F = 129.43, R = 0.92, N = 24, P > 0.0001; (B) Distance to the continuous forest: $\mathrm{asin}(\sqrt{y}/100) = 2.35 + 0.02\sqrt{x}$, F = 0.15, R = 0.08, N = 24, P = 0.7028

which includes revenues of alternative crops on former forest land. Finally, we show the impact of a more extensive coffee management on CBB infestation in case that land owners are faced with declining pollination services due to deforestation.

4.1 Combined effect of fruit set and berry weight

In contrast to previous pollination service studies on coffee (Roubik 2002, Klein et al. 2003a, Ricketts et al. 2004), we did not exclusively focus on fruit set or berry weight, but combined our findings concerning fruit set and berry weight and related them to net revenues in an integrated approach. Note that a variation in soil fertility might have an impact on yields, too. This effect would possibly have an influence on berry weight, if at all. However, a separate analysis without taking berry weight into account showed a similar relation between forest distance and fruit set, which motivated us not to include soil fertility as a decisive factor in our analysis.

Calculations based on fruit set alone may either over- or underestimate pollination service values. An overestimation might result from the assumption that fruit set corresponds directly to the quantity of coffee beans harvested later. However, Ricketts et al. (2004) showed that - combining fruit set and seed mass - an adequate pollination on sites far from forest would increase coffee yields by 20.8%, in contrast to 11.5% when using fruit set, only. Combining both effects results in yield estimates that build a broader basis for the evaluation of pollination services. However, as shown below, a comprehensive evaluation procedure should include quite some more aspects than just multiplying the respective yields by market prices.

4.2 Net revenues and market price changes

Coffee growers are interested in net revenues (i.e., profits) rather than gross revenues, because the former reflect the actual net contribution to the households' income (Gobbi 2000). This should be taken into account when evaluating bee pollination as an ecosystem service (Ricketts et al. 2004). In our Indonesian case, an analysis based on gross revenues would lead to average pollination values of USD 43 to 57 per ha of deforested area (depending on scenario), which reflects an overestimation of about 23%. In Ecuador, the distorting effect would be even stronger and result in pollination values of USD 57 to 76 per ha deforested area, i.e., about 57% higher than in case of a calculation based on net revenues.

Additionally, the final effect on the producers' profits depends on the demand side of the product market, where possible price changes have to be considered. In an early article, Southwick and Southwick (1992) emphasized this aspect, which, nevertheless, is often neglected when analyzing the positive impact of pollination services (see Ricketts et al. 2004).

Although coffee is a commodity traded on a global level and changes in the supply of a small country will not influence the global coffee price, local prices could still be affected: in case that more and more coffee growers reduce their supply when faced with declining revenues, the traded quantity decreases, which might lead to a price increase on the local market (Kevan and Phillips 2001).

Our market analyses showed that coffee markets in our study regions can be characterized as local monopsonies with many producers on the supply side, most of them holding coffee sites of less than five hectares and contributing just a small part of the traded quantity. On the demand side there is often only one single buyer, who has the power to set the price and buys coffee beans or berries directly from producers at the farm gate. We found that supply reductions corresponding to our deforestation scenarios are unlikely to have an impact on local coffee market prices (Olschewski et al. 2006). However, deforestation on a larger scale might cause prices to increase due to a stronger supply reduction; in that case, a calculation based on constant coffee prices would overestimate the net revenue cutbacks caused by reduced pollination services.

4.3 Regional net welfare effects

Destroying forests adjacent to coffee systems reduces pollination services. This effect has been evaluated by previous studies (De Marco and Coelho 2003, Klein et al. 2003a, Ricketts et al. 2004), however, it is often neglected that former forest land will not be abandoned but is normally used for other crops or pasture. Our study takes possible revenues of such alternatives into account in order to calculate the regional net welfare effect of land-use changes.

Provided that the alternative crop production is sustainable, our results show that it is reasonable for landowners to convert their forest land to other crops, because forest conservation would lead to a welfare loss (Olschewski et al. 2006). If we assume that the farmers have the right to convert forest land, negatively affected coffee growers could think of compensating them for maintaining the forest. However, the respective payments would not be sufficient to change the land owners' behavior: an incentive for deforestation remains. Note that this conclusion is based on the (local) pollination service value, alone. Taking additional (global) ecosystem services of forests, such as biodiversity conservation, into account could lead to the result that forest conservation would generate a welfare improvement. In that case it has to be determined, how an operational payment scheme could be implemented, which combines both local and global services.

4.4 Shade-coffee certification

Given this situation, the question arises whether coffee systems themselves can take over conservation tasks within a landscape mosaic. Biodiversity-rich land-use systems like extensive agroforestry under complex shade are often practiced by smallholders in developing countries (Perfecto et al. 1996), using small patches of land to diversify their income sources (Oxfam America 2005). However, the recent price decrease on the world coffee market has caused many producers to abandon their coffee fields and convert this land to produce economically more attractive crops (O'Brian and Kinnaird 2003, Benítez et al. 2006, Oxfam America 2005).

Philpott and Dietsch (2003) and Dietsch et al. (2004) show that shaded coffee systems can contribute significantly to conservation goals, especially when combined with organic and fair trade certification. Organic certification norms include the selective harvesting of ripe berries, the collection of berries fallen off the plant and the recycling of production residuals in form of organic fertilizer. This procedure is suitable to prevent plants from CBB infestation, where berries should be harvested selectively when ripe, and overripe berries should be collected and destroyed to interrupt infestation cycles (Kalshoven 1981, COFENAC 2002). This requires a more intensively managed system with a higher labor input. In our study regions, about 8 additional working days per hectare and year have to be taken into account, and, as a consequence, net revenues per hectare would decline from USD 100 to 92 in Sulawesi, whereas in Manabí a more drastic reduction of net revenues from USD 52 to 28 would take place.

Our results show that fulfilling certification requirements causes a relatively low net revenue decline of about 8 % in Sulawesi, whereas in Manabí net income decreases by 46 %. In both cases, farmers would have no incentive to intensify their coffee management as long as coffee yields and prices remain constant. To avoid net income reductions (assuming constant coffee prices), the additional pest control would have to increase coffee yields by at least 6 % (from 269 to 285 kg per ha) in Indonesia, and by 23 % (from 853 to 1050 kg per ha) in Ecuador. According to our findings, the respective yield increase could be achieved by reducing the infestation rate under the assumption that the above described precautionary harvest management is applied (compare Figure 2A). However, Perfecto et al. (2005) argue that an additional high price premium that goes directly to the producers would be necessary to make conservation through shade-coffee systems a viable option for land owners. This goes a long way with the results found by Benítez et al. (2006) who compare land uses including price and yield risks. They conclude that without considerable payments by certification programs, it would be questionable whether biodiversity-rich shade-coffee systems could be prevented from being converted to economically more attractive land uses.

5 Conclusions

As a natural habitat of bee populations, forests provide pollination services to adjacent coffee areas. The value of these services can be determined by comparing forest destruction scenarios with a reference situation, where coffee yields depend on the distance to forests. We found that the main economic impact, measured as net revenues per ha, takes place within a distance of 400 m to forest patches, and is highly influenced by the management system.

Our scenarios in both countries show that most land-use alternatives generate higher net revenues per hectare than the value of pollination services assigned to the deforested area. Consequently, from a landowner's point of view, there is a strong incentive to convert forests into cropland. Even a payment for ecosystem services, if based on pollination services alone, would hardly be sufficient to reduce pressure on forest margins.

Given this situation, conservation on privately owned and used land becomes more and more important. The production of certified shade coffee shows a possible solution for coffee growers confronted with the impact of adjacent forest sites destruction, infestation problems and falling world coffee market prices. In comparison with traditional production systems, special management norms generate higher costs, but at the same time yields and revenues can be increased. This creates incentives for landowners to maintain their shade coffee production systems and to conserve their ecological functions within a landscape mosaic, which are of special importance in highly fragmented areas where only small patches of natural forest remain.

References

Aizen MA, Feinsinger P (2003) Bees not to be? Responses of insect pollinator faunas and flower pollination to habitat fragmentation. in Bradshaw GA, Marquet PA (eds) How landscapes change: Human disturbance and ecosystem fragmentation in the Americas, Ecological Studies: Volume 162. Springer-Verlag Berlin, Germany, pp 111-129

Benítez PC, Kuosmanen T, Olschewski R, van Kooten GC (2006) Conservation payment under risk – a stochastic dominance approach. Am J Agr Econ, 88(1): 1-15

Blanche R, Cunningham SA (2005) Rain forest provides pollinating beeles for atemaya crops. J Econ Entom 98: 1193-1201

Blanche KR, Cunningham SA, Ludwig JA (2006) Proximity to rainforest enhances pollination and fruit set in macadamia and longan orchards in north Queensland, Australia. J Appl Ecol in press

Chacoff NP, Aizen MA (2006) Edge effects on flower-visiting insects in grapefruit plantations bordering premontane subtropical forest (Unique ID: 217568-18476). J Appl Ecol 43(1): 18-27

COFENAC (2002) Café orgánico – Guia del cafecultor. Consejo Cafetalero Nacional. Manta, Ecuador

Cunningham SA (2000) Depressed pollination in habitat fragments causes low fruit set. Proc Royal Soc London. Series B 267: 1149-1152

Daily GC, Alexander S, Ehrlich PR, Goulder L, Lubchenco J, Matson PA, Mooney, Postel S, Schneider SH, Tilman D, Woodwell GM (1997) Ecosystem Services: Benefits Supplied to Human Societies by Natural Ecosystems, Issues Ecol 2: 1-18

De Marco Pjr, Coelho FM (2003) Services performed by the ecosystem: forest remnants influence agricultural cultures pollination and production. Biodiv Conserv 13: 1245-1255

Dietsch T, Philpott SM, Rice R, Greenberg R, Bichier P (2004) Policy alternatives for conservation in coffee landscapes. Science 303: 625

Ghazoul J, Liston KA, Boyle TJB (1998) Disturbance-induced density - dependent seed set in Shorea siamensis(Dipterocarpaceae), a tropical forest tree. J Ecol86:462-473

Gillison AN, Liswanti N, Budidarsono S, van Noordwijk M, Tomich TP (2004) Impact of cropping methods on Biodiversity in coffee agroecosystems in Sumatra, Indonesia. Ecol Soc 9(2): 7 [online] URL: http://www.ecologyandsociety.org/vol9/iss2/art7

Gobbi JA (2000) Is biodiversity-friendly coffee financially viable? An analysis of five different coffee production systems in western El Salvador. Ecol Econ33: 267-281

Heard T, Exley E (1994) Diversity, Abundance and distribution of insect visitors to Macadamia Flowers. Environ. Entomol. 23: 91-100

IISD (2005) Biodiversity: Science and Governance Bulletin. A summary report of the International Scientific Conference 24.-28. January 2005 in Paris. [online] URL: http://www.iisd.ca/sd/icb http://www.iisd.ca/sd/icb

Kalshoven LGE (1981) Pest of crops in Indonesia. P.T. Ichtiar Baru, Van Hoeve, Jakarta, 701 pp.

Kevan PG, Phillips TP (2001) The economic impact of pollinator declines: an approach to assessing the consequences. Cons Ecol 5(1): 8 [online] URL: http://www.consecol.org/vol5/iss1/art8 http://www.consecol.org/vol5/ iss1/art8

Klein AM, Steffan-Dewenter I, Tscharntke T (2003a) Fruit set of highland coffee increases with the diversity of pollinating bees. Proc Royal Soc London Series B 270: 955-961

Klein AM, Steffan-Dewenter I, Tscharntke T (2003b) Pollination of Coffea cenephora in relation to local and regional agroforestry management. J Appl Ecol 40: 837-845

Kremen C (2005) Managing ecosystem services: what do we need to know about their ecology. Ecol Lett 8: 468-479

Kremen C, Williams NM, Bugg RL, Fay JP, Thorp RW (2004) The area requirements of an ecosystem service: crop pollination by native bee communities in California. Ecol Lett 7: 1109-1119

Le Pelley RH (1968) Pests of coffee. Tropical Science Series. Longmans Press, Green and Co. Ltd., London

Manrique AJ, Thimann RE (2002) Coffee (Coffea arabica) pollination with africanized honeybees in Venezuela. Intercienca 27: 414-416

Mathieu F, Brun LO, Frerot B, Suckling DM, Frampton C (1999) Progression in field infestation is linked with trapping of coffee berry borer, Hypothenemus hampei (Col., Scolytidae). J Appl Entom 123: 535-540

Morandin LA, Winston ML (2006) Pollinators provide economic incentive to preserve natural land in agroecosystems. Agric Ecosyst Environ. in press

O'Brian TG, Kinnaird MF (2003) Caffeine and Conservation. Science 300: 587

Olschewski R, Tscharntke T, Benítez PC, Schwarze S, Klein AM (2006) Economic evaluation of pollination services comparing coffee landscapes in

Ecuador and Indonesia. Ecol Soc 11(1): 7 [online] URL http://www.ecologyandsociety.org/vol11/iss1/art7/

Oxfam America (2005) The coffee crisis continues. Situation assessment and policy recommendations for reducing poverty in the coffee sector. [online]URL http://www.oxfamamerica.org/crisis_continues

Perfecto I, Rice R, Greenberg R, van der Voort M (1996) Shade coffee as refuge of biodiversity. Bioscience 46: 598-608

Perfecto I, Vandermeer J, Mas A, Soto Pinot L (2005) Biodiversity, yield, and shade coffee certification. Ecol Econ 54: 435-446

Philpott SM, Dietsch T (2003) Coffee and conservation: a global contextand the value of farmer involvement. Conserv Biol 17: 1844-1846

Ricketts TH (2004) Tropical forest fragments enhance pollinator activity in nearby coffee crops. Conserv Biol 18: 1262-1271

Ricketts TH, Daily GC, Ehrlich PR, Michener CD (2004) Economic value of tropical forest to coffee production. Proc Natl Acad Sci USA 101: 12579-12582

Roubik DW (2002) Tropical agriculture - The value of bees to the coffee harvest. Nature 417: 708

SICA (2003) Estadísticas Agropequarias. Precio Finca y Rendimientos. Ministerio de Agricultura y Ganadería, Quito

Southwick EE, Southwick LJr (1992) Estimating the Economic Value of Honey Bees (Hymenoptera: Apidae) as Agricultural Pollinators in the United States. J Econ Entom 85(3): 621-633

Tscharntke T, Klein AM, Kruess A, Steffan-Dewenter I, Thies T (2005) Landscape perspectives on agricultural intensification and biodiversity-ecosystem service management. Eco Lett 8: 857-874

Vijayan V, Rehman P, Vijayalakshmi CK (1999) Influence of shade on coffee berry borer. Insect Environ 5: 132

Wegbe K, Cilas C, Decazy B, Alauzet C, Dufour B (2003) Estimation of production losses caused by the coffee berry borer (Coleoptera: Scolytidae) and calculation of an economic damage threshold in Togolose coffee plots. J Econ Entom 96: 1473-1478

Zeller M, Schwarze S, van Rheenen T (2002) Statistical Sampling Frame and Methods Used for the Selection of Villages and Households in the Scope of the Research Program on Stability of Rainforest Margins in Indonesia (STORMA). STORMA Discussion Paper Series Subprogram A (SDPS-A) No. 1, Göttingen, Germany, and Bogor, Indonesia

Insect diversity responses to forest conversion and agroforestry management

Merijn M. Bos[1]*, Patrick Höhn[1], Shahabuddin Saleh[2], Boris Büche[1], Damayanti Buchori[3], Ingolf Steffan-Dewenter[1], and TejaTscharntke[1]

[1] Agroecology, University of Göttingen, Waldweg 26, D-37073 Göttingen, Germany
[2] Faculty of Agriculture, University of Tadulako, Kampus Bumi Tadulako Tondo, Palu-Indonesia
[3] Department of Pest and Plant Diseases, Faculty of Agriculture IPB, Bogor Agricultural University, Jl. Kamper Kampus, IPB Darmaga, Bogor 16680, Indonesia

*corresponding author: Merijn Bos, Email: boommarter@hotmail.com

Summary

The ongoing loss of pristine tropical rainforests increases the potential importance of agroforestry systems for the conservation of tropical arthropod diversity. Shaded agroforestry systems can still support high levels of biodiversity, even resembling those supported by undisturbed forests, but intensively managed open agroforestry systems may cause severe losses in insect diversity. In this study we evaluate the conservation value of agroforestry systems for species richness and diversity (Simpson's index) of four insect groups at natural forest sites and three different types of cacao-dominated agroforestry systems in Central Sulawesi, Indonesia. The agroforestry systems were characterised by low, intermediate and high diversity of shade trees. Each habitat type was studied with 4 replicates, i.e. 16 study sites altogether. We compared responses of solitary bees and wasps, dung beetles and lower canopy dwelling beetles and ants. These taxa represent diverse and functionally important insect groups: solitary bees and wasps act as crop pollinators or pest predators, dung beetles as decomposers of mammalian excrements and canopy dwelling beetles and ants include abundant herbivores and predators. High percentages of forest species did not occur in agroforestry systems, but diversity and species richness in agroforests remained as high as or even higher than in the forest lower canopy. Diversity, species richness and abundance of the functionally important dung beetles and canopy ants showed strong resilience against both forest conversion and changes in agroforestry management. Diversity, species richness and abundance of solitary bees and wasps as well as

Tscharntke T, Leuschner C, Zeller M, Guhardja E, Bidin A (eds), The stability of tropical rainforest margins, linking ecological, economic and social constraints of land use and conservation, Springer Berlin 2007, pp 279-296

canopy beetles even seemed to profit from the effects of opening the upper canopy that was related to forest conversion and changes in shade tree compositions. On the latter two groups the effects of opening the upper canopy were stronger than effects of reduced shade tree richness. Based on our results, we recommend the inclusion of agroforestry systems with a diversity of shade trees in tropical conservation plans in addition to pristine forest reserves. Furthermore, regional differences in local agroforestry management contributed to between 31% (for dung beetles) and 58% (for canopy beetles) of the total species richness, which stresses the importance of conservation policies aimed at a diversity of habitat types on a broader landscape scale.

Keywords: agroforestry, ants, beetles, biodiversity, lower canopy, conservation, habitat preference, knockdown fogging, management, pit-fall traps, solitary bees , solitary wasps, trap nests

1 Introduction

The global-scale conversion of natural ecosystems is a major cause of biodiversity loss and threatens ecosystem functioning, sustainable land use and economies (Hoekstra et al. 2005). Tropical rainforests are one of the most species-rich and functionally important terrestrial ecosystems (Myers et al. 2000). In the past 50 years an estimated 32% of these tropical rainforests have been converted to human-dominated systems and a further loss of 10-15% has been projected by 2050 (Millenium Ecosystem Assessment 2005). Hot spots of tropical rainforest conversion are Southeast Asia, with annual deforestation rates of 2.5 million hectares (0.91%) and Central America, with 2.5 million ha (0.38%) per year (Achard et al. 2002). With the continuing loss of tropical forests, cultivated areas are gaining interest for their potential value for conserving tropical biodiversity. Diversified agricultural systems such as agroforestry are suggested to serve as tools in nature conservation policies (Rice and Greenberg 2000; Putz et al. 2001; Donald 2004; McNeely 2004; Schroth et al. 2004).

1.1 The conservation potential of agroforestry systems

Agroforestry systems are generally characterized by a canopy cover of shade trees below which a wide range of crop plants can be grown (Schroth et al. 2004). In human-dominated, deforested landscapes, agroforestry systems provide the only remaining habitat type with a substantial tree cover (Schroth et al. 2004). Agroforestry supports some of the most important tropical cash crops including oil palm (*Elaeis* spp), rubber (*Ficus elastica*), cacao (*Theobroma cacao*) and coffee (*Coffea* spp.). Additionally, various kinds of timber, local fruit trees and annual crops are grown in agroforestry systems, thereby providing building material, firewood and food for local communities (e.g., Rice and Greenberg 2000, Siebert et al. 2002),

Agroforestry systems range widely in land-use intensity from extensively managed, patchy and densely shaded plantations owned by large numbers of smallholders to large scale, non-shaded and monotonous plantations owned by few farmers (e.g., Perfecto et al. 1997, Siebert 2001, Jones et al. 2003, McNeely 2004, Foley et al. 2005). Acknowledging the intrinsic and economic values of biodiversity (e.g., Meffe 1998, Altieri 1999, Sodhi 2004, Foley et al. 2005), studies increasingly focus on biodiversity as supported by various agroforestry types (e.g., Perfecto et al. 1997, Lawton et al. 1998, Siebert 2002, Jones et al. 2003, Klein et al. 2002, 2004, Schulze et al. 2004, Shahabuddin et al. 2005).

1.2 Land-use management and the conservation of biodiversity

The entomofauna of agroforestry systems is very diverse (Kalshoven 1950, Room 1971, Entwistle 1972, Campbell 1984) but has mainly been studied in the context of pest management. Increasing intensity of agroforestry management means simplification of the habitat by reducing the diversity of canopy trees, partial or complete removal of shade tree layers, increasing chemical inputs and more frequent cleaning of dead wood and leaf litter (Perfecto and Snelling 1995, Perfecto et al. 1997, Klein et al 2002, Siebert 2002, Jones et al. 2003). Land-use intensification might fail to increase crop productivity because it is paralleled by the loss of functionally important species groups resulting in reduced pollination services or increased pest problems (Klein et al. 2003, Rice and Greenberg 2000, McNeely 2004, Perfecto et al. 2005). Many studies indicate that increasing land-use intensity leads to biodiversity loss. For example, Chung et al. (2000) found that species richness of canopy beetles remained in extensive timber plantations more similar to nearby forest sites than beetle richness in intensive oil palm plantations. Jones et al. (2003) showed a decrease of termite species richness in rubber plantations from extensively managed rubber plantations, preserving 62% of the level of forest richness, to intensively managed plantations which preserved 44%.

Agroforestry in tropical America recently underwent a significant intensification of coffee plantations (Perfecto and Snelling 1995, Perfecto et al. 1997, Perfecto et al. 2005, Philpott 2005). Rustic, shaded and small scale agroforestry was largely replaced by large, monotonous coffee plantations with little or no shade. In comparison to extensive, well shaded agroforests, beetle species richness decreased by 26% but increased in abundance by up to 17% in more intensively managed coffee agroforestry systems (Perfecto et al. 1997). Similarly, pruning and removal of shade trees changed community structure and caused losses in ant diversity (Perfecto and Snelling 1995, Philpott 2005). Rice and Greenberg (2000) and McNeely (2004) emphasise the importance of remaining well developed stands of shade trees for land-use sustainability, preserving soil fertility, reducing soil erosion and increasing resistance to disturbances by extreme climatic events. For example, farmers in Africa successfully increased amounts of available nitrogen in the soil by replanting leguminose shade trees, thus forming a basis for land rehabilitation (Sanchez 1999). Additionally, Ramirez et al. (2001) showed how intercropping of shade providing timber trees in cacao plantations can lead to lower sensitivity to fluctuating market-prices.

1.3 Landscape heterogeneity and the overall supported biodiversity

At a local scale the species richness of agroforestry systems can be expected to depend on their structure, including the diversity of crop plants and canopy trees and the management intensity (e.g., Chung 2000, Jones et al. 2003, Schulze et al. 2004). On a landscape scale the spatial turnover in the species composition of local communities can significantly contribute to regional biodiversity. Thus, landscape heterogeneity can have important effects on biodiversity. For example, regional landscape heterogeneity contributed 56% of regionally recorded plant species (Wagner et al. 2000), 39% of regional solitary bee and wasp species (Tylianakis et al. *in press.*) and about 25% of regional beetle and butterfly species (Gering et al. 2002, Summerville et al. 2003).

Generally, landscapes composed of a mosaic of different land use types, including extensively managed agroforestry systems, maintain highest overall species numbers. In addition, distance to undisturbed 'source habitats' such as undisturbed rainforest play a role for local species richness (Klein et al. *in press.*). With high biodiversity at a landscape scale, local disturbances are likely to be buffered by recolonisation from diverse neighbouring communities, resulting in less vulnerable faunal communities (Peterson et al. 1998).

2 Cacao management and the conservation of biodiversity in Indonesia

Here we present results from a study focussing on agroforestry systems in Central Sulawesi, Indonesia. The island underwent a 'cacao boom' throughout the 1990s (Potter 2001, Belsky and Siebert 2003) in which coffee-dominated agroforestry was largely replaced by cacao-dominated agroforestry systems, turning Indonesia into the world's third most important contributor to the international cacao production.

Indonesia is experiencing the world's highest rates of deforestation, the majority of which is driven by agricultural expansion (Achard et al. 2004) and results in unprecedented biodiversity losses (Sohdi et al. 2004). In the light of the disappearing primary rainforest habitats, agroforestry systems are very likely to play a role in the conservation of tropical biodiversity, especially in intensively cultivated regions of Sulawesi, where 80% of Indonesia's cacao is grown (Vingerhoets 2002).

Cacao is originally a shade dwelling tree species from the lowland forests of Central America (Urquhart 1955). In agroforestry, cacao is initially grown under shade which acts as buffer against major microclimatic changes, most important for younger cacao trees, but for adult trees highest productivity can be expected without shade (Urquhart 1955, Entwistle 1972). Shaded, cacao-dominated agroforestry systems can still support high levels of biodiversity, for some insect groups even resembling that of tropical rainforests (Schulze et al. 2004). Furthermore, in Central Sulawesi shade trees are increasingly removed (Belsky and Siebert 2003) posing a potential threat to the species rich flora and fauna supported by cacao dominated agroforestry (Siebert 2001), similar to the conversion of shade to sun coffee in tropical America. Unshaded agroforestry has not only negative effects on biodiversity but also results in higher dependence of cacao trees on watering and fertilizers (Urquhart 1955, Entwistle 1972).

The few studies that directly compared insect diversity between agroforestry systems and natural forest sites did not reach consensus on whether agroforestry systems support significantly lower insect diversity than forest sites or not (Lawton et al. 1998, Chung et al. 2000, Schulze et al. 2004). Effects of management intensification on biodiversity have shown to be much stronger than the initial conversion of forest to diversely shaded agroforestry systems (Perfecto et al. 1997, Chung et al. 2000, Jones et al. 2003). We tested the hypothesis that

1. *not forest conversion but increasing land use intensity results in reduced insect diversity*

On a larger spatial scale agroforestry systems differing in management regime, soil and water condition and landscape context comprise a heterogeneous landscape that can be the basis of a high, regionally supported biodiversity (as in Wagner et al. 2000, Gering et al. 2002, Summerville et al. 2003, see above). Therefore, we hypothesise that

1. *large spatial species turnover between different land-use systems contributes to regional species richness of insects.*

We tested the hypotheses using data on the insect fauna of 12 agroforestry systems and four natural forest sites. The agroforestry systems were classified according to three different compositions of shade trees: shade trees remaining from natural forest stands, a heterogeneous stand with a diversity of planted shade trees and a homogenous stand with one or two species of planted legume trees. The sequence from forest shade to homogenous shade represents a decline in structural diversity of the agroforestry systems as a habitat for insects. The study sites were situated around the Toro village about 100km south of Central Sulawesi's capital city Palu. The village is situated at the western border of the Lore Lindu National park. All study sites had a core area of 30 by 50 meters and were at least 300 meters away from each other. All forest sites were part of the national park.

The different insect groups we focus on here are solitary bees and wasps (Hymenoptera, Aculeata), dung-feeding scarabaid beetles (Coleoptera: Scarabaeidae, subfamilies Scarabaeinae, Coprinae and Aphodiinae, hereafter dung beetles), lower canopy ants (Hymenoptera: Formicidae) and lower canopy beetles (Coleoptera). Solitary bees are important pollinators of cultivated crops (Free 1993, Roubik 1995), solitary wasps can act as pest predators (O'Neill 2001) and dung beetles are important decomposers of mammal excrements (Klein 1989, Hanski and Cambefort 1991). Ants form a major component of tropical arthropod diversity and, as the most abundant group, play an important role in tropical ecosystems as predators (Hölldobler and Wilson, 1990). Lastly, the canopy dwelling beetles are the most species rich insect group found in tropical forested areas (e.g., Erwin 1982, Davies et al. 1997, Lawton et al. 1998) but also in tropical agroforestry systems (Perfecto et al. 1997, Chung et al. 2000).

The five selected insect groups have been poorly studied in the context of cacao agroforestry (e.g., Room 1971, Klein et al. 2002, Schulze et al. 2004, Shahabuddin 2005) and rarely in a comparison between differently managed cacao plantations and nearby located undisturbed forest sites.

3 Results

3.1 Hypothesis 1: Not forest conversion but increasing land use intensity results in reduced insect diversity

The change from natural forest to agroforestry systems was accompanied by a significant increase in temperature and cover of the herb layer (Figure 1 a and d). Contrastingly, relative humidity and upper canopy cover dropped significantly (Figure 1 b and c). In a factor analysis we combined the environmental variables in a single factor because of strong auto-correlation (Spearman rank-correlations, p 0.05). The resulting factor explained 71.7% of all variance in the variables and was positively related to temperature and herb layer cover but negatively to relative humidity and canopy cover.

In a direct comparison of abundance, species richness and a diversity index (combining abundance and species richness) between forest and agroforestry sites we quantified faunal turnover and conservation potential of agroforestry systems for different insect groups. We show how different management of agroforestry systems may play a role in the conservation of insect diversity and forest specific insect fauna.

We used different methods to sample the involved insect groups: Bees and wasps were sampled with trap nests (n=8 per site, see Tscharntke et al. 1998 for methodology), dung beetles by baited pit fall traps (n=10 per site) and canopy beetles and ants by insecticide fogging (n=4 shade dwelling trees at forest sites and 4 cacao trees at the agroforests). We collected 2854 solitary bees and wasps belonging to 32 species, 761 dung beetles of 25 species, 3247

Table 1. The average and total abundances (N) and species richness (Spp.)collected per insect group and study site. Arithmetic means ± standard error per land-use type and ANOVA results are given. Means per land-use type are taken from the total values per site. Differences between means indicated with [a] and [b] are significant according to Tukey's HSD post-hoc test.

	Natural forest	Cacao under forest remnants	Cacao under diverse, planted shade	Cacao under homogeneous, planted shade	ANOVA results	Total
Abundance:						
Dung beetles	61.5 ± 10.76	39.8 ± 6.81	47.8 ± 19.41	41.3 ± 6.71	$F_{(3, 12)}=0.67$, $p=0.58$	761
Bees and wasps	12.5 ± 6.89[a]	180.0 ± 46.47[b]	216.0 ± 27.46[b]	305.0 ± 28.20[b]	$F_{(3, 12)}=15.96$, $p<0.001$	2854
Canopy ants	176.75 ± 51.48	187 ± 76.10	231.5 ± 90.58	216.5 ± 37.44	$F_{(3, 12)}=0.14$, $p=0.93$	3247
Canopy beetles	17.8 ± 6.38[a]	34.5 ± 10.49[ab]	42.5 ± 11.24[ab]	63.5 ± 7.98[b]	$F_{(3, 12)}=4.24$, $p<0.05$	633
Species richness:						
Dung beetles	11.5 ± 1.04	10.0 ± 1.08	7.5 ± 1.66	8.8 ± 0.25	$F_{(3, 12)}=2.32$, $p=0.13$	25
Bees and wasps	2.3 ± 0.63[a]	8.3 ± 1.18[b]	9.3 ± 1.55[b]	9.8 ± 0.75[b]	$F_{(3, 12)}=10.16$, $p<0.01$	32
Canopy ants	8.8 ± 1.60	11.0 ± 1.08	8.8 ± 1.70	12.3 ± 0.85	$F_{(3, 12)}=2.02$, $p=0.16$	44
Canopy beetles	11.3 ± 3.68	22.8 ± 5.42	19.3 ± 5.62	25.5 ± 0.87	$F_{(3, 12)}=1.64$, $p=0.23$	208

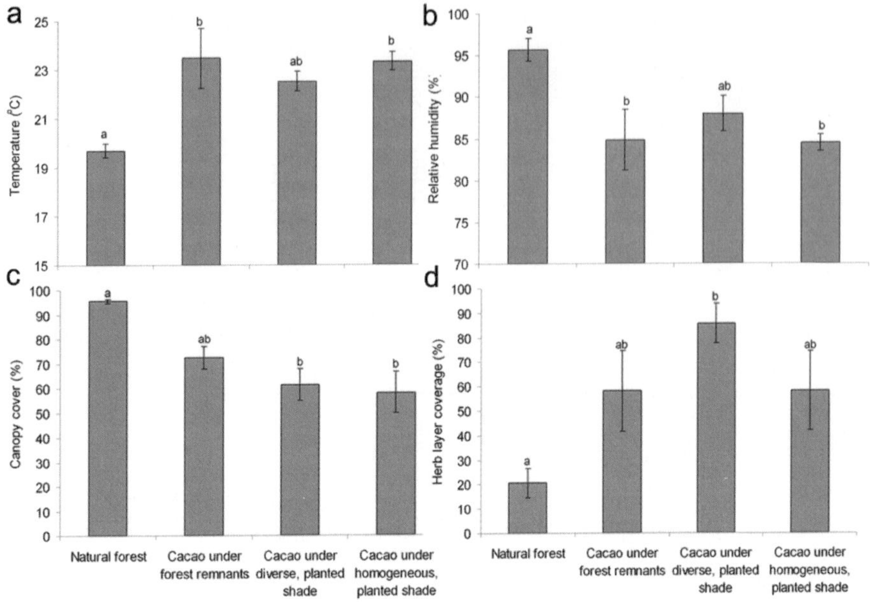

Fig. 1. The effects of forest conversion on a) Temperature (°C, ANOVA: $F_{(3, 12)}=6.52$, $p<0.01$) averages were taken from three morning measurements (by digital data-loggers), and averaged per site, b) Relative humidity (%RH, ANOVA: $F_{(3, 12)}=5.39$, $p<0.05$) was measured simultaneously with temperature, c) Canopy cover (ANOVA:$F_{(3, 12)}=5.39$, $p<0.05$), estimated with a spherical densiometer at 4 spots per site and averaged, and d) Herblayer coverage (%, ANOVA: $F_{(3, 12)}=4.43$, $p<0.05$) was estimated at four spots per site and averaged. All variables were averaged per land-use type, bars indicate standard errors between the four sites per land-use type, differences between values indicated with a and b are significant.

ants of 44 species and 633 canopy beetles belonging to 208 species. Identifications were made by specialists (see acknowledgements), on the basis of literature and on the basis of previously collected material from the same study region (by Klein et al. 2004, Shahabuddin et al. 2005). Unidentified species were sorted to morphospecies under supervision of specialists.

Per site abundances of solitary bees and wasps and of lower canopy beetles were higher in homogeneously shaded agroforests than in natural forests; however, abundances of dung beetles and canopy ants were not affected by land-use type (see Table 1). Species richness and abundance of solitary bees and wasps were positively related to the factor that was calculated from the environmental variables (species richness: $r=0.64$, $F_{(1,14)}=9.65$, $p=0.008$, abundance: $r=0.59$, $F_{(1, 14)}=7.58$, $p=0.02$). This means that solitary bee and wasp densities and species richness generally increased towards the least shaded agroforestry systems with dense herb layers, with a dryer and warmer microclimate than that of forest sites. Density and species richness of the other insect groups and diversity of all groups were not significantly affected by the factor ($p>0.05$).

For species richness, first order Jackknife estimations were calculated from observed values per site (EstimateS v7.00, Colwell 2004). The observed numbers of species covered between 63±0.9% (canopy beetles) and 67%±1.3% (canopy ants) of the estimates. Species richness saturation for solitary bees and wasps was significantly higher for the agroforestry systems than for the natural forest sites (ANOVA: $F_{(3, 12)}=4.47$, $p<0.05$). The observed number of species of all four insect groups were strongly correlated with estimated values for species richness ($r^2>0.90$, $p<0.00001$), so we used observed species richness for our statistical analyses, but acknowledge that actual species richness was presumably higher.

The species richness of solitary bees and wasps differed significantly between land-use types, with highest values in the three types of shaded agroforests (see table 1). Species richness of dung beetles,canopy beetles and ants showed no significant response to differences in land-use (see table 1).

Although species richness of canopy beetles increased towards the homogeneous agroforests, diversity (Simpson's reciprocal index for diversity insensitive to rare species) was significantly higher at agroforests shaded by forest remnants (27.1 ± 3.47) than at the forest sites (13.5±1.16) or the homogeneously shaded cacao agroforestry systems (13.0±2.40), see Figure 2d. The diversity of the other three insect groups also tended to be highest in cacao shaded by forest remnants, but differences remained insignificant (Figure 2a-c).

Accordingly, our results support the hypothesized negative effects of management intensification on diversity and that the effects are stronger than that of forest conversion to agroforestry systems.

3.2 Hypothesis 2: Large spatial species turnover between different land-use systems contributes to regional species richness of insects

We distinguished between species richness at the α, β and γ level in an additive partitioning approach (Lande 1996, Veech et al. 2002). Between each of the spatial scales in our set-up we calculated the β-diversity (among-scale richness) as the difference between total species richness (γ) and mean species richness (α) within one spatial scale. The scales in our study were: sample unit (trap nest or tree), study-sites, land-use types and total study area. This way we explain total species richness by the differences in number of species between the spatial levels (see also Wagner et al. 2000, Gering et al. 2003, Summerville 2003). Species recorded only once were excluded from the calculations. The data on dung beetles was lumped per site and was not available per trap.

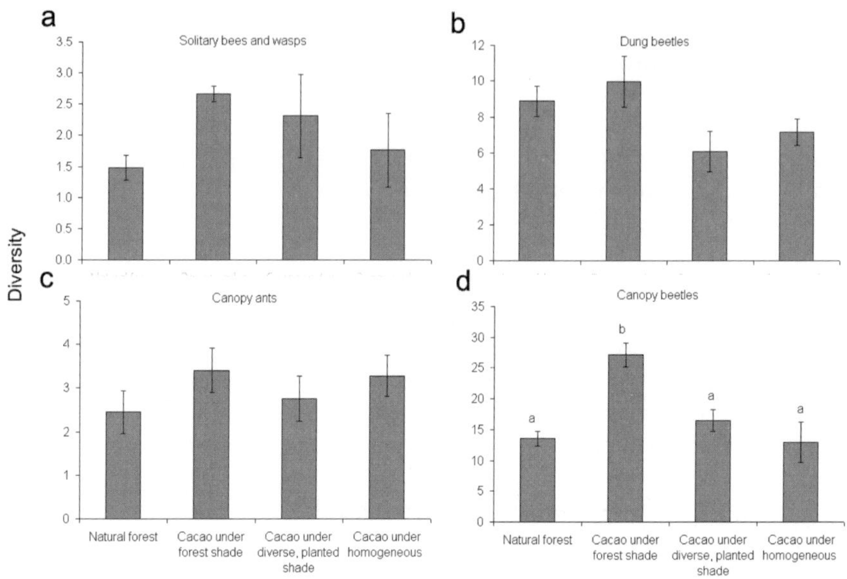

Fig. 2. Diversity (Simpson's reciprocal index) of a) Solitary bees and wasps (ANOVA: $F_{(3,\ 12)}=1.73$, p=0.21), b) Dung beetles (ANOVA: $F_{(3,\ 12)}=2.67$, p=0.09), c) Canopy ants (ANOVA: $F_{(3,\ 12)}=0.87$, p=0.48) and d) Canopy beetles(ANOVA: $F_{(3,\ 12)}=5.7$, p=0.01). Higher values stand for higher diversity. Bars indicate standard errors. Differences between letters a and b mentioned above the standard error are significant (based on Tukey's HSD post-hoc test). The index is calculated as 1 divided by the sum of the squared relative abundances of each species. The minimum values of 1 would stand for a community consisting of 1 species and the maximum possible value for a community cannot exceed the number of species.

Bees and wasps comprised a total of 32 species, of which seven occurred at forest sites. Three of those (42.9% of all forest species) were recorded only at the forest sites. The dominant eumenid wasp species *Rhynchium haemorrhoidale* (Hymenoptera: Eumenidae, 66% of all records) was the only species trapped at all land-use types but showed a pronounced preference for the monotonously shaded agroforestry systems. In the additive partitioning of species richness, the differences between study sites contributed 45% of the regional solitary bee and wasp species richness (Figure 3).

Fifteen of the 20 ant species (75.0%) found at the forest sites were also recorded at agroforestry sites. Remarkably, only three of these 20 species showed higher abundance at the forest than at the agroforest sites. Forty-four percent of regional species richness in the study area depended on the differences between the land-use types (Figure 3).

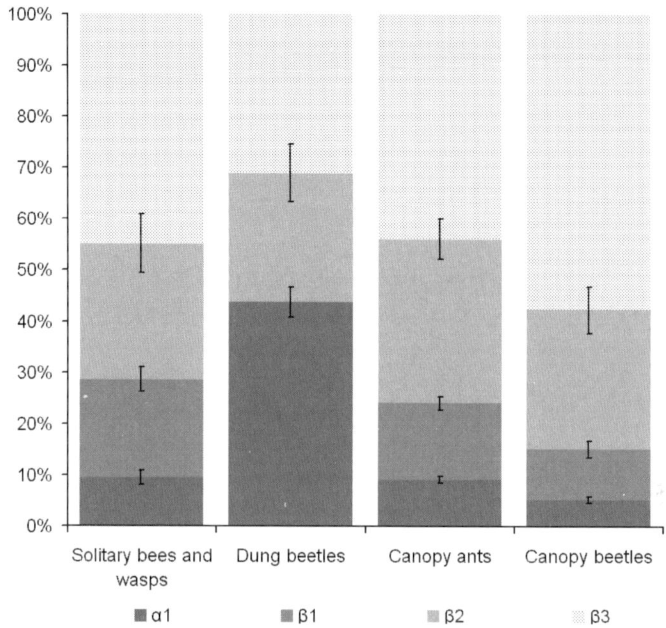

Fig. 3. The additive partitioning of species richness (species that were observed only once were excluded) showing the importance of each spatial scale for total species richness. Where values were averaged, bars indicate standard errors. α_1 = species richness per trap nest or tree (dung beetle data was lumped per study site, hence α_1 and β_1 could not be calculated); β_1 = the change in species richness from trap nest or tree to study site (α_2), β_2 = the change in species richness from study site to land-use type (α_3) and β_3 = the change in species richness from land-use type to the total species richness (α_4, or γ). Species richness at a spatial scale n equals $\alpha_n = \alpha_{n-1} + \beta_{n-1}$ in which the highest possible α_n equals total species richness γ.

Of all insect groups, dung beetles showed highest faunal overlap between forest and agroforestry sites. Of the 17 species recorded at the forest sites, 13 (76.5% of all forest species) were also found in agroforestry systems and all species were relatively common. The additive partitioning of the species richness showed that 31% of dung beetle species richness could be explained by differences between land-use types (Figure 3).

From the highly diverse canopy beetles, only five of the 40 species (12.5%) recorded at the forest sites were also found in the agroforestry systems, and, thus, they showed the smallest faunal overlap between the land-use types. Because of this high dissimilarity between agroforests and forest sites, 58% of the overall species richness was explained by the differences between land-use types (Figure 3).

Supporting the second hypothesis, the dissimilarity among the habitats (three agroforestry types and the forest) largely contributed to overall insect richness (31-58%). Further, we found high percentages of forest species that did not occur in agroforests.

4 Discussion

The cacao–dominated agroforestry systems still supported surprisingly high levels of insect diversity and species richness. Although forest conversion led to the loss of 24.5-87.5% of the species recorded at the forest sites, species richness and diversity in the managed agroforestry systems were similar to that found in the natural forest sites or even higher. Increasing management intensity of agroforestry systems with a reduction of shade-tree richness did have negative effects on the diversity (species richness and abundance combined in Simpson's index) of solitary bees and wasps, and canopy beetles, but not on the diversity of canopy ants and dung beetles. These results support the hypothesis that in addition to forest conversion land-use intensification is a major cause for decreases in insect diversity.

When we considered species richness and abundance separately, lower canopy beetles and solitary bees and wasps appeared to even profit from land-use intensification. Solitary bees and wasps had highest species richness and abundance in the homogeneous agroforests, and lower canopy beetle species richness and abundances were also highest in the homogeneously shaded agroforests. The differences in the patterns for species richness or abundance compared to the diversity index indicate that the insect communities of homogeneously shaded sites were dominated by relatively few abundant species whereas less intensively managed sites with a diverse shade canopy supported a more even species-abundance distribution.

We showed that responses of insects to forest conversion and agroforestry management depended strongly on the group studied and whether species richness or diversity was compared. Dung beetles showed no response to forest conversion and shade management. Most species at the forest sites were also able to survive in agroforestry systems. This is probably related to the high dispersal capability of dung beetles when they search for fresh feeding patches and indicates their adaptability to a broad amplitude of habitat conditions (Peck and Forsyth 1982, Hanski and Cambefort 1991). Lower canopy beetles and solitary bees and wasps showed stronger responses to both forest conversion and changes in agroforestry management but the overall diversity seemed to profit from intermediate human disturbance. We do not know how the studied insect groups would respond to further intensification and complete removal of shade trees. Perfecto et al. (1997) reports drastic declines in beetle and ant species richness from well-shaded coffee agroforestry systems to zero-shade coffee.

The conversion from tropical rainforest to agroforestry systems was accompanied by a significant decrease in the density of the upper canopy and an increase in herb-layer density and species richness of herbaceous plants (R. Pitopang, *pers. comm.*). Although we could not find significant relationships with environmental variables, the denser undergrowth vegetation of agroforestry systems presumably supported higher herbivore and pollinator densities (but not necessarily diversity) explaining the observed increase in solitary bees and lower canopy beetles. In our study, 52% of the collected lower canopy dwelling beetles represented herbivorous species and that guild showed an extreme increase towards the monotonous agroforestry systems (data not shown).

The increases in the herb layer and concomitant increases in herbivores could also have led to the higher densities of predaceous solitary wasps in the trap nests. Klein et al. (2004) observed the most common solitary wasp, *Rhynchium haemorrhoidale*, preying on cacao leaf consuming caterpillars. This wasp species accounted for 66% of all collected solitary wasps and increased sharply towards the homogenous agroforests.

As argued above, our results indicate that some insect groups from the lower canopy can profit from lower shade levels and the related increase in cover and diversity of the herb layer. However, insect communities from higher canopy layers are likely to respond oppositely to what we have found for bees, wasps and beetles in lower canopy. For example Lawton et al. (1998) and Chung et al. (2004) found a significant reduction in upper canopy beetle diversity from natural forest to agroforestry systems. Similarly, the diversity of butterflies and birds in agroforesty systems was lower than in natural forests (Schulze et al. 2004). Contrasting responses by insects in upper and lower canopy to habitat disturbance have been suggested before by Davis and Sutton (1998) and Fermon et al. (2005).

Whereas diversity remained high, species compositions changed drastically from natural forest sites to intensively managed agroforestry systems. Especially forest communities of solitary bees and wasps and lower canopy dwelling beetles were decimated after forest conversion. Canopy ants and dung beetles seemed to depend less on locally specific factors as species compositions remained more similar to those found at the forest sites.

The drastic changes of insect communities in response to forest conversion and changes in agroforestry management, led to large faunal turnover (i.e., β diversity). The differences between land-use types explained between 31% and 58% of the total recorded species richness. These results support the hypothesis that differences in local management can contribute significantly to the insect richness at the landscape level. The insect group that was most strongly affected by differences in management (lower canopy beetles) had the largest faunal turnover; more than half of their regional species richness could be explained by differences in land-use.

5 Conclusions and Outlook

Agricultural intensification and high disturbance levels due to intensified management practices have negative effects on faunal diversity in that a few 'winner' species make up a less diverse community at the cost of many 'loser' species (McKinney and Lockwood 1999). However, disturbance must extend beyond a threshold in order to cause a decline in biodiversity and before that threshold is reached, disturbance can lead to higher diversity (Connell 1978). In Europe, for example, extensive early 20th century agriculture has lead to an increase in rural biodiversity, whereas the agricultural intensification since the 1950s led to sharp declines (Tscharntke et al. 2005).

Although different insect groups respond in different ways, forest species generally declined after forest conversion. As reported here, however, overall species richness can remain high, even in agroforestry systems with homogeneous shade trees. Because species richness puts large emphasis on rare species, we combined species richness and abundance in a diversity index that was relatively insensitive to large numbers of rare species. This diversity index did not increase towards the homogeneous plantations and was for most insect groups highest at the intermediate agroforestry systems shaded by forest remnants. Hence, the most valuable type of agroforestry system from a diversity conservation point of view, are systems with a canopy of remaining forest trees where in other cases shade trees have been planted.

We did not detect effects of agroforestry intensification as strong as in tropical America where shade tree stands were completely removed in many landscapes. Under further intensification we predict similar negative effects on the still highly diverse insect fauna of the cacao-dominated agroforestry in Indonesia. Leaving shade tree stands contributes to the conservation of high insect species richness. Diversification of shade trees and the maintenance of natural forest remnants are recommended tools in the conservation of insect diversity in tropical agroforestry systems.

Faunal turnover as a result of land-use dependent differences in community structure greatly contributes to regional insect richness. The effect of different insect communities in different land-use types made up major parts of the regional insect in our study. To preserve the conservation value of cultivated areas with agroforestry, conservation policies should aim at heterogeneity at the landscape level and promote a variety of forested land-use systems.

Acknowledgements
This study has been funded by the Deutsche Forschungsgemeinschaft (DFG) through the multidisciplinary German-Indonesian research project STORMA (Stability of Rainforest Margins, SFB 552) coordinated by Adam Malik, Wolfram Lorenz, Surya Tarigan and Daniel Stietenroth. We are thankful to Christian H. Schulze and Jason Tylianakis for their useful comments during the progress of the work. Thanks to Akhmad Rizali, Christoph Bayer, Otto Merkl,

Chris Reid, Wolfgang Schawaller and other taxonomists for their taxonomic assistance. Lastly, thanks to our assistants Arifin, Armand and Pak Arsath and to Kusnadi, Rina Yusuf and all drivers hired by STORMA for their help in Indonesia.

References

Achard F, Eva HD, Stibig H-J, Mayaux P, Callego J , Richards T, Malingreau J-P (2002) Determination of deforestation rates of the world's humid tropical forests. Science 297: 999-1002

Altieri MA (1999) The ecological role of biodiversity in agroecosystems. Agriculture, Ecosystems and Environment 74: 19-31

Belsky JM, Siebert SF (2003) Cultivating cacao: Implications of sun-grown cacao on local food security and environmental stability. Agriculture and Human Values 20: 270-285

Campbell CAM (1984) The influence of overhead shade and fertilizers on the Homoptera of mature Upper-Amazon cocoa trees in Ghana. Bulletin of Entomological Research 74: 163-174

Chung AYC, Eggleton P, Speight MR, Hammond PM, Chey VK (2000) The diversity of beetle assemblages in different habitat types in Sabah, Malaysia. Bulletin of Entomological Research 90: 475-496

Colwell RK (2004) EstimateS: Statistical estimation of species richness and shared species from samples. Version 7. User's Guide and application. http://purl.oclc.org/estimates

Connell JH (1978) Diversity in tropical rainforests and coral reefs: high diversity of trees and corals is only maintained in a nonequilibrium state. Science 199: 1302-1309

Davies JG, Stork NE, Brendell MJD, Hine SJ (1997) Beetle species diversity and faunal similarity in Venezuelan rainforest tree canopies. In: Stork NE, Adis J, Didham RK (eds), Canopy Arthropods. London: Chapman & Hall, pp. 27-52

Davis AJ, Sutton SL (1998) The effects of rainforest canopy loss on arboreal dung beetles in Borneo: implications for the measurement of biodiversity in derived tropical ecosystems. Diversity and Distributions 4: 167-173

Donald PF (2004) Biodiversity impacts of some agricultural commodity production systems. Conservation Biology 18: 17-37

Entwistle PF (1972) Pests of Cocoa, Longman, London

Erwin TL (1982) Tropical forests: their richness in Coleoptera and other arthropod species. The Coleopterists Bulletin 36: 74-75

Fermon H, Waltert M, Mühlenberg M (2003) Movement and vertical stratification of fruit-feeding butterflies in a managed West African rainforest. Journal of Insect Conservation 7: 7-19

Foley JA, DeFries R, Asner GP, Barford C, Bonan G, Carpenter SR, Chapin FS, Coe MT, Daily GC, Gibbs HK, Helkowski JH, Holloway T, Howard

EA, Kucharik CJ, Monfreda C, Patz JA, Prentice IC, Ramankutty N, Snyder PK (2005) Global consequences of land use. Science 309: 570-574

Free JB (1993) Insect pollination of crops, Academic Press, London

Gering JC, Crist TO (2002) The alpha-beta-regional relationship: providing new insights into local-regional patterns of species richness and scale dependance of diversity components. Ecology Letters 5: 433-444

Hanski I, Cambefort Y (1991) Dung Beetle Ecology, Princeton, Princeton University Press, New Jersey

Hoekstra JM, Boucher TM, Ricketts TH, Roberts C (2005) Confronting a biome crisis: global disparities of habitat loss and protection. Ecology Letters 8: 23-29

Hölldobler B, Wilson EO (1990) The Ants, Harvard University Press, Cambridge

Jones D, Susilo TFX, Bignell DE, Hardiwinoto S, Gillison AN, Eggleton P (2003) Termite assemblage collapse along a land-use intensification gradient in lowland central Sumatra, Indonesia. Journal of Applied Ecology 40: 380-391

Kalshoven LGE (1950) De plagen van cultuurgewassen in Indonesie, Uitgeverij van Hoeve, 's-Gravenhage/Bandoeng

Klein A-M, Steffan-Dewenter I, Tscharntke T (2004) Foraging trip duration and density of megachilid bees, eumenid wasps and pompilid wasps in tropical agroforestry systems. Journal of Animal Ecology 73: 517-525

Klein A-M, Steffan-Dewenter I, Tscharntke T (2003) Fruit set of highland coffee increases with the diversity of pollinating bees. Proceedings of the Royal Biological Society of London 270: 955-961

Klein A-M, Steffan-Dewenter I, Tscharntke T (2002) Predator-prey ratios on cacoa along a land-use gradient in Indonesia. Biodiversity and Conservation 11: 683-693

Klein A-M, Steffan-Dewenter I, Tscharntke T (2006) Rainforest promotes trophic interactions and diversity of trap-nesting Hymenoptera in adjacent agroforestry. Journal of Animal Ecology (in press)

Klein BC (1989) Effects of forest fragmentation on dung and carrion beetle communities in Central Amazonia. Ecology 70: 1715-1725

Lande R (1996) Statistics and partitioning of species diversity, and similarity among multiple communities. Oikos 76: 5-13

Lawton JH, Bignell DE, Bolton B, Bloemers GF, Eggleton P, Hammond PM, Hodda M, Holt RD, Larsen TB, Mawsley NA, Stork NE, Srivastava DS, Watt AD (1998) Biodiversity inventories, indicator taxa and effects of habitat modification in tropical forest. Nature 391: 72-76

McKinney ML, Lockwood JL (1999) Biotic homogenization: a few winners replacing many losers in the next mass extinction. Tree 14: 450-453

McNeely JA (2004) Nature vs. nurture: managing relationships between forests, agroforestry and wild biodiversity. Agroforestry Systems 61: 155-165

Meffe GK (1998) The potential consequences of pollinator declines on the conservation of biodiversity and stability of food crop yields. Conservation Biology 12: 8-17

Millennium Ecosystem Assessment (2005) Ecosystem and human well-being: synthesis, Island Press, Washington, DC.

Myers N, Mittelmeier RA, Mittelmeier CG, da Fonseca GAB, Kent J (2000) Biodiversity hotspots for conservation priorities. Nature 493: 853-858

O'Neill KM (2001) Solitary wasps: behavior and natural history, Cornell University Press, New York

Peck SB, Forsyth A (1982) Composition, structure and competitive behavior in a guild of Ecuadorian rain forest dung beetles (Coleoptera, Scarabaeidae). Canadian Journal of Zoology 60: 1624-1634

Perfecto I, Vandermeer J, Hanson P, Cartín V (1997) Arthropod biodiversity loss and the transformation of a tropical agro-ecosystem. Biodiversity and Conservation 6: 935-945

Perfecto I, Vandermeer J, Mas A, Pinto LS (2005) Biodiversity, yield, and shade coffee certification. Ecological Economics 54: 435-446

Perfecto I, Snelling R (1995) Biodiversity and the transformation of a tropical agroecosystem: Ants in coffee plantations. Ecological Aplications 5: 1084-1097

Peterson G, Allen CR, Holling CS (1998) Ecological resilience, biodiversity, and scale. Ecosystems 1: 6-18

Philpott SM (2005) Changes in arboreal ant populations following pruning of coffee shade-trees in Chiapas, Mexico. Agroforestry Systems 64: 219-224

Potter L (2001) Agricultural intensification in Indonesia: outside pressures and indigenous strategies. Asia Pacific Viewpoint 42: 305-324

Putz F, Blate EGM, Redford KH, Fimbel R, Robinson J (2001) Tropical forest management and conservation of biodiversity: an overview. Conservation Biology 15: 7-20

Ramirez OA, Somarriba E, Ludewigs T, Ferreira P (2001) Financial returns, stability and risk of cacao-plantain-timber agroforestry systems in Central America. Agroforestry Systems 51: 141-151

Rice RA, Greenberg R (2000) Cacao cultivation and the conservation of biological diversity. Ambio 29: 167-172

Room PM (1971) The relative distribution of ant species in Ghana's cocoa farms. Journal of Animal Ecology 40: 735-751

Roubik DW (1995) Pollination of cultivated plants in the Tropics, FAO Agricultural Services Bulletin 118, Rome.

Sanchez PA (1999) Delivering on the promise of agroforestry. Environment, Development and Sustainability 1: 275-284

Schroth G, da Fonseca GAB, Harvey CA, Gascon C, Vasconcelos HL, Izac A-MN (2004) Agroforestry and biodiversity conservation in tropical landscapes, Island Press, Washington DC

Schulze CH, Waltert M, Kessler PJA, Pitopang R, Shahabuddin, Veddeler D, Muhlenberg M, Gradstein SR, Leuschner C, Steffan-Dewenter I, Tscharntke T (2004) Biodiversity indicator groups of tropical land-use systems: comparing plants, birds and insects. Ecological Applications 14: 1321-1333

Shahabuddin, Schulze CH, Tscharntke T (2005) Changes of dung beetle communities from rainforests towards agroforestry systems and annual cultures in Sulawesi (Indonesia). Biodiversity and Conservation 14: 863-877

Siebert SF (2002) From shade- to sun-grown perennial crops in Sulawesi, Indonesia: implications for biodiversity conservation and soil fertility. Biodiversity and Conservation 11: 1889-1902

Sodhi NS, Koh LP, Brook BW, Ng PKL (2004) Southeast Asian biodiversity: an impending disaster. Trends in Ecology and Evolution 18: 654-660

Summerville KS, Boulware MJ, Veech JA (2003) Spatial variation in species diversity and composition of forest Lepidoptera in Eastern deciduous forests of North America. Conservation Biology 17: 1045-1057

Tscharntke T, Gathmann A, Steffan-Dewenter A (1998) Bioindication using trap-nesting bees and wasps and their natural enemies: community structure and interactions. Journal of Applied Ecology 35: 708-719

Tscharntke T, Klein A-M, Kreuss A, Steffan-Dewenter I, Thies C (2005) Landscape perspectives on agricultural intensification and biodiversity-ecosystem service management. Ecology Letters 8: 857-874

Tylianakis JM, Klein A-M, Tscharntke T (2005) Spatiotemporal variation in the diversity of Hymenoptera across a tropical habitat gradient. Ecology 86: 3296-3302

Urquhart DH (1955) Cocoa, Longmans, London

Veech JA, Summerville KS, Crist TO, Gering JC (2002) The additive partitioning of species diversity: recent revival of an old idea. Oikos 99: 3-9

Vingerhoets, J (2002) International Cocoa Organization annual report 2001-02. ICCO.

Wagner HH, Wildi O, Ewald KC (2000) Additive partitioning of plant species diversity in an agricultural mosaic landscape. Landscape Ecology 15: 219-227

Plant diversity in homegardens in a socio-economic and agro-ecological context

Katja Kehlenbeck[1*], Hadi Susilo Arifin[2], and Brigitte L. Maass[1]

[1] Department of Crop Sciences: Tropical Crops, Georg-August-University, Grisebachstr. 6, D-37077 Göttingen, Germany
[2] Department of Landscape Architecture, Faculty of Agriculture, IPB, Bogor, Indonesia

*corresponding author: Katja Kehlenbeck, phone +49-(0)511-393750, Email: katja_kehlenbeck@yahoo.de

Summary

Homegardens are generally regarded as a very complex, species-rich agroforestry system managed in a sustainable manner over decades or even centuries. In many densely populated tropical regions, homegardens appear to be the last forest-like islands surrounded by increasingly extended, uniform staple crop fields. With their multi-layered vegetation structure, homegardens serve as an important habitat for wild flora and fauna in these areas. They fulfil not only important ecological, but also many social and cultural functions. However, the major purposes of homegardens are subsistence production and income generation, particularly in rural areas. At forest margins, high production levels in homegardens might help to reduce deforestation. Furthermore, homegardens should be considered as a model for sustainable agroforestry systems, integrating both economic and ecological advantages. Plant diversity, as a basis for homegarden productivity and sustainability, is influenced by a combination of agro-ecological as well as socio-economic factors. The complex interactions of all these factors are not yet fully understood. This paper presents an overview of the existing knowledge and identifies gaps regarding the factors determining plant species diversity and composition in homegardens. We further illustrate this with two case studies from Indonesia (Central Sulawesi and West Java), in which temporal and spatial variations were investigated. In conclusion, plant diversity was mainly influenced by elevation as well as commercialisation, urbanisation, and fragmentation. It was fairly dynamic over time, particularly, when commercialisation was possible. To preserve the sustainability of homegardens and their suitability for *in situ*

Tscharntke T, Leuschner C, Zeller M, Guhardja E, Bidin A (eds), The stability of tropical rainforest margins, linking ecological, economic and social constraints of land use and conservation, Springer Berlin 2007, pp 297-319

conservation of plant genetic resources, any promotion to intensify production in homegardens should consider the overall ecological functioning of the system in a landscape context.

Keywords: agro-biodiversity, agro-ecosystem, agroforestry, commercialisation, function, homegarden, *in situ* conservation, plant diversity, plant genetic resources, species composition, sustainability, urbanisation, vegetation structure, vegetation dynamics

1 Introduction

Homegardens are one of the most complex and diverse agricultural systems worldwide. As their appearance is highly variable, there are several definitions of this system. Homegardens are mostly defined as a piece of land surrounding a homestead that is cultivated with a diverse mixture of perennial and annual plant species arranged in a multi-layered vertical structure, often in combination with raising livestock (Christanty 1990, Kumar and Nair 2004, Soemarwoto 1987). Homegarden systems have existed for millennia (Kumar and Nair 2004, Soemarwoto and Conway 1992) in many tropical regions, where they played an important role towards the development of early agriculture and domestication of crops and fruit trees (Miller and Nair 2005). Individual homegardens have been continuously cultivated for many decades and even centuries, for example, in Sri Lanka (Hochegger 1998). For this reason, homegardens are generally regarded as sustainable production systems (Christanty 1990, Kumar and Nair 2004, Landauer and Brazil 1990, Soemarwoto and Conway 1992, Torquebiau 1992). However, quantitative support for this statement is rare, as most of the published homegarden studies are rather descriptive. In addition, no long-term quantitative study of the same homegarden has been reported. This might be due to the difficulties in measuring sustainability *per se*, resulting in an indirect assessment by using more or less widely accepted sustainability indicators (Kumar and Nair 2004). Torquebiau (1992) and Huxley (1999) suggested several sustainability indicators related to the resource base (e.g., soil, climate, biodiversity), the system's functioning (e.g., management, outputs), and the impact of homegardens on other systems (e.g., forests, wildlife). Kumar and Nair (2004) present extensive data of several recent homegarden studies that mainly describe soil-related and socio-economic aspects of sustainability. Biodiversity, particularly plant species diversity, is the aspect probably most frequently assessed in homegarden research. Numerous publications described plant species and their uses in homegardens worldwide, e.g., in Indonesia (Abdoellah et al. 2002, Arifin et al. 1998, Kehlenbeck and Maass 2004, Soemarwoto 1987), Vietnam (Trinh et al. 2003), India (Peyre et al. 2005), Sri Lanka (Hochegger 1998), Ethiopia (Tesfaye Abebe 2005), Ghana (Bennett-Lartey et al. 2002), Tanzania (Hemp 2005), Mexico (Vogl et al. 2002), Brazil (Albuquerque et al. 2005), Peru (Coomes and Ban 2004), and

Cuba (Castiñeiras et al. 2002, Wezel and Bender 2003). Because of their large crop species and varietal diversity, homegardens are also regarded as an ideal production system for *in situ* conservation of plant genetic resources (Trinh et al. 2003, Watson and Eyzaguirre 2002), crucial for the long-term sustainability of agro-ecosystems. In addition, homegardens can contribute substantially to the conservation of forest tree species, particularly where natural forests have largely been replaced by agricultural fields (Hemp 2005).

In Indonesia and many other tropical regions, the conversion of primary forest to frequently unsustainable agricultural lands has increased dramatically. Forest margins are particularly concerned due to easy access, even in protected areas such as national parks. Sustainable and productive agricultural systems urgently need to be promoted in such agricultural frontier areas to reduce the pressure on further forest conversion. Agroforestry systems like homegardens and forest gardens could serve as a model for such systems because they offer an alternative measure for sustainable use of natural resources. However, neither the functioning nor the potential of homegardens have been sufficiently studied. Research is needed, particularly, concerning nutrient and water balances, carbon sequestration, the value of non-conventional products and services, system productivity, and sustainability, including temporal changes (Kumar and Nair 2004). Especially, changes of plant diversity and its driving factors need to be emphasised. This paper presents firstly an overview of the existing knowledge on roles and functioning of homegardens. Factors determining plant species diversity and composition in homegardens are discussed in more detail. Secondly, we illustrate the complex influence of certain agro-ecological and socio-economic factors on plant diversity with two case studies from Indonesia (Central Sulawesi and West Java), in which temporal and spatial variations were investigated.

2 State of the art in homegarden research

2.1 The multiple roles of tropical homegardens

The basic function of homegardens, especially in rural areas, is subsistence production (Figure 1) (Kumar and Nair 2004, Soemarwoto and Conway 1992). Because of the high plant species diversity existing in homegardens, a wide spectrum of multiple-use products can be generated with relatively low labour, cash or other external inputs (Christanty 1990, Hochegger 1998, Soemarwoto and Conway 1992). Homegardens generally serve as a complement to staple crop fields by producing mainly fruits, vegetables, spices, and many non-food products (Figure 1) (Albuquerque et al. 2005, Karyono 1990, Kehlenbeck and Maass 2004, Kumar and Nair 2004, Michon and Mary 1994, Peyre et al. 2005). However, in densely populated or heavily degraded areas without sufficient staple crop fields, homegardens may also provide large portions of staple food (Soemarwoto and Conway 1992, Tesfaye Abebe 2005). Homegarden products, including those from animals reared in the gardens, have a rather high nutritional value in terms of protein, minerals, and vitamins (Soemarwoto and Conway 1992). As these diverse products are available year-round, homegardens also contribute to food security in times or seasons of scarcity (Christanty 1990, Karyono 1990). Therefore, the importance of homegardens for countering malnutrition and food insecurity has attracted increasing attention (Kumar and Nair 2004). This, for example, has resulted in several manuals for the promotion of growing vegetables in tropical homegardens, as compiled by Helen Keller International (2004).

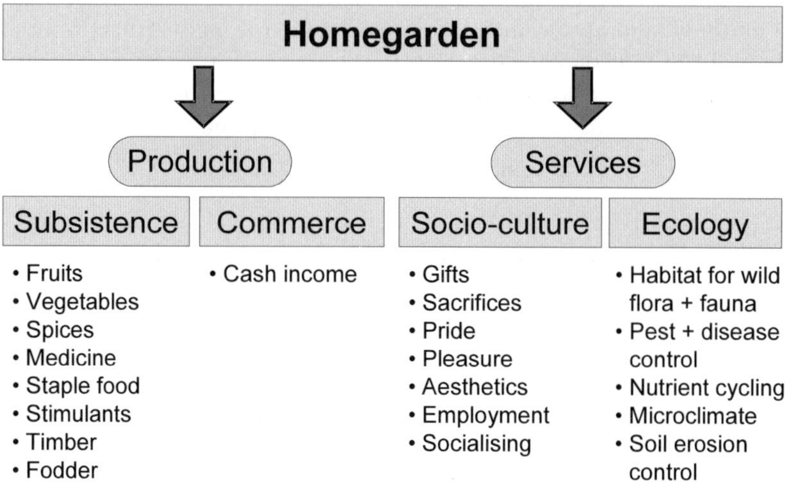

Fig. 1. Main functions of homegardens and selected products/outputs.

The second important function of homegardens is the generation of cash income, particularly in regions with good market access (Figure 1) (Christanty et al. 1986, Michon and Mary 1994, Tesfaye Abebe 2005, Trinh et al. 2003). Most of the income is said to be derived from perennials such as fruit trees, cacao, and coffee, but in peri-urban areas or tourist centres, also vegetables and ornamentals are frequently grown as cash crops (Abdoellah et al. 2002, Soemarwoto and Conway 1992). The portion of income from a homegarden may vary from only 1–7% (Arifin et al. 2005) to more than 50% of the household's total income (Trinh et al. 2003), depending on market access, among other factors.

In addition to the production, homegardens have important social and cultural functions (Abdoellah et al. 2002, Christanty 1990, Soemarwoto and Conway 1992). They are mostly 'open' for everyone, thus, providing a place for children to play and for the neighbourhood to meet and chat. The exchange of homegarden products and planting material is common in many traditional societies. Homegardens also serve as status symbol and the aesthetic purpose partly outweigh the productive function, especially in urban areas and better-off households (Arifin et al. 1998, Karyono 1990). Some plant species in homegardens are believed to have a magical value (Abdoellah et al. 2002), others are necessary for religious ceremonies, e.g., Hindu Balinese families need their homegardens as source and place for making sacrifices (Arifin et al. 2002, Kehlenbeck unpublished).

Homegardens also fulfil ecological functions (Figure 1), particularly in landscapes where large, monotonous, and monofunctional agricultural fields dominate (Christanty 1990). The multi-layered vegetation structure of homegardens is said to resemble natural forests and offers a habitat for a diverse community of wild plants and animals (Albuquerque et al. 2005, Hemp 2005, Karyono 1990, Michon and Mary 1994). This structure appears to contribute substantially to the sustainability of homegarden systems.

2.2 Biodiversity as a sustainability indicator

The multi-layered vegetation structure in homegardens, created by the wide spectrum of cultivated and wild plants, appears to be responsible for many other benefits/advantages of this system. The contribution of (agro-) biodiversity to (agro-)ecosystem functioning and sustainable production is more and more recognised both for man-made and natural systems (Atta-Krah et al. 2004, Clergue et al. 2005, Main 1999, Schwartz et al. 2000). On the one hand, intra-specific diversity is not only a key source for breeding, but also important for sustainability because it enables individual species to adapt to a changing environment and, therefore, ensures their long-term survival (Atta-Krah et al. 2004, Main 1999). Inter-specific diversity, on the other hand, leads to important synergistic ecological processes and enables ecosystem functioning. This refers to efficient, complementary resource utilisation, efficient nutrient recycling, and a low risk of soil erosion (Clergue et al. 2005, Main 1999, Kumar and Nair 2004, Soemarwoto and Conway 1992, Torquebiau 1992, Wiersum 2004).

In addition to ecological aspects of sustainability, biodiversity also contributes to socio-economic sustainability of agro-ecosystems. Productivity of a species-rich agro-ecosystem is generally rather high and stable because the multi-species system forms a buffer against biotic (pests and diseases) and abiotic (storms and droughts) stress (Atta-Krah et al. 2004, Clergue et al. 2005, Wiersum 2004). Recent studies demonstrated the importance and monetary value of plant diversity for pollination services and pest control in agroforestry systems (Bos et al., this volume, Olschewski et al., this volume). A diverse, multi-species production system reduces the risk of total crop failure and provides year-round available products of high nutritional value. Thus, plant diversity contributes to sustainability in the aspect of household food security (Atta-Krah et al. 2004, Huxley 1999, Kumar and Nair 2004, Main 1999).

Wiersum (2004) stressed a further potential of multi-species agroforestry systems towards sustainability: a diverse range of useful plant species in a system enables its effective adjustment to changing socio-economic conditions and demands of future generations. Nevertheless, the suitability of biodiversity as an indicator to assess sustainability might be critically examined because there is no threshold value for an ideal number of species in a sustainable system (Main 1999). Biodiversity also seems to be highly variable over time, while homegarden research has so far neglected to quantify such changes.

2.3 Factors influencing plant diversity in homegardens

Garden diversity varies depending on a combination of external and intrinsic factors that are related to garden features, agro-ecology, and socio-economy (Christanty et al. 1986, Hodel et al. 1999). However, plant diversity is not only influenced by clearly identifiable single factors but rather by a complex interaction between several of them. This interaction is not yet understood, and additional intrinsic factors, like individual preferences and practices of the gardener, may even play an overriding role for the composition and level of plant diversity (Abdoellah et al. 2002, Castiñeiras et al. 2002, Hodel et al. 1999). In addition, these factors may vary in their relative importance over time and, thus, affect plant diversity (Figure 2). A better understanding of these interrelationships and the processes leading to them would help to assess the sustainability of the system as well as its suitability for *in situ* conservation of plant genetic resources. How selected factors influence plant diversity in homegardens is presented below in more detail.

Agro-ecological factors

Agro-ecological factors such as elevation, climate, or soil fertility might limit plant diversity in homegardens (Figure 3). Many homegarden studies have highlighted the effect of elevation on plant diversity, which, in terms of species richness, generally decreases with increasing elevation (0–1000 m) due to decreasing mean temperature (Hodel et al. 1999, Karyono 1990). Tesfaye Abebe (2005) also reported lower diversity indices in higher elevations (1500–2000 m). Less fruit tree species, but more vegetables and medicinal plant species were cultivated in homegardens of higher elevations (Castiñeiras et al. 2002, Shrestha et al. 2002, Soemarwoto and Conway 1992). However, research results of Quiroz et al. (2002) in Venezuela suggested that the highest diversity is found in homegardens of intermediate elevation (600–1300 m), where gardeners have the opportunity to cultivate both tropical and subtropical useful plants.

Not only temperature, but also precipitation influences plant diversity. Homegardens in West Java harboured higher plant diversity in the wet than in the dry season (Soemarwoto and Conway 1992). Plant diversity of Ghanaian homegardens was higher in the humid forest ecozones than in the hot and

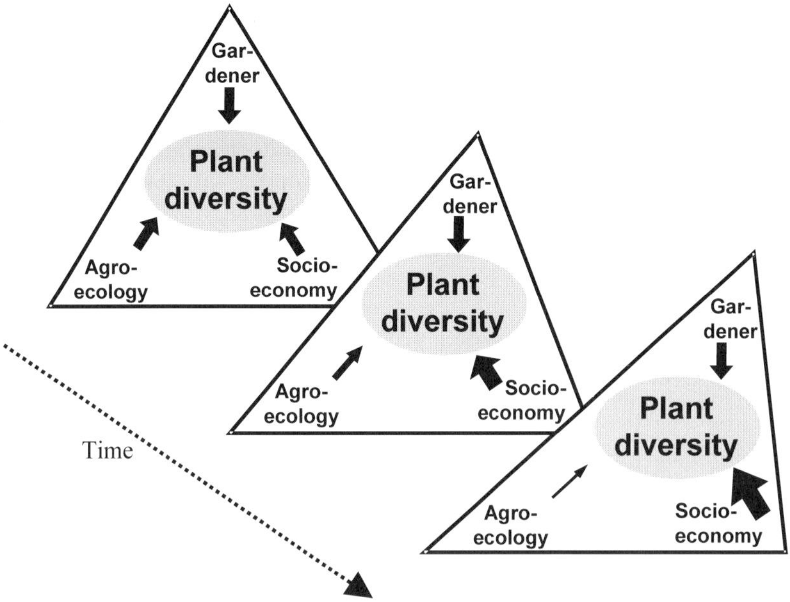

Fig. 2. Schematic illustration of the relative importance of different factors on plant diversity in homegardens and their temporal changes. Here, the importance of socio-economic factors (e.g., commercialisation) increases over time, while that of agro-ecological characteristics (e.g., infertile soil) decreases, for example due to the use of industrial fertiliser.

dry savanna zones (Bennett-Lartey et al. 2002). In contrast, Cuban homegardens harboured higher diversity under semiarid conditions as compared to humid conditions, caused though by irrigation (Wezel and Bender 2003). However, variation in plant diversity might occur also due to small-scale climatic variation, like droughts caused, for example, by El Niño events (Figure 3).

Soil fertility is another agro-ecological factor that generates variation of plant diversity, but its influence has not yet been studied in detail. Hodel et al. (1999) simply assumed an influence of soil factors on diversity without quantifying this. In forest gardens, Kaya et al. (2002) reported lower species diversity on marginal soils compared to more fertile soils. Many cultivated plant species, particularly vegetables and spices, do not give adequate yield under unfavourable soil conditions such as low pH value or P content. Therefore, gardeners might stop cultivating these species while switching to a reduced set of crops that can cope with low soil fertility.

Garden features

Within the major factors influencing plant diversity, garden size is one of those frequently analysed. Among others, Abdoellah et al. (2002), Hodel et al. (1999), and Quiroz et al. (2002) reported a positive relationship between garden size and plant species richness. In small homegardens, particularly tree species richness decreased, resulting in a poor vertical vegetation structure. However, diversity indices such as the Shannon index decreased with increasing garden size, probably due to more uniform planting patterns and dominance of a few species in large gardens (Peyre et al. 2005). Homegarden age is thought to influence plant species richness positively (Coomes and Ban 2004, Quiroz et al. 2002). After setting up a new homegarden, gardeners start planting with a rather small set of crops. Over time, more and more species may be introduced by gardeners or resprout from the former vegetation, while established, reliable species remain (Figure 3). However, Hodel et al. (1999) did not find any relationship between plant diversity and garden age when surveying homegardens of Vietnam.

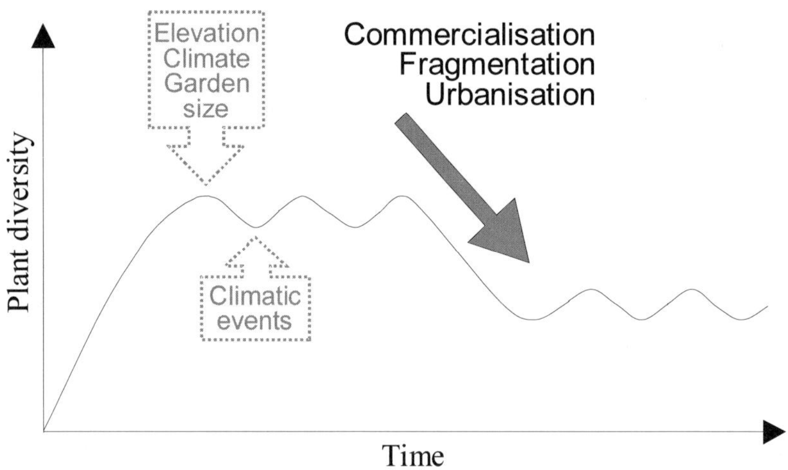

Fig. 3. Schematic development of plant diversity (except ornamentals) in homegardens over time under the influence of changing socio-economic conditions. Agroecological factors (e.g., elevation) may limit plant diversity, while short-term/small-scale climatic events might cause a certain fluctuation.

Socio-economic factors

Among socio-economic factors, the negative influence of market proximity and market-oriented production on plant diversity in homegardens has frequently been recorded (Figure 3) (Abdoellah et al. 2002, Christanty 1990, Michon and Mary 1994, Quiroz et al. 2002, Shrestha et al. 2002, Soemarwoto and Conway 1992, Tesfaye Abebe 2005). In remote areas, traditional subsistence homegardens provide the owner families with a wide spectrum of products to meet their daily needs, thus, resulting in a high plant diversity. Good market access, on the other hand, might push gardeners from subsistence to semi-commercial or commercial production. Peyre et al. (2005) and Shrestha et al. (2002) argued that cultivation of cash crops, particularly of annual vegetables, leads to decreasing numbers of perennials such as fruit and timber trees. Therefore, commercial homegardens often lack the upper vegetation strata. The focus of development agencies on improving nutrition and income through the promotion of mostly exotic, annual vegetables, might lead to such negative effects (Shrestha et al. 2002). But also a slightly positive relationship of plant species richness and the amount of cash income generated in the homegarden is reported (Hodel et al. 1999, Quiroz et al. 2002, Trinh et al. 2003). This effect was explained by higher labour investment in the commercial homegardens, from which also subsistence crops could profit. Possibly, a well-balanced mix of subsistence and cash crop production might lead to higher plant diversity.

Besides commercialisation, the scarcity of land generally also reduces the biodiversity in homegardens. Families with insufficient staple crop fields are forced to grow high proportions of staples in their homegardens (Soemarwoto and Conway 1992). As many staple food crops are light-demanding, perennials like fruit or timber trees disappear from such homegardens (Karyono 1990). Urbanisation is also said to reduce plant species diversity (Hodel et al. 1999, Karyono 1990, Michon and Mary 1994). In peri-urban regions with good access to large markets in the city, many traditional homegardens are converted into commercial fruit tree or vegetable gardens, thereby losing most of the less productive subsistence plants. At the highest urbanisation level, homegardens are generally rather small and dominated by ornamentals, giving priority to the aesthetic function instead of subsistence production.

Gardeners' characteristics

Certain characteristics of the gardener and his/her household are known to influence plant diversity in homegardens. A gardener's age can influence plant diversity positively (Quiroz et al. 2002), possibly because, over the years, gardeners try to cultivate new crops while they continue to plant well-tried species. In addition, older gardeners often have more time for homegardening and are supported by their grown-up children. Consequently, higher time allocation to homegardening leads to higher plant diversity (Hodel et al. 1999, Quiroz et al. 2002). Therefore, also large households with large labour force generally maintain a higher species richness in their homegardens as compared to small, labour force-constrained households (Quiroz et al. 2002, Tesfaye Abebe 2005). Farmers as compared to non-farmers might maintain a higher plant diversity in their homegardens due to higher time allocation and experience of the farmers (Arifin et al. 1997).

How formal education or sex of the gardener influence plant diversity in homegardens is still uncertain (Castiñeiras et al. 2002, Hodel et al. 1999, Quiroz et al. 2002). Similarly, the influence of a household's wealth status on plant diversity is debated controversially. Generally, homegardens of well-off households are said to harbour fewer food-producing plant species because such households purchase food and prefer ornamentals (Hodel et al. 1999, Karyono, 1990). In other cases, higher plant diversity found in homegardens of wealthy families was related to larger garden sizes and larger landholdings suitable for staple crop production (Coomes and Ban 2004, Shrestha et al. 2002).

Ethnicity of the gardener may also be a factor explaining variation in plant diversity of homegardens (Hodel et al. 1999, Soemarwoto and Conway 1992). Ethnic and cultural influences are particularly important for plant species composition. Different ethnic groups prefer different plant products and, therefore, cultivate for example more vegetables or more medicinal plants in their homegardens (Christanty et al. 1986, Shrestha et al. 2002, Soemarwoto and Conway 1992, Trinh et al. 2003). Migration can, thus, have a positive effect on plant diversity in homegardens, as long as the plant species brought from the migrants' home regions establish successfully in the new environment. On the other hand, migrants also adopt useful plants from indigenous gardeners. However, plant diversity in migrant homegardens could be rather low due to poverty and discrimination (Hodel et al. 1999), e.g., by assigning land of poor soil quality for settlement to such groups. Besides, shortage of labour for homegarden management and poor access to suitable agricultural land for staple food crops might further decrease plant diversity in migrant homegardens.

In conclusion, no individual factor alone determines the plant diversity found in homegardens, but rather a complex combination of agro-ecological, socio-economic, cultural, and political factors causes spatial and temporal variation of plant species. In the following two sections, this is illustrated by recent results of homegarden research in Sulawesi and Java.

3 Case study 1: Plant diversity in rural homegardens of Central Sulawesi, Indonesia

In association with the collaborative research program STORMA, selected homegardens on the island of Sulawesi were studied, focussing on the dynamics in diversity of cultivated plants over time and the specific influences of selected factors on plant diversity. The study was conducted from 2001 to 2004 in the Napu Valley (elevation about 1100 m), Central Sulawesi, described in more detail by Kehlenbeck and Maass (2004, 2006).

Ten households with homegardens were randomly selected from each of five villages differing in their market access and origin of inhabitants (Table 1). The study included interviews with gardeners, measurement of homegarden size, and complete inventories of useful plants (i.e., excluding ornamentals and weeds). In three villages, homegardens were studied over time (Kehlenbeck and Maass 2006), in two villages only in 2004. Besides plant species number and abundance, also density of plant individuals (no. of individuals/100 m^2), Shannon index (H'), and Sørensen's similarity coefficient were determined

(Magurran 1988). For assessing changes of plant species composition over time, a principal component analysis (PCA) was carried out, based on the mean abundance of cultivated plant species per village in the 30 homegardens revisited.

Table 1. Characteristics of five villages studied in the Napu valley, Central Sulawesi, in 2004.

	Wuasa	Rompo	Wanga	Siliwanga	Tamadue-Trans
Year of foundation	1892	1915	1923	1992	1991
Inhabitants (no.)	2,600	400	350	600	700
Ethnicity	90% locals	90% locals	75% locals	95% migrants	99% migrants
Distance to paved road	0 km	5 km	0 km	0 km	5 km
Market access	Good	Poor	Medium	Medium	Poor

Sources: STORMA survey data (subproject A4 'Economic Analysis of Land Use Systems of Rural Households') and Kehlenbeck (unpublished).

In 2004, homegarden sizes ranged from 250 m^2 to 2,400 m^2 (medians per village are given in Table 2), and they were established 2–40 years ago (median = 11 years). In the 50 homegardens, a combined total of 206 plant species were grown, including about 42 wild species (mainly used as fuel wood/timber, medicine, or vegetable) and many underutilised species (mainly used as vegetable). In addition to the 206 useful plant species, 162 ornamental and 58 weed species were identified in the homegardens.

3.1 Variation of plant diversity over time

Principal Component Analysis based on plant species composition resulted in a clear distinction of the three villages along the ordination axes (Figure 4). Despite temporal changes, villages remain clearly separated in the ordination space. Axis 1 explains about 50% of the total variability and is positively correlated with some staple crops, particularly paddy rice (*Oriza sativa*), cassava (*Manihot esculenta*), and sweet potato (*Ipomoea batatas*). Axis 2, explaining about 20% of the total variability, is negatively correlated with certain traditional vegetable species, such as *Clerodendron minahassae*, eggplant (*Solanum macrocarpon*) and yard-long bean (*Vigna unguiculata*), but positively with the cash crop species groundnut (*Arachis hypogaea*), vanilla (*Vanilla planifolia*), and spring onion (*Allium fistulosum*). Thus, axis 2 can be interpreted as reflecting the continuum from traditional subsistence crops towards modern cash crops.

In Wuasa, homegardens were characterised by a rather small portion of staples, but a high portion of cash crops, recently growing strongly (Figure

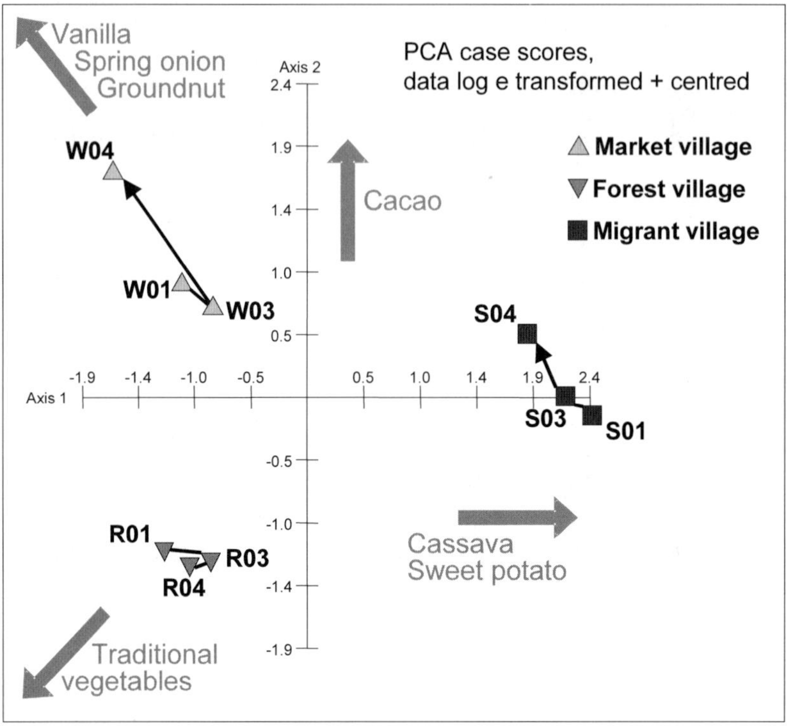

Fig. 4. Changes of plant species composition (except ornamentals) from 2001 to 2004 in 30 homegardens of three villages in the Napu valley, Central Sulawesi. Results of a Principal Component Analysis, based on mean abundance data per village of 196 plant species cultivated in the homegardens surveyed over time; data log e transformed and centred before analysis.

4). This pattern reflects the increasing importance of commercial crops in this particular village with its rather good market access. In Siliwanga, a migrant village with intermediate market access, a similar, but less pronounced trend was found. Its starting position was different from Wuasa, due to a markedly different plant species composition, characterised by the dominance of staple crops in Siliwanga. However, the abundance of these staples had already decreased over the short time span of this investigation. Partly, they were replaced, for example, by the cash crop cacao (*Theobroma cacao*). In the remote village Rompo with rather poor market access, no change towards more cash crops has been detected. Homegardens in this village were still characterised, for example, by traditional vegetables, whereas abundance of cash crops and staples was rather low. Market access, therefore, seems to play a major role towards commercialisation of homegardens. Contrary to findings in the literature, this commercialisation has not yet led to a decrease of the overall relatively high plant species diversity in the homegardens studied (Figure 5).

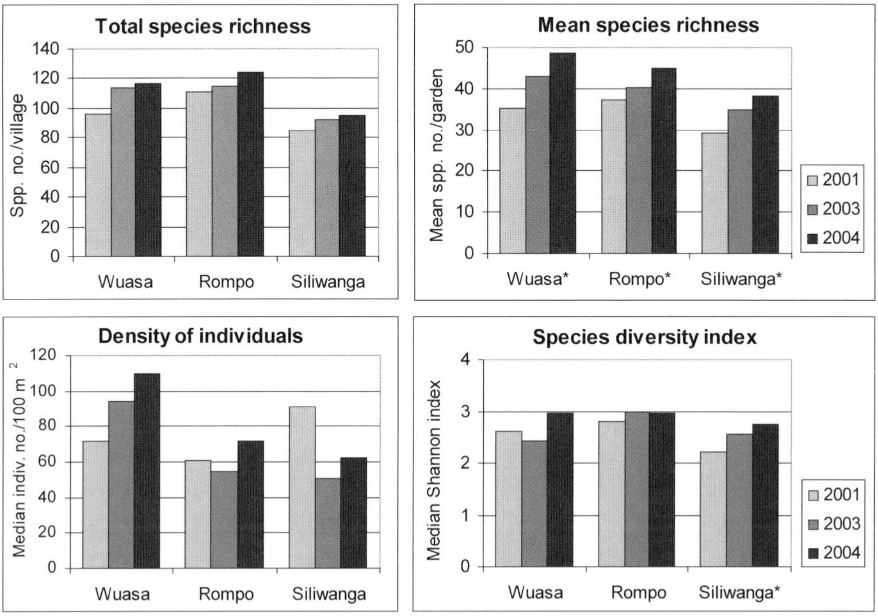

Fig. 5. Diversity parameters of plant species (except ornamentals) of 30 homegardens in three villages of the Napu valley, Central Sulawesi, from 2001 to 2004. (In villages followed by an asterisk, changes of the respective variable over time were significant at $P \leq 0.05$ by Friedman test).

Plant species richness in the homegardens revisited even increased markedly over time both per village and per garden (Figure 5). In the three villages, a combined total of 152, 171, and 178 useful plant species were identified in 2001, 2003, and 2004, respectively. Mean density of plant individuals increased over time only in the market village. Changes in Shannon diversity index were not clear except in the migrant village Siliwanga, where the index increased significantly. In the latter village, however, plant species richness and Shannon index continued to be markedly lower than in the two local villages.

3.2 Spatial differences of plant diversity

When comparing homegardens in all five research villages, marked differences in their plant diversity became apparent. In homegardens of the two migrant villages, Siliwanga and Tamadue-Trans, diversity parameters were mostly lower than in those of the three local villages that did not differ significantly from each other (Table 2). In addition to diversity parameters, plant species composition was clearly different between the five villages studied in 2004. Sørensen's coefficient showed a higher similarity between the three local

Table 2. Median size, mean plant species (except ornamentals) richness, median density of plant individuals, as well as median Shannon diversity index (ranges in brackets) of 48 homegardens in five villages of the Napu valley, Central Sulawesi, in 2004.

	Median HG size (m^2)	Mean plant species number		Median no. of plant individuals/100 m^2		Median Shannon index H'	
Wuasa (local)	740	48.5a	(23–68)	69.0a	(29–394)	3.0ab	(1.5–3.5)
Rompo (local)	660	44.9ab	(28–65)	74.5a	(39–96)	3.0a	(2.4–3.3)
Wanga (local)	600	38.1ab	(13–56)	37.0ab	(20–117)	3.1ab	(1.5–3.3)
Siliwanga (migrant)	900	35.9ab	(22–44)	42.5ab	(13–110)	2.6ab	(2.3–3.0)
Tamadue (migrant)	2250	33.0b	(22–50)	22.5b	(7–46)	2.3b	(2.1–2.9)

Medians were used, when variables were not normally distributed. Means and medians in a column followed by different letters are significantly different at $P \leq 0.05$ (by Tukey HSD for plant species number; Nemenyi test for density of plant individuals and Shannon index).

villages (0.72–0.76) than between these and the two migrant villages (0.58–0.63) or between the two migrant villages (0.64).

3.3 Identification and discussion of main factors influencing plant diversity in homegardens of Central Sulawesi

The influences of several agro-ecological, garden-based, and socio-economic factors as well as gardeners' characteristics on plant species richness and Shannon diversity index were analysed by multivariate regression analysis, based on data of all 48 homegardens studied in 2004.

Agro-ecological parameters

As elevation and climate were roughly constant throughout the research area, only the influence of the factor 'soil fertility' on plant diversity was investigated in this study. The following soil fertility parameters were tested: sand content, content of C_{total} and plant available P (Bray I), pH value (H_2O), effective cation exchange capacity ($CEC_{eff.}$), and base saturation. Plant species richness was not affected by the tested parameters. C content influenced the Shannon index negatively. This unexpected negative influence of C content (which is comparable to organic matter content) is related to the fact that in the migrant villages with their rather young and large homegardens, the soil had a relatively high C content, but a low P content and pH value. At the same time, plant diversity in many of these homegardens was rather low (Table 2), caused by the dominance of staple food or cash crops in many migrant homegardens (see below).

Garden features

Plant species richness was positively influenced mainly by garden size (Figure 6). Influence of garden age on plant diversity was not detected.

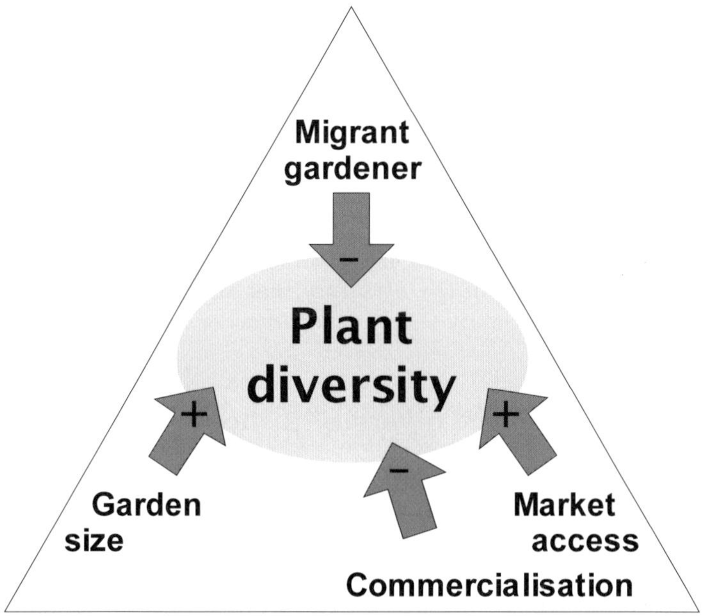

Fig. 6. Schematic diagram how different factors influence diversity of cultivated plants (except ornamentals) in 48 homegardens in five villages of the Napu valley, Central Sulawesi, in 2004.

Socio-economic factors

Generally, a better market access is said to cause a decrease of plant diversity in homegardens. In 2001, this tendency was detected in the study region, where plant species richness and Shannon diversity index were highest in the remote village Rompo (Figure 5). In 2004, however, results showed a positive influence of market access on plant diversity (Figure 6). This could be caused by the activity of a new village development project that promoted medicinal plants in homegardens. In the market village Wuasa, the project was relatively successful, resulting in a conspicuous increase of medicinal plant species richness from 3 (mean in 2001) to about 8 species per garden in 2004. Commercialisation, on the other hand, influenced plant diversity negatively. In homegardens dominated by cash crops like cacao or coffee, plant diversity was markedly lower than in subsistence homegardens.

Gardeners' characteristics

Neither household-related features (e.g., household size or wealth status) nor gardeners' characteristics (e.g., age, sex, level of formal education) played a significant role for plant diversity, apart from the origin of the gardener. In homegardens of migrants, significantly lower plant diversity could be expected than in homegardens of locals (Figure 6). Due to economic constraints after arrival, many migrants in the study area focussed strongly on staple food (in Siliwanga) or cash crop production (in Tamadue) in their homegardens. Plant diversity might increase in the future, when migrant families establish staple crop fields or plantations in the surroundings of their villages. In Siliwanga, such an increase of plant diversity already occurred over time (Figure 5). However, the situation in migrant homegardens of the Napu valley was quite far away from statements in the literature that such homegardens generally harbour higher plant diversity than homegardens of locals.

4 Case study 2: Plant diversity in homegardens of West Java, Indonesia

In West-Java, changes of homegardens' plant diversity were studied along an urban-rural continuum as well as along an elevation gradient. For the investigation of urbanisation effects, the vegetation structure and composition of 115 homegardens in six villages were studied (Arifin et al. 1998). The six villages differed in their urbanisation level. One was a rural village, three were characterised as intermediately urbanised, and two as urban villages, founded in the 1940s, the 1950s, and the 1980/1990s, respectively. In each homegarden, both useful and ornamental plants were inventoried. Overall, homegarden sizes ranged from 30 m^2 to 4000 m^2, with means per village from 70 m^2 to 530 m^2 (Table 3). A combined total of 440 plant species were grown in the 115 homegardens, about half of the species being ornamentals.

For the study of elevational effects on plant diversity, 30 homegardens each were investigated in three villages in the same watershed. The villages were located at different elevations (low: 300 m, intermediate: 950 m, high: 1300 m), and were founded in the 1940s, the 1950s, and the 1960s, respectively. Vegetation composition (both useful and ornamental plants) and structure of homegardens, as well as management patterns were studied in detail.

4.1 Spatial differences of plant diversity

Rural–urban continuum

Plant species number varied largely among the 115 homegardens studied (Table 3). Mean species number per homegarden did not differ markedly between the rural, the intermediate, or the urban homegardens. However, urban gardens harboured a markedly higher number of ornamental plant species per homegarden than rural gardens. In addition, the proportions of both ornamental species and individuals increased with a higher level of urbanisation (rural: 40% of total species and 47% of total individuals were ornamentals; urban: 70% of total species and > 80% of total individuals were ornamentals). Homegarden size decreased continuously from rural to urban sites (Table 3). Vertical vegetation structure decreased in complexity from rural to urban. Consequently, urban homegardens mostly lacked the strata above 5 m, and strata above 2 m were represented by only few individuals.

Table 3. Mean size and mean diversity characteristics of useful and ornamental plant species (range in brackets) of 115 homegardens studied in six villages with different urbanisation levels in Cianjur and Bogor, West Java, in 1995.

Village urbanisation level		HG size (m^2)	Total plant spp. no. per HG	No. of ornamental spp. /HG	No. of total plant individuals per 100 m^2	No. of ornamental individuals per 100 m^2
Rural	(N = 30)	530	36 (6–82)	14	63	29
Intermediate 1	(N = 21)	380	49 (32–79)	25	72	40
Intermediate 2	(N = 10)	190	32 (7–85)	14	107	59
Intermediate 3	(N = 20)	130	20 (2–53)	11	58	34
Urban 1	(N = 10)	70	34 (18–63)	24	280	233
Urban 2	(N = 24)	90	37 (16–78)	26	240	201

Rural homegardens were distinguished from the urban ones by their dominant function in subsistence production, expressed by the rather high number and abundance of useful plants, particularly fruit trees, vegetables, and staple crops. In urban homegardens with their representative, aesthetic function, plant diversity was composed mainly of ornamentals.

Elevation gradient

Plant species richness in homegardens decreased continuously from low to high elevations (Table 4). In the intermediate and high elevations, homegardens were markedly smaller than in the lowlands. Despite their rather small size, homegardens of intermediate elevation harboured much more plant individuals than homegardens in the high- or lowlands. Density of individuals was markedly lower in homegardens of the lowlands than in higher elevations. However, the highest density was found in homegardens of intermediate altitude. Plant species composition was different along the elevation gradient. Homegardens in the lowlands were dominated by perennial fruit trees, and in the highlands by vegetables.

Table 4. Mean size and mean diversity characteristics of useful and ornamental plant species (range in brackets) of 90 homegardens studied in three villages of different elevation in Cianjur watershed, West Java, in 2000.

Village elevation level	HG size (m^2)	Total plant spp. no. per HG	No. of total plant individuals per HG	No. of total plant individuals/ 100 m^2
High (1300 m; N = 30)	190	27 (14–36)	280 (107–670)	150
Intermediate (950 m; N = 30)	220	40 (27–64)	490 (225–771)	220
Low (300 m; N = 30)	560	44 (26–74)	350 (182–867)	60

4.2 Discussion of main factors influencing plant diversity of homegardens in Java

Agro-ecological parameters

Elevation negatively influenced plant species richness in homegardens. However, the lower plant diversity in highland homegardens might also be caused by their relatively smaller size as compared to the lowland ones. This positive relationship between homegarden size and plant species richness is well documented in the literature (Section 2.3). The statement by Quiroz et al. (2002) that the highest plant diversity is found in the homegardens of intermediate elevations (i.e., 600–1300 m) is only partly confirmed by this study. Although density of plant individuals was highest in homegardens of intermediate elevation, species richness was not.

Socio-economic factors

Urbanisation affected the richness and abundance of useful plant species negatively, but those of ornamentals positively. As in the elevation gradient, these results might be confounded with the decrease of homegarden size along the rural-urban continuum. However, the shift in plant species composition from useful plants to more ornamentals due to urbanisation confirms statements by Hodel et al. (1999) and Karyono (1990).

Fragmentation of rural homegardens was also found to change plant diversity in homegardens (Arifin et al. 1997). Where the formerly rather large, rural homegardens were divided into small plots because of the traditional inheritance system, the upper vegetation layers tended to disappear. The proportion of a few cash crops as well as ornamental plants increased in disfavour of many subsistence crops.

5 Common issues for the maintenance of homegardens' plant diversity in Indonesia

Plant diversity in Indonesian homegardens is mainly threatened by commercialisation, fragmentation, and urbanisation. This loss of plant diversity may be accompanied by a disruption of the agro-ecosystem's functioning and, as a consequence, the loss of its sustainability, e.g., due to the interruption of nutrient cycles or the impoverishment of the multi-layered vegetation structure. Maintaining the complex and species-rich, subsistence-oriented character of traditional homegardens should, thus, be promoted. Local knowledge on management and use of the diverse plant community in homegardens should be preserved, enlarged, and spread both within and among regions. If homegarden size falls below a critical minimum of about 100 m^2, plant diversity and structure as well as a garden's production potential suffer strongly (Arifin et al. 1997). Households should, therefore, avoid to divide their homegardens, particularly the back and side yard gardens, where most of the useful plants occurred (Arifin unpublished). However, there is an increasing need for income generation from homegardens. To meet this demand, the efficiency of garden cash crop production could be improved. Many semi-commercial homegardens give only unsatisfactory yields because the gardeners lack specific knowledge of how to manage the cash crop properly. Besides, only cash crops that need few external inputs, but obtain a rather high, stable price and, simultaneously, do not disturb biodiversity, structure, and functioning of homegardens should be promoted, e.g., shade-tolerant, climbing spices such as vanilla (*Vanilla planifolia*) or black pepper (*Piper nigrum*).

6 Conclusions and Outlook

Traditional homegardens may serve as a model for a sustainable agroforestry system. Plant diversity is seen as a major factor towards sustainability and productivity of the system. However, plant diversity in homegardens is very dynamic over time and prone to marked reduction due to a complex interaction of different socio-economic factors, e.g., commercialisation. A holistic research and development approach is needed to meet both the livelihood goal (better nutrition and income of gardeners) and the conservation goal (preserving agro-biodiversity and ecosystem functioning) in homegarden management

(Shrestha et al. 2002). The suitability of homegardens for *in situ* conservation of plant genetic resources should not be taken for granted due to the temporal changes of plant diversity, small population sizes per species, and strong selection pressure by gardeners. Therefore, gardeners should be closely involved in the whole conservation process (Kehlenbeck and Maass 2006). Additionally, not single homegardens, but rather an aggregate of the homegardens of whole villages or regions should be considered as a conservation unit (Trinh et al. 2003).

Acknowledgements
Many thanks to all the people in the research villages, especially to the respondent families, for their cooperation, kindness, and ready help during the field surveys. The support provided by members and staff of STORMA (SFB 552) in Bogor, Palu, and Göttingen is also gratefully acknowledged. The German Academic Exchange Service (DAAD) and German Science Foundation (DFG) provided financial support for the study of homegardens in Sulawesi. The research on Javanese homegardens was funded by the Core University Program in Applied Biosciences Tokyo University/Bogor Agricultural University, the Japan Society for the Promotion of Science (JSPS), and the Directorate General of Higher Education, Indonesian Ministry of Education and Culture.

References

Abdoellah OS, Parikesit, Gunawan B, Hadikusumah HY (2002) Home gardens in the upper Citarum watershed, West Java: a challenge for *in situ* conservation of plant genetic resources. In: Watson JW, Eyzaguirre PB (eds) Home gardens and *in situ* conservation of plant genetic resources in farming systems. Proceedings of the Second International Home Gardens Workshop, 17-19 July 2001, Witzenhausen, Germany. IPGRI, Rome, pp 140-147

Albuquerque UP, Andrade LHC, Caballero J (2005) Structure and floristics of homegardens in Northeastern Brazil. Journal of Arid Environments 62: 491-506

Arifin HS, Sakamoto K, Chiba K (1997): Effects of the fragmentation and the change of the social and economical aspects on the vegetation structure in the rural home gardens of West Java, Indonesia. Landscape Research–Journal of the Japanese Institute of Landscape Architecture 60: 489-494

Arifin HS, Sakamoto K, Chiba K (1998) Effects of urbanisation on the vegetation structure of home gardens in West Java, Indonesia. Japanese Journal of Tropical Agriculture 42: 94-102

Arifin HS, Arifin NHS, Suryadarma IGP (2002) Integrating the value of local tradition and culture in ecological landscape planning in Indonesia.

Landscape Research–Journal of the Japanese Institute of Landscape Architecture 65: 196-200 [In Japanese, English version available from the senior author]

Arifin HS, Chozin MA, Sarma M, Sakamoto K (2005) The farming system of Indonesian home garden (*pekarangan*) in Cianjur watershed, Cianjur district–West Java. Japanese Journal of Tropical Agriculture (submitted)

Atta-Krah K, Kindt R, Skilton JN, Amaral W (2004) Managing biological and genetic diversity in tropical agroforestry. Agrofor Syst 61: 183-194

Bos M, Höhn P, Saleh S, Büche B, Buchori D, Steffan-Dewenter I, Tscharntke T (this volume) Insect diversity responses to forest conversion and agroforestry management. In: Tscharntke T, Leuschner C, Zeller M, Guhardja E, Bidin A (eds) The stability of tropical rainforest margins, linking ecological, economic and social constraints of land use and conservation. Springer, Berlin Heidlberg New York

Bennett-Lartey SO, Ayernor GS, Markwei CM, Asante IK, Abbiw DK, Boateng SK, Anchirinah VM, Ekpe P (2002) Contribution of home gardens to *in situ* conservation of plant genetic resources farming systems in Ghana. In: Watson JW, Eyzaguirre PB (eds) Home gardens and *in situ* conservation of plant genetic resources in farming systems. Proceedings of the Second International Home Gardens Workshop, 17-19 July 2001, Witzenhausen, Germany. IPGRI, Rome, pp 83-96

Castiñeiras L, Fundora Mayor Z, Shagarodsky T, Moreno V, Barrios O, Fernández L, Cristóbal R (2002) Contribution of home gardens to *in situ* conservation of plant genetic resources in farming systems - Cuban component. In: Watson JW, Eyzaguirre PB (eds) Home gardens and *in situ* conservation of plant genetic resources in farming systems. Proceedings of the Second International Home Gardens Workshop, 17-19 July 2001, Witzenhausen, Germany. IPGRI, Rome, pp 42-55

Christanty L (1990) Home gardens in tropical Asia, with special reference to Indonesia. In: Landauer K, Brazil M (eds) Tropical home gardens. The United Nations University, Tokyo, Japan, pp 9-20

Christanty L, Abdoellah OS, Marten GG and Iskandar J (1986) Traditional agroforestry in West Java: the *pekarangan* (homegardens) and *kebun-talun* (annual-perennial rotation) cropping systems. In: Marten GG (ed) Traditional agriculture in Southeast Asia. Boulder, Colorado, USA, pp 132-158

Clergue B, Amiaud B, Pervanchon F, Lasserre-Joulin F, Plantureux S (2005) Biodiversity: function and assessment in agricultural areas. A review. Agron Sustain Dev 25: 1-15

Coomes OT, Ban N (2004) Cultivated plant species diversity in home gardens of an Amazonian peasant village in Northeastern Peru. Econ Bot 59: 420-434

Helen Keller International (2004) Research and publications. <http://www.hki.org/research/index.html> [03.11.2005]

Hemp A (200) The banana forests of Kilimanjaro: biodiversity and conservation of the Chagga homegardens. Biodivers Conserv 15: 1193-1217

Hochegger K (1998) Farming like the forest - Traditional home garden systems in Sri Lanka. Margraf, Weikersheim, Germany, 203 pp

Hodel U, Gessler M, Cai HH, Thoan VV, Ha NV, Thu NX and Ba T (1999) *In situ* conservation of plant genetic resources in home gardens of Southern Vietnam. IPGRI, Rome, Italy, 106 pp

Huxley P (1999) Tropical agroforestry. Blackwell Science, Oxford, UK, 371 pp

Karyono (1990) Home gardens in Java. Their structure and function. In: Landauer K, Brazil M (eds) Tropical home gardens. The United Nations University, Tokyo, Japan, pp 138-146

Kaya M, Kammesheidt L, Weidelt HJ (2002) The forest garden system of Saparua island, Central Maluku, Indonesia, and its role in maintaining tree species diversity. Agrofor Syst 54: 225-234

Kumar BM, Nair PKR (2004) The enigma of tropical homegardens. Agrofor Syst 61: 135-152

Kehlenbeck K, Maass BL (2004) Crop diversity and classification of homegardens in Central Sulawesi, Indonesia. Agrofor Syst 63: 53-62

Kehlenbeck K, Maass BL (2006) Are tropical homegardens sustainable? Some evidence from Central Sulawesi, Indonesia. In: Kumar BM, Nair PKR (eds) Tropical homegardens: A time-tested example of sustainable agroforestry. Advances in Agroforestry 3. Springer, Germany pp 339-354

Landauer K, Brazil M (eds) (1990) Tropical home gardens. The United Nations University, Tokyo, Japan, 257 pp

Magurran AE (1988) Ecological diversity and its measurement. Croom Helm, London, UK, 179 pp

Main AR (1999) How much biodiversity is enough? Agrofor Syst 45: 23-41

Michon G, Mary F (1994) Conversion of traditional village gardens and new economic strategies of rural households in the area of Bogor, Indonesia. Agrofor Syst 25: 31-58

Miller RP, Nair PKR (2005) Indigenous agroforestry systems in Amazonia: from prehistory to today. Agrofor Syst 66: 151-164

Olschewski R, Tscharntke T, Benitez PC, Schwarze S, Klein A-M (this volume) Economic evaluation of ecosystem services as a basis for stabilizing rainforest margins? The example of pollination services and pest management in coffee landscapes. In: Tscharntke T, Leuschner C, Zeller M, Guhardja E, Bidin (eds) The stability of tropical rainforest margins, linking ecological, economic and social constraints of land use and conservation. Springer, Berlin Heidlberg New York

Peyre A, Guidal A, Wiersum KF, Bongers F (2005) Dynamics of homegarden structure and function in Kerala, India. Agrofor Syst 66: 101-115

Quiroz C, Gutiérrez M, Rodríguez D, Pérez D, Ynfante J, Gámez J, Pérez de Fernandez T, Marquez A, Pacheco W (2002) Home gardens and *in situ* conservation of agrobiodiversity–Venezuelan component. In: Watson JW, Eyzaguirre PB (eds) Home gardens and *in situ* conservation of plant genetic resources in farming systems. Proceedings of the Second International

Home Gardens Workshop, 17-19 July 2001, Witzenhausen, Germany. IPGRI, Rome, pp 73-82

Schwartz MW, Brigham CA, Hoeksema JD, Lyons KG, Mills MH, van Mantgem PJ (2000) Linking biodiversity to ecosystem function: implications for conservation ecology. Oecologia 122: 297-305

Shrestha P, Gautam R, Rana RB, Sthapit B (2002) Home gardens in Nepal: status and scope for research and development. In: Watson JW, Eyzaguirre PB (eds) Home gardens and *in situ* conservation of plant genetic resources in farming systems. Proceedings of the Second International Home Gardens Workshop, 17-19 July 2001, Witzenhausen, Germany. IPGRI, Rome, pp 105-124

Soemarwoto O (1987) Homegardens: a traditional agroforestry system with a promising future. In: Steppler HA, Nair PKR (eds): Agroforestry: a decade of development. International Council for Research in Agroforestry (ICRAF), Nairobi, Kenia, p 157-170

Soemarwoto O, Conway GR (1992) The Javanese homegarden. Journal for Farming Systems Research-Extension 2: 95-118

Tesfaye Abebe (2005) Diversity in homegarden agroforestry systems of Southern Ethiopia. Tropical Resource Management Paper No. 59, Erosion and Soil & Water Conservation Group. Dept. of Environmental Sciences, Wageningen University and Research Centre, Wageningen, Netherlands, 119 pp

Torquebiau E (1992) Are tropical agroforestry home gardens sustainable? Agric Ecosyst Environ 41: 189-207

Trinh LN, Watson JW, Hue NN, De NN, Minh NV, Chu P, Sthapit BR, Eyzaguirre PB (2003) Agrobiodiversity conservation and development in Vietnamese home gardens. Agric Ecosyst Environ 97: 317–344

Vogl CR, Vogl-Lukasser BN, Caballero J (2002) Homegardens of Maya migrants in the district of Palenque, Chiapas, Mexico. Implications for sustainable rural development. In: Stepp JR, Wyndham FS, Zarger RK (eds) Ethnobiology and biocultural diversity. University of Georgia Press, Athens, USA, pp 631-647

Watson JW, Eyzaguirre PB (eds) (2002) Home gardens and *in situ* conservation of plant genetic resources in farming systems. Proceedings of the Second International Home Gardens Workshop, 17-19 July 2001, Witzenhausen, Germany. International Plant Genetic Resources Institute (IPGRI), Rome, Italy, 184 pp

Wezel A, Bender S (2003) Plant species diversity of homegardens of Cuba and its significance for household food supply. Agrofor Syst 57: 39-49

Wiersum KF (2004) Forest gardens as an 'intermediate' land-use system in the nature-culture continuum: characteristics and future potential. Agrofor Syst 61: 123-134

Tree species diversity relative to human land uses in tropical rain forest margins in Central Sulawesi

S. Robbert Gradstein[1*], Michael Kessler[1], and Ramadhanil Pitopang[2]

[1] Institute of Plant Sciences, University of Göttingen, Untere Karspüle 2, 37073 Göttingen, Germany
[2] Department of Forest Managament and Herbarium Celebense, Tadulako University, Palu, Indonesia

*corresponding author: S. Robbert Gradstein, phone +49 (0)551-392229, Email: sgradst@uni-goettingen.de

Summary

The large-scale exploitation and conversion of tropical forests causes growing concern about the continued existence of the rich biodiversity of these forests. In the framework of the interdisciplinary STORMA project in Lore Lindu National Park area in Central Sulawesi, we studied tree diversity in six different land use types in the margins of submontane rain forest: undisturbed forest, forest with rattan extraction, selectively logged forest, cacao forest gardens, cacao plantations with mixed canopy of planted trees, and cacao plantations with a monospecific canopy. By analyzing such a finely subdivided use gradient it was attempted to answer the question how human usage and biodiversity conservation in the study area may be reconciled. Tree species (dbh > 10 cm) were sampled in 24 plots of 0.25 ha in all six land use types (4 replicates each). In total, 251 tree species (143 genera, 59 families) were recorded. Number of tree species per 0.25 ha was 51-63 in primary forest and gradually decreased towards the studied cacao systems. However, when native and cultivated tree species were considered separately, significant differences were detected among plantation types in terms of tree diversity. Tree endemism in forest plots totalled ca. 15% and was in good accordance with endemism in woody plants of Sulawesi. The number of endemic species was strongly reduced in cacao systems, although percentage endemism did not decline significantly in cacao forest gardens. Roughly one third of tree species in the forest plots were of economic importance as commercial timber trees; timber diversity was little affected by moderate human use of the forest but was significantly reduced in

Tscharntke T, Leuschner C, Zeller M, Guhardja E, Bidin A (eds), The stability of tropical rainforest margins, linking ecological, economic and social constraints of land use and conservation, Springer Berlin 2007, pp 321-334

cacao forest gardens and dropped to near zero in other plantation types. The mean basal area of 57 m^2 (36-80 m^2) per ha in natural forest was lower than the previously recorded value from the study area but is still almost double as high as the mean value typical for tropical lowland forests in Southeast Asia.

The results of this study support the notion that tree diversity in the submontane forests of Central Sulawesi is unusually high and rich in large-sized timber trees, although tree size varies locally. Moderate human use of the forest ecosystems does not significantly affect tree diversity. We conclude that conservation of tropical tree diversity is compatible with human exploitation of tropical forest as long as a canopy of native trees is maintained. Future conservation policies in rain forest margin areas should therefore focus on developing measures aimed at sustainable use of the natural resources. Promotion of such activities may help to stabilize tropical rain forest margins in Central Sulawesi.

Keywords: endemism, human impact, land use changes, Indonesia, species richness, Sulawesi, sustainable use, timber diversity, tree diversity, tropical rain forest margins

1 Introduction

Throughout the tropics natural forests are increasingly exploited and converted into pastures or cropland. The large-scale forest conversion causes growing concern about the continued existence of the rich biodiversity of these forests. At the same time, the number of studies focusing on the effect of tropical forest conversion on biodiversity is increasing. The majority of these studies show that forest modification and clearance have negative impacts on biodiversity, leading to reduction of species richness and diversity (e.g. Bawa and Seidler 1998, Beck et al. 2002, Krömer and Gradstein 2004, Lawton et al. 1998, Parthasarathy 1999, Schulze et al. 2004, Turner et al. 1997). In some taxa, however, species richness increases in secondary forests and open habitats (Fujisaka et al. 1998, Gradstein 1992, Kappelle et al. 1995, Klein et al. 2002, Nöske 2005). Secondary forests and agroforestry systems may help to maintain a certain portion of species diversity but not of all taxonomic groups (Schulze et al. 2004). In particular, species richness of trees seems to be suffering seriously from forest modification (Kessler et al. 2005, Turner et al. 1997).

In the framework of the interdisciplinary research project STORMA of the German Research Foundation, set up to study the biological and ecological degradation of rainforest margin areas and to develop strategies to stabilize these margins (Gerold et al. 2004), we have analyzed tree diversity in different land use types in the region of Lore Lindu National Park, Central Sulawesi, Indonesia. Lying between Wallace's and Weber's biogeographic lines and taking a central position in Malesia, Sulawesi has been isolated from the mainland of Southeast Asia since the end of the Miocene (Audley-Charles 1983). The long-term isolation of Sulawesi has allowed the development of a characteristic fauna and flora with a unique composition. Of an estimated 5000 recorded species of flowering plants, including more than 2000 woody ones, about 15 % are endemic (Roos et al. 2004). Our preliminary work on woody plant diversity in Lore Lindu Park showed that tree diversity in the park is unusually high, reaching to 150 tree species ($\geq$10 cm dbh) per hectare at 900-1200 m elevation (Kessler et al. 2005). We also found that basal area in primary forest was 140 m^2 ha^{-1}, one of the highest values recorded in tropical forests worldwide. Species richness declined from primary forest to forest gardens, secondary forests, and cacao plantations, with species of Lauraceae, Meliaceae, and Euphorbiaceae being predominant in primary forests, Euphorbiaceae, Rubiaceae and Myristicaceae in forest gardens, and Euphorbiaceae, Urticaceae, and Ulmaceae in young secondary forests. Cacao plantations were composed almost exclusively of cacao trees and two species of legume shade trees (Kessler et al. 2005). Comparative studies of birds and butterflies demonstrated parallel declines of species richness (Schulze et al. 2004).

In this chapter we present new data on tree diversity in different land use types in a previously unstudied part of Lore Lindu National Park and with a more detailed differentiation of land use types than has previously been possible. Our main objective was to determine the impact of land use changes in the area on the taxonomic composition and structural diversity of the tree diversity of the forest. By analyzing a finely subdivided use gradient it is attempted to answer the question how human usage and biodiversity conservation in the study area may be reconciled.

2 Materials and Methods

The study area was located in the surroundings of Toro, a village at the southwestern margin of Lore Lindu National Park about 100 km south of the town Palu. The park is mountainous, ranging from 200 m to 2700 m in elevation. The study sites are between 800 m and 1100 m in elevation and belong to the submontane forest zone (Kessler et al. 2005). Mean annual rainfall in the study area varied from 1200-2200 mm during the period from 1997 to 2001. Generally, forest habitats in the interior of Lore Lindu National Park are still relatively undisturbed but the margins of the park are characterized by a mosaic of near-primary forests, degraded forests, forest gardens, and plantations with cacao, coffee, maize, and paddy rice as the most important crops (Gerold et al. 2004).

Tree diversity was studied in six different land use types, three of forest and three of agroforest systems. Further description of the study sites are provided in the chapters by Bos et al., Dietz et al, and Migge-Kleian et al. in this volume.

- A. Natural forest with traditional use only; human activities restricted to collecting of medicinal plants and extensive hunting; rattan palms abundantly present.
- B. Natural forest subject to rattan extraction.
- C. Selectively logged forest, containing small to medium sized gaps, disturbance of ground vegetation, and increased abundance of lianas following the selective removal of canopy trees and rattan.
- D. Cacao forest garden: cacao cultivated under natural shade trees (= remaining forest cover) at the forest margin.
- E. Cacao cultivated under a mixed canopy of planted shade trees at the forest margin.
- F. Cacao cultivated under a monospecific canopy of planted shade trees at the forest margin.

Each land use type was sampled in plots with 4 independent replicates. Plots were selected as much as possible at similar elevation to avoid macroclimate-induced differences in plant composition, but unavoidably plots of land use types A-C were located at slightly higher elevation (hill tops) than land use types D-F (lower slopes). Plot size was 50 x 50 m. Each plot was subdivided into 25 subplots of 10 x 10 m^2 each and all trees dbh $\geq$10 cm were recorded, individually numbered with aluminum tags, their position in the plot mapped, their dbh measured, and trunk as well as total height estimated.

All recognizable morphospecies of trees were collected in sets of at least seven duplicates. Identification of the plant specimens was done at Herbarium Celebense – CEB (Universitas Tadulako, Palu) and Herbarium Bogoriense – BO (Bogor). Economic value and rarity of species was determined by using *Prosea* (Soerianegara & Lemmens 1993, Lemmens et al. 1995, Hong 2004), *Flora Malesiana* (Steenis 1950-) and information from specialists. Vouchers were deposited in BIOT, BO, CEB, GOET, and L.

We compared taxonomic and structural composition, rarity, and economical importance of tree species between the habitat types. Structural composition was analyzed by comparing the distribution of tree height and diameter classes. Rarity was expressed by the percentages of endemic species and economic importance by the percentages of major and minor timber species in the different land use types.

3 Results

We recorded a total of 251 tree species (143 genera, 59 families) in twenty four 0.25 ha plots in six land use types, including 52 species with economic importance as timber trees, 33 endemic species, and 23 cultivated species. In the natural forest (type A) we recorded 51-63 tree species $\geq$10 cm dbh per plot, belonging to 20-29 families (Table 1, Figure 1). Species richness of

Table 1. Summary statistics of tree species richness and structural characteristics in plots of 0.25 ha each in six land use types with four replicates each. A = Natural forest; B = Natural forest subject to rattan extraction; C = Selectively logged forest; D = Cacao forest garden; E = Cacao cultivated under a mixed canopy of planted shade trees; F = Cacao cultivated under a monospecific canopy of planted shade trees. Different lower case letters designate significantly different values (post hoc Tukey tests).

	A		B		C		D		E		F		ANOVA	
Number of	Mean ± SD		Mean ± SD		Mean ± SD		Mean ± SD		Mean ± SD		Mean ± SD		F	*p*
species	55.8 ± 5.5	a	58.0 ± 8.0	a	48.3 ± 4.0	a	20.8 ± 7.8	b	19 ± 7.5	b	9.0 ± 4.5	b	43.7	<0.001
genera	43.0 ± 7.3	a	44.0 ± 6.1	a	39.5 ± 5.9	a	19.5 ± 5.5	b	18.0 ± 6.2	b	8.8 ± 4.2	c	26.3	<0.001
families	27.0 ± 3.4	a	27.0 ± 4.5	a	24.8 ± 3.8	a	14.3 ± 3.6	b	13.5 ± 4.5	b	6.8 ± 3.3	b	19.3	<0.001
native species	55.8 ± 5.5	a	58.0 ± 8.0	a	48.3 ± 4.0	a	18.5 ± 7.9	b	10.3 ± 8.8	bc	5.5 ± 3.7	c	51.8	<0.001
endemic species	8.0 ± 0.8	a	9.0 ± 2.9	a	7.8 ± 1.0	a	2.3 ± 2.6	b	0.8 ± 1.5	b	0	b	20.3	<0.001
cultivated species	0	a	0	a	0	a	2.3 ± 1.0	b	8.8 ± 1.3	c	3.5 ± 1.0	b	80.7	<0.001
major timber trees	5.0 ± 1.2	a	4.0 ± 4.5	a	4.8 ± 1.5	a	0.8 ± 0.5	b	0	b	0.3 ± 0.5	b	28.3	<0.001
minor timber trees	14.8 ± 2.6	a	13.3 ± 4.3	a	10.5 ± 3.1	a	4.5 ± 2.4	b	1.8 ± 2.2	bc	0.3 ± 0.5	c	19.9	<0.001
stems ≥ 10 cm dbh	140.5 ± 17.3	a	143.5 ± 22.0	a	129.0 ± 10.3	a	77.5 ± 21.1	b	70.0 ± 22.1	b	81.0 ± 55.5	b	5.7	0.002
% of endemic species	14.5 ± 2.0	a	16.1 ± 6.8	a	16.2 ± 3.1	a	10.4 ± 8.4	ab	3.3 ± 6.5	b	0	c	6.6	0.001
basal area ($m^2 ha^{-1}$)	56.7 ± 18.2	a	51.0 ± 1.8	a	33.7 ± 9.3	ab	20.5 ± 8.4	bc	14.9 ± 9.7	c	11.9 ± 6.3	c	13.8	<0.001

trees in forest with rattan extraction and in lightly logged forest was not significantly different from that in natural forest; cacao plantations, however, had significantly lower species numbers (Table 1, Figure 1). There were no significant differences between the plantations in terms of total tree diversity. However, when only native tree species were considered, tree species declined significantly from the cacao forest gardens with 12-30 species, across cacao plantations with a mixed canopy of shade trees with 3-23 species to cacao plantations with a monospecific shade canopy with only 1-10 species. Thus, the number of native tree species per 0.25 ha was highest in natural forest and gradually decreased from land use types A to F. Unsurprisingly, this decrease was partly offset by the cultivated trees, which were most species rich in the cacao plantations with a mixed canopy of shade trees with 7-10 species per plot, significantly less diverse in cacao plantations with a monospecific shade canopy (2-4 species) and in cacao forest gardens (1-3 species), and lacking in the three forest types.

In contrast, endemic trees were best represented in the three forest types with 6-12 species per plot, and declined strongly in the three cacao agroforestry systems (0-6 species). This pattern was partly a result of the lower overall native tree diversity in the cacao agroforestry systems. When the percentage of endemic species was considered, then the cacao forest gardens did not differ significantly (0-20% endemic species) from the three forest types (10-20%), and only the two cacao systems with planted trees had significantly reduced percentages of endemic trees (0-13%).

Roughly one third of the tree species in the forest plots (15-20 spp.) were of economic importance as commercial timber trees; of these, 4-5 were major timber species and the rest minor ones. Timber diversity was little affected by moderate human use of the forest but was significantly reduced in cacao forest gardens and dropped to near zero in cacao plantations.

At the family level, the taxonomic composition of the land use types showed major differences. Canopies in natural forests were dominated by Meliaceae, Lauraceae, Sapotaceae, and Fagaceae, in forest with rattan extraction and selectively logged forest by Rubiaceae, Fagaceae, and Myristicaceae, in cacao forest gardens by Moraceae, Myristicaceae, and Melastomataceae, and in the remaining cacao plantation types by planted Fabaceae, Lauraceae, and palms. Predominant canopy tree species in the natural forest were *Palaquium quercifolium* (Sapot.) followed by *Chionanthus laxiflorus* (Oleac.), *Dysoxylum densiflorum* (Meliac.), *Castanopsis acuminatissima* (Fagac.), *Lithocarpus celebicus* (Fagac.) and *Ficus* sp (Morac.), in the forest with rattan extraction *Pandanus polycephalus* (Pandan.), *Palaquium obovatum*, *Neonauclea intercontinentalis* (Rubiac.), and *Meliosma sumatrana* (Sabiac.), and in the selectively logged forest *Neonuclea intercontinentalis*, *Pandanus polycephalus, Oreochnide rubescens* (Urtic.), *Canarium hirsutum* (Burser.), and *Castanopsis acuminatissima*. A very different composition of canopy tree species was found in the cacao systems, with *Artocarpus elasticus* (Morac.) *A. vrieseanus, Turpinia sphaerocarpa* (Staphyl.), *Horsfieldia costulata* (Myrist.) and *Astro-*

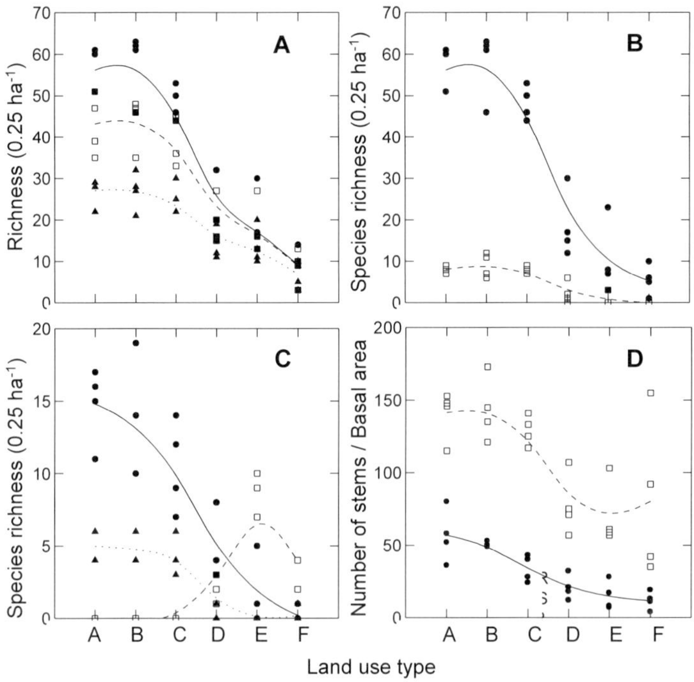

Fig. 1. Species richness and structural parameters of trees ≥10 cm dbh on 24 plots in six different land use types (see text and Table 1 for explanation) around Toro, Central Sulawesi. **A.** Richness of species (closed circles, continuous line), genera (open squares, dashed line), and families (closed triangles, dotted line). **B**. Species richness of native trees (closed circles, continuous line) and endemic species (open squares, dashed line). **C**. Richness of minor (closed circles, continuous line) and major timber tree species (closed triangles, dotted line), and of cultivated species (open squares, dashed line). **D**. Number of stems per 0.25 ha (open squares, dashed line) and basal area per ha (closed circles, continuous line). Trend lines were fitted by distance-weighted least squares smoothing with tension 0.5.

nia macrophylla (Melast.) being the most important forest species used as shade trees in cacao forest gardens, *Syzygium aromaticum* (Myrtac.), *Horsfieldia costulata* (Myrist.), *Erythrina subumbrans* (Fabac.) *Bischoffia javanica* (Euphorb.), *Arenga pinnata* (Arec.), *Durio zibethinus* (Bombac.), *Cocos nucifera* (Arec.), *Persea americana* (Laurac.) in cacao plantations with a mixed shade canopy, and *Erythrina subumbran*, *Cinnamomum burmanni (*Laurac.), *Arenga pinnata*, *Glyricidia sepium* (Fabac.) and *Cocos nucifera* in cacao plantations with a monospecific shade canopy.

Tree height decreased gradually towards cacao plantations and tree abundance (= individuals per plot) varied considerably but was generally lower in cacao plantations than in forest (Table 1). Basal area of trees was highest in natural forest (36-80 m^2 per ha), which had more large-sized trees than other land use types. Basal area in forest with rattan extraction plots (49-53 m^2) and in selectively logged forest (24-43 m^2) was lower, but these differences were not significant. Cacao agroforestry systems had generally lower values, indicating that many large trees had been extracted.

4 Discussion and conclusions

The natural forests analyzed in this paper are slightly less species rich (51-63 species per 0.25 ha) than those on the opposite, eastern and northeastern sides of Lore Lindu Park, which have up to 148 tree species (dbh $\geq$10 cm) in one hectare and 66-76 in 0.25 ha (Kessler et al. 2005; Table 2). The species numbers reported here are much higher, however, than those reported by Mansyur (2003) and Brodbeck et al. (2004) from Lore Lindu Park (Table 2). The low figures obtained in the latter studies are difficult to explain and may be due to differences in the field methodologies applied.

Table 2. Number of tree species per hectare in natural tropical rain forests of Southeast Asia and Australia. Approximate species numbers per 0.25 hectare are estimated at ca. 45% of those reported for one hectare based on tree species accumulation curves in primary forest of Sulawesi in Kessler et al. (2005), where 0.25 ha corresponded to 44.6% of the values at 1 ha.

Site	No. of species per ha	No. of species per 0.25 ha	Elevation	Source
Gunung Mulu NP, Sarawak, Malaysia	214	ca. 96	200	Proctor et al. 1983
Pasoh FR, Malaysia	210	ca. 95	200	Kochummen et al. 1990
Queensland, Australia	108	ca. 49	730	Philips et al. 1994
CMBRS, Papua New Guinea	228	ca. 103	900	Wright et al. 1997
Gunung Halimun NP, West Java	116	ca. 52	1000	Simbolon et al. 1997
Negros Island, Phillipines	92	ca. 41	1000	Hamann et al. 1991
Lore Lindu NP, Central Sulawesi	–	57	1050	this study
Lore Lindu NP, Central Sulawesi	148	66	1100	Kessler et al. 2005
Lore Lindu NP, Central Sulawesi	89	ca. 40	1100	Brodbeck et al. 2004

Comparison with lowland rain forests elsewhere in Southeast Asia and in Australia shows that tree species numbers vary considerably among sites, for climatic, edaphic, or historical reasons (Table 2; Davidar et al. 2005). Although tree species per hectare was not recorded in the present study, our data clearly support the notion that tree diversity in the submontane rain forests of Central

Sulawesi is high and not depauperate relative to the rest of Southeast Asia, in contrast to animal groups such as mammals, birds, and butterflies, where the long-term isolation of Sulawesi has restricted immigration (Kessler et al. 2005, Roos et al. 2004, Whitten et al. 2002). At the same time, the available data document a considerable variability of tree species richness in natural forest along the margins of Lore Lindu National Park. The causes for this variability are still unknown and are subject to further study.

Tree species number in the different land use types near Toro gradually decreased with increased forest disturbance and towards agroforest systems, and was lowest in cacao plantations with a monospecific canopy of shade trees. A similar decrease in species diversity was recorded from the eastern and northeastern side of Lore Lindu Park by Schulze et al. (2004) and Kessler et al. (2005), who recorded highest species diversity in natural forests, followed by forest gardens and cacao plantations. Data on disturbed forest and on different types of cacao plantations were not retrieved in the latter study and are provided here for the first time.

Not surprisingly, cacao plantations, representing agroforestry systems, had by far the lowest tree species richness. The contrast is especially conspicuous relative to the forest gardens, where cacao trees are planted under the natural forest canopy and where higher levels of tree diversity are maintained. Cacao forest gardens near Toro had 15-32 tree species per 0.25 ha and are almost as species rich as those on the eastern margin of the park, which have 19-35 tree species per 0.25 ha (Kessler et al. 2005). Much lower values (32-40 spp. per hectare!) reported from Lore Lindu Park by Brodbeck et al. (2004) are presumably due to different field methodologies applied (see above).

Sperber et al. (2003) found that parasitoid butterfly diversity was higher in cacao plantations with high diversity of shade trees than in those with low tree diversity. Similar results were obtained by Bos et al. (this volume) and Sporn and Gradstein (2005) from studies on insect and epiphyte diversity in cacao plantations in the study area in Sulawesi. These data show that proper management of shade tree diversity plays a vital role not only in maintaining the sustainability of cacao agroforestry production systems but also in helping maintaining high biodiversity values in these locations. The importance of these agroforestry systems for plant conservation is particularly obvious in regions where native forest vegetation has almost disappeared (Andersson and Gradstein 2005, Bandeira et al. 2005).

Secondary forests, regenerating after clear-felling, were not included in the present study but were treated by Kessler et al. (2005). These forests stand out by their total lack of large trees and the abundance of thin-stemmed trees. The high richness of trees $\geq$5 cm in secondary forests showed that this forest type has the potential to recover a considerable richness, if allowed to mature. In terms of taxonomic composition, the abundance of Meliaceae, Lauraceae, and Moraceae appeared to be considerably reduced in secondary forests relative to primary forests, whereas in Urticaceae, Ulmaceae, and Asteraceae it was increased. The latter families are typical fast-growing pioneer taxa of early successional stages throughout the tropics (Turner 2001) that are of little economic interest.

Endemic species are of considerable conservation concern and represent about 15% of the tree flora of Sulawesi (Keßler et al. 2002). This overall value is in good accordance with the percentages recorded by us at the plot level, with 10-20% of the native tree species recorded in the three forest types and the cacao forest gardens being endemic to the island. The representation of endemic species declined strongly to 0-13% in the two types of cacao plantations, however, showing that endemic species are more susceptible to severe habitat disturbances than widespread taxa. This is in accordance with general hypotheses on the vulnerability of endemic plants to habitat modifications (Kruckeberg and Rabinovitz 1985, Major 1988).

Our previous study in the forest on the northeastern side of the park yielded a basal area of ca. 140 m^2 per hectare, one of the highest values ever recorded in tropical forests (Kessler et al. 2005). The unusually high figure was partly due to the presence of numerous large fig trees, with diameters up to 6 m. However, even if these giant figs were excluded from the calculation, the basal area of the forest remained at about 108 m^2. The mean basal area of 57 m^2 (36-80 m^2) per ha in natural forest near Toro agrees with the results of Dietz et al. (this volume) from Toro and is much lower than the previously recorded value, but is still almost double the mean value of ca. 30 m^2 typical for tropical lowland forests in Southeast Asia (Dawkins 1959, Turner 2001). Part of this difference may be attributed to the general increase of basal area values with elevation, but it is likely that a strong geographical component is also included, with Sulawesi (or at least Lore Lindu National Park) having atypically high values. The causes for this geographical patterns are still unknown, but may involve the unique taxonomic composition of the tree flora of Sulawesi.

In conclusion, we found that moderate human use of the forest ecosystems by rattan and selected timber extraction did not result in significant decreases of tree biodiversity. Even forest gardens with cacao planted under natural shade trees and a canopy closure of over 80% (Dietz et al., this volume) supported considerable levels of tree biodiversity taking into account the lower tree density characteristic of these land use systems. Indeed, individual-based rarefaction showed that tree diversity in cacao forest gardens was somewhat lower, but not significantly so, than in the natural forest (M. Kessler unpubl. data). The conservation of tropical tree biodiversity is thus largely compatible with human exploitation of tropical forest as long as a canopy of native trees is maintained. Our study indicates that future conservation policies in rain forest margin areas should focus on developing measures aimed at sustainable use of the natural resources rather than attempting to exclude human activities. It shows that moderately intensive forest use in these areas is not necessarily in conflict with conservation of tree diversity (but may be so for other groups of organisms). Promotion of sustainable human activities may thus help to stabilize tropical rain forest margins in Central Sulawesi.

Acknowledgements
This study was carried out in the framework of the interdisciplinary research program "Stability of Rainforest Margins in Indonesia" (STORMA) funded by the German Research Foundation (DFG-SFB 552, grant to SRG). We gratefully acknowledge logistic support from STORMA's Indonesian partner universities in Bogor and Palu, Institut Pertanian Bogor (IPB) and Universitas Tadulako (UNTAD), the Ministry of Education in Jakarta (DIKTI), the authorities of Lore Lindu National Park, our Indonesian counterparts S. Tjitrosoedirdjo (SEAMEO-BIOTROP) and J. Mogea (Herbarium Bogoriense), and STORMA's coordinating teams in Göttingen, Bogor and Palu. Finally, we thank the local people of Toro, especially to Pak Said, Umar, Iskandar, Berwin, Abraham, and Radjab, for their hospitality and help in the field.

References

Andersson MS, Gradstein SR (2005) Impact of different management intensities on non-vascular epiphyte diversity in cacao plantations in western Ecuador. Biodiversity and Conservation 14: 1101-1120

Audley-Charles MG (1983) Reconstruction of eastern Gondwanaland. Nature 306: 48-50

Bandeira FP, Martorell C, Meave JA, Caballero J (2005) The role of rustic coffee plantations in the conservation of wild tree diversity in the Chintanec region of Mexico. Biodiversity and Conservation 14: 1225-1240

Bawa K, Seidler R (1998) Natural forest management and the conservation of biological diversity in tropical forests. Conservation Biology 12: 46-55

Beck J, Schulze CH, Linsenmair KE, Fiedler K (2002) From forest to farmland: diversity of geometer moths along two habitat gradients on Borneo. Journal of Tropical Ecology 18: 33-51

Bos M, Höhn P, Shahabuddin S, Buchori D, Steffan-Dewenter I, Tscharntke T (this volume) Insect responses to forest conversion and agroforestry management. In: Tscharntke T, Leuschner C, Zeller M, Guhardja E, Bidin A (eds) The stability of tropical rainforest margins, linking ecological, economic and social constraints of land use and conservation. Springer, Berlin Heidlberg New York

Brodbeck F, Weidelt HJ, Mitlohner R (2004) Traditional Forest Gardens in Central Sulawesi: A Sustainable Land Use System? In: Gerold G, Fremery M, Guhardja E (eds) Land Use, Nature Conservation and the Stability of Rainforest Margins in Southeast Asia. Springer, Berlin Heidelberg New York, pp 445-459

Davidar P, Puyravaud JP, Leigh EG (2005) Changes in rain forest tree diversity, dominance and rarity across a seasonality gradient in the Western Ghats, India. Journal of Biogeography 32: 493-501

Dawkins HC (1959) The volume increment of natural tropical high-forest and limitations to improvements. Empire Forestry Review 38: 175-180

Dietz J, Hölscher D, Leuschner C, Malik A, Amir MA (this volume) Forest structure as influenced by different types of community forestry in a lower montane rainforest of Central Sulawesi, Indonesia. In: Tscharntke T, Leuschner C, Zeller M, Guhardja E, Bidin A (eds) The stability of tropical rainforest margins, linking ecological, economic and social constraints of land use and conservation. Springer, Berlin Heidlberg New York

Fujisaka S, Escobar G, Veneklaas GE (1998) Plant community diversity relative to human land uses in an Amazon forest colony. Biodiversity and Conservation 7: 41-57

Gerold G, Fremery M, Guhardja E (2004) Land Use, Nature Conservation and the Stability of Rainforest Margins in Southeast Asia. Springer, Berlin Heidelberg New York

Gradstein SR (1992) The vanishing tropical rain forest as an environment for bryophytes and lichens. In: Bates JW, Farmer AR (eds) Bryophytes and lichens in a Changing Environment. Clarendon Press, Oxford, pp 232-256

Hamann A, Barbon EB, Curio E, Madulid DA (1999) A botanical inventory of a submontane tropical rainforest on Negros Island, Philippines. Biodiversity and Conservation 8: 1017-1031

Hong LT, Lemmens RHMJ, Prawirohatmodjo I, Soerianegara I, Sosef MSM, Wong WC (2004) Plant Resources of South-East Asia – Timber trees. ETI, Workingham, UK

Kappelle M (1996) Los Bosques de Roble (*Quercus*) de la Cordillera de Talamanca, Costa Rica: Biodiversidad, Ecología, Conservación y Desarollo. InBio, Heredia Costa Rica

Kappelle M, Kennis PAF, De Vries RAJ (1995) Changes in diversity along a successional gradient in a Costa Rican upper montane *Quercus* forest. Biodiversity and Conservation 4: 10-34

Keßler PJA, Bos MM, Sierra Daza SEC, Willemse LPM, Pitopang R, Gradstein SR (2002) Checklist of woody plants of Sulawesi, Indonesia. Blumea, Supplement 14: 1-160

Kessler M, Keßler PJA, Gradstein SR, Bach K, Schmull M, Pitopang R (2005) Tree diversity in primary forest and different land use systems in Central Sulawesi, Indonesia. Biodiversity and conservation 14: 547-560

Klein A, Steffan-Dewenter I, Buchori D, Tscharntke T (2002) Effects of land-use intensity in tropical agroforestry systems on coffee flower-visiting and trap-nesting bees and waps. Conservation Biology 16: 1003-1014

Kochummen KM, Lafrankie Jr JV, Manokaran N (1990) Floristic Composition of Pasoh Forest Reserve, a Lowland Rain Forest in Peninsular Malaysia. Journal of Tropical Forest Science 3: 1-13

Krömer T, Gradstein SR (2004) Species richness of vascular epiphytes in montane forests and fallows in Bolivia. Selbyana 25: 190-195

Kruckeberg AR, Rabinovitz D (1985) Biological aspects of endemism in higher plants. Annual Review of Ecology and Systematics 16: 447-479

Major J (1988) Endemism: a botanical perspective. In: Myers AA, Giller PS (eds) Analytical biogeography. Chapman and Hall, New York, pp 117-146

Lawton JH, Bignell DE, Bolton B, Bloemers GF, Eggleton P, Hammond PM, Hodda M, Holt RD, Larsen TB, Mawdsley NA, Stork NE, Srivastava DS, Watt AD (1998) Biodiversity inventories, indicator taxa and effects of habitat modification in tropical forest. Nature 391: 72-76

Lemmens RHMJ, Soerianegara I, Wong WC (1995) Timber trees: Minor commercial timbers. Prosea 5, 2: 1-655.

Mansyur M (2003) Analisis Vegetasi Hutan Di Desa Salua dan Kaduwaa, Taman Nasional Lore Lindu, Sulawesi Tengah. Proseding Workshop dan Promosi Flora Kawasan Timur. Denpasar, Indonesia

Migge-Kleian S, Woltmann L, Anas I, Schulz W, Steingrebe A, Schaefer M (this volume) Impact of forest disturbance and land use change on soil and litter arthropod assemblages in tropical rainforest margins. In: Tscharntke T, Leuschner C, Zeller M, Guhardja E, Bidin A (eds) The stability of tropical rainforest margins, linking ecological, economic and social constraints of land use and conservation. Springer, Berlin Heidlberg New York

Mori S, Boom BM, de Carvalho AM, dos Santos TS (1983) Southern Bahian Moist Forests. The Botanical Review 49: 155-232

Nöske N (2005) Effekte anthropogener Störung auf die Diversität kryptogamischer Epiphyten (Flechten, Moose) in einem Bergregenwald in Südecuador. online: http://webdoc.sub.gwdg.de/diss/2005/noeske/

Parthasarathy N (1999) Tree diversity and distribution in undisturbed and human-impacted sites of tropical wet evergreen forest in the southern Wastern Ghats, India. Biodiversity and Conservation 8: 1365-1381

Phillips OL, Hall P, Gentry AH, Sawyer SA, Vasquez R (1994) Dynamics and species richness of tropical rain forest. Proc. Natl. Acad. Sci. U.S.A. 91: 2805-2809

Proctor JJ, Anderson M, Chai P, Vallack HW (1983) Ecological studies in four contrasting lowland rain forest in Gunung Mulu National Park, Sarawak.Journal of Ecology 71: 237-260

Roos M, Keßler PJA, Gradstein SR, Baas P (2004) Species diversity and endemism of 5 major Malesian islands: diversity-area relationships. Journal of Biogeography 31: 1893-1908

Schulze C, Waltert M, Keßler PJA, Pitopang R, Shahabuddin, Veddeler D, Mühlenberg M, Gradstein SR, Leuschner C, Steffan-Dewenter I, Tscharntke T (2004) Biodiversity indicator taxa of tropical land-use systems: comparing plants, birds and insects. Ecological Applications 14: 1321-1333

Soerianegara I, Lemmens RHMJ (1993) Timber trees: Major commercial timbers. Prosea 5, 1: 1-610.

Simbolon H, Suzuki E, Yoneda M (1997) Study on the biodiversity and dynamic of forest ecosystem in Gunung Halimun National Park, West Java, Indonesia. Workshop on Biodiversity and Dynamics of Forest Ecosystems in Western Pacific and Asia, Kyoto, Japan, 10-11 November 1997

Sperber CF, Nakayama K, Valverde MJ, Neves FD (2004) Tree species richness and density affect parasitoid diversity in cacao agroforestry. Basic and Applied Ecology 5: 241-251

Sporn S, Gradstein SR (2005) Diversity comparison of non-vascular epiphytes in cacao plantations with different shade canopies. In: Stietenroth D, Lorenz W, Tarigan S, Malik A (eds), Proceedings International Symposium "The Stability of Tropical Rainforest Margins." Universitätsverlag, Göttingen, p 148

Steenis CGGJ van (1950-) Flora Malesiana [continuing series]. Kluwer (Noordhoff-Kolff), Dordrecht, Netherlands

Turner IM (2001) The Ecology of the Tropical Rainforest. Cambridge University Press, Cambridge, UK

Turner IM, Wong YK, Chew PT, bin Ibrahim A (1997) Tree species richness in primary and old secondary tropical forests in Singapore. Biodiversity and Conservation 6: 237-543

Whitmore TC, Sayer JA (eds) (1992) Tropical Deforestation and Species Extinction. Chapman & Hall, London New York Tokyo

Whitten AJ, Henderson GS, Mustafa M (2002) The Ecology of Sulawesi, 2nd Edition. Periplus, Jakarta Tokyo Singapore

Wright DD, Jessen JH Burke P, Gomez de Silva Garza H (1997) Tree and Liana Enumeration and Diversity on a One-Hectare plot in Papua New Guinea. Biotropica 29: 250-260

Alternatives to slash-and-burn in forest-based fallow systems of the eastern Brazilian Amazon region: Technology and policy options to halt ecological degradation and improve rural welfare

Jan Börner[1*], Manfred Denich[1], Arisbe Mendoza-Escalante[1], Bettina Hedden-Dunkhorst[2], and Tatiana Deane de Abreu Sá[3]

[1] Center for Development Research (ZEF), University of Bonn, Walter-Flex-Str. 3, 53113 Bonn, Germany
[2] Federal Agency of Nature Conservation (BfN), Konstantinstr. 110, 53179 Bonn, Germany
[3] Empresa Brasileira de Pesquisa Agropecuária, Embrapa Sede, Parque Estação Biológica - PqEB s/n., Brasília-DF, - Brazil - CEP 70770-901

* corresponding author: Jan Börner, present address: Technische Universität München, Lehrstuhl für Vegetationsökologie, Am Hochanger 6, 85350 Freising, Germany, E-mail: janboerner@uni-bonn.de

Summary

In many smallholder farming systems in the humid tropics, the slash-and-burn practice is used for land preparation. Increasing land-use intensity by shortening fallow periods often contributes to the degradation of the natural resource base of the fallow system, i.e. the fallow vegetation and soil. In the eastern Amazon region of Brazil, we therefore searched for ways to maintain the sustainability of the traditional fallow system and to adapt it to changing agro-ecological and economic conditions.

We identified two major agro-ecological constraints of the traditional fallow system with slash-and-burn: (1) high losses of nutrients and organic matter during the burn, and (2), if land-use intensity increases, fallow regeneration capacity declines.

As alternatives to slash-and-burn, we studied modifications to those practices recognized to be harmful to the ecological sustainability of the fallow system, i.e. mulching for the management of soil organic matter and fire-free land clearing with bush choppers to transform fallow vegetation into mulch.

Tscharntke T, Leuschner C, Zeller M, Guhardja E, Bidin A (eds), The stability of tropical rainforest margins, linking ecological, economic and social constraints of land use and conservation, Springer Berlin 2007, pp 335-363

Mulching allows extending the cropping period, planting crops off-season, and modifying crop rotations. Additionally, biomass and nutrient accumulation of degraded fallow vegetation can be improved by enrichment plantings using fast-growing leguminous tree species.

Our socio-economic analysis focused on the implications of technology change on income and land-cover change at farm and field levels. Based on farm-household data collected from 270 randomly selected households, a farm-household level bio-economic model was developed to analyze the consequences of improved access to mechanized plowing and mechanical mulching for typical smallholdings in the study area.

Model simulations suggest that the costs of mulching are still very high compared to other mechanized land preparation technologies that provide similar economic benefits from the farmers' point of view. Among the recommendations to reduce mulching costs is the use of simpler mulching equipment on areas with young fallows.

Technology scenarios indicate that countervailing policy measures are necessary if mechanized chopping and mulching is to bring about the desired ecological benefits.

Taxes on ecologically undesirable forms of land preparation, e.g. slash and burn, are promising policy options to promote chopping-and-mulching or other fire-free land preparation techniques given that they can be provided at costs that range between 60 to 110 Euros ha^{-1}. Tax revenues could be used for financing environmental conservation payments and/or a crop yield insurance that applies to crops that are produced using environmentally friendly production technologies.

Keywords: mulch, bush chopper, fallow management, bio-economic modeling, tax, yield insurance

1 Introduction

Deforestation is one of the key challenges faced by the global community, not only because of the associated emissions of greenhouse gases, loss of biodiversity and other ecosystem services (United Nations Millennium Declaration 2000), but also because of soil degradation and the pollution of air and water resources, especially in tropical regions. These problems usually affect the rural poor more than other groups of society.

According to Alencar et al. (2004), between 18 and 30% of the deforestation in the Brazilian Amazon region is caused by smallholders with farms smaller than 100 ha. Using the example of settlements in the states of Acre and Rondônia, Vosti et al. (2002) have shown that the current agricultural frontier areas are likely to become almost completely deforested within the next 25 years. In this context, decision-makers are particularly interested in knowing (1) what kind of land-use systems are likely to emerge on agricultural frontier areas in the future; (2) how economic, technological, and environmental change will affect rural livelihoods in the absence of a valuable natural resource, such as primary forests; and (3) whether migration to new forest margins is an inevitable consequence.

The elaboration of these policy issues in this chapter is based on key findings of ecological, agronomic, and socio-economic studies carried out in the context of the bilateral German-Brazilian research program "Studies on Human Impact on Forests and Floodplains in the Tropics - SHIFT" in the eastern Brazilian Amazon. Our study area, the Bragantina region in the northeast of Pará state is one of the oldest colonization areas in the Amazon region. By 1950, most of the arable land in the Bragantina region had been cleared of primary forests, and since then a stable smallholder sector has evolved contrary to the pessimistic expectations of early observers (Penteado 1967). In fact, the history of the Bragantina region confirms the position put forward by Diogenes S. Alves (Part III of this book) that the "land degradation / abandonment followed by economic stagnation" paradigm is not always adequate to describe agricultural frontier development in the long-run.

The debate around sustainable agriculture in tropical forest areas frequently deals with land-use technologies. Slash-and-burn is the most common technology used by farmers both in the Amazon region and worldwide to convert primary or secondary forests into agricultural land. The use of fire and extended fallow periods has often been regarded as economically unproductive and environmentally harmful (Penteado 1967, Fearnside 1986, Nepstad et al. 1999). Consequently, many technological alternatives to slash-and-burn are being promoted by different stakeholders, often without a thorough understanding of the ways in which technology change affects rural livelihoods and the environment.

More than 150 years of agriculture in the Bragantina region have shown that it is the combination of agro-environmental policies with economic growth and technology change that has shaped the region's landscape (Hurtienne 2001), where many economic and ecological functions of primary forests have been taken over by secondary forests. Nevertheless, there is evidence suggesting that both traditional and modern agricultural activities gradually reduce the vitality of the secondary forests, which are of both ecological and agronomic importance.

In this chapter we therefore focus on technology and policy options for improving the economic and ecological sustainability of the fallow-based production system in the Bragantina region and juxtapose our findings to the situation at the forest margins. Specifically, we focus on the following research questions:

1. What is the impact of slash-and-burn and alternative mechanized land preparation technologies on key bio-physical and agronomic indicators at plot level?

2. What is the short and medium-term impact of using slash-and-burn and mechanized alternatives on land-cover change, carbon sequestration and income at farm level?
3. How can policy instruments be used to simultaneously address environmental degradation and rural welfare?
4. How can policy instruments be used to enhance technology adoption?
5. What implications do these findings have for the future of smallholder agriculture at the forest margins?

While 1, 2, and 5 directly address the three policy issues raised above, answering questions 3 and 4 may help to improve existing and develop new policy instruments to reconcile environmental conservation and technology change in the old colonization areas of the eastern Amazon region.

2 Methodology and transdisciplinary research

For the sake of space we omit a detailed documentation of the methodologies used to generate specific findings and refer instead to the respective publications. What follows is a brief overview of the methodological steps taken by the involved natural scientists and agricultural economists.

Agronomic research focused on the bio-physical analysis of the fallow system, e.g. by conducting vegetation surveys (Denich 1989, Vieira 1996, Baar 1997), biomass studies (Denich 1989, Nunez 1995, Schmitt 1997, Brienza 1999, Tippmann 2000), nutrient flux analyses (Hölscher 1994, Sommer 2000, Wickel 2004), and studies on the regeneration capacity of fallow vegetation after the cropping period (Wiesenmüller 1999). During the agronomic research phase, an alternative to the slash-and-burn technology was developed that replaces manual slashing and the use of fire by mechanical chopping of the fallow vegetation and mulching (Denich et al. 2004, Denich et al. 2005). The "chop-and-mulch" technology has been tested in various forms ranging from manual chopping to the use of heavy machinery that transforms secondary forests of up to 300 Mg ha^{-1} of fresh biomass into a homogeneous mulch layer (Block 2005). The most extensive experience was gained with tractor-driven bush choppers that handle fallows of up to 12 years old (or 150 Mg biomass ha^{-1}; Figure 1).

Researcher and farmer-managed cropping experiments were conducted to compare the agronomic performance of mulching versus slash-and-burn. Moreover, different forms of enriching fallows with fast-growing leguminous trees were tested as a means to enhance fallow regrowth and shorten fallow periods (Kato et al. 1999, Bünemann 1998, Brienza 1999, Sommer 2000).

Most of the agronomic research was conducted in the municipality of Igarapé-Acu in the center of the study region (Figure 2).

The agronomic research phase was followed by socio-economic research that aimed at improving the understanding of the determinants of smallholder

Fig. 1. Chop-and-mulch (a) and harrowing (b) using tractor power in the Bragantina Region

production and technology choice at field (plot), farm, and regional levels. In this research phase, the agro-ecological and socio-economic diversity of the Bragantina region was assessed. For this purpose, large-scale farm-household and plot level surveys were carried out and accompanied by an analysis of institutions and agriculture-related organizations at village, municipality, and regional levels (Mendoza-Escalante 2005, Börner 2006). During the plot-level survey, soil samples were taken from randomly selected plots with annual crops. To obtain a general indicator of soil fertility on these plots, we conducted a bio-assay using rice as an indicator plant.

The farm-household data were used in an econometric profit function analysis to determine the monetary value of fallowing among other economic determinants of smallholder production (Mendoza-Escalante 2005). Furthermore, a farm-household model that incorporates bio-physical relationships was developed using primarily linear mathematical programming and stochastic simulation techniques. The model (hereafter mathematical programming model) was used to analyze land-use decisions and technology choice of a representative type of farm household under different policy scenarios at farm level (Börner 2006). The model uses standard methods to account for aversion to production and price risks, such as the MOTAD and the 'Expected-Value Variance' approach described in Hardaker et al. (2004). However, following Hardaker et al. (2004), risk aversion was not measured via the direct elicitation of utility functions. Instead, a factor of constant absolute risk aversion was parameterized, such that modeled crop mix in the first period corresponds to the crop mix observed in the study area in 2002/2003. For all model simulations, a 25-year planning horizon was assumed.

Both the econometric and the mathematical programming models assume that farm households maximize some indicator of economic utility. The mathematical programming model maximizes non-essential consumption subject to agronomic and socio-economic constraints, such as the availability of land, fallow vegetation, family labor, basic consumption needs. This allows attaching an economic value (shadow price) even to resources that do not have a market value, e.g. fallow vegetation. The mathematical programming model also assumes that smallholders are averse to risks, such as varying prices and yields depending on their individual degree of risk aversion.

For a detailed description of the two methodologies, we refer to Hazell and Norton (1986) and Intriligator (1971) for mathematical programming and to Chambers (1988) and Diewert (1973) for econometric profit function analysis.

Stochastic simulation techniques were used following Berg (1998) to assess the impact of fertilizer on the variance of yields. Therefore, expected yield and yield-variance response functions to fertilizer application were generated for each crop. Since crop yields are often correlated, modeling production risk requires accounting explicitly for gross margin covariances. Hence, the yield response functions were fed into a quadratic version of the linear mathematical programming model. Most of the results presented below, however, were obtained using the linear version of the model, as it requires considerably less solving time and allows for a longer planning horizon.

Both econometric and mathematical programming models account for the key findings of previous bio-physical research, e.g. fallow regrowth was estimated from biomass surveys, and fallow length / yield relationships are based on experiment station results (Kato 1998a, b).

All monetary figures presented in this chapter are expressed in 2002 Brazilian Reais (1 R$ = 0.38696 Euro).

3 Smallholder agriculture and ecological limitations in the Bragantina region

3.1 Overview

The Bragantina region, defined as the micro-regions Castanhal and Bragantina, covers 12,410 km^2 east of Belém, the capital of the state of Pará (Figure 2). With 44.7 inhabitants km^{-2}, it is the most densely populated rural area in the Amazon region, which also stands out in terms of a good transport infrastructure and regional market integration (IBGE 2000).

Sixty percent of the land cover of the Bragantina region is made up of secondary vegetation of different ages, showing that fallow-based agricultural production systems dominate. In 1995/1996, 76% of the annual and perennial crops grew on smallholdings (IBGE 1998). Typical annual crops that serve for both home consumption and sale are cassava and beans. One of the main income sources for farmers is the on-farm processing of cassava roots into cassava flour. Corn is a common intercrop and mainly used for home consumption and as animal fodder. Passion fruit, black pepper, coconut, and oranges are widespread perennial cash crops on smallholder farms. A small proportion of smallholders that is well endowed with land has invested in extensive livestock operations.

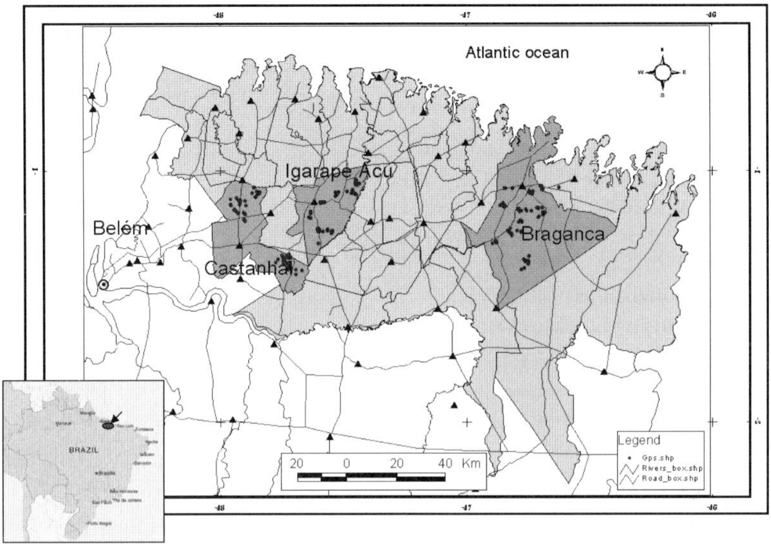

Fig. 2. Brazil, the study area, and the sample municipalities. The scattered dots in the highlighted municipalities are the locations of sample households.

Table 1 presents farm-household characteristics according to per capita income quartiles. The category 'Location' indicates the number of households located in a given sampling municipality with Castanhal being the western-most municipality close to the urban center of Belém, and Bragança, the remotest municipality in the eastern part of the study area.

Low-income households concentrate in Bragança, whereas most high-income households are located in Castanhal. The survey data further suggest that households in the forth income quartile are also generally wealthier, achieve higher education levels and tend to be better endowed with land and fallow vegetation than households in the lower two quartiles. No significant differences in soil quality were found between income quartiles. Hence, wealthier and land endowed farms do not conserve soil fertility better than small and low-income farm households.

The deep-rooting woody species in the secondary vegetation of the Bragantina region are the backbone of the traditional fallow system, not only because they regenerate quickly by vegetative re-sprouting after slash-and-burn (Denich 1989). Soil and vegetation studies have also shown that the woody vegetation recycles nutrients and water from the subsoil via its extensive root system and thus fulfills important agro-ecological functions (Hölscher 1994, Sommer et al. 2000, Wickel 2004).

Table 1. Farm-household characteristics by per capita income quartiles (Qut.)*

	1.Qut.	2.Qut.	3.Qut.	4.Qut.	Means
Total per capita income	285	792	1651	7599	2585
Agricultural per capita income					
Crop production (R$ year^{-1})	158[a]	500[b]	1083[c]	4844[d]	1649
Livestock production (R$ year^{-1})	11[a]	-8[a]	24[a]	1944[a]	495
Total agricultural production (R$ year^{-1})	169[a]	493[b]	1107[c]	6788[d]	2143
Share of commercialized production value (%)	46[a]	60[b]	63[c]	73[d]	60
Off-farm per capita income					
Non-agricultural off-farm employment (R$ year^{-1})	49[a]	103[a]	166[a]	342[a]	165
Transfers (e.g. pensions) (R$ year^{-1})	47[a]	187[b]	350[c]	440[c]	255
Agricultural off-farm employment (R$ year^{-1})	21[a]	10[a]	28[a]	29[a]	22
Physical endowment					
Total farm land (ha)	17.4[a]	19.6[a]	15.8[a]	31.9[b]	21.2
Old fallows in % of farm land	13.6[a]	14.4[a]	14.2[a]	19.6[b]	15.4
[average fallow age (years)]	[12[a]]	[13[a]]	[15[b]]	[19[c]]	[17]
Young fallows in % of farm land	46.1[a]	41.8[a]	32.2[b]	36.8[a]	39.3
[average fallow age years]	[5[a]]	[5[a]]	[5[a]]	[6[a]]	[6]
Cropland (ha)	2.4[a]	3.1[b]	3.4[b]	6.7[c]	3.9
Soil fertility (grams of above-ground biomass)**	2.24[a]	2.05[a]	2.13[a]	2.09[a]	2.29
Wealth index***	-1.43[a]	-0.04[b]	0.15[b]	1.33[c]	0.00
Labor use					
Family labor (adult equivalent)	3.4[a]	3.0[a]	2.9[b]	2.7[b]	3.0
Hired labor (mandays year $^{-1}$)	29[a]	40[b]	47[b]	115[c]	63
Household composition					
Familiy size (individuals)	7.5[a]	6.3[b]	6.2[b]	4.6[c]	6.2
Education of the head of the household (years)	2.5[a]	2.9[a]	2.3[a]	4.0[b]	2.9
Location					**Sum**
Households (HH) from Castanhal (west)	11	19	24	36	90
HH from Igarapé-Açu (center)	18	25	25	22	90
HH from Bragança (east)	39	24	18	10	91
Total N	68	68	67	68	271

Source: SHIFT/NAEA field data

a,b,c,d: Mean values that are not significantly different ($p < 0.05$) from each other are assigned the same letter

* E.g. The 1. Quartile represents the 25% of the sample with the lowest per capita income

**Result from bio-assay with rice as an indicator plant, high dry weight of above-ground biomass indicate high soil fertility

***Based on household durables (Sahn and Stifel 2000)

Soils in the Bragantina region are predominantly Ultisols and Oxisols with low organic matter and nutrient contents, low cation exchange capacities, low pH, and high aluminum concentrations.

3.2 Ecological degradation, economic performance, and agricultural technologies

This section shows how agricultural technologies contribute to ecological degradation in the Bragantina region and how the chop-and-mulch technology could help to reconcile environmental conservation and agricultural productivity.

Agricultural census data from the municipalities in the study area show that the total area covered by secondary vegetation decreased by 31% between 1980 and 1995. However, the area under old secondary vegetation increased by roughly 32% (Figure 3).

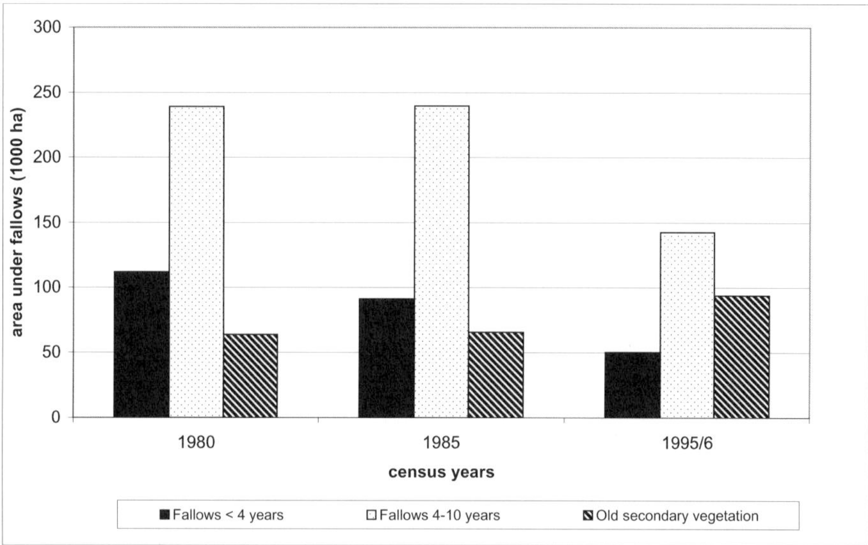

Fig. 3. Absolute area covered by different types of secondary vegetation (fallows) in the districts of the study area in 1980, 1985, and 1995 (IBGE, various agricultural censuses).

Apart from the overall reduction in secondary forest cover with the associated carbon and biodiversity losses, ecological degradation occurs in two different ways, both of which are related to the use of land preparation technologies: first, as a gradual decline in the regeneration potential of the secondary forests and, second, through highly negative nutrient balances, which, in the absence of external nutrient supplies, lead to reduced soil fertility.

Under slash-and-burn, fallow re-growth depends on the length and the frequency of cropping cycles, and vegetation studies have shown that intensively and repeatedly used fallows have a lower biomass and woody species density than less exploited fallows of the same age (Vieira 1996, Baar 1997, Denich et al. 2005).

Mechanized plowing and harrowing, which is associated with the destruction of tree roots, is the most prominent technological alternative to slash-and-burn in the Bragantina region. Our research shows that woody species are significantly reduced after mechanical land preparation (Table 2). After plowing or harrowing, fallows temporarily lose important agronomic functions, such as fast biomass accumulation, deep soil nutrient recycling, weed suppression, and their private and social value as an extraction resource (Hedden-Dunkhorst 2003). Sommer et al. (2000) demonstrate, moreover, that the destruction of roots and stumps through mechanization can lead to a loss of up to 15 Mg ha^{-1} of below-ground carbon.

Table 2. Biomass (t ha^{-1}) on fields after different land preparation methods

	No tillage		Plowing and harrowing	
	woody	herbaceous	woody	herbaceous
Cropping period	0.174^{a}	0.173^{A}	0.006^{b**}	0.116^{A}
	(0.042)	(0.042)	(0.003)	(0.019)
2-year fallow	10.1^{a}	4.4^{A}	4.3^{b*}	6.2^{A}
	(2.0)	(1.3)	(1.8)	(1.1)
5-year fallow	22.8^{a}		15.0^{a}	
	(5.4)		(4.4)	

Source: Denich et al. (2005)
Number of repetitions: 6; Plot size 120 m^2 (with sub-plots of 12 m^2)
Values in brackets are standard errors
a,b (A): Mean values that are not significantly different (t-Test; ** $p < 0.05$, * $p < 0.1$) from each other are assigned the same letter (lower case for woody and upper case for herbaceous biomass)

Especially the slash-and-burn technology is characterized by high nutrient losses. On average, 96% of the nitrogen, 47% of the phosphorus (the limiting nutrient for plant growth in the study region), and 48% of the potassium stored in the vegetation is lost through volatilization during burning (Mackensen et al. 1996).

The economic consequences of ecological degradation in the Bragantina region are difficult to quantify. Comparing yield levels for typical crops since 1920 is inconclusive, as yields have been influenced by unobserved technology change (Börner 2006). A bio-assay of 542 plots from 270 farms has shown that soil quality is significantly and positively correlated with fallow length. However, cassava yields and gross margins obtained on fields with lower soil fertility are significantly higher due to the use of external farm inputs, e.g. in the western Bragantina region (Table 3). This indicates that short fallow periods do not necessarily have a negative influence on the profitability of agriculture. Instead, economic performance seems to be determined by access to improved technology, which is linked to the provision of extension and the amount of machinery available in a given municipality.

Table 3. Soil fertility, technology use, and cassava yields in municipalities Castanhal, Igarapé-Açu and Bragança

	Castanhal (west)	Igarapé-Açu (center)	Bragança (east)	Mean
Average yield cassava roots (t ha^{-1})	21.7[a]	17.9[a]	10.4[b]	16.2
Fertilizer use (% of farmers)	90	86	43	73
Tractor use (% of farmers)	76	49	18	47
Extension contact (% of farmers)	44	17	11	24
Tractors per agricultural establishment[1]	0.26	0.09	0.01	0.05
Plows per agricultural establishment[1]	0.15	0.03	0.01	0.03
Soil fertility (grams of above-ground biomass)[2]	2.27[a]	2.01[b]	2.57[c]	2.29

Source: SHIFT/NAEA field data
[1]IBGE agricultural census 1995/6
[2]Result from bio-assay with rice as an indicator plant, high dry weight of above ground biomass indicate high soil fertility
a,b,c: Mean values that are not significantly different ($p < 0.05$) from each other are assigned the same letter

The chop-and-mulch technology was designed with the intention to avoid the negative aspects of conventional mechanization in terms of its devastating effects on the vitality of the secondary vegetation. It also favours a more efficient utilization of plant material and nutrients than the slash-and-burn system. This allows extending the traditional cropping period from 2 to 4 years, as nutrients are only gradually released by the mulch layer. However, nutrient-demanding crops are affected by a temporal immobilization of nutrients in the mulch layer, which requires additional fertilizer applications.

If chop-and-mulch substitutes slash-and-burn, the external costs associated with the use of fire for land preparation could be reduced, e.g. annual material damages from accidental fires amounting to an average of 50 R$ per farm (Börner 2006).

4 Technologies and valuation of resources at the farm level

This section describes the three land preparation technologies under study, namely slash-and-burn, mechanical plowing and harrowing, and mechanical chop-and-mulch. Introducing mechanized alternatives to slash-and-burn shifts the valuation of resources at the farm level. Understanding the forces at work helps to explain how improved access to these alternatives, e.g. in the course of economic development or government initiatives, translates into land-use and land-cover change. Most of the findings presented in this section are based on econometric or mathematical programming models. Table 4 gives an overview of some key characteristics of slash-and-burn and the mechanized alternatives.

4.1 Slash-and-burn

Slash-and-burn remains the most important land preparation technique for smallholders in the Amazon region due to the ease of application and the low requirements of capital-demanding external inputs. In the Bragantina region, fallow periods typically range from 4 to 10 years and vary with crop species and land endowment. At plot level, Mendoza-Escalante (2005) shows econometrically that fallowing contributes on average 22% (or 385 R$ ha^{-1}) to the profit of plots planted with annual crops. This implies that converting fallow vegetation has relatively high opportunity costs, i.e. the benefit foregone if fallow is converted for other purposes than annual cropping. It also explains that most farms maintain considerable amounts of secondary forests as a valuable farming input. At farm level, the value of fallow vegetation is lower (Table 4), because woody vegetation is not beneficial for all farm operations. For example, for the establishment of pastures and perennial plantations, fallow vegetation represents a cost, as it has to be removed and thus does not contribute directly to the economic benefits derived from subsequent land uses. In Table 4 the values in brackets show that the valuation of both labor and fallow increases if farmers use chemical fertilizers in the slash-and-burn system. Apart from the provision of nutrients, fallowing and ashes from burning provide two other important services, namely weed suppression and soil pH regulation. A fertilized cropping system without mechanical land preparation and chemical pH control therefore requires at least two to three years of fallow between cropping periods. Increased productivity due to fertilizer use thus also increases the valuation of services from fallow.

Despite the high population density, model simulations suggest that the conversion of secondary forests is also limited by labor availability during the slash-and-burn season. While the value of male family labor for vegetation clearing is below the average wage rate during most of the year, its valuation peaks during the land preparation season (Table 4). In other words, labor in the slash-and-burn season is particularly scarce, and hence expensive, which limits the amount of land that can be prepared.

Table 4. Cassava yields and resource valuation using different land preparation technologies

	Average cassava yield in the study area[1] [kg ha^{-1}]	Average cassava yield without fertilization[2] [kg ha^{-1}]	Average cassava yield with fertilization and optimal treatment[2,3] [kg ha^{-1}]	Land preparation costs[1] [R\$ ha^{-1}]	Valuation of fallow at the farm level[4] [R\$ ha^{-1}]	Valuation of labor at the farm level[4] [R\$ (peak season)]	Weeding requirement[1] [times/crop cycle]
Slash-and-burn	14085	13820	30680	150	128 [283][6]	184 [214][6]	1-3
Plowing/harrowing	20288	5507	35230	180	38	235	>4
Chop-and-mulch	n.a.	15130	28020	935 (576)[5]	185	228	2

1 SHIFT/NAEA field data
2 Kato 1998a, b
3 Embrapa Amazônia Oriental personal communication 2003
4 Mathematical programming model
5 Scenario assuming lower cost mulching equipment, low biomass fallows, and reduced mulching intensity (i.e. higher coarseness of mulch layer)
6 Assuming that farms use chemical fertilizers

4.2 Conventional mechanization

The relative valuation of labor and fallow vegetation changes dramatically if mechanical land preparation technologies are available. For example, plowing and harrowing practically decouples fallow length from crop yields by substituting fallow services (e.g. nutrient supply and liming effect of the ashes) external inputs, such as lime and chemical fertilizers. As a consequence, fallow vegetation represents a cost to production, as it has to be removed prior to cultivation. This results in a reduction by 71% of the valuation of fallows at farm level.

Once cleared from vegetation, stumps, and surface roots, the land can be prepared mechanically, which releases labor for preparing additional land using the slash-and-burn technique. The overall valuation of labor, however, increases due to the higher profitability of annual crop production in the mechanized system, but also as a consequence of higher weeding requirements. Plowing and harrowing can be offered at a cost only slightly higher than the estimated cost of manual vegetation clearing (Table 4).

4.3 Chop-and-mulch

Experiments have shown that a variety of soil characteristics, such as soil moisture and organic matter content, improve significantly under mulch (Sommer 2000, Kato 1998a, b). This can have very different implications for agricultural crops. Without fertilization, cassava yields are about the same as those after slash-and-burn in the first cropping period following land preparation and tend to be higher in the second period as nutrients are then released from the mulch by decomposition. Bean yields are hardly affected by land preparation, and both rice and corn require fertilization due to the immobilization of nutrients in the mulch (Kato 1998a, b).

Since it is unrealistic to expect farmers to use fertilizers only under chop-and-mulch, the figures in Table 4 are based on the assumption that fertilizers are also used under slash-and-burn (see values in brackets). In such a scenario, using chop-and-mulch reduces the economic value of fallows by 35 % from 283 to 185 R$ ha^{-1}: first, because it is cheaper to use fertilizers than fallows as a plant nutrient source and, second, because older fallows are more expensive to chop than younger fallows, which is not necessarily the case with manual slashing. Similar to conventional mechanization, chop-and-mulch increases the valuation of family labor, since more output can be produced per unit of labor.

Data from three different sites suggest that chopping costs tend to increase with the amount of vegetation biomass and the availability of massive tree stumps. Large fields are significantly cheaper to prepare than smaller fields, because field size determines the time lost maneuvering. To achieve homogeneous mulch particles, most fields have to be chopped twice. Theoretically, chopping costs could be reduced from 935 to 576 R$ ha^{-1} using light and less expensive equipment on plots with young fallow vegetation, which are chopped only once.

Our results suggest that inducing the average farmer in the research area to adopt the chop-and-mulch technology requires a reduction of the average chopping costs (935 R$ ha^{-1}) by 70%. If, however, this farmer had also access to mechanical plowing and harrowing, chopping costs would have to be reduced even further, which shows that the two mechanical alternatives to chop-and-mulch compete.

In summary, both chop-and-mulch and mechanical plowing/harrowing lead to a lower valuation of fallow vegetation as a production factor, whereas the valuation of family labor increases. All three technologies can potentially produce yields almost 100% higher than the yields obtained on most of the fields in the traditional slash-and-burn system (Table 4). What determines yield levels, however, is rather the amount of fertilization than the choice of land preparation technology. In fact, the question arises, why do farmers actually not use fertilizers to enhance productivity.

4.4 Determinants of fertilizer use

Economists usually expect fertilizer use to be determined by relative prices, relative scarcity of production factors, and risk preferences. Relative prices favor the use of fertilizer for almost all common crops in the study region. An investment of 1 R$ per cropping period in NPK increases the net present value of slash-and-burn over 18 years by at least 7 R$. Moreover, previous research suggests that net per capita income, labor availability, soil fertility, social engagement in producer organizations, and other farm-household characteristics have low explanatory power with regard to fertilizer use (Börner 2006). Instead, fertilizer use seems to depend on the type of crop that is to be planted. For example, water melons, cucumbers, and beans are frequently fertilized. Apart from perennial cash crops, which are also fertilized, these crops are among the most nutrient-demanding annual cash and food crops and produce practically no fruit without fertilizers (Kato 1998a, b, Kato et al. 1999). Cassava and corn, on the other hand, produce low but sufficient yields without external nutrient supply and hence represent a minimum risk alternative to activities that require up-front cash outlays.

It seems that little prevents farmers from making rational use of fertilizers. In the following, we will therefore explore the impact of risk preferences on using fertilizers for specific crops.

5 Sustainability of current agricultural practices

In this section we explore the ecological and economic sustainability of smallholder agriculture in the Bragantina region under the assumption that the current economic and technological frame conditions remain constant over time. In the context of our study, economic sustainability requires that actual consumption does not compromise future consumption, i.e. farm-household income is at least non-decreasing over time. Accordingly, ecological sustainability means that the environmental services provided by natural resources, e.g. carbon sequestration, are non-decreasing over time. The results presented here and in the subsequent sections were obtained using the mathematical programming model.

Figure 4 shows the main land-use categories and how they develop over time. Most of the categories remain stable, but degraded fallow fluctuates in regular cycles. These cycles are caused by the abandonment of black pepper plantations in 7-year intervals (not shown). A marked decrease over time can be observed in the average fallow age, indicating that both carbon sequestration and woody species diversity are reduced.

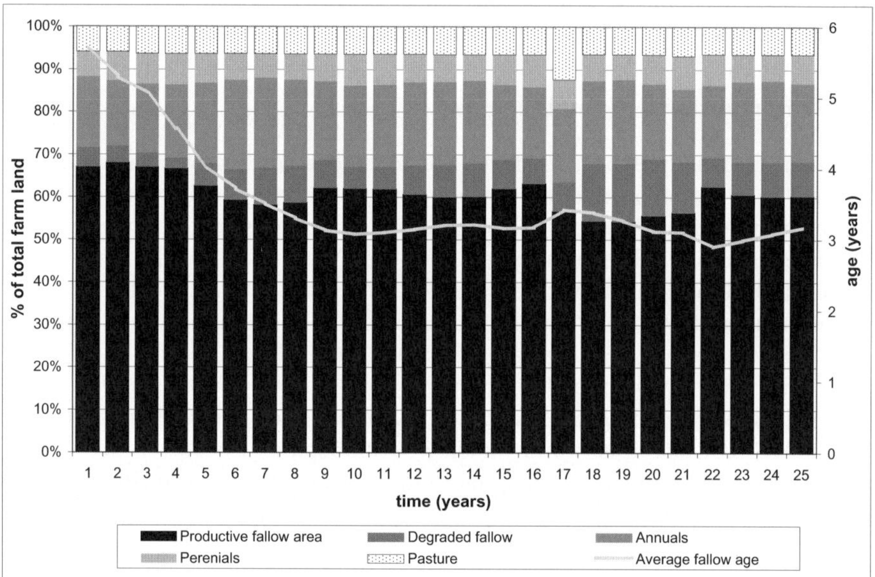

Fig. 4. Main land-use categories and average fallow age in a 25-year simulation under status quo conditions using the mathematical programming model

Three categories of indicators that could be called 'indicators of sustainability' in the context of the 'critical triangle" of sustainable agricultural development (cf. Vosti and Reardon 1997), will be looked at in more detail: (1) net household income as a welfare indicator; (2) average fallow age, secondary forest cover, and the total of stored below and above-ground carbon as indicators of environmental degradation; and (3) the amount of hired labor as a stimulus for the local rural economy. Table 5 shows the estimated trends in these indicators over the 25-year planning horizon as a means to evaluate their average change over time.

Table 5. Time trend estimates of 'sustainability indicators' at the farm level over a 25-year simulation using the mathematical programming model

	Trend coefficient	Standard error	R
Welfare (R\$)			
Net household income	3.396	29.25	0.00
Environment			
Average fallow age (years)	-0.078 **	0.01	0.56
Secondary forest cover (ha)	-0.010	0.01	0.08
Below- and above-ground carbon (t)	-3.843 **	0.64	0.61
Economic growth			
Hired labor (days year^{-1})	-0.198	0.16	0.06
Product mix			
Value of cassava flour production (R\$)	-15.851 *	8.59	0.13
Value of bean production (R\$)	3.630 *	1.65	0.24

N=25
* $p < 0.05$
** $p < 0.01$

The trends in net income and hired labor are not statistically significant, indicating that they remain stable over time. This implies that, *ceteris paribus,* the average type of farm in the Bragantina operates economically sustainable in the sense that today's decisions do not compromise future income opportunities. However, they also indicate that there is little scope for local economic growth under the assumption that economic and technological frame conditions do not change over time.

In contrast to the economic indicators, the environmental measures decrease over time. Especially the average age of fallows and the total carbon stored on the model farm show negative trends. In the case of carbon, the reduction is a classical negative externality, as carbon sequestration has no economic value for the farmer as an economic agent. However, the change in average fallow age does affect the farmer, since younger fallows provide fewer nutrients through slash-and-burn. To compensate this effect, a substitution takes place from strongly fallow-dependent cassava production to rather fertilizer-dependent bean production (see 'product mix' in Table 5). This indicates that farmers do substitute nutrients from fallow vegetation with chemical fertilizers, but it does not explain why this substitution is associated with the switch from cassava to bean production. Figure 5 shows how risk aversion influences decisions on crop mix and fertilizer use.

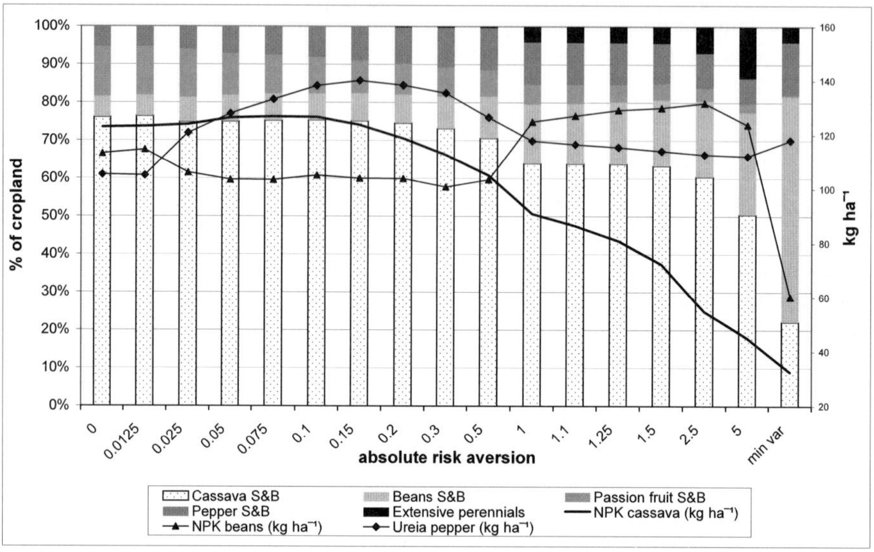

Fig. 5. Risk aversion (measured as constant absolute risk aversion following Hardaker et al. 2004), fertilizer use, and crop mix using the quadratic version of the mathematical programming model.

Especially in the case of cassava production, fertilizer application decreases with increasing risk aversion (Figure 5, from left to right). However, risk aversion must not necessarily induce low fertilizer use intensity. For example, beans appear to be produced at higher intensities at high levels of risk aversion, whereas moderate risk aversion induces higher fertilizer use intensity in black pepper production relative to risk neutrality, i.e. absolute risk aversion is 0. This suggests that fertilizer use intensity depends on the farmers' individual degree of risk aversion and the extent to which crop yield variability is affected by fertilizer application.

One implication of these findings is that highly risk-averse farmers will be more reluctant to adopt technologies that require the use of fertilizers in cassava production, while they might be more interested in doing so for the production of bean and perennial cash crops. Interestingly, in the less developed eastern part of the study area, mechanical land preparation has been primarily adopted in combination with bean production, which is in line with this interpretation of the results.

The main conclusion from this section is that current agricultural practices allow farmers to maintain, albeit not improve, their current income level. Nonetheless, key indicators of environmental sustainability, e.g. carbon sequestration and fallow age, decline over time, because they are either of no value to the farmer (carbon) or the related services (nutrients from fallow) can be partly substituted by external inputs such as fertilizers. Substitution with fertilizers is associated with a change in crop mix in favor of fertilizer-dependent annual crops, e.g. beans, or perennial cash crops, e.g. black pepper or passion fruit, depending on the degree of risk aversion of the individual farmer.

6 Indicators of sustainability and mechanical land preparation

This section addresses the second research question by looking at the impact of the introduction of mechanized land preparation. For this purpose, results from simulations with access to mechanical land preparation are compared to the results from the previous section.

Considerable trade-offs exist with respect to the economic and environmental benefits and costs of technology change (Figure 6). It shows the percentage change in the indicators of sustainability relative to the simulation in section 5, if the farm uses fertilization in the slash-and-burn system and has access to mechanical plowing as well as chop-and-mulch at reduced costs.

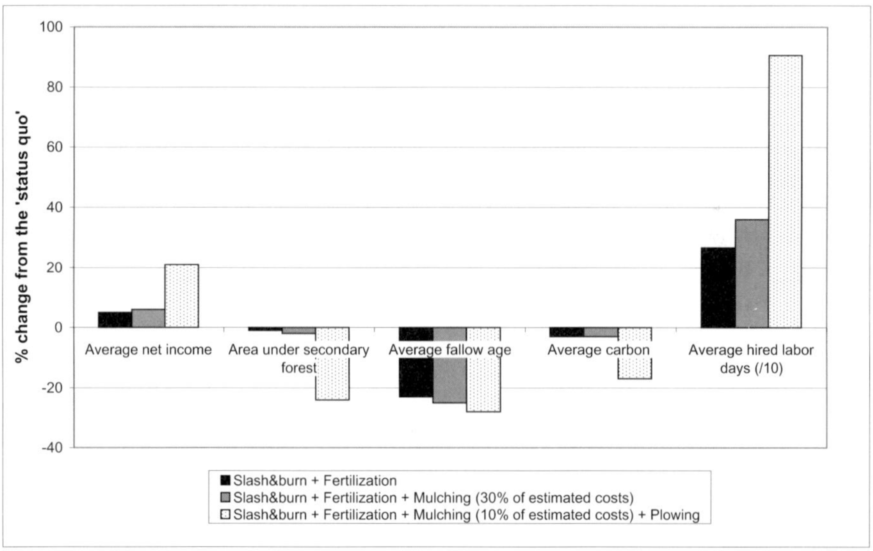

Fig. 6. Percental changes in sustainability indicators relative to the status quo simulation if the farm has access to fertilization and mechanical land preparation technologies. Hired labor days were divided by 10 for expositional reasons, because changes were greater than 100%.

Fertilizer use in combination with mechanical land preparation technologies has clear positive impacts on household income. Especially continuous agriculture with mechanized plowing greatly increases the productivity of both land and labor. This is at the cost of more than 20% of the area under fallow vegetation and 17% of stored carbon relative to the simulation in section 5, as mechanical plowing reduces the valuation of fallow vegetation in comparison to pure fallow-based agriculture (see also section 4).

Mechanical land preparation technologies do not substitute for slash-and-burn. Instead, 53% of the area cleared each year is prepared using slash-and-burn because, in the presence of mechanical land preparation, labor can still profitably be allocated to manual vegetation clearing. Consequently, far more land is prepared than in a scenario without access to mechanical land preparation, which is reflected in a large reduction of secondary forest cover. As shown in section 4, the valuation of labor increases in the presence of mechanical land preparation technologies. This explains that considerably more labor is hired even at wages higher than the average wage in the study region.

In general, the socio-economic analysis is rather disencouraging with regard to mechanical land preparation as a means to improve the environmental sustainability of smallholder agriculture in the study area. However, if available at a service cost around 180 R$ ha^{-1} (market price for plowing and harrowing), both mechanical plowing and chop-and-mulch have the potential to enhance rural welfare and economic development.

At least in the case of mechanical plowing, the example of the western part of the Bragantina region shows that the technology will increasingly be available to smallholders in the course of economic development. The following section will therefore explore how instruments of agro-environmental policy can be used to reduce the negative environmental effects of technological change without neutralizing its positive effects on rural welfare.

7 Policy scenarios for sustainable agriculture

The current set-up of agro-environmental policies in the Amazon region involves policies both for environmental protection and economic growth through technology provision. A very restrictive environmental protection measure, the *Reserva Legal*, requires farms in the Amazon region to protect a large proportion of forested land on their property as a nature reserve. At least in the Bragantina region, the *Reserva Legal* is hardly enforced and model simulations suggest that enforcement would have severe negative income effects for smallholders.

Some local governments in the Bragantina region promote mechanized plowing through the provision of a subsidized machinery service (so-called "*patrulhas mecanizadas*"), which is widely accepted by small and large farms in the respective municipalities.

Proambiente is an upcoming program, which will provide farmers that adopt environmentally friendly production technologies with subsidized credits and lump sum payments for environmental services. Chop-and-mulch is one of the technologies currently considered eligible under the program and, until today, has been offered free of charge to interested farmers.

To answer the third and fourth research questions, policy scenarios essentially addressing the same policy objectives were analyzed using the mathematical programming model. Apart from propositions of alternative policy measures, useful messages for the improvement of existing policy action can be extracted from the results.

7.1 Set-aside payments for the protection of secondary forests

The formulation of the *Reserva Legal* shows that the government has recognized the need to protect forests from conversion into other uses. A more flexible option to conserve primary and secondary forests is a permanent subsidy on areas that are taken out of the agricultural crop/fallow cycle. Contrary to an environmental standard, such as the *Reserva Legal*, a subsidy has the advantage that farms can set aside forests according to their individual resource endowment. For example, setting productive land aside is less costly for large farms than for smaller ones, such that the value of the subsidy can be adjusted depending on farm size. Figure 7 shows that in the presence of mechanical alternatives to slash-and-burn, both carbon sequestration and secondary forest conservation are achieved at lower rates of a hypothetical subsidy. This is a consequence of the reduced valuation of fallow on farms that have access to mechanical land preparation (see section 4).

The set-aside payment increases the value of fallow such that a side-effect is a reduction of the area annually prepared by slash-and-burn by 65% at a payment of 100 R$ ha^{-1} on farms with full technology access. Reduced material damages from accidental fires are another external benefit of these payments.

7.2 Taxing slash-and-burn and providing alternatives

The high positive income effect of mechanical plowing suggests that the local government initiatives to enhance the use of mechanical plowing and harrowing by subsidized service provision overshoot the mark. In fact, if plowing and harrowing is offered at prices that reflect real costs, the private economic gains are still very high. As a result we propose to levy a tax on the use of slash-and-burn on farms that benefit from the availability of machinery at the municipality level. This considerably reduces the profitability of preparing land manually in addition to the land that has been prepared mechanically (Figure 8).

In the tax scenario, hypothetical tax rates from 50 to 500 R$ were tested. Farmers would radically reduce the area annually prepared by slash-and-burn

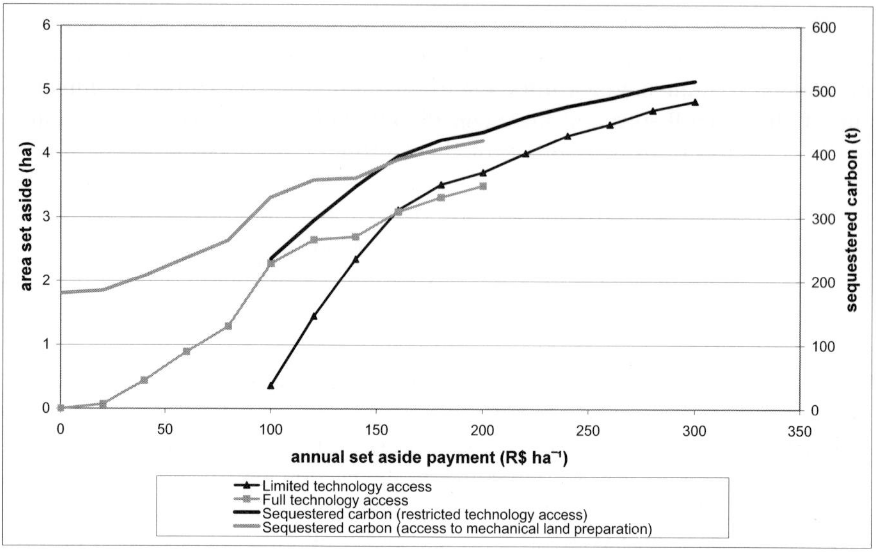

Fig. 7. Set-aside payments and associated carbon sequestration on farms with and without access to mechanical land preparation technologies.

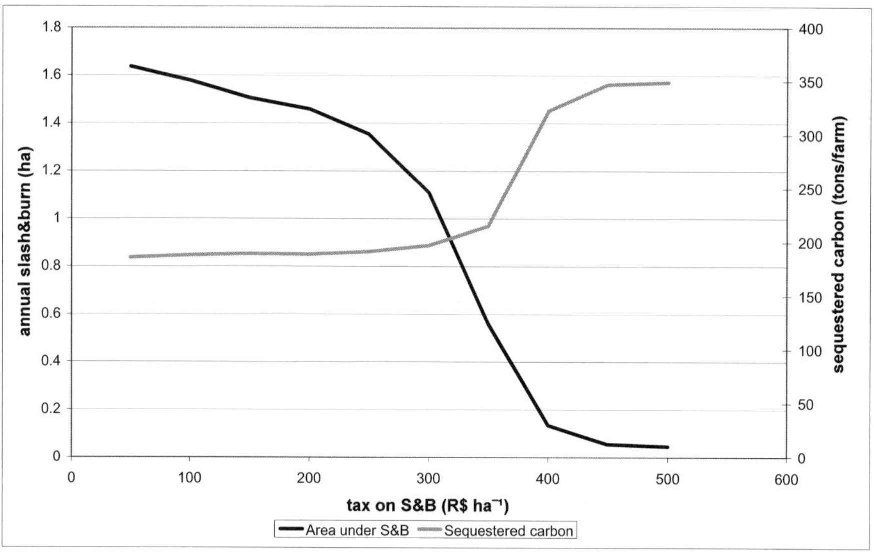

Fig. 8. Tax on slash-and-burn and associated changes in annually prepared land and average carbon storage.

at tax rates between 250 and 400 R$ ha^{-1} (Figure 8). The threshold for considerable increases in carbon sequestration lies at a tax rate of approximately 350 R$ ha^{-1}, while the maximum tax revenue, i.e. area burned multiplied by the tax rate, is generated at a rate of 250 R$ ha^{-1}.

Even with the maximum tax rate of 500 R$ ha^{-1}, the average annual household income (not shown here) is still 4% above the simulation with no access to mechanical land preparation (see section 5).

7.3 A technology-specific crop-yield insurance

The third policy instrument that was tested is a technology-specific crop-yield insurance. The concept of crop-yield insurances is not new to agricultural policy making in Brazil. The *Programa de Garantia da Atividade Agropecuária* (PROAGRO) has been offering yield insurances for credit financed activities since 1976. During the 1980s, much research was devoted to the issue of crop-yield variability and the viability of crop-yield insurances (see for example Hazell et al. 1986 and Anderson and Hazell 1989). At the time, experience with crop-yield insurances was rather negative, mainly due to high transaction costs, moral hazard, and adverse selection. However, since PROAGRO introduced a new premium scheme in 1996 combined with modern harvest monitoring and agricultural zoning systems, transaction costs and moral hazard have been reduced such that PROAGRO is no longer in deficit.

For environmental credit programs, in particular *Proambiente*, it appears sensible to couple crop-yield insurances to specific agricultural technologies, e.g. chop-and-mulch. An insurance is likely to enhance adoption especially by farmers that are reluctant in view of the risks associated with switching from fertilizer-free slash-and-burn to fertilizer-based chop-and-mulch (see also section 4). Moreover, transfers of an insurance-based subsidy are lower than in the case of a full subsidy of mulching costs, because the insurance sets in only if yields are lower than expected. The threshold yield level upon which the insurance comes into effect depends on desired degree of adoption, individual risk aversion, and the development of market prices for agricultural crops. Our simulation experiment is therefore based on an assumed threshold level.

It was assumed that mulching can be offered at a price about 70% lower than the estimated average price of 935 R$ ha^{-1} (see section 4). The insurance is offered free of charge providing farmers with a full compensation of yield losses greater than 10% of the expected yield due to pest, accidental fires, climatic variability, etc.

At medium to high levels of risk aversion, insured chop-and-mulch is adopted at considerably higher rates than uninsured chop-and-mulch (Figure 9). With the insurance, the area under mulch increases at the expense of both slash-and-burn and mechanical plowing, while it leaves the forested area almost unaffected (result not shown). The latter also applies to carbon sequestration, such that the external benefits of a technology-specific crop-yield

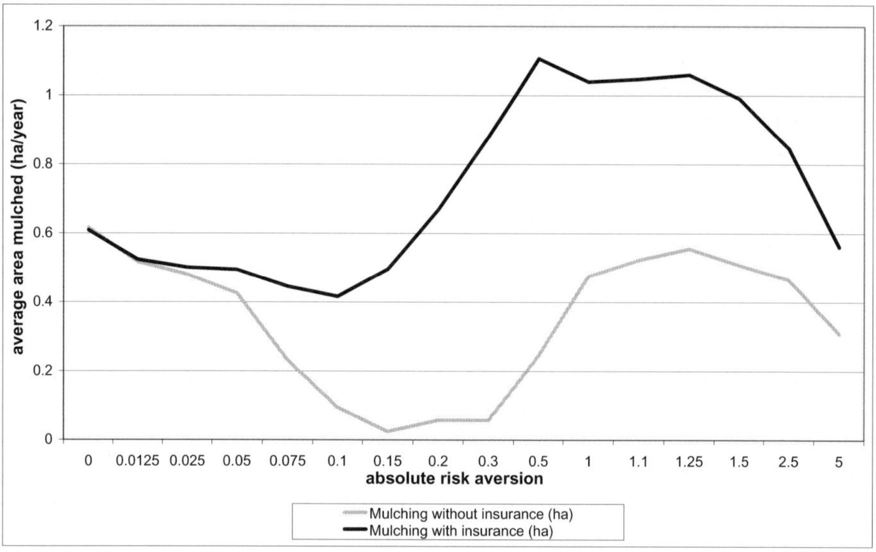

Fig. 9. Average annually chopped area at different degrees of constant absolute risk aversion with and without a technology-specific crop-yield insurance.

insurance are limited to a slight reduction in the negative effects of slash-and-burn, e.g. accidental fires and non-carbon greenhouse gas emissions.

8 Conclusions and implications for policy

We have studied an old colonization region in the eastern Amazon region in Brazil in order to improve our understanding of the interactions between smallholder agriculture and secondary forests in a fallow-based farming system.

Our findings confirm that secondary forests provide a variety of ecological functions relevant for the performance of agricultural crops. In the traditional slash-and-burn system, burning vegetation provides nutrients and fallowing reduces weed pressure. Economically, this is reflected in the very high share of fallowing (22%) in the total profit obtained from annual crops under slash-and-burn. Despite the high value of fallow, repeated slash-and-burn cycles gradually reduce the regeneration capacity of fallow vegetation and remove soil nutrients.

The introduction of mechanized land preparation is becoming more and more popular, but it is dramatically altering the valuation of labor, land, and fallow vegetation at farm level. As a result, more land is cultivated than under slash-and-burn, which also contributes to a decline in the areas covered with secondary forests, the regeneration capacity of fallow vegetation, and the provision of environmental services such as carbon sequestration.

Mechanized chop-and-mulch turned out to be better suited to maintain fallow regeneration capacity and below-ground carbon than plowing and harrowing, as it is not based on the destruction of tree roots. However, service costs need to be reduced by at least 70% to induce adoption by an average smallholder in the study area.

Findings from both economic and ecological approaches indicate that crop performance depends less on the choice of land preparation technology than on fertilizer use. One key determinant of fertilizer use appears to be aversion to production risks. Since the yield variance of some crops, e.g. cassava, increases rapidly with fertilizer application, this can represent an adoption barrier to technologies that depend on fertilizer use.

Three types of policy instruments were analyzed with regard to their effectiveness in promoting secondary forest conservation, farm-household income and the adoption of environmentally friendly technologies. Both a tax on slash-and-burn for farmers that benefit from improved technology access and set-aside payments for forest conservation turned out to be efficient measures in reducing the negative externalities of technological change.

A scenario of technology-specific crop-yield insurance for fields prepared by chop-and-mulch suggests that such a measure would increase the likelihood of adoption. While it would have little impact on forest conservation, it could contribute to reducing damage through accidental fires.

Finally, we found that agriculture at the forest margins must not necessarily result in land abandonment as a consequence of ecological degradation. The example of more than 120 years of smallholder agriculture in the eastern Amazon region of Brazil shows that agriculture and considerable proportions of vital secondary forests can coexist over extended periods of time. Abandonment and migration to new forest margins is thus not a necessary consequence of the deforestation of primary forests for agricultural purposes.

Nevertheless, economic and technological changes represent a challenge for policy makers concerned with the future of agriculture at the present forest margins. Without appropriate action, these areas are likely to be the source of substantial carbon emissions even beyond the removal of the remaining primary forests. Moreover, the long-term ecological consequences of the aggressive use of conventional mechanical land preparation technologies are still unknown and might well be irreversible with regard to the persistence of tree-dominated secondary forests. Our research suggests that efficient instruments exist to reduce the negative environmental effects of technological change while at the same time preserving its economic benefits. These instruments can either complement or substitute existing policy measures, as their elements are not entirely new to the agro-environmental policy environment in the Amazon region.

Acknowledgements
The project was a cooperation of the Center for Development Research (ZEF, University of Bonn, Germany), the Institute for Agricultural Engineering (University of Göttingen, Germany), EMBRAPA Amazônia Oriental (Belém-PA, Brazil), and the Núcleo de Altos Estudos Amazônicos (NAEA, Belém-PA, Brazil), under the Governmental Agreement on Cooperation in the field of Scientific Research and Technological Development between Germany and Brazil. It was financed by the bilateral research program "Studies on Human Impact on Forests and floodplains in the Tropics - SHIFT" (No. 01LT0012 and 01LT0015).

9 References

Alencar A, Nepstad D, McGrath D, Moutinho P, Pacheco P, Carmen M do, Diaz V, Soares Filho BS (2004) Desmatamento na Amazônia: indo além da "Emergência Crônica". IPAM, Belém, Brazil

Anderson JR, Hazell PBR (eds,1989) Variability in Grain Yields. Implications for Agricultural Research and Policy in Developing Countries. John Hopkins University Press, Baltimore and London

Baar R (1997) Vegetationskundliche und –ökologische Untersuchungen der Buschbrache in der Feldumlagewirtschaft im östlichen Amazonasgebiet. Göttinger Beiträge zur Land- und Forstwirtschaft in den Tropen und Subtropen, Heft 121

Berg E (1998) Der Einfluß von Unsicherheit und Risikoeinstellung auf die Intensität und Produktionsprogramm im Ackerbau. Published in Polish. Zagadnienia ekonomiki rolnej 4-5, 37-57

Block A (2005) Göttinger Mähhäcksler Tritucap, und Forstmulcher – Nicht Brennende Flächenvorbereitung am Beispiel der Zona Bragantina, Nord-Ost-Amazonien, Brasilien. Ph.D. Thesis. University of Göttingen

Börner J (2006) A bio-economic model of small-scale farmers' land use decisions and technology choice in the eastern Brazilian Amazon. Ph.D. thesis. University of Bonn (1/6/2006)
at: http://hss.ulb.uni-bonn.de/diss_online/landw_fak/2006/boerner_jan-christoph URN: urn:nbn:de:hbz:5N-07475

Brienza Jr. S (1999) Biomass dynamics of fallow vegetation enriched with leguminous trees in the Eastern Amazon of Brazil. Göttinger Beiträge zur Land- und Forstwirtschaft in den Tropen und Subtropen, Heft 134

Bünemann, E (1998) Einfluß von Mulch und mineralischem Dünger auf *Zea mays* und *Vigna unguiculata* in der Feldumlagewirtschaft Ostamazoniens. Diploma Thesis, University of Göttingen

Chambers RG (1988) Applied Production Analysis: A Dual Approach. Cambridge University Press, Cambridge

Denich M (1989) Untersuchungen zur Bedeutung junger Sekundärvegetation für die Nutzungssystemproduktivität im östlichen Amazonasgebiet, Brasilien. Göttinger Beiträge zur Land- und Forstwirtschaft in den Tropen und Subtropen. Heft 46

Denich M, Vielhauer K, Kato MS de A, Block A, Kato OR, Sá TD de Abreu, Lücke W, Vlek PLG (2004) Mechanized land preparation in forest-based fallow systems: The experience from Eastern Amazonia. Agroforestry Systems 61, 91-106

Denich M, Vlek PLG, Sá TD de Abreu, Vielhauer K, Lücke W (2005): A research concept for the development of fire-free fallow management techniques in the Eastern Amazon region, Brazil. Agriculture, Ecosystems and Environment 110 (1-2), 43-58

Diewert WE (1973): Functional Forms for Profit and Transformation Functions. Jounal of Economic Theory 6, 284-316

FAO (2003): State of the World's Forest. Rome (1/1/2006) at: http://www.fao.org/

Fearnside PM (1986) Human Carrying Capacity of the Brazilian Rainforest. Columbia University Press, New York

Hardaker JB, Huirne RBM, Anderson JR, Lien G (eds, 2004) Coping with Risk in Agriculture (2nd edition). CABI Publishing, Oxfordshire Cambridge

Hazell PBR, Norton RD (1986) Mathematical Programming for Economic Analysis in Agriculture. Macmillan Publishing Company, New York

Hazell PBR, Pomareda C, Valdéz A (1986) Crop Insurance for Agricultural Development: Issues and Experience. John Hopkins University Press, Baltimore and London

Hedden-Dunkhorst B, Denich M, Vielhauer K, Sousa Filho FR de, Sá TD de Abreu, Hurtienne T, Costa F de Assis, Mendoza Escalante A, Börner J (2003): Forest-Based Fallow Systems as a Safety Net for Smallholders in the Eastern Amazon. Paper presented at International Conference on Rural Livelihoods, Forests and Biodiversity 19-23 May, Bonn, Germany at:http://www.cifor.cgiar.org/publications/corporate/cd-roms/bonn-proc/pdfs/papers/T1_FINAL_Hedden-Dunkhorst.pdf (verified April 4, 2006)

Hölscher D (1994) Wasser- und Stoffhaushalt eines Agrarökosystems mit Waldbrache im östlichen Amazonasgebiet. Göttinger Beiträge zur Land- und Forstwirtschaft in den Tropen und Subtropen, Heft 106

Hurtienne T (2001) Agricultura Familiar e Desenvolvimento Rural Sustentável na Amazônia. In: Coelho CMN, Castro E, Mathis A, Hurtienne T (eds): Estado e Políticas Públicas na Amazônia. Editora Cejup Ltda, Belém-PA

IBGE (1998): Censo Agropecuário 1995/6 (1/1/2004) at: http://www.ibge.gov.br/

IBGE (2000): Censo Demográfico 2000 (1/1/2004) at: http://www.ibge.gov.br/

Intriligator M (1971) Mathematical Optimization and Economic Theory. Prentice Hall, Englewood Cliffs

Kato MS de Andrade (1998a) Fire-free land preparation as an alternative to slash-and-burn agriculture in the Bragantina region, Eastern Amazon: Crop performance and phosphorus dynamics. Ph.D. Thesis. Cuvillier Verlag, Göttingen

Kato OR (1998b) Fire-free land preparation as an alternative to slash-and-burn agriculture in the Bragantina region, Eastern Amazon: Crop performance and nitrogen dynamics. Ph.D. Thesis. Cuvillier Verlag, Göttingen

Kato MS de Andrade, Kato OR, Denich M, Vlek PLG (1999) Fire-free alternatives to slash-and burn for shifting cultivation in the Eastern Amazon region: The role of fertilizers. Field Crops Research 62, 225-237

Mackensen J, Hölscher D, Klinge R, Folster H (1996) Nutrient transfer to the atmosphere by burning of debris in eastern Amazonia. Forest Ecology and Management 86(1), 121-128

Mendoza-Escalante A (2005) Analysis of Smallholder Agricultural Production in the Eastern Amazon: Empirical Evidence and Policy Prospects for the Bragantina Region. Ph.D. Thesis. Cuvillier Verlag, Göttingen

Nepstad DA, Moreira A, Alencar A (1999) Flames in the Rainforest: Origins, Impacts and Alternatives to Amazon Fire. Pilot Program for the Conservation of the Rainforests of Brazil, World Bank, Washington

Nunez JBH (1995) Fitomassa e estoque de bioelementos das diversas fases da vegetação secundária, provenientes de diferentes sistemas de uso da terra no nordeste paraense, Brasil. Master Thesis, Federal University of Pará, Belém

Penteado AR (1967) Problemas de colonização e uso da terra na região Bragantina do Estado do Pará, I+II. Ph.D. Thesis. Universidade Federal do Pará, Belém

Sahn DE, Stifel D (2000) Poverty Comparisons Over Time and Across Countries in Africa. World Development 28, 2123-2155

Schmitt D (1997) Untersuchungen über nicht-destruktive Methoden zur Bestimmung von Biomasse junger Sekundärvegetation im östlichen Amazonien. Diploma Thesis. Institute of Agronomy in the Tropics, University of Göttingen

Sommer R (2000) Water and nutrient balance in deep soils under shifting cultivation with and without burning in the Eastern Amazon. Ph.D. Thesis. Cuvillier Verlag, Göttingen

Sommer R, Denich M, Vlek PG (2000) Carbon storage and root penetration in deep soils under small-farmer land-use systems in the eastern Amazon region, Brazil. Plant and Soil 219, 231-241

Tippmann R (2000) Assessment of Carbon Sequestration in Landscape under the Clean Development Mechanism of the Kyoto Protocol. Diploma Thesis, University of Bonn

United Nations (2000) United Nations Millennium Declaration. (1/1/2006) at: http://www.un.org/millenniumgoals/

Vieira IC (1996) Forest Succession after Shifting Cultivation in Eastern Amazonia. Ph.D. Thesis. University of Stirling

Vosti SA, Reardon T (eds, 1997) Sustainability, Growth, and Poverty Alleviation. John Hopkins University Press, Baltimore

Vosti SA, Carpentier CL, Witcover J (2002) Agricultural intensification by smallholders in the western Brazilian Amazon: From deforestation to sustainable land use. RR 130. International Food Policy Research Institute, Washington

Walker R (2000) Deforestation and Cattle Ranching in the Brazilian Amazon: External Capital and Household Processes. World Development 28 (4), 683-699

Wickel B (2004) Water and nutrient dynamics of a humid tropical agricultural watershed in Eastern Amazonia. Ecology and Development Series 21. Cuvillier Verlag, Göttingen

Wiesenmüller, J. (1999) Einfluß landwirtschaftlicher Flächenvorbereitung auf die Dynamik des Wurzelsystems und die oberirdische Regeneration der Sekundärvegetation Ostamazoniens, Pará, Brasilien. PhD Thesis. University of Göttingen

Protected Area Management and Local Benefits – A case study from Madagascar

Klas Sander[1*] and Manfred Zeller[2]

[1] Forest Economist, Environment Department, The World Bank, 1818 H Street NW, Washington DC, USA
[2] Institute of Rural Development, University of Goettingen, Waldweg 26, 37073 Goettingen, Germany

*corresponding author: Klas Sander, Email: Klas.Sander@web.de

Summary

This paper analyzes the comparative advantage of alternative forest management policies taking into account the subjective welfare judgment of individual rural households. The methodological approach chosen for this study is a financial farm household analysis combining both, economic data at the household level and natural science data of forest resources obtained in a research area in the Northwest of Madagascar. This interdisciplinary approach takes account of the interdependence between ecosystem dynamics and economic decision-making processes. By applying a poverty index, it is possible to make a further distinction between groups of rural households that are inherently different with regard to their economic potential and activities. The results of this study demonstrate that, due to their livelihood strategies, it is especially the poorest households that suffer most from a strict conservation approach, while better-off households would benefit more due to an improved provision of indirect forest services, particularly watershed protection. We also analyze the potential for compensating opportunity costs of restricted forest utilization through international transfer payments and the potential of implementing economic incentives for afforestation at the household level. In this context, we extend our analysis beyond mere economic aspects of resource

The research was funded by the Tropical Ecology Support Program (TÖB) of the German Agency for Technical Cooperation (GTZ). However, the views and opinions expressed in this paper as well as any errors or omissions remain the sole responsibility of the authors. This paper resulted from Mr. Sander's PhD research at the Institute of Rural Development, University of Göttingen.

Tscharntke T, Leuschner C, Zeller M, Guhardja E, Bidin A (eds), The stability of tropical rainforest margins, linking ecological, economic and social constraints of land use and conservation, Springer Berlin 2007, pp 365-387

management and provide further information how alternative management strategies can meet the increasing direct demand on forest resources, while also achieving the objective of biodiversity conservation. Lastly, the potential of forest resources to contribute to local and regional economic development in developing countries is discussed.

Keywords: Africa, Madagascar, forest management, financial farm household analysis, charcoal production, poverty alleviation

1 Introduction

The ongoing destruction of global forest resources and its negative ecologic, economic, and social consequences is a challenge to the global community. Since many of the goods and services provided by forests are public goods, market mechanisms generally fail to provide the socially optimal amount of these goods and services. As a result, governments and other actors are frequently called in to design appropriate policies and programs to correct for market failure. In this context, the most commonly proposed policy approach is the creation of protected areas (PA). This policy strategy is often based on macro-economic analyses, which – representing the viewpoint of a social planner – emphasize the considerable economic benefits that can be gained from the conservation of forests compared to alternative, extractive management options of natural resources (Bruner et al. 2001, Pagiola et al. 2002, Pearce and Moran 1994, IIED 2003). The monetary values that are applied in these analyses are commonly obtained from studies using hypothetical markets – at the national and global level. When the success of such conservation strategies is measured against progress made in reducing the actual rate of deforestation, but not against indicators of poverty alleviation, it is also difficult to assess the contributions of such policies to achieve the objectives expressed in the Millennium Development Goals.

In contrast, it is the purpose of this study to carry out a financial analysis of alternative forest management strategies from the viewpoint of a rural household. We apply data from field surveys executed in the Northwest of Madagascar. We follow an analytical approach as formulated by Lutz et al. (1994, p.276) who emphasized that farmers consider only costs and benefits that actually accrue to them from the decisions they make about how to use their resources. They value the costs and benefits at the prices they actually face and decisions about land use are ultimately made by the farmers themselves and not by social planners or government agencies. Consequently, we did not consider any social values such as biodiversity or climate regulation in our analysis. Alternatively, the research project placed significant interest on a detailed exploration of the complex socio-economic and socio-cultural environment framing the decision-making process of natural resource utilization by rural households. However, this analysis not only considers economic aspects of resource management at the micro-level, but combines them with natural science data of forest resources to analyze the interdependence of ecosystem dynamics and economic decision-making processes. This extends the analysis beyond mere economic aspects of forest utilization and it provides further information how alternative management strategies can meet the increasing direct demand for forest resources while also achieving the provision of global public goods. Even though this study seems to be rather case study specific at a first glance, we believe that our results have implications for forest management strategies beyond the country level, especially when a large share of the population directly depends on forest resources for their livelihoods.

The structure of this paper is as follows: Following, the current policy framework of forest resources management in Madagascar is briefly described. Section 3 looks at the conceptual framework and the data. Descriptive statistics of the household and forest ecosystems are discussed in Section 4. Part 5 is devoted to a discussion of the modeling procedure applied in the analysis. The results of the financial analysis are presented in Section 5. We finish with conclusions in Section 6.

2 The Policy Framework of Forest Management in Madagascar

Madagascar's biotic resources are characterized by an extremely high level of endemism of plants and animals (EU 1999; Ganzhorn & Sorg 1996). 90% of its 250 species of reptiles, 29 of its lemur species and 80% of its plant species are unique to the island (O'Connor 1996). Thus, Madagascar is often classified as a megadiversity country of highest conservation priority representing one of the most important reservoirs for biological diversity (Larson 1994).

However, deforestation is one of the biggest problems in Madagascar. Shifting agriculture and the collection of woodfuel[1] as the main source of energy are among the main factors driving deforestation (World Bank 2004). According to recent estimates (Steiniger et al. 2003), average annual deforestation lies around 0.86% for Madagascar as a whole. However, these numbers may increase to up to 7-8% if small areas are analyzed individually (Ackermann 2003, Oberlé 2001). Over the past decades Madagascar has fallen deeper into poverty, with its GDP per capita declining from US$ 383 to US$ 246 between 1960 and today (UNDP 2001, World Bank 2003a and 2003b). Annual population growth rates reported for Madagascar are generally beyond the African average and they vary between 2.7% and 3.8%. 70% of the total population lives in rural areas.

As a response to the problem of deforestation, the Government of Madagascar (GoM) has adopted a strict conservation policy with biodiversity conservation as the guiding principle. Since 1991, the GoM has invested about

[1] Woodfuel encompasses both firewood and charcoal

US$ 75 million in the establishment of a protected area (PA) network (Carret and Loyer 2003). As of 2003, the PA network already consisted of 18 national parks, 5 "integrated" reserves, 23 special reserves, and 2 marine reserves (World Bank 2003). The GoM has the objective to increase areas under protection from currently 1.7 million hectares to 6,0 million hectares (World Bank 2004). This objective is complemented by the goal to improve institutional efficiency and accountability, to facilitate dedicated efforts to improve sector governance, and to support law enforcement efforts in the natural resource management sector.

It is argued that the conservation approach to forest resource management benefits especially the poorest people as opposed to making use of its extractive production potential. This argument is supported by several macroeconomic studies that focus on forest values such as biodiversity, recreation, and watershed protection (Kramer et al. 1995, Carret and Loyer 2003). The monetary values that are applied in these analyses are often obtained from studies using hypothetical markets – at the national and global level – but so far without being able to translate these economic values into real resource flows, i.e. compensation payments that target the opportunity costs faced at the micro-level.

In contrast to the ambitious objectives of the GoM, it remains unclear how increasing management tasks and development programs can jointly be financed in a sustainable manner. At present, only 7% of management costs can be refinanced through park fees, while 70 – 80% are contributed by international donors and 15 – 20% by the GoM. This casts some uncertainty on the plan to generate compensation payments from a 50% share of park entrance fees. In addition, Carret and Loyer (2003) estimate that management costs of PA will increase from US$ 2.5 per hectare per year to about US$ 5.0 per hectare per year.

3 Conceptual Framework and Research Design

3.1 Conceptual framework

We follow a conceptual framework as presented in Figure 1. It represents the economic decision-making process of a farm household (HH) in a two period model with time periods T and $T+n$. It assumes that farm households can generate income through returns in their physical or working capital including land (K), labor (L), human capital (H), and social capital (S). Therefore, in period T the income generation process of household i is given by

$$Q_T^i = f(K_T^i, L_T^i, H_T^i, S_T^i), T = 1, \ldots, n \tag{1}$$

Forest resources are not privately owned by individual households, but they are included as physical capital K in the production function of the individual household i, because in Madagascar – as often observed in developing countries – property rights over forest resources are poorly defined leading to *de facto* open access resources.

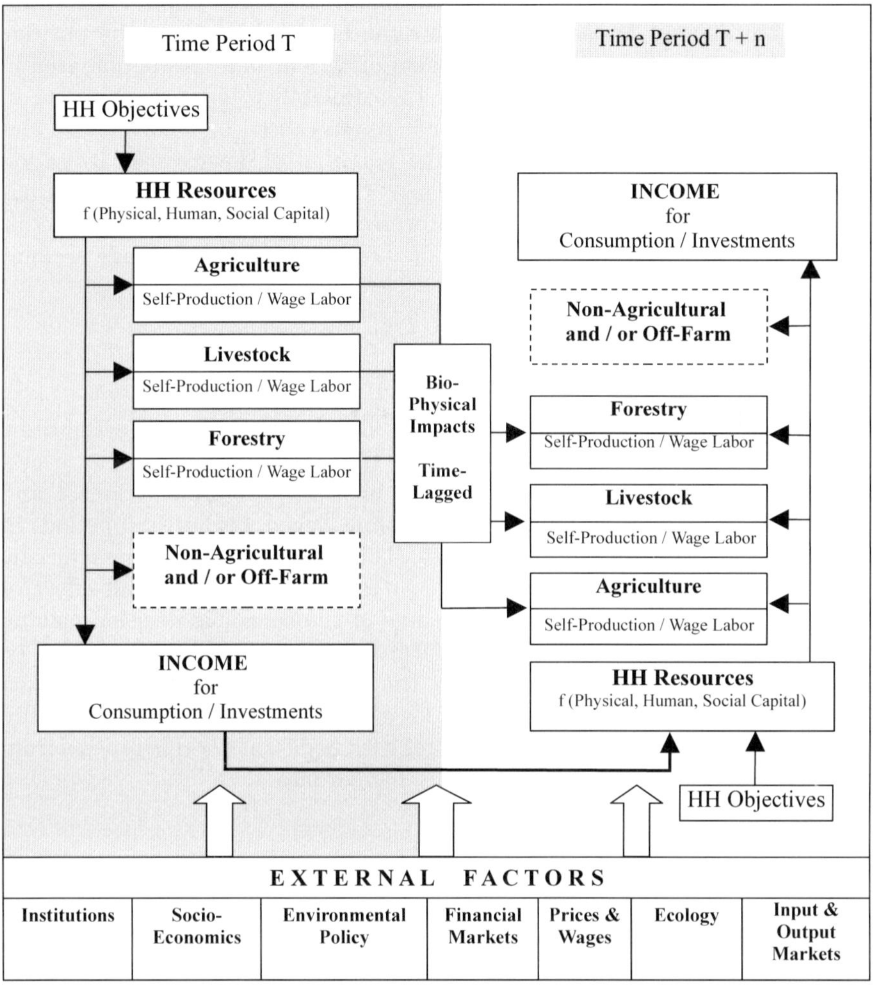

Fig. 1. The Decision-Making Framework of a Rural Household in a Two-Period Model

In each time period, this income generation process is constrained by external factors, such as the production potential of soils, the ecological characteristics of forest resources, the socio-cultural environment, input and output markets, and – most importantly in the context of this study – the political framework. As these external factors change, so does the strategy of each individual household i to generate income.

Furthermore, we assume in our model that in later periods time-lagged biophysical impacts have an effect on physical capital and, thus, income generation processes of individual households. Such time-lagged impacts are either created by different households or the same household and can be positive or negative. As an example, the extension of agricultural production area in a watershed may negatively influence the availability of forest products in the subsequent time period, or increased erosion caused by deforestation and increased cattle breeding may cause the potential of downstream irrigation agriculture to decline in the future. These time-lagged impacts become additional factors in future production processes, playing an important role for the economic decision-making process of a rural household. We take account of this by extending equation (1) into

$$Q^i_{T+n} = f(K^i_{T+n}(Q^{\sum i}_T), L^i_{T+n}, H^i_{T+n}, S^i_{T+n}). \tag{2}$$

Q^i_T represents production processes of all households living in the same area in any preceding time period T. These time lagged impacts do not necessarily have to be immediate, i.e. occurring in subsequent time periods T and T+1 or T+23 and T+24, but can also be time-lagged with several periods in between, e.g. in period T+2 and T+15. This depends on the characteristic of the physical capital and the nature of the impact. The overall effect of those time-lagged impacts on the outcome of the economic decision-making process heavily depends on how such future economic events in periods $T{+}n$ are valued by individuals in the present time period T. This problem is taken account of by applying a discount rate for calculating present values of future net benefit or income streams. Therefore, with equation (2) this process from the viewpoint of individual household i changes into

$$PV^1 = Q^i_T + \sum Q^i_{T+n}(1+r)^{-(T+n)} \tag{3}$$

From (3) the considerable influence of discounting behavior on the decision making processes of individual households can be seen. With increasing discount rates, future economic events – and, thus, time lagged economic effects influencing these events – are valued less compared to present events.

Following this model, individual households compare the present value of alternative income generation processes and they choose the strategy that provides them with the highest PV given their specific economic constraints.

3.2 Data

The research area is located in the region of Marovoay in the Northwest of Madagascar. 479 households were randomly selected from two communities representing a sampling intensity of about 15% for each community. A standardized questionnaire was applied to collect detailed data on all economic activities of the households over the past 12 months. Complementary questions about migrational aspects were also included. The household survey was executed from June to August 2001.

The sample was stratified according to the geographical location of households with regard to a irrigated rice production area. The stratum for households living in or near this area is labeled "RPA", while the stratum for households living further away is labeled "Periphery".

The households of the sample were further categorized into three poverty groups – the poorest, the less poor, and the better-off. This categorization was accomplished by creating a poverty index using Principal Component Analysis (PCA) as outlined by Henry et al. (2003). The variables that were identified by PCA with significant factor loadings are: Expenses for Clothing[2], Family Size, Number of Children (age < 14), House Type, Ownership of Sewing Machine, Number of Lamabe[3] owned, Ownership of Ox-Cart, Ownership of Radio, Total Value of Livestock owned, and Surface of all Rice Field owned. This additional stratification does not refer to the absolute poverty of households, but only allows distinguishing between relative poverty groups of households living in the research area. The relevance for considering multidimensional aspects of poverty in the context of forest management is discussed by Angelsen and Wunder (2003).

For modeling the ecological-economic interdependence of the savannah ecosystem, circular inventory plots with a diameter of 30m were used for the forestry inventory. Data containing information on the diameter at breast height (dbh), the height, the local name, and its local uses was obtained for all trees with a dbh exceeding 5cm. In addition, data on frequency of tree stumps was recorded to analyze the anthropogenic impact of wood use. In total, 71 plots corresponding to an inventory area of about 20 hectares was covered. For other forest ecosystems, secondary data obtained from Ackermann (2003) and Lopez (2003) were applied in the modeling process.

4 The Socio-Economic and Ecological Framework

The economic potential of the study area is constituted by the second largest rice production area in Madagascar encompassing a total of about 20,000 hectares of irrigated rice production. In the south, the region is delimited by the Ankarafantsika National Park with a total area of approximately 130,000 hectares.

Applying a purchasing power parity exchange rate, about 52% of all households live below the internationally defined poverty line of US$ 1 per person

[2] This is the benchmark indicator for the PCA

[3] A mattress that is locally produced from natural fibers used during the harvest period for rice threshing

per day. If the subsistence use of forest products is valued at shadow prices, this number decreases to about 42%. With regard to relative poverty, Table 1 shows that even though relative poverty groups were established using multidimensional aspects of poverty, these categorization also reflects well differences in total income – adding subsistence as well as cash income – with the poorest households generating about 50% less income compared to better-off households.

Table 1. Mean total income per household per year

		Mean Total Income per HH	
		FMG = Franc Malagasy	PPP-US$
Rice production area	Poorest	2,930,989	1,288
	Less Poor	3,817,733	1,678
	Better Off	5,981,492	2,629
	Sub-Total	4,247,358	1,867
Periphery	Poorest	2,586,398	1,137
	Less Poor	4,033,110	1,773
	Better Off	4,905,830	2,156
	Sub-Total	3,841,779	1,689
Total		4,122,890	1,812

Source: Own Household Survey 2001; PPP-US$ (Worldbank 2003); US$1 = 2275 FMG

Demographic trends in the research area are characterized by high population growth rates, with two main factors contributing to this development: a) significant migration dynamics into the area, and b) natural reproduction rates beyond the country's average due to the high share of young families moving to the area. Migration dynamics are highest in the periphery of the RPA, with 39% of all households having actively migrated into the area compared to 30% of households living near the RPA. In total, about 50% of all households consider themselves as being migrants[4]. More than one-third of all migrants have arrived in the study area in the decade prior to the survey and another third in the period 1980-90. Educational levels among households are very low, with the poorest households having almost never received any kind of formal education, compared to about 85% of the better-off households. Furthermore, levels of education are significantly lower for household living in the periphery of the RPA.

Even though agricultural activities in the research area are dominated by rice production, significant differences exist between the two geographical

[4] It has to be noticed that individuals in Madagascar commonly define their migrational status with regard to the location of the family grave. Consequently, even persons that were born in the area where they live may still consider themselves as a migrant.

strata. Households near the RPA focus on rice production, while households in the periphery concentrate on dry farming systems converting savannah and forest land into agricultural production area. Quantities of rice available to a household are on average about twice as high near the RPA, while the quantity of dry agricultural products per household in the periphery exceeds those near the RPA about ten times. Comparing relative poverty groups, the better-off households near the RPA have about twice as much rice available than the poorest households.

In contrast to their lower agricultural production potential, households living in the periphery are characterized by intensive livestock breeding activities – particularly cattle. On average, the number of cattle owned by households in the periphery exceed those owned by household living near the RPA by 46%. This aspect represents an important finding for the modeling process of the financial farm household analysis, because if cattle grazing goes beyond the carrying capacity of the ecosystem in the periphery of the rice perimeter, negative effects on agricultural production, e.g. erosion, have to be expected in future time periods.

Due to relatively high market integration, households in the research area also have access to non-agricultural and off-farm income sources. Agricultural labor represents the main source for this type of income, with increasing importance the poorer households are. Especially the poorest households living near the RPA depend on this source of income.

The utilization of forest resources contributes significantly to the livelihood of the households in the research area, mainly for satisfying energy consumption needs in form of fuelwood and charcoal and for construction purposes. HH in the periphery – especially the poorest – depend on forest products more compared to HH living near the RPA. Over the past year before the survey, 11% and 16% of households living near the RPA and in the periphery respectively used wood from nearby forest resources for construction purposes, e.g. housing and cattle parks. Only 4% of all households bought construction wood on local markets. Wood based energy sources are used by all households, with no household possessing the equipment for using alternative energy sources for cooking[5]. Charcoal – which is exclusively sold on the market – is produced by only 2% of the households near the RPA compared to 12% in the periphery.

Other forest based activities, such as the collection of honey, edible plants, roots, etc. are only of minor importance, except for the collection of wild fruits, which are collected by 47% of all households. In contrast, 43% of households near the RPA also harvest cultivated fruits, compared to only 26% in the periphery. Medicinal plants are collected by 10% of households in the periphery, but only 3% of households living near the RPA use these products, with poorer households following this activity more than better-off households. Not all of these minor uses take place in forests, but are frequently observed in the

[5] It is estimated that about 80% of all energy consumed in Madagascar is coming from wood, i.e. fuelwood and charcoal (World Bank 2004).

periphery of forests and also near fields and on river banks. Therefore, it may be spurious to link the economic importance of these subsistence activities exclusively to forest resources.

Table 2. Energy Consumption Patterns among Households in the Research Area [N = 477]

		collect FW	collect FW & buy CC	buy FW	buy FW & CC	buy CC
Rice production area	Poorest	80.0%		7.3%	1.8%	10:9%
	Less Poor	64.5%	11.8%	9.1%	8.2%	6.4%
	Better Off	61.8%	15.5%	9.1%	6.4%	7.3%
	Sub-Total	68.8%	9.1%	8.5%	5.5%	8.2%
Periphery	Poorest	100.0%				
	Less Poor	95.9%	2.0%			2.0%
	Better Off	91.8%	6.1%			2.0%
	Sub-Total	95.9%	2.7%			1.4%
Total		77.1%	7.1%	5.9%	3.8%	6.1%
Annual quantity of FW collected per capita [m³]		1.50	0.78	0.03	0.02	0.00

Source: Own Household Survey; FW = Fuelwood, CC = Charcoal

In contrast, the energy consumption patterns of households require further analysis to evaluate their impacts on the utilization of forest resources in the research area. Despite the above mentioned fact that all households rely on wood to satisfy their consumption demands for energy, only 79% of the households living near the rice perimeter collect fuelwood themselves compared to 98% of households in the periphery. The poorer households are the more they depend on this type of subsistence activity. As Table 2 indicates, HH near the RPA buy wood based energy more frequently on the market compared to households living in the periphery. This can be attributed to increasing opportunity costs of fuelwood collection. About 85% of household near the RPA judge that collecting fuelwood has become more difficult over the past year, compared to 67% in the periphery. According to the opinions of the respondents in the survey, more than 50% attribute that decline to the increase in population, while about 15% think that the establishment of the adjacent National Park leads to decreasing availability of forest resources. On average, mean collection time is more than ninety minutes per collection for households near the RPA, while households in the periphery only allocate about fifty minutes for this activity if fuelwood is collected on a daily basis. The time per collection increases with lower frequencies of collection. Similarly, the distance to the place of collection increases with lower frequencies of collection. Fuelwood collection is predominantly carried out by the male members of the household. More than 10% of households collect fuelwood within the national park, while about two-third collect in its vicinity.

Lastly, credit transactions have to be analyzed for estimating discount rates applied in the subsequent financial analysis. For the surveyed household, credit transactions in cash and kind have mean interest rates of about 70%. The highest interest rates were observed for credits taken for immediate health expenses. Loan periods are short, averaging only 6.3 months and never exceeding 14 months. This indicates that credits are mainly used for balancing consumption expenses over the year, but hardly for financing any productive investments.

5 The Financial Farm Household Analysis

For modeling the financial costs and benefits of farm households under alternative policy scenarios, those policy options had to be identified first. According to the analysis of current policy trends in Madagascar, three scenarios were identified and formulated for the analysis as presented in Table 3.

Table 3. Description of Scenarios analyzed in the Financial Analysis

Base Scenario	Policy Option 1	Policy Option 2	Policy Option 3
No policy intervention at all; all forest resources are accessible; land-use dynamics may lead to a full conversion of forest land for agricultural production	Strict conservation strategy of all forest resources; only savannah is accessible to satisfy demand for wood products and for extending agricultural production area	This option is identical to PO 1; in addition, compensation is paid to households as 50% of revenues generated from entry fees paid to the National Park	Economic incentives for tree-planting activities are created; only access to closed forests is restricted; 50% of all HH that are living in the research area after 30 years are assumed to participate in tree-planting; on average 0.5ha are assumed to be planted

From the viewpoint of the global social planner, the base scenario results in significant social costs i.e. the loss of biodiversity, recreational value, watershed protection, climatic regulation, and CO_2 sequestration. These costs are not considered for our specific model unless directly impacting farm HH production as specified below.

In contrast, Policy Option (PO) 1 and 2 represent the most preferable alternatives for managing forest resources from a non-local[6] social point of

[6] Non-local encompasses indirect economic values at the regional (e.g. watershed protection) as well as global (e.g. biodiversity) level.

view. The forest cover in the research area remains at 24%, representing the ecosystems that are currently stratified as closed and degraded forests. High non-local benefits are expected to be derived from the conservation of biodiversity, the protection of watersheds, climatic regulation, carbon sequestration, and maintaining the recreational potential of forest resources in Madagascar.

Under PO3 the model results in a final forest cover at the end of the simulation period of 11%. However, this excludes the area under plantation forestry, which is highly variable depending on the number of HH that engage in afforestation and the average area planted. This policy alternative represents a compromise from the viewpoint of the social planner, since only a very minimum area under natural forest is conserved and, consequently, non-local social benefits are reduced compared to PO1 and PO2. For reasons of simplicity it is assumed in our model that even the afforestation efforts with exotic, fast-growing tree species cannot off-set these effects and, hence, we only concentrate on their contribution to satisfy household demands for wood products, i.e. construction timber and wood based energy.

The biophysical interactions of model components are taken account of in our analysis as presented in Figure 2. For our analysis we developed a dynamic simulation model using MS-Excel software applications. Population growth, growth rates of forest ecosystems, and visitor rates are considered as exogenous variables in our model. An increasing number of households extend their agricultural production area, which leads to the conversion of forests. The absolute number of households that could eventually live in the research area was calculated as a function of accessible forest area and the mean amount of agricultural production area per household. Accordingly, deforestation rates were calculated as a function of initial accessible forest area and population growth rates. It was assumed that only dry farming systems were extended since the establishment of irrigation farming systems requires certain terrestrial conditions that are already fully exploited.

In addition to absolute forest land being converted into agricultural production area, the relative proportion of different forest ecosystems with regard to total forest area was also assumed to change, with natural closed forests being converted into degraded forests, and degraded forests being converted into savannah. The re-growth potential of forest ecosystems was equally considered in the modeling process (see Table 4). On the basis of this data, the benefits and costs of wood consumption of local households was calculated with regard to decreasing wood volumes available from accessible forest resources, while the availability of non-wood products was related to the spatial decline of forest cover.

Three different types of forest ecosystems can be categorized for the research area: a) closed natural dry forests (13%), b) degraded natural dry forests (15%), and c) savannah woodlands (72%). Considering the remarkably high frequency of about 360 tree stumps per ha in the savannah woodland, it can be concluded that the natural wood stock would be much higher without human harvest activities.

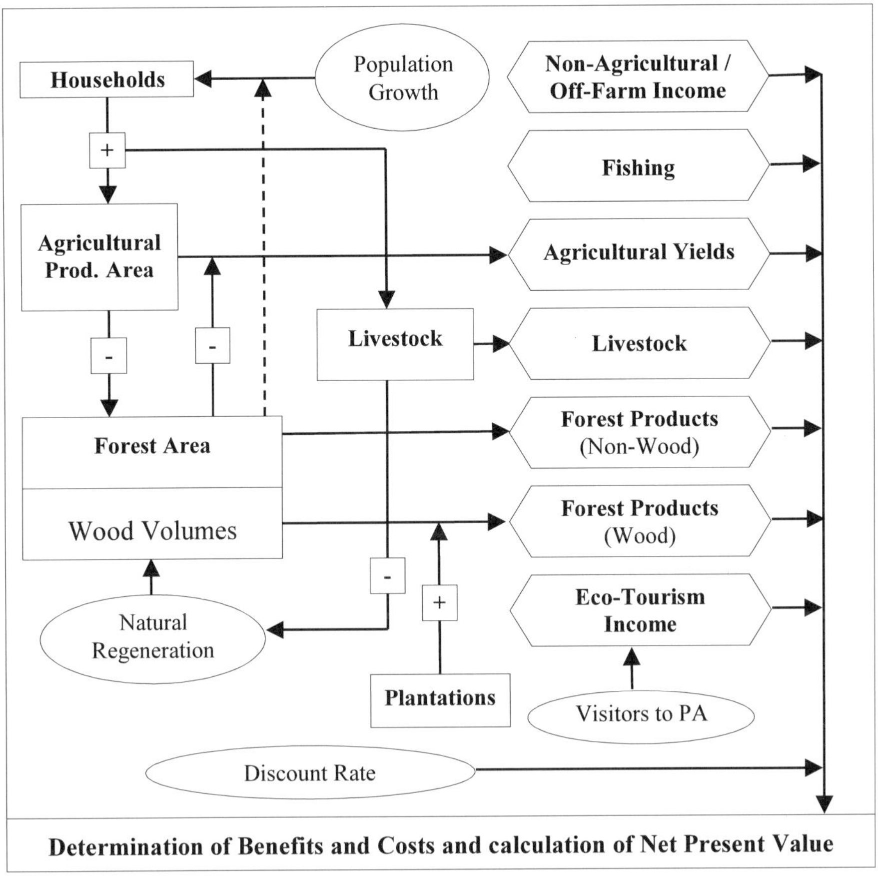

Fig. 2. Interactions between Model Components

To establish a link between different rates of deforestation under alternative policy options and their impact on agricultural production levels, point estimates were obtained from production function analyses recently carried out by other research studies in Madagascar. For example, one of the production function analyses reports that a doubling of deforestation in a watershed from 26% to 52% would cause production levels of irrigated rice to decline by about 8% (Minten et al. 2002). Furthermore, Raharinjanahary (2004) calculated a reduction of agricultural production levels around 40% when fields are affected by erosion and sedimentation. With this data, a logistic function was established for calculating reduction factors of agricultural production with regard to the percentage of forest cover remaining in the research area. Table 5 summarizes the underlying equations of socio-economic and biophysical interdependences on which the model is based.

Table 4. Volume and Growth Rates of Forest Ecosystems

Forest Type	Volume [m^3/ha]	Growth Rate with no Disturbance [m^3/ha/a]	Source
Forest Closed	87.2	2.4	LOPEZ 2003
Forest Degraded	25.7	1.2	LOPEZ 2003
Savannah Woodland	5.1	1.0	Own Inventory Data 2003 Clement 1982

According to data presented by Ackermann (2003), the negative effects of cattle grazing on deforestation and, consequently, agricultural production, was modeled by calculating an overgrazing factor. As the overgrazing factor increases, the re-growth in m^3/ha/a declines and may even go down to zero.

For modeling afforestation activities by households, data was obtained from projects that work in the field of afforestation at the household level (Herzog 2001, PLAE 2003, Sepp 2003). It was assumed that the main input such as plants and other technical equipment is provided free of charge to the households, while households only have to contribute labor.

A modeling period of 30 years was chosen for the analysis. We consider this period as appropriate, because a shorter period of analysis would not sufficiently account for future benefits or costs, e.g. benefits from plantation forestry or costs attributable to deforestation dynamics, while a longer period of analysis is judged unnecessary due to the ambiguity for predicting future events.

6 Results of Financial Cost-Benefit-Analysis

At first, the results of the financial cost-benefit calculations are analyzed considering a zero discount rate, i.e. cost and benefits for the farm household that occur in later time periods receive the same weight as cost and benefits in earlier periods. Even though this may be unrealistic taking into account the real discounting constraints faced by households in the research area as presented earlier, it gives full weight to any future losses attributable to deforestation, while also reflecting future benefits from afforestation activities. Subsequently, a sensitivity analysis is applied to analyze the influence of changing discount rates on the outcome of cost-benefit calculations.

Several important conclusions can be drawn from the results presented in Table 6. Firstly, the Base Scenario (BS) displays positive net present values for all poverty groups over all geographical strata. This shows clearly that – despite the often phrased "irrational" resource utilization by local households – this resource use pattern is absolutely rational given the specific socio-economic framework in which rural households take their decision

Table 5. Equations of the Simulation Model

Exogenous Variables		
Population Growth	=	population growth rate is assumed exogenous absolute number of people living in the area is limited by the land area (forest / savannah) that can be converted to dry farming agriculture which is different for each policy scenario (compare table 3)
Visitors to Protected Areas	=	obtained from real visitors rates to adjacent protected area
Discount Rate	=	obtained from analysis of credit transactions of surveyed HH
Endogenous Variables		
Households (HH)	=	f(population growth)
Agricultural Production Area	=	f(population growth) average area of dry farming system per HH
Deforestation Rate	=	f(extension of agricultural production area)
Livestock	=	f(population growth) average number of livestock owned per HH
Growth Rate of Forests	=	f(ecosystem type; carrying capacity of ecosystem for livestock) for each eco-system different growth rates from secondary sources, see Table 1
Wood Volume	=	f(deforestation rate; natural growth rate)
Plantation	=	f(population growth) size of area planted per HH is assumed to be fixed at 0.5 ha
Benefits / Costs[1]		
Agricultural Yields	=	f(deforestation rate) Logistic relationship between deforestation and level of agricultural yields simulated with a logistic function developed by point estimates from production function analyses: $$Y = \frac{A}{1+e^{\frac{FC-B}{C}}} \quad (4)$$ with Y = Level of agricultural yield (%), FC = Forest cover (%), A, B, C = Constants
Benefits from Wood FP	=	f(decrease in wood volume) if aggregated consumption of wood < wood stock in year_i => benefit (valued at market prices adjusted as farmgate price) if aggregated consumption of wood > wood stock in year_i => cost (valued at market prices)
Benefits from Non-Wood FP	=	f(deforestation rate) harvested quantities of non-wood products are assumed to decline linear with disappearance of forests, i.e. deforestation rate
Compensation Payments	=	f(visitors to protected areas; number of HH in research area)

[1]Benefits from Non-agricultural / Off-Farm Income and Fishing are assumed to be not influenced by model components

Table 6. Net Present Value of Base Scenario and Net Increment of Alternative Policy Options assuming a Discount Rate of Zero [in PPP-US$]

		Base Scenario total NPV	annual NPV (30 years)	% of present annual income	**PO1**	**PO2** total net incremental benefit	**PO3**
Rice production area	Poorest	33,405	1,114	86.5%	-1,739	-1,537	758
	Less Poor	46,345	1,545	92.1%	-399	-196	1,146
	Better Off	76,628	2,554	97.2%	1,731	1,934	1,571
	Sub-Total	52,081	1,736	93.0%	-133	70	1,159
Periphery	Poorest	27,885	930	81.8%	-2,629	-2,427	918
	Less Poor	48,170	1,606	90.6%	-3,781	-3,578	401
	Better Off	59,888	1,996	92.6%	-219	-17	1,320
	Sub-Total	45,690	1,523	90.2%	-2,210	-2,008	879
Total		50,026	1,668	92.0%	-770	-567	1,073

Source: Own Analysis; PO = Policy Option

on natural resource management[7]. Furthermore, having the advantage of calculating NPV under zero discount rates, by dividing the total NPV with the number of years of the model period, we derive the average annual income to households. For the poorest households, income will decrease most under the BS, which is due to decreasing availability of forest products caused by ongoing deforestation. Particularly the households in the periphery suffer from this development. Again, these numbers do not include any social benefits, but only the benefits – and costs – as they accrue directly to the households[8]. As a result of the high discounting rates of households in the research area, these future losses in income are hardly valued by households in the present time period.

Secondly, looking at the net incremental benefits of the alternative policy options with regard to the base scenario, this demonstrates that PO1 and PO2 are judged negatively from the viewpoint of most households in the study area. It is especially the poorest households that suffer most from the restrictions imposed on forest utilization under these policy approaches and even a compensation scheme as envisaged under PO2 does not at all compensate for foregone income. Furthermore, as shown by the results, PO1 and PO2 have a particularly strong impact on the households in the periphery due to their more pronounced dependence on forest based activities. In contrast, for better-off households living near the RPA, the economic situation improves mainly due to improved watershed protection and, thus, constant yields of

[7] The same results apply for all discount rates above zero.

[8] These costs include the time lagged environmental effects of collective land use that directly feedback into household income.

paddy rice. These results demonstrate that forest conservation would – as proposed – benefit households that depend on indirect forest services, but that it is to the disadvantage for those households that base their livelihood strategies on extractive forest utilization. Considering our preceding analysis, it has to be highlighted that it is mainly the extractive use of wood products – but not Non-Timber Forest Products – that make up the largest share of subsistence and cash income for these households.

Still, even for the better-off households this analysis may actually be too optimistic, since we did not attempt to model any price increases in wood based energy sources resulting from forest use restrictions. Such price shifts are likely to take place considering the dependence and demand of all households in the research area on wood based energy sources (compare Woodwell 2002). In addition, no substitutes for wood based energy sources are readily available and they would also require significant initial investments, e.g. in new stoves. However, the better-off households still have the advantage of having endowments in form of agricultural yields that – even though their relative value would decrease if market prices for wood based energy increased – can be exchanged for energy sources with the lowest marginal costs, i.e. at the market price. In contrast, the poorest household would lose an important share of their cash as well as in-kind income sources and, thus, this would make it much harder for them to satisfy their demands for energy and any other consumption goods in exchange for these endowments when restrictions are imposed on forest use.

Considering these arguments, PO3 represents the best policy alternative, even though better-off households do not gain as much as under PO1 or PO2. Instead, even the poorest households gain under this policy scenario compared to the BS. The latter effect is mainly attributable to a less restricted access to forest resources and additional future benefits coming from afforestation efforts. As a consequence, PO3 does not further increase the relative poverty gap between households in the research area as opposed to PO1 and PO2.

Following, the sensitivity analysis with regard to changing discount rates (DR) is carried out. For reasons of graphical illustration we only present eight situations from PO2 and PO3 in Figure 3. Only for the better-off households in the periphery it its observed that with increasing DR the financial evaluation becomes inferior compared to the poorest households near the RPA. This change can be attributed to the changes of present values of overlapping future events, i.e. relative changes in forest product availability and agricultural production. Among households living near the RPA, only for the better-off is the situation positive over the whole range of DR applied in the sensitivity analysis. In contrast, for all households PO3 is superior compared to the base scenario. Moreover, as already discussed above, the difference between relative poverty groups and geographical strata is much less compared to PO2, thus, indicating that varying interests of different household with regard to their livelihood strategies are better balanced under this policy option. We regard the results of this analysis as very robust, because we did not even attempt

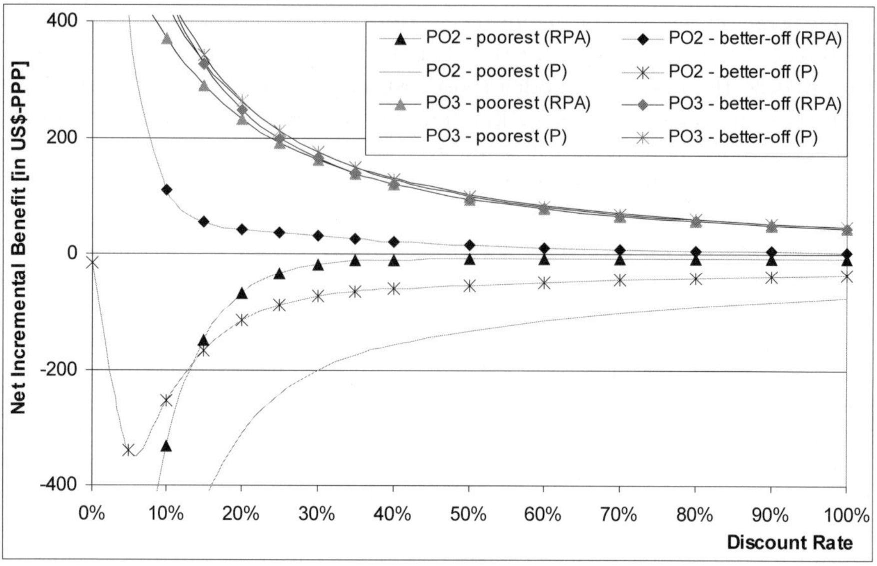

Fig. 3. Sensitivity Analysis of Changing Discount Rates

to model an increase in real market prices for wood based energy sources as well as improved secondary effects such as watershed protection that are attributable to plantation forestry.

In addition to the economic analysis, land use dynamics as they occur according to the assumptions made in the model are also analyzed to assess their implications for policy recommendations. For the Base Scenario, it will take 10 years until only savannah woodlands exists as sources for harvesting forest products in the research area. After 28 years all land is eventually converted for agriculture.

For PO1 and PO2 land use changes are different. As the area under closed and degraded forest remains unchanged, the land under savannah will decline more rapidly and all accessible land will be converted into agriculture after 21 years. For PO3, degraded forest will disappear after 6 years of the modeling period, which is about 4 years earlier compared to the Base Scenario. Furthermore, land under savannah will be completely converted for agricultural purposes in year 26, which is only about 2 years earlier compared to the Base Scenario. Most importantly, however, is the increase in area dedicated to plantation forestry rising to about 1300 ha after 30 years.

Comparing supply and demand of wood shows that only under Policy Option 3 the increasing demand for fuelwood, charcoal, and construction wood can be met by forest resources growing in the study area. Most drastically, demand will exceed supply already after less than 14 years under PO 1 and 2 and about 22 years under the Base Scenario. This underlines further that the need for afforestation to meet future needs of forest products cannot be underestimated and has to be seriously considered for developing appropriate forest resource management policies. This also illustrates that by not modeling future price increases in real terms underestimates the benefits attributable to plantation forestry. If such changes had been taken account of in our model, PO3 is hypothesized to display higher net benefit flows and net present values.

The results of our analyses allow important inferences: Firstly, it is highly insufficient to regard rural households as a homogenous group, when the impact of forest policies on rural livelihood strategies is evaluated. Strong differences exist between relative poverty groups with the poorest suffering most when use restrictions are imposed. In this context, it has to be acknowledged that the poorest households hardly posses any other endowments they could rely on for supporting their livelihoods if access to forest resources was significantly restricted.

Secondly, it was shown that it is not the use of Non-Timber Forest Products, but the extraction of wood that constitutes an important incomes source to rural households (compare Southgate 1998). Thus, the proposition that rural households could build their livelihood strategies merely on the extraction of NTFP may be false and only the use and marketing of wood based products can in fact contribute to the overall goal of poverty alleviation. Therefore, the substitution of these products – particularly wood based energy – needs to be critically discussed with regard to its impact on local and regional economic development. As our analysis has demonstrated, the production and marketing of wood based energy can be viewed as a promising tool to support income generation by the poorest households, while the wealthier households focus on agricultural production. Taking account of the fact that with an increase in incomes even higher quantities of wood will be required to satisfy energy demands, it is necessary to develop policy strategies how this increasing demand can be met in a sustainable manner, but not to further restrict the use of these products. In addition, if energy sources are substituted for imported fuels, e.g. petroleum based substances or natural gas, these have to be purchased by households imposing further restrictions on already tight budgets, particularly for the poorest households. According to Boberg (2000), negative implications for food consumption of rural households can be expected. In addition, while extending local markets and, hence, income generation through taxes at the local and regional level when the consumption of locally produced energy is supported, substitutes often have to be imported resulting in further constraints for balance of payments at the national level.

Consequently, policy recommendations based on the results of this analysis should focus on the development and practical implementation of economic incentives at the local level with the objective to facilitate tree-planting activities among rural households. Such incentives are mainly the transfer of clearly defined property and use rights of trees and the land they are planted on to rural households and input subsidies for tree-planting activities (Kaimowitz 2002, Woodwell 2002). In this context, it has to be emphasized that supporting afforestation and utilization of trees at the household level in rural areas of developing countries bears several advantages, particularly for reducing risk and vulnerability (compare Dubois 2003).

From the analysis it can be hypothesized that due to the economic disadvantages that are created especially for the poorest households living near forest resources by imposing strict conservation policies, the approach of establishing protected areas may be inappropriate considering the demand for wood based products, which represent an economic necessity to the majority of rural households in developing countries. Any successes of such policy approaches in conserving biodiversity may be attributable to mere serendipity. The results of this analysis also expose the bias that can be a problem when (global) macro-economic analyses are applied. It was shown that if these values are not transferred into real resource flows those values create a strong bias in favor of global benefits compared to local costs.

7 Conclusions

Our analysis showed that rural households do not necessarily gain from the strict conservation approaches of forest resources management as often proposed by economic analyses that are carried out at the macro-level. At the micro-level, the direct use potential of forests, e.g. for fuelwood or even the conversion of forests into agricultural production area, often outweighs the potential of indirect benefits, e.g. watershed protection, and, consequently, no net gain can be achieved for many individual households if forest resources are managed under strict conservation policies. Similar findings are presented by Marcus (2001) and Cernea and Schmidt-Soltau (2003). Moreover, it was shown that relative poverty groups are differently affected by implemented policy measures, with the poorest being worst off.

From our results, we propose that further research should concentrate on analyzing real opportunity costs of forest utilization and its contribution to support local livelihoods. Until present, most research efforts have devoted their resources for the monetary valuation of global benefits obtained from conservation of natural resources[9]. In this context, it is proposed to reconsider the logic for applying the results of certain valuation methods for economic analyses. If significantly high numbers – which ought to represent real economic values – are applied to justify the implementation of policy strategies without ever being able to translate these economic values into real resource flows that arrive at the household level of forest dependent households, it must be expected that those policies are not successful. This inconsistency is then not merely attributable to some kind of market, government, or community failure, but is also linked to the methodology applied for obtaining these values (compare Grainger 1997). In contrast, the potential for extractive uses of

[9] Barkmann et al. (this volume) give an example that stated preference studies can also be used for assessing the local monetary value of conservation.

forest should not be underestimated with respect to its contribution for long-term economic development. This is especially important when the objective of poverty alleviation is seriously contemplated.

References

Ackermann K (2003) Entwicklung von Bewirtschaftungsempfehlungen für Sekundärwälder in Nordwestmadagaskar. TÖB Abschlussbericht, Deutsche Gesellschaft für Technische Zusammenarbeit (GTZ) Eschborn

Angelsen A, Wunder S (2003) Exploring the Forest-Poverty link: key concepts, issues and research implications. CIFOR occasional paper No. 40

Barkmann J, Glenk K, Handi H, Sundawati L, Witte J-P, Marggraf R (this volume) Assessing economic preferences for biological diversity and ecosystem services at the Central Sulawesi rainforest margin – a choice experiment approach. In: Tscharntke T, Leuschner C, Guhardja E, Zeller M, Bidin A (eds) The stability of tropical rainforest margins, linking ecological, economic and social constraints of land use and conservation. Springer, Berlin Heidelberg New York

Boberg, J (2000) Woodfuel markets in developing countries – a case study of Tanzania. Ashgate

Bruner AG, Gullison RE, Rice RE, da Fonseca GAB (2001) The effectiveness of parks to protect biodiversity. Science 291: 125 - 128

Carret JC, Loyer D (2003) Madagascar protected area network sustainable financing – economic analysis perspective. Contribution to the World Park's Congress, Durban South Africa, September 2003

Cernea MM, Schmidt-Soltau K (2003) Biodiversity conservation versus people resettlement: risks to nature and risks to people. Paper presented at the CIFOR international conference on "Rural Livelihoods, Forests, and Biodiversity" in Bonn, Germany, May 19-23, 2003

Dubois O (2003) Forest-based poverty reduction: a brief review of facts, figures challenges and possible ways forward. In: Oksanen T **et al)** (eds) Forests in poverty reduction strategies, EFI proceedings No. 47, 2003

EU (1999) Collecte et analyse de données pour l'aménagement durable des forêts – joindre les efforts nationaux et internationaux. Programme de partenariat CE-FAO (1998-2000), Actes de l'atelier sous-regional pour les pays du Bassin du Congo + Madagascar sur la collecte et analyse des des données forestières - perspectives pour le secteur forestière en Afrique, Lambarene, Gabon, 27 septembre – 1 octobre 1999 http://www.fao.org/montes/fon/fons/outlook/africa/acp/lamb/Lamb.htm

Ganzhorn JU, Sorg JP (1996) Synopsis. In: Ganzhorn JU, Sorg JP (eds) Ecology and Economy of a Tropical Dry Forest in Madagascar. Primate Report Special Issue, Verlag Erich Goltze, Göttingen

Grainger A (1997) Compensating for opportunity costs in forest-based global climate change. Critical reviews in environmental science and technology 27: 163-176

Henry C, Sharma M, Lapenu C, Zeller M (2003) Assessing the Relative Poverty of Microfinance Clients - an operational tool; Consultative Group to Assist the Poorest (CGAP), Washington DC, USA

Herzog F (2001) Analyse socio-économique d'un programme de reboisement: un exemple de la région d'Antsiranana au Nord de Madagascar. Deutsche Gesellschaft für Technische Zusammenarbeit (GTZ), Eschborn

IIED (2003) Valuing forests – a review of methods and applications in developing countries. Environmental Economics Programme, International Institute for Environment and Development IIED, January 2003

Kaimowitz D (2002) Not by bread alone... forests and rural livelihoods in Sub-Saharan Africa. Mimeo, CIFOR 2002

Kramer R, Sharma A, Munashinge M (1995) Valuing Tropical Forests: Methodology and Case Study from Madagascar. The Worldbank Environment paper No. 13

Larson B (1994) Changing the economics of environmental degradation in Madagascar: lessons from the National Environmental Action Plan process. World Development 22 (5): 671-689

Lopez P (2003) Sekundäre Trockenwälder in Nordmadagaskar – Eine Fallstudie über ihre Strukturen, Potentiale und Nutzungsformen. Göttinger Beiträge zur Land- und Forstwirtschaft in den Tropen- und Subtropen, Heft 148, Göttingen 2003

Lutz E (1994) The costs and benefits of soil erosion conservation: the farmers' viewpoint. The World Bank Research Observer 9 (2), 273-295.

Marcus RR (2001) Seeing the forest for the trees: integrated conservation and development projects and local perceptions of conservation in Madagascar. Human Ecology 29 (4): 381-397

Minten B, Brand J, Randrianarisoa JC (2002) Etude sur l'impact de la déforestation sur la riziculture irriguée – cas des petits bassins versants de Maroantsetra, Nord-Est de Madagascar. Cahiers d'études et de recherches en économie et sciences sociales, Programme ILO / Cornell University, Office National d'Environnement (ONE), PAGE

Oberlé A (2001) Intégration d'une approche cartographique dans la gestion et l'aménagement de bassins versants – au sein d'un programme de lutte anti-érosive au Nord-Ouest de Madagascar. DEA Systèmes Spatiaux et Environnement, Faculté de Géographie et d'Aménagement, Université Louis Pasteur Strasbourg

O'Connor A (1996) What are the historical causes of deforestation in Madagascar and what is the situation today? http://www.colby.edu/personal/thtieten/defor-mad.html

Pagiola S, Bishop J, Landell-Mills N (2002) Selling forest environmental services – market-based mechanisms for conservation and development. Earthscan

Pearce D, Moran D (1994): The economic value of biodiversity. IUCN, Earthscan

PLAE (2003): Internal project data base. Projet Lutte Anti-Erosive (PLAE), AHT-International, Essen, Germany

Raharinjanahary H (2004) Evaluation des pertes agricoles dues à l'érosion et à l'ensablement- Etude de cas dans la région de Marovoay. Rapport final TÖB, POLFOR-GTZ, Antananarivo, Madagascar

Sepp S (2003) Les interventions forestières : premier bilan et propositions d'orientations pour le future. Rapport de mission, projet Gestion Rationnelle de l'Energie et de l'Environnement (GREEN-MAD), Février 2003

Steiniger M, Harper G, Juhn D, Hawkins F (2003) Analyse de Changement de Couverture Forestière Nationale: ˜ 1990 et ˜ 2000 ; Conservation International (CI); Center for Applied Biodiversity Science (CABS); U.S. National Air and Space Administration (NASA)

Southgate D (1998) Tropical forest conservation – an economic assessment of the alternatives in Latin America. Oxford University Press, Oxford

UNDP (2001) World Development Report, UNDP, New York

Woodwell JC (2002) Fuelwood and land use in West Africa: understanding the past to prepare for the future. Presented at the workshop in Ouagadougou, Burkina Faso, of Comité Inter-Etats de lutte contre la sécheresse au Sahel and USAID, September 2002

Worldbank (2003a) Madagascar – third environmental program support project. Updated project information document, report no. PID11561, The Worldbank, January 23rd, 2003

Worldbank (2003b) Madagascar at a glance – summary of development indicators. The Worldbank, August 2003

World Bank (2004) Madagascar – third environmental program support project. Updated project information document (PID), Antananarivo, 2004

Part III

Integrated concepts of land use in tropical forest margins

Potentials to reduce deforestation by enhancing the technical efficiency of crop production in forest margin areas

Alwin Keil[1]*, Regina Birner[2], and Manfred Zeller[1]

[1] Institute of Agricultural Economics and Social Sciences in the Tropics and Subtropics, University of Hohenheim (490), D-70593 Stuttgart, Germany
[2] International Food Policy Research Institute (IFPRI), Washington, D.C., USA

*corresponding author: Alwin Keil, phone: +49 (0)711 4593301, fax: +49 (0)711 4592582, Email: Alwin.Keil@uni-hohenheim.de

Summary

The establishment of tree crop plantations such as oil palms, coffee, or cocoa considerably contributes to the loss of tropical forests. Taking the case of cocoa production in Central Sulawesi as an example, this chapter investigates whether there is a potential for reducing deforestation by improving the productivity of tree crop plantations in rainforest margin areas by better crop management. Increased productivity would enable farmers to earn a living from a smaller area of land; thus, the expansion of low-productivity perennial cropping systems into forest land can be viewed as a waste of forest resources. In order to assess whether the productivity of the existing cocoa plantations in the research area could potentially be improved, the level of technical efficiency attained is estimated. A farmer is technically inefficient if he fails to produce the maximum output attainable for the level of inputs he uses. Technical inefficiency is caused by lacking know-how, for example with respect to the timeliness of agricultural operations such as weed control or crop hygienic measures.

After introducing the concept of technical efficiency (TE) and describing the method of Stochastic Frontier analysis to empirically estimate the level of TE, we show how this method was applied to our Sulawesi case. Using data collected in a sample of 202 farm households, we estimate separate Stochastic Frontier production functions for the two most important crops in the research area: cocoa, the primary cash crop, the cultivation of which has become widespread only in the 1990s, and irrigated rice (paddy) that has been grown for generations. Apart from estimating the levels of TE attained, we

Tscharntke T, Leuschner C, Zeller M, Guhardja E, Bidin A (eds), The stability of tropical rainforest margins, linking ecological, economic and social constraints of land use and conservation, Springer Berlin 2007, pp 391-416

also investigate the influencing factors of efficiency in one single statistical procedure.

The estimated average TE in rice production is 77% as opposed to only 37% in cocoa cultivation; thus, the potential to increase production by improving crop management (not by increasing the level of input use!) is particularly large in the case of cocoa. The analysis of the factors influencing TE shows that poverty and illiteracy have an efficiency reducing effect in both rice and cocoa production. Agricultural extension services significantly increase efficiency in rice cultivation while this effect is not observed in cocoa production.

Acknowledging that increasing the productivity of perennial cropping systems in forest margin areas may also create incentives for deforestation, we then discuss the conditions under which enhanced productivity can be expected to have a forest saving or a forest clearing effect. We finally assess the likely outcome of improved TE in the case of Central Sulawesi and conclude that the comparatively low efficiency level currently found indicates a considerable potential for reducing deforestation by increasing farm incomes on already converted forest land, thus meeting both environmental and economic objectives. Policy interventions aimed at realizing this potential should include improved agricultural extension focusing on technical advice on the proper management of cocoa, but, at the same time, they need to control the influx of migrants attracted by the profitability of cocoa cultivation.

Keywords: technical efficiency, stochastic frontier analysis, cocoa, deforestation, Indonesia

1 Introduction

In Southeast Asia, tropical forest is to a considerable extent being replaced by perennial crops such as oil palms, cocoa, and coffee, which are grown both by smallholders and large-scale commercial enterprises. A question of particular policy relevance in this context refers to the relation between the level of technical efficiency (TE) attained in these perennial cropping systems and the rate of deforestation. Policy interventions such as the provision of extension services that help farmers improve TE and, hence, derive higher incomes from already existing perennial crops may reduce the incentive to convert additional forest land by enabling farmers to earn a living from a smaller area of land. Thus, the expansion of low-efficiency perennial cropping systems into forest land can be viewed as a waste of forest resources. Even though there is a huge body of literature on tropical deforestation (see Kaimowitz and Angelsen (1998) for a review), surprisingly little attention has been paid to the link between TE and deforestation. Even though increasing the productivity of crop cultivation may also create new incentives for deforestation, there is a need to study the TE of the perennial cropping systems in rainforest margin areas in order to determine whether and to which extent these systems are managed in a technically inefficient way.

The literature on TE has largely neglected perennials; as a review of empirical studies of TE in developing country agriculture by Thiam et al. (2001) reveals, most studies have focused on food crops, or have conducted a whole farm analysis without differentiating between farm enterprises. In Asian countries, rice production has been the subject of numerous efficiency studies, whereas the cultivation of perennial cash crops such as cocoa and coffee has received very little attention. In the mountainous province of Central Sulawesi, Indonesia, the valleys are predominantly occupied by irrigated rice (paddy), while the mountains are still largely covered with forest. However, the area has become subject to a cocoa boom in the past 15 years, which has attracted migrants and caused a rapid loss of the natural forests. Cocoa is grown by family farms, many of which also cultivate paddy. Since cocoa, unlike paddy, is a comparatively new crop in the area, the question arises whether the farmers are able to cultivate this crop in a technically efficient way.

Against this background, the objectives of this chapter are (1) to show how the TE of perennial cropping systems in rainforest margin areas can be analyzed, and (2) to discuss the potential to reduce deforestation by enhancing the level of TE in agricultural production in forest margin areas in general, and in our research region in particular. Even though the chapter is motivated by the link between TE and deforestation, the analytical focus is placed on the TE analysis. The remainder of the chapter is structured as follows: A brief introduction to the concept of TE and the method used for its estimation is provided in section 2, while section 3 specifies the models applied for our Sulawesi case study in detail. Results are presented in section 4 and discussed in section 5. Finally, conclusions and policy implications are derived in section 6.

2 Measuring efficiency and its determinants

2.1 The concepts of technical, allocative, and economic efficiency

Farrell (1957) proposed a measure of the efficiency of a firm that consists of two components: Technical efficiency (TE), which reflects the firm's ability to obtain maximal ('frontier') output from a given set of inputs, and allocative efficiency (AE), which represents its ability to choose the optimal proportions of inputs and outputs, given their respective prices. These two measures are then combined to provide a measure of total economic efficiency (EE):

$$\mathrm{EE} = \mathrm{TE} \times \mathrm{AE}, \quad \text{where } \mathrm{TE}, \mathrm{AE} \in [0,1] \tag{1}$$

The distinction between these measures is illustrated in Figure 1, which shows two possible production functions for a single output and a single variable input. The outer bound production function, TVP_1, displays higher levels of output for all positive levels of input use than the inner production function, TVP_2. Therefore, a farm that operates at any point on TVP_1 is more technically efficient than a farm operating at any point on TVP_2. The optimum level of input use is attained when marginal returns equal marginal costs. In Figure 1, marginal returns are represented by the slope of the production functions, and marginal costs by the slope of the cost function, TFC. Hence, optimal levels of input use are achieved in points A and C on the production functions TVP_1 and TVP_2, respectively, indicating full allocative efficiency. Assuming that TVP_1 represents the frontier output, the distinction between technical and allocative efficiency allows to identify four possible scenarios: First, a farm may display both technical and allocative inefficiency (point D); second, it may achieve allocative efficiency but be technically inefficient (point C); third, the firm may be technically efficient but display allocative inefficiency (point B) and, fourth, the farm may achieve both technical and allocative efficiency (point A). The term economic efficiency is reserved for this last situation (summarized from Ellis 1993: 67-68).

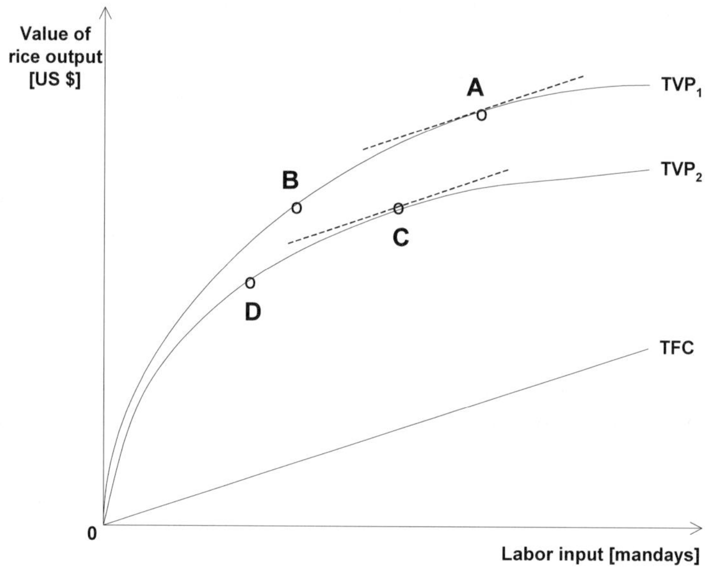

Fig. 1. The concept of technical, allocative, and economic efficiency. (TVP = Total value product at technically efficient $_1$ and inefficient $_2$ levels of production; TFC = Total factor costs) Source: Adapted from Ellis (1993) p.67)

2.2 Measuring technical efficiency using Stochastic Frontiers

The decomposition of economic efficiency estimates into allocative and technical efficiency components suffers from various methodological, analytical, and data problems (Chambers 1988). We will therefore confine our discussion to the estimation of technical efficiency, for which two reliable methods exist, acknowledging that the level of economic efficiency will be lower than (or, theoretically, equal to) our TE estimates (see Equation (1)).

A production function is defined to represent the maximum output attainable for a given set of inputs, i.e., the 'frontier' output (Chambers 1988 p.8). Since the production frontier is never known in practice, Farrell (1957) suggested that the function be estimated from sample data using either a non-parametric piece-wise linear technology or a parametric function. The first approach resulted in the development of Data Envelopment Analysis (DEA), which uses linear programming to construct the frontier (see Seiford 1996, for a comprehensive review of the methodology). The chief advantage of DEA is that no explicit functional form needs to be imposed on the data. Its main disadvantage is that DEA attributes all deviations from the frontier to inefficiency. However, especially in developing country agriculture, empirical data may be heavily contaminated by statistical noise due to measurement errors, variability in climatic and edaphic conditions, and affliction of crops with pests and diseases, resulting in an underestimation of TE levels attained (Coelli et al. 1998 p.219).

In the parametric approach, i.e., the direct estimation of a production function, it is assumed that the decision-makers select the levels of inputs and output to maximize *expected* rather than actual profit. Notably in agricultural production, where farmers face uncertainty with respect to the amount of output to be produced and the output prices to be received, this assumption seems appropriate (Coelli et al. 1998 p.209). In this instance, Zellner et al. (1966) showed that the single-equation production function does not suffer from simultaneous-equation bias. The parametric approach evolved from the estimation of deterministic frontiers to the use of Stochastic Frontiers for efficiency analysis. While deterministic frontiers, same as DEA, do not account for random deviations, Stochastic Frontiers accommodate statistical noise. DEA will therefore generally yield lower efficiency estimates than Stochastic Frontiers (Brümmer 2001).

Because of its advantage of accounting for statistical noise, in the following we will concentrate on the Stochastic Frontier approach that was independently proposed by Aigner et al. (1977) and Meussen and van den Broeck (1977). The Stochastic Frontier production function is given by:

$$Y_i = F(X_i, \beta) \exp(V_i - U_i), \quad i = 1, 2, \ldots, N, \tag{2}$$

where

$$Y_i = \text{Quantity (or value) of output of the i-th firm}$$
$$F(\cdot) = \text{Suitable production function}$$
$$X = \text{Vector of input quantities}$$
$$\beta = \text{Vector of parameters to be estimated}$$
$$V = \text{Random error term}$$
$$U = \text{Non-negative error term representing technical inefficiency}$$

V is a random variable, assumed to be independently and identically distributed as $N(0, \sigma_V)$. U, which accounts for systematic departures from the frontier, i.e., technical inefficiency, is assumed to follow a particular one-sided distribution. A number of different distributions have been proposed in the literature, namely the half-normal and exponential (e.g. by Aigner et al. 1977), the truncated normal (Stevenson 1980), and the two-parameter Gamma distribution (Greene 1990). The technical efficiency (TE) measure for the i-th household $\text{TE}_i = \exp(-U_i) \in [0, 1]$ is the ratio of the observed output and the maximum attainable output at the frontier. The maximum likelihood (ML) estimation of Equation (2) yields estimates of β and γ where $\gamma = \sigma_U^2/\sigma^2 \in [0, 1]$, and $\sigma^2 = \sigma_U^2 + \sigma_V^2$.

The basic structure of the Stochastic Frontier model is depicted in Figure 2, in which the productive activities of two firms, denoted i and j, are considered. Firm i uses inputs with values given by x_i and obtains the output Y_i. The frontier output Y_i^* exceeds the value on the deterministic production function, $f(x_i; \beta)$, because its productive activity is associated with particularly favorable conditions leading to a positive random error V_i. Firm j uses inputs of the value x_j and obtains the output Y_j. Contrary to firm i, the associated frontier output Y_j^* lies below the value of the deterministic production function, $f(x_j; \beta)$, because of particularly unfavorable conditions, which lead to a negative random error V_j. Hence, in both cases the observed production values are less than the corresponding frontier values, but the (unobservable) frontier production value lies above the deterministic production function in one case, and beneath it in the other (summarized from Battese 1992).

2.3 Investigating efficiency determinants

A number of empirical studies, such as Pitt and Lee (1981) and Kalirajan (1981) have investigated the determinants of technical inefficiency by regressing the efficiency estimates on firm-specific characteristics in a second-stage analysis. However, this approach is inconsistent in its distributional assumptions: In the first stage, the efficiency effects are required to be independently and *identically* distributed (Jondrow et al. 1982), whereas in the second stage they are assumed to be a function of firm-specific factors, which implies that they are *not* identically distributed. Kumbhakar et al. (1991) and Reifschneider and Stevenson (1991) noted this inconsistency and specified Stochastic Frontier models in which the inefficiency effects were defined to be explicit

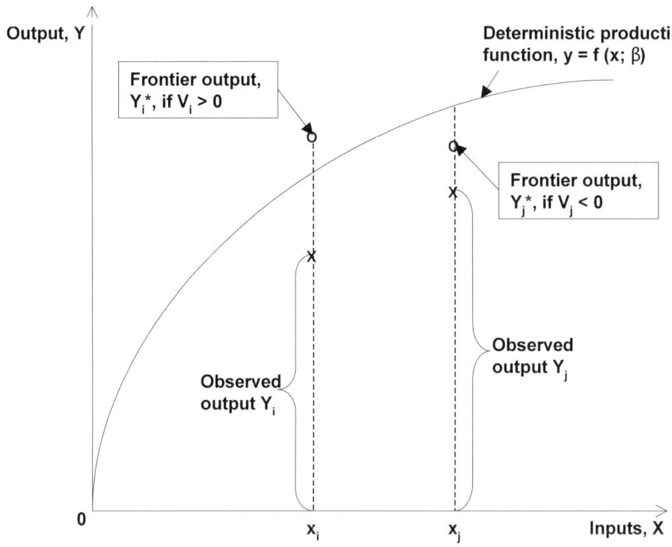

Fig. 2. The concept of a Stochastic Frontier production function. Source: Adapted from Battese (1992) p.191.

functions of firm-specific factors, and all parameters were estimated in a single-stage ML procedure. Battese and Coelli (1995) extended these approaches to accommodate panel data, which, in addition to efficiency determinants, permits the investigation of technical change and technical efficiency change over time. In their model, the TE effects, U, are obtained by truncation (at zero) of the normal distribution with mean μ_{it} and variance σ_U^2, such that $\mu_{it} = Z_{it}\delta$, where Z_{it} is a vector of firm-specific explanatory variables, and δ is a vector of parameters to be estimated. In our case study, the Battese and Coelli (1995) model was applied.

3 Technical efficiency and its determinants in cocoa and rice cultivation in Central Sulawesi

3.1 Research area and data collection

The research area is the Palu River watershed in Central Sulawesi province, Indonesia, extending from 1° to 1° 30' S and around 120° E. Its mountainous topography results in a distinct rainfall gradient, with the coastal zone receiving only 600 mm of rain p.a., while precipitation rates exceed 3,000 mm at higher elevations (WWF 1981). Therefore, agricultural land use is very location specific, depending on local rainfall, topography and soil properties. In general, irrigated rice and cocoa are the two most important crops grown.

To account for the variation in local climatic conditions, data were collected in a stratified random sample of 202 farm households in eight villages located at different elevation levels, comprising 148 farmers who grow irrigated rice and 105 farmers who grow cocoa (51 farmers grow both crops). Using farmer recall, in early 2003 plot-level production data were collected for the years 2002 and 2001, whereby in the case of rice two cropping seasons were covered per year. For both crops, data were also collected for the most serious drought period experienced in recent years. Production data were then aggregated to the household level to be consistent with the investigation of farm-specific efficiency determinants.

3.2 Model specifications

The Stochastic Frontier for irrigated rice is specified as a transcendental logarithmic ('Translog') function of the following general form:

$$\ln Y_{it} = \beta_0 + \sum_{m=1}^{n} \beta_{0m} D_{mit} + \sum_{k=1}^{5} \beta_k \ln(X_{kit}) + \frac{1}{2} \sum_{k=1}^{5} \sum_{j=1}^{5} \beta_{kj} \ln(X_{kit}) \ln(X_{jit}) + V_{it} - U_{it} \quad (3a)$$

In the case of cocoa, the Translog specification yielded unstable parameter estimates due to multicollinearity. Therefore, the simpler Cobb-Douglas functional form was preferred which does not include the second-order terms of the input variables and, hence, avoids the problem of multicollinearity[1]. Thus, the Stochastic Frontier model for cocoa is specified as follows:

$$\ln Y_{it} = \beta_0 + \sum_{m=1}^{n} \beta_{0m} D_{mit} + \sum_{k=1}^{5} \beta_k \ln(X_{kit}) + V_{it} - U_{it}. \quad (3b)$$

The notation used in Equations (3a) and (3b) is defined as follows:

$\ln Y$ = Natural logarithm (ln) of the output
i = Household index ($i = 1, \ldots, N$; $N = 148$ for rice, $N = 105$ for cocoa)
t = Time index ($t = 1, \ldots, T$; $T = 5$ [cropping seasons] for rice, $T = 3$ [years] for cocoa)
β = Vector of parameters to be estimated
D_m = Vector of dummy variables ($m = 1, \ldots, n$; $n = 6$ for rice, $n = 4$ for cocoa)
$\ln X_k$ = ln of the input vector
V = $N(0, \sigma_v)$ distributed random error term
U = Non-negative error term representing technical inefficiency

[1] The Translog and Cobb-Douglas specifications produced very similar TE estimates (see section 4.4).

The non-negative error component is specified as a function of potential firm-specific determinants of technical inefficiency:

$$U_{it} = \delta_0 + \sum_{r=1}^{s} \delta_r Z_{rit} + W_{it} \tag{4}$$

where

$\delta =$ Vector of parameters to be estimated

$Z_r =$ Vector of potential inefficiency determinants ($r = 1, \ldots, s$; $s = 11$ for rice, $s = 9$ for cocoa)

$W = N(0, \sigma_U)$ distributed random variable, where σ_U is defined such that $U_{it} \geq 0$

The computer program Frontier 4.1 was used to obtain maximum likelihood estimates of the parameters in the models defined by Equations (3a) and (4) for irrigated rice and Equations (3b) and (4) in the case of cocoa (see Coelli 1996).

Table 1 lists the variables used in the Stochastic Frontiers and inefficiency models, and their summary statistics. Whilst the specification of most variables in the Stochastic Frontier functions is straightforward, a brief explanation of the following variables seems appropriate: The variable *time* was included to account for technical change which is hypothesized to have a yield increasing effect over time. Since a drought period experienced before 2001 varies between households, *time* is specified as a dummy variable that differentiates between the years 2002 or 2001 on the one hand, and a prior drought period on the other. In 11% of the observations for rice and 47% of the observations for cocoa, no fertilizers, herbicides, or pesticides were applied. It may be that the positive and zero cash-input cases differ with regard to the partial production elasticities of the other inputs and their error variances because the technological structures for the two cases are different; the dummy variable *dmat* accounts for these zero-observations and, hence, avoids biased parameter estimates (see Battese 1997). In the case of cocoa, production data were aggregated to the household level from up to six different plots. The edaphic and micro-climatic conditions of different cocoa plots of one household may be quite heterogeneous in the research area, since in many cases long-used plots are located in the valley, while others are situated at higher elevations where the natural forest has relatively recently been cut. The variable *sevplots* is included in the model to account for this heterogeneity. Apart from variables capturing the input of land, capital and labor, also the adequacy of water supply to the crops is accounted for. This is crucial because, as indicated above, the mountainous research area is characterized by high temporal and spatial rainfall variability. Since rainfall data were not available at a local resolution, the variable *water* reflects the farmers' perception of the sufficiency of water supply on a scale from 1 (= totally insufficient) to 5 (=

Table 1. Definition of variables used in the Stochastic Frontier models for rice and cocoa production in Central Sulawesi, and their summary statistics

Variable	Variable definition	Irrigated rice[a] Mean	Irrigated rice[a] St. Dev.	Cocoa[b] Mean	Cocoa[b] St. Dev.
Dependent variable[c]					
ln Y	= Logged total output of husked rice (dried cocoa seeds) in kg	1093.57	1145.12	939.08	2438.70
Dummy variables					
yieldred	= Dummy, if yield was drastically reduced by pests/diseases (= 1, else = 0)	0.19	0.39	0.18	0.38
redsoil	= Dummy, if soil color is redish/yellowish (= 1, else = 0)	0.22	0.41	N/A	
lowelev	= Dummy, if altitude is < 100 m a.s.l. (= 1, else = 0)	0.49	0.50	N/A	
onlyherb	= Dummy, if weeding is done solely by herbicides (= 1, else = 0)	0.13	0.33	N/A	
time	= Dummy, if season/year was 2001 or 2002 (= 1, else = 0)	0.95	0.22	0.96	0.20
dmat	= Dummy, if no fertilizer/herbicides/pesticides were used (= 1, else = 0)	0.11	0.31	0.47	0.50
sevplots	= Dummy, if prod. data were aggregated from several cocoa plots (= 1, else = 0)	N/A		0.34	0.48
Input variables[c]					
land	= Logged area of irrigated rice (cocoa) in ares	76.35	57.42	104.08	93.41
seed	= Logged amount of rice seed used in kg	102.27	79.57	N/A	
material	= Logged expenses for fertilizer, herbicides, and pesticides in thousands of IDR[d]	162.98	206.64	564.29	1317.12
labor	= Logged labor input for weed control (in rice)/maintenance (in cocoa) in hours	102.35	130.55	391.28	851.74
water	= Logged average water supply score on a scale from 1 (= totally insufficient) to 5 (= ideal)	3.76	1.01	3.30	0.61
age	= Logged age of the cocoa plantation in years	N/A		6.98	3.44

(continued)

Table 1. Definition of variables used in the Stochastic Frontier models for rice and cocoa production in Central Sulawesi, and their summary statistics (continued)

Variable definition		Summary statistics			
		Irrigated rice[a]		Cocoa[b]	
		Mean	St. Dev.	Mean	St. Dev.
Inefficiency effects					
headsex	= Dummy, if HH head is female (= 1, else = 0)	0.08	0.27	N/A	
headage	= Age of the HH head	46.74	14.13	48.58	14.52
headlit	= Dummy, if HH head is literate (= 1, else = 0)	0.89	0.31	0.81	0.39
povindex	= Poverty index, derived from asset, food, and housing related indicators	0.04	0.95	0.40	1.06
extension	= Number of extension visits in the last 12 months	1.90	4.11	1.36	3.05
offinc	= Dummy, if off-farm income was available in the last 12 months (= 1, else = 0)	0.63	0.48	0.55	0.50
special	= Dummy, if HH only cultivates rice (cocoa) (= 1, else = 0)	0.63	0.48	0.53	0.50
palolo	= Dummy, if HH resides in Palolo sub-district (= 1, else = 0)	0.19	0.39	0.52	0.50
kulawi	= Dummy, if HH resides in Kulawi sub-district (= 1, else = 0)	0.33	0.47	0.22	0.42
time	= Dummy, if season/year was 2001 or 2002 (= 1, else = 0)	0.95	0.22	0.96	0.20
irriresp	= Dummy, if HH head/close relative/close friend was responsible for irrigation water distribution in the last 12 months (= 1, else = 0)	0.48	0.46	N/A	

[a] Unbalanced panel of 423 observations obtained from 148 different rice growing farmers over a maximum of five cropping seasons.
[b] Unbalanced panel of 209 observations obtained from 105 different cocoa growing farmers over a maximum of three years.
[c] Summary statistics are given for the unlogged variables.
[d] IDR = Indonesian Rupiah. 1 US$ = 8,900 IDR (February 2003).
N/A = Not applicable; HH = Household.

ideal). In the case of cocoa, the varying productivity over the lifespan of the plantation is an additional factor to be considered; hence, the variable *age* was included in the model.

The variables in the inefficiency models are based on Bravo-Ureta and Pinheiro's (1993) review of the empirical literature on farm-specific determinants of efficiency in developing country agriculture, and our own considerations. Female household heads (variable *headsex*) are usually widows; therefore, labor availability for the timely accomplishment of agricultural tasks may be problematic, entailing a lower level of TE[2]. The age of the household head (*headage*) may have an ambiguous influence on TE: on the one hand, due to their longer farming experience, older farmers may have developed a higher level of TE than younger farmers; on the other hand, older farmers are likely to be more conservative and therefore less willing to adopt new practices, implying a negative influence on TE. Literacy of the household head (*headlit*) is expected to have a positive effect on TE, since it greatly facilitates the acquisition and processing of information related to agricultural production. The variable *povindex* is a poverty index which was constructed from nine asset, food, and housing related indicators by Principal Component Analysis. Since wealthier households are likely to have better access to information (Feder et al. 1985) and credit (Eswaran and Kotwal 1989), they are hypothesized to attain higher levels of TE than poorer households. The agricultural extension service provides advice and information to farmers to improve their technical competence in farming operations. Thus, the number of extension visits received (*extension*) is expected to have an efficiency enhancing effect. The existence of off-farm income sources (*offinc*) can have a positive or negative impact on TE in agriculture: if off-farm income sources are utilized during slack labor periods in crop production in order to finance agricultural inputs, this could potentially have an efficiency enhancing effect (Kalirajan 1990); however, if off-farm income opportunities compete for labor during critical farming activities, or off-farm income is generally of greater importance to the household, this may reduce TE in agricultural operations (Ali and Flinn 1989). The variable *special* indicates whether a farmer cultivates both rice and cocoa or is specialized on one of these two primary crops in the research area (apart from growing minor crops for home consumption). Specialization is hypothesized to enhance efficiency because labor resources and expertise can be focused on the cultivation of one main crop. The three adjacent sub-districts where data were collected are characterized by differences in altitude, distance to the provincial capital, and quality of physical infrastructure such as roads and irrigation systems. These differences are likely to have implications on TE in crop production, such as the timely availability of fertilizer and pesticides and the quality of irrigation; hence, two respective dummy variables were included in the model. The variable *time* also appears in the inefficiency models

[2] This variable does not appear in the cocoa model since only three out of the 105 HHs that grow cocoa (2.9%) are female headed.

to capture a temporal change in TE; due to growing farming experience, TE is expected to increase over time. The variable *irriresp* is applicable only to rice production since cocoa is not being irrigated: it indicates whether or not the farmer or a socially close person was involved in the distribution of irrigation water. There is evidence of frequent disputes about water distribution at the village level; we hypothesize that the 'political capital' of either being the person responsible for water distribution or having a close relationship with the person in charge has an efficiency enhancing effect due to positive implications on the amount and/or timeliness of irrigation water supplied.

4 Results

4.1 Hypotheses tests of model applicability

The applicability of the Stochastic Frontier method can easily be tested based on the likelihood ratios (LR) produced by different model specifications; Table 2 presents the results of several LR tests for the Sulawesi case. In the rice model, the null hypothesis that the simplification of the Translog model to a Cobb-Douglas form is an appropriate representation of the data is strongly rejected (first row). The second row tests the null hypothesis that the one-sided error term accounting for inefficiency is actually zero, i.e., that the sample farmers are fully technically efficient. The LR tests compare the Stochastic Frontier specifications with their ordinary least squares counterparts. It should be noted that the test statistic for this hypothesis follows a mixed χ^2 distribution since the test involves inequality and equality restrictions; upper and lower bounds for the mixed distribution are provided by Kodde and Palm (1986). The hypothesis of no inefficiency is rejected at $P < 0.001$ in both cases. Thirdly, the null hypothesis that the coefficients of the inefficiency determinants are jointly zero is also strongly rejected for both models. The variable measuring labor input for weeding in the rice model and its squared and interaction terms did not yield any statistically significant regression coefficients; therefore, the hypothesis was tested that the coefficients of weeding labor and the related second-order terms are jointly zero (bottom row of Table 2). The critical χ^2value for a total of six restrictions and an error probability of 10% is 10.64, which was not attained by the LR test value λ. Hence, the restricted model was accepted, and the results presented refer to this statistically optimized model[3].

[3] The two specifications produce very similar estimates of the partial production elasticities with respect to the variables *land*, *seed*, *material*, and *water*. Furthermore, the estimates of mean TE are very similar at 0.766 and 0.771 for the unrestricted and the restricted models, respectively. The Pearson correlation coefficient between the household-specific TE estimates produced by the two models is 0.996.

Table 2. Results of likelihood ratio tests of hypotheses related to the specification of the Stochastic Frontier models for rice and cocoa production in Central Sulawesi

Null hypothesis	Irrigated rice Crit. χ^2 value	Test value λ	Cocoa Crit. χ^2 value	Test value λ
H_0: $\beta_{ij} = 0$; $i \leq j = 1 \ldots 5$	29.59	57.70**	-	
H_0: $\gamma = 0$	33.82[a]	132.49**	30.54[a]	64.56**
H_0: $\delta_1 = \ldots = \delta_s = 0$	24.73	30.16*	27.88	45.30**
H_0: $\beta_{Labor} = 0$	10.64[b]	7.12	-	

*(**) The Null hypothesis is rejected at the 1% (0.1%) level of error probability.
[a] Mixed χ^2distribution (see Table 1 in Kodde and Palm, 1986).
[b] Critical χ^2 value for a total of six restrictions and the 10% level of error probability.

4.2 Parameter estimates of the Stochastic Frontiers

The results of the parameter estimates of the Stochastic Frontiers for irrigated rice and cocoa are presented in the upper section of Table 3. The signs of the coefficients of the dummy variables *yieldred*, and *redsoil* are as expected. *Lowelev* has a significant negative coefficient. The coefficients of *time* are statistically not significant, implying that a yield increasing effect of technical change over time is not supported by the data. While in the Cobb-Douglas production function for cocoa the estimated coefficients of the input variables are identical with their partial production elasticities, in the Translog function the individual coefficients of the input variables are not directly interpretable, but the elasticities of mean output with respect to the four input variables are calculated at the means of the logs of the input variables and their second-order coefficients as follows:

$$\frac{\partial \ln Y}{\partial \ln X_k} = \beta_k + \beta_{kk}\overline{\ln X_k} + \sum_{j \neq k}^{4} \beta_{kj}\overline{\ln X_j}, \quad k = 1, \ldots, 4 \tag{5}$$

All partial output elasticities of the input variables are positive and highly significant. The elasticity of scale was computed as the sum of the partial production elasticities of those input variables that are under the effective control of the entrepreneur; their standard errors were derived from the variance-covariance matrices of the respective models. The estimated elasticities of scale are not significantly different from unity; hence, in neither model the hypothesis of constant returns to scale can be rejected (bottom section of Table 3).

4.3 Parameter estimates of the efficiency determinants

Table 4 presents the results of the parameter estimates of the inefficiency models for rice and cocoa cultivation. For the interpretation it is crucial to

Table 3. Maximum likelihood (ML) estimates of the parameters in the Stochastic Frontiers for rice and cocoa production in Central Sulawesi

	Irrigated rice			Cocoa		
Variable	Parameter	ML estimate[a]		Parameter	ML estimate[a]	
Coefficients in the Stochastic Frontier production functions						
Constant	β_0	3.253 (1.179)	***	β_0	0.030 (0.146)	
yieldred	β_{01}	- 0.264 (0.042)	***	β_{01}	- 0.644 (0.050)	***
redsoil	β_{02}	- 0.087 (0.039)	**	-		
lowelev	β_{03}	- 0.108 (0.049)	**	-		
onlyherb	β_{04}	0.325 (0.045)	***	-		
time	β_{05}	0.087 (0.094)		β_{02}	- 0.002 (0.083)	
dmat	β_{06}	- 0.175 (1.128)		β_{03}	1.194 (0.261)	***
sevplots	-			β_{04}	0.401 (0.101)	***
land	β_1	-1.280 (0.324)	***	β_1	0.721 (0.063)	***
seed	β_2	1.241 (0.364)	***	-		
material	β_3	0.019 (0.202)		β_2	0.118 (0.023)	***
labor	-			β_3	0.152 (0.032)	***
water	β_4	0.915 (0.548)	*	β_4	0.803 (0.179)	***
age	-			β_5	0.485 (0.091)	***
$\frac{1}{2}$(land)^2	β_{11}	- 0.445 (0.128)	***	-		
$\frac{1}{2}$(seed)^2	β_{22}	- 0.780 (0.128)	***	-		
$\frac{1}{2}$(material)^2	β_{33}	0.014 (0.018)		-		
$\frac{1}{2}$(water)^2	β_{44}	0.301 (0.322)		-		
land × seed	β_{12}	0.756 (0.115)	***	-		
land × material	β_{13}	- 0.010 (0.012)		-		
land × water	β_{14}	0.415 (0.143)	***	-		
seed × material	β_{23}	- 0.004 (0.012)		-		
seed × water	β_{24}	- 0.476 (0.155)	***	-		
material × water	β_{34}	- 0.019 (0.014)		-		
Partial output elasticities						
land		0.634 (0.053)	***		0.721 (0.063)	***
seed		0.248 (0.055)	***		-	
material		0.099 (0.022)	***		0.118 (0.023)	***
labor		-			0.152 (0.032)	***
water		0.697 (0.066)	***		0.803 (0.179)	***
age		-			0.485 (0.091)	***
Returns-to-scale		0.982 (0.027)			0.991 (0.047)	

*(**)[***] Significant at the 10% (5%) [1%] level of error probability.
[a] Values in parentheses are standard errors.

keep in mind that a negative sign of a coefficient stands for a negative influence on inefficiency, i.e., an efficiency enhancing factor, whereas a positive sign must be interpreted as efficiency reducing.

Table 4. Maximum likelihood (ML) estimates of the parameters in the inefficiency models for rice and cocoa production in Central Sulawesi

Variable	Parameter	Irrigated rice ML estimate[a]		Parameter	Cocoa ML estimate[a]	
Inefficiency model						
Constant	δ_0	- 0.556 (0.761)		δ_0	- 0.299 (0.667)	
headsex[b]	δ_1	0.930 (0.416)	**	-		
headage	δ_2	- 0.017 (0.006)	***	δ_1	0.018 (0.006)	***
headlit	δ_3	- 2.088 (0.609)	***	δ_2	- 0.396 (0.228)	*
povindex	δ_4	- 0.865 (0.289)	***	δ_3	- 0.091 (0.080)	
extension	δ_5	- 0.117 (0.043)	***	δ_4	0.030 (0.032)	
offinc	δ_6	2.114 (0.700)	***	δ_5	0.419 (0.172)	**
special	δ_7	- 0.757 (0.241)	***	δ_6	- 0.395 (0.203)	*
palolo	δ_8	- 2.160 (0.722)	***	δ_7	- 0.080 (0.240)	
kulawi	δ_9	- 0.464 (0.226)	**	δ_8	0.328 (0.248)	
time	δ_{10}	- 0.373 (0.515)		δ_9	0.736 (0.501)	
irriresp[b]	δ_{11}	- 1.411 (0.491)	***	-		
Variance parameters						
σ^2		1.331 (0.377)	***		0.694 (0.104)	***
γ		0.971 (0.010)	***		0.999 (0.000)	***
Mean efficiency		0.771			0.371	

*(**)[***] Significant at the 10% (5%) [1%] level of error probability.
[a] Values in parentheses are standard errors.
[b] Variable is not applicable to the cocoa model.

The signs of five explanatory variables are consistent both in the case of rice and cocoa: First, literacy of the household head (*headlit*) is found to be efficiency enhancing. Second, the negative significant coefficient of *povindex* in the case of rice means that wealth is an efficiency enhancing factor in rice production; the sign of this coefficient is consistent in the model for cocoa, but it is statistically not significant. Third, the availability of off-farm income was found to be efficiency reducing in both models, as indicated by the positive significant coefficient of *offinc*. Fourth, as hypothesized, farmers who specialize on one major crop (variable *special*), i.e., who grow either rice or cocoa, are found to be more efficient in the cultivation of the respective crop than those farmers who cultivate both crops. And, fifth, the coefficient of the location dummy variable *palolo* is negative in both models, but statistically significant only in the case of rice.

Two efficiency determinants have opposite signs in the models for rice and cocoa, but their coefficients are statistically significant only in the rice model: The number of extension visits (*extension*) is efficiency enhancing in the case of rice production, which is consistent with the literature. The location dummy variable *kulawi* also has a negative significant coefficient in the rice model. The coefficient of the *time* dummy is negative in the rice model, and positive in the case of cocoa. However, neither of the coefficients is statistically significant, thus neither supporting nor rejecting the hypothesis of increasing TE over time.

The age of the household head (*headage*) was found to have opposite effects on TE in the two models: in the case of rice cultivation it has an efficiency enhancing effect, whereas in the case of cocoa it is efficiency reducing.

Two variables are only applicable to the rice growing farmers: The coefficient of the variable *headsex* is positive, meaning that female headed households are technically less efficient in the cultivation of rice than male headed households. Finally, rice farmers who are either involved in the distribution of irrigation water themselves or are closely related to the person responsible (variable *irriresp*) attain higher levels of TE than rice farmers for whom this is not the case.

4.4 Technical efficiency estimates

Mean TE is estimated to be 77% in the case of rice cultivation, and 37% in the case of cocoa (bottom section of Table 4). It is worth noting that, at 38%, the tested Translog specification yielded a very similar estimate of mean TE in cocoa cultivation. Furthermore, the firm-specific TE estimates produced by the Translog and the Cobb-Douglas specifications are highly correlated at 0.86, and the conclusions derived with respect to efficiency determinants are identical.

Figure 3 shows the frequency distributions of the TE estimates in rice and cocoa cultivation at the farm level, based on the mean TE over all observed periods. TE in rice production ranges from 0.34 to 0.94, with the majority of farmers (55%) exhibiting TE above 0.80. In cocoa production, TE ranges from 0.04 to 0.98, and is below 0.50 for 73% of farmers. Those farmers who grow both rice and cocoa attain a mean TE of 78% in rice cultivation, as opposed to 30% in the case of cocoa. In 90% of mixed farms TE estimates in rice production are at least 0.20 points higher than in cocoa production, the maximum difference being 0.83.

5 Discussion

5.1 Estimated production functions

While most parameter estimates and deduced partial output elasticities in the production functions conform to our expectations, some coefficients re-

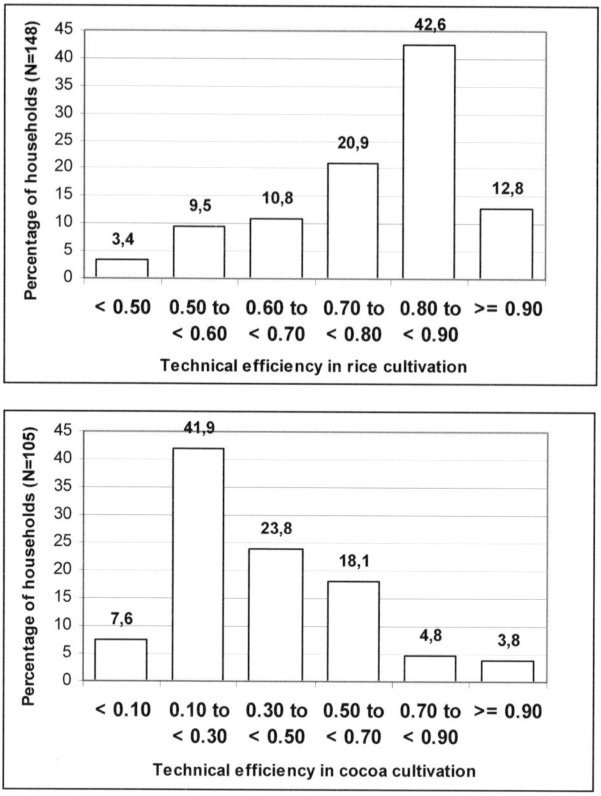

Fig. 3. Distribution of technical efficiencies in the production of irrigated rice (top) and cocoa (bottom) in Central Sulawesi.

quire more detailed comments: In the research region, a negative coefficient of *lowelev* is not surprising, considering that the maximum altitude of rice production is approximately 1000 m a.s.l. and low-lying areas suffer from excessively high temperatures especially in the dry season. In the rice model, the variable *labor* is defined as the total labor input for weed control since this is assumed to be the most influential activity with regard to yield levels. However, in cases of serious weed pressure, yields may be reduced despite a high labor input for weed control, resulting in a very small and statistically insignificant partial production elasticity of *labor*, which led to the exclusion of this variable from the final model. It is interesting to note that both in the rice and cocoa models the partial production elasticity of the *water* variable is highly significant, thus confirming the importance of the inclusion of climate-related variables in agricultural production functions. A large positive partial production elasticity of *age* in the case of cocoa is not surprising, considering that the mean age of cocoa plantations in the sample is only seven years; hence, a large share of cocoa plantations in the research area is still in the stage of increasing productivity.

5.2 Factors influencing efficiency

The LR tests confirm that both the production of rice and cocoa in the research area are subject to technical inefficiency and that, taken together, the explanatory variables in the inefficiency models have a significant impact on the level of TE (Table 2, second and third rows). Our results show that farmer levels of TE can vary greatly between different crops (Figure 3); therefore, the factors influencing efficiency are of particular interest.

The signs of most efficiency determinants are consistent with our hypotheses. In rice production, female headed households attain lower levels of TE than male headed households: female household heads are usually widows and due to labor scarcity they may not be able to accomplish certain agricultural tasks in an optimal manner. Literacy is efficiency enhancing, which is consistent with our expectations since it greatly facilitates the access to information. Poverty is efficiency reducing since it has implications on a household's ability to acquire information and credit. This factor is statistically significant only in the case of rice, which may partly be due to the larger number of observations as compared to the cocoa model. The availability of off-farm income has an efficiency reducing effect on agricultural production. This may be due to competition for labor resources during critical farming activities, such as weed control, or to a general lack of determination in farming if the off-farm income source is of primary importance to the household. Our result is consistent with the findings of Ali and Flinn (1989). Specialization is efficiency enhancing because resources and expertise can be focused on one major crop. Agricultural extension is found to be efficiency enhancing in the case of rice cultivation, but no such effect is supported by the cocoa data. A weakness of this variable is that it does not contain any information regarding the contents of the extension message. It may be that in the case of mixed farms the professional advice focused on rice rather than cocoa, but it can also be deduced that agricultural extension regarding the cultivation of cocoa needs to be improved.

An efficiency influencing factor can also have opposite effects in different activities, which is observed in the case of farmer age: while older farmers are more efficient in rice production than younger farmers, they are less efficient in the cultivation of cocoa. In the case of rice, older farmers are likely to benefit from their experience. Since cocoa is a relatively new crop in the research area, older farmers do not necessarily have more experience in its cultivation than younger farmers. In cases of new technologies, older farmers may be less adaptable and less determined or capable to improve their skills than younger farmers, leading to lower levels of TE.

As hypothesized, the involvement of the household head or a socially close person in water distribution at the village level has positive implications on the level of TE attained in rice production. This indicates that, within a village, the distribution of irrigation water is not carried out in an entirely fair and equal manner, but is also influenced by social relations and 'political capital'. This finding is of particular importance in view of the high partial output elasticity of the *water* variable (bottom section of Table 3).

5.3 Levels of technical efficiency

The results clearly indicate that substantial levels of technical inefficiency impede agricultural production in Central Sulawesi. Thiam et al. (2001) reviewed three studies which estimated TE for rice production in India and the Philippines using Stochastic Frontiers and panel data, the estimates being 89%, 69%, and 70%. Our estimate of 77% is therefore consistent with other comparable studies in Asia. One of the rare attempts found in the literature to estimate TE for cocoa production is a study by Gimbol et al. (1995), using data from three major cocoa producing provinces in Papua New Guinea (PNG). In a sample of 30 smallholder cocoa producers mean TE was estimated to be 57%, with a range from 15% to 96%. At 37%, our estimate of mean TE in cocoa cultivation is considerably lower. In Central Sulawesi, a low TE estimate is supported by field observations which show obvious deficiencies in the management of cocoa plantations, such as inappropriate pruning practices, failure to remove diseased fruits etc.. In the three PNG provinces, cocoa production started in the 1950s to 1960s. Therefore, the higher mean TE may be attributable to the longer experience of PNG smallholders with this crop. Gimbol et al. (1995) did not attempt to identify influencing factors of TE.

5.4 Technical efficiency and deforestation

The mean estimated TE in the cultivation of irrigated rice of 77% implies that production could be considerably increased using the same input vector and technology. However, far more significant increases in production by improved know-how could be achieved in the case of cocoa, the primary cash crop in the research area, for which mean TE is estimated to be merely 37%, and of which area expansion is the main cause of the on-going conversion of forest land (Ruf 2001). In order to conceptualize the conditions under which enhanced productivity would have a forest saving or a forest clearing effect, and what the likely outcome in the case of Central Sulawesi would be, we will start with the analytically simplest case, where farmers are integrated into perfect output and input markets. Under this assumption, we can employ a von Thünen approach linking the net profit obtained from the cultivation of a given cocoa plot (the 'land rent') with the distance between this plot and the village center (von Thünen 1826): when farmers move inputs and outputs between a village and the field, they incur transport costs leading to a diminishing land rent with increasing distance (Figure 4). Equation 6 shows the variables determining the land rent in this simple model (Angelsen et al. 2001 p.23):

$$r = py - wl - qk - vd \tag{6}$$

where

$r =$ land rent per ha
$p =$ output price
$y =$ yield per ha
$w =$ agricultural wage rate
$l =$ labor requirements per ha
$q =$ price of capital
$k =$ capital requirements per ha
$v =$ transport costs per km
$d =$ distance between the homestead and the cultivated plot [km]

The outer limit of agriculture is indicated by the distance d* in Figure 4 where the land rent becomes zero. Enhanced TE will increase yields, y, resulting in an upward shift of the rent gradient and, hence, an outward shift of the agricultural frontier, d*; in the perfect-markets scenario it would thus clearly aggravate deforestation.

The above assumption of a perfect output market may be justified in the case of cocoa: this is an export crop whose price is largely determined by world market conditions and not by the local supply and demand[4]. However, the assumption of perfect input markets is unrealistic in the context of rural areas in developing countries where, more often than not, farmers have limited access to credit and face poorly functioning labor markets (Zeller et al. 1997, p.6). Family labor often has few alternative uses outside the farm, and many households cannot afford to hire in labor; in such circumstances, the amount of family labor available will limit the area that can be cultivated, as indicated by the dotted vertical line in Figure 4 (Angelsen et al. 2001 p.24). With regard to our case study, we find that 16% of farmers hired labor regularly to maintain their cocoa plots, 7% did so in one year only, and 77% did not hire any labor in the period under consideration; this suggests that also in Central Sulawesi the area under cocoa cultivation is likely to be limited by the amount of family labor available.

Given that farmers in Central Sulawesi are to a certain extent labor constrained, it is, furthermore, of interest to know whether there is room to absorb more labor productively in the cultivation of the already existing cocoa plots.

[4] Note that this would not apply in the case of a locally marketed food crop with a relatively inelastic demand.

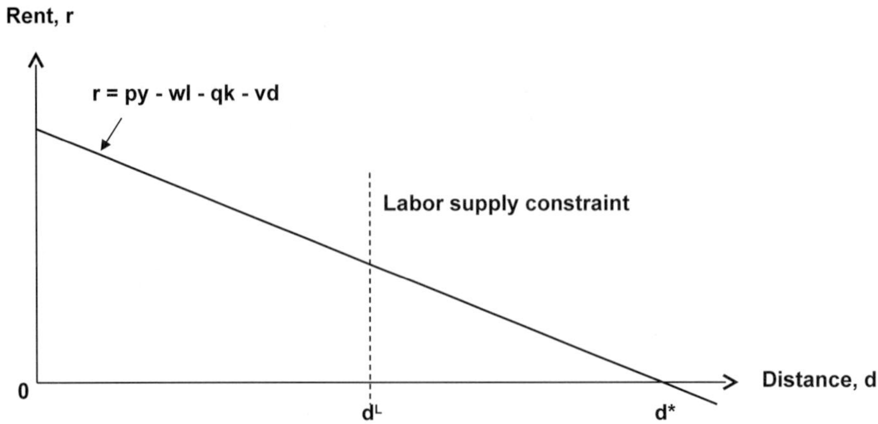

Fig. 4. The agricultural frontier as a function of land rent and distance. Source: Adapted from Angelsen et al. (2001) p.24.

The literature on the impact of technical change on deforestation suggests that labor-using technical change is likely to reduce deforestation by limiting the amount of family labor available for area expansion (Angelsen and Kaimowitz 2001 p.9; Maertens 2003 p.80). The marginal productivity of labor in cocoa production in the sample households was IDR (Indonesian Rupiah) 22,883 per personday, while the agricultural wage rate was merely IDR 15,000; since the marginal return exceeds the marginal cost by 53%, it would be economically justified to intensify labor use on the already existing cocoa plantations, for example by consistently practicing crop hygienic measures such as the removal of diseased fruit from the plantation. Therefore, from a micro-economic perspective, enhanced TE in cocoa production coupled with an intensification of labor use has the potential to reduce deforestation and increase agricultural incomes, thus achieving environmental and economic goals at the same time. However, at the macro-economic level, the labor supply constraint that limits area expansion for a given household can be offset by the in-migration of farm households attracted by the profitability of cocoa production, which can lead to massive deforestation (Ruf 2001). In this volume, the issue of forest conversion and rural migration is discussed by Abdulkadir-Sunito and Sitorus (Part I) and by Weber et al. (Part III).

6 Conclusions, policy implications, and outlook

In summary, we conclude that substantial yield increases could be achieved by higher levels of TE in both rice and cocoa production in Central Sulawesi. The potential increase is particularly large in the case of cocoa. Since this crop is predominantly grown at the rainforest margin and its area expansion exacerbates deforestation, efficiency enhancing measures clearly have the potential to contribute to the protection of the remaining forest by raising incomes from existing cocoa plots. Respective policy measures should aim at reducing illiteracy and alleviating poverty by increasing farm income. Measures focusing on off-farm income opportunities may have an efficiency reducing effect on agricultural production. Moreover, agricultural extension should be intensified and technical advice specifically concerning the management of cocoa should be emphasized. However, it is crucial to acknowledge that the profitability of cocoa cultivation is likely to continue attracting migrants, and that forest conservation policies need to control their influx into forest margin areas if they are to be effective.

With regard to future research, we conclude that policy-oriented analyses of efficiency in agricultural production have to be farm enterprise specific to yield results which allow for an efficient targeting of policy measures. In areas characterized by climate variability, TE analyses should take water availability into account. Our results also indicate that further research on the link between TE and deforestation is justified, especially with regard to the question under which conditions improved efficiency will fulfill the hope of reducing deforestation.

References

Abdulkadir-Sunito M, Sitorus MTF (this volume) From ecological to political buffer zone: ethnic politics and forest encroachment in Upland Central Sulawesi. In: Tscharntke T, Leuschner C, Zeller M, Guhardja E, Bidin A (eds) The stability of tropical rainforest margins, linking ecological, economic and social constraints of land use and conservation. Springer, Berlin Heidelberg New York

Aigner D, Lovell CAK, Schmidt P (1977) Formulation and estimation of stochastic frontier production function models. Journal of Econometrics 6: 21-37

Ali M, Flinn JC (1989) Profit efficiency among basmati rice producers in Pakistan Punjab. American Journal of Agricultural Economics 71: 303-310

Angelsen A, Kaimowitz D (eds) (2001) Agricultural technologies and tropical deforestation. CABI Publishing, in association with Center for International Forestry Research (CIFOR), Wallingford, UK

Angelsen A, van Soest D, Kaimowitz D, Bulte E (2001) Technological change and deforestation: a theoretical overview. In: Angelsen A and Kaimowitz D (eds) Agricultural technologies and tropical deforestation. CABI Publishing, in association with Center for International Forestry Research (CIFOR), Wallingford, UK, pp 19-34

Battese GE (1992) Frontier production functions and technical efficiency: A survey of empirical applications in agricultural economics. Agricultural Economics 7: 185-208

Battese GE (1997) A note on the estimation of Cobb-Douglas production functions when some explanatory variables have zero values. Journal of Agricultural Economics 48: 250-252

Battese GE, Coelli TJ (1995) A model for technical inefficiency effects in a stochastic frontier production function for panel data. Empirical Economics 20: 325-332

Bravo-Ureta BE, Pinheiro AE (1993) Efficiency analysis of developing country agriculture: A review of the frontier function literature. Agricultural and Resource Economics Review 22: 88-101

Brümmer B (2001) Estimating confidence intervals for technical efficiency: The case of private farms in Slovenia. European Review of Agricultural Economics 28: 285-306

Chambers RG (1988) Applied production analysis: A dual approach. Cambridge University Press, Cambridge, UK

Coelli TJ (1996) A guide to FRONTIER Version 4.1: A computer program for stochastic frontier production and cost function estimation. CEPA Working Paper 96/07. Centre for Efficiency and Productivity Analysis, University of New England, Armidale, NSW, Australia

Coelli TJ, Rao DSP, Battese GE (1998) An introduction to efficiency and productivity analysis. Kluwer Academic Publishers, Dordrecht, The Netherlands

Ellis F (1993) Peasant economics. Farm households and agrarian development. 2nd Edition. Cambridge University Press, Cambridge, UK

Eswaran M, Kotwal A (1989) Credit as insurance in agrarian economies. Journal of Development Economics 31: 37-53

Farrell MJ (1957) The measurement of productive efficiency. Journal of the Royal Statistical Society 120: 253-290

Feder G, Just R, Silberman D (1985) Adoption of agricultural innovations in developing countries: A survey. Economic Development and Cultural Change 33: 255-298

Gimbol KC, Battese GE, Fleming EM (1995) Technical efficiencies of smallholder cocoa producers in Papua New Guinea: A Stochastic Frontier Analysis. Quarterly Journal of International Agriculture 34: 337-357

Greene WH (1990) A Gamma-distributed stochastic frontier model. Journal of Econometrics 46: 141-163

Jondrow J, Lovell CAK, Materov IS, Schmidt P (1982) On the estimation of technical inefficiency in the stochastic frontier production function model. Journal of Econometrics 19: 233-238

Kaimowitz D, Angelsen A (1998) Economic models of tropical deforestation: A review. Center for International Forestry Research (CIFOR), Bogor, Indonesia

Kalirajan K (1981) An econometric analysis of yield variability in paddy production. Canadian Journal of Agricultural Economics 29: 283-294

Kalirajan K (1990) On measuring economic efficiency. Journal of Applied Econometrics 5: 75-85

Kodde DA, Palm FC (1986) Wald criteria for jointly testing equality and inequality restrictions. Econometrica 54: 1243-1248

Kumbhakar SC, Ghosh S, McGuckin JT (1991) A generalized production frontier approach for estimating determinants of inefficiency in U.S. dairy farms. Journal of Business and Economic Statistics 9: 279-286

Maertens M (2003) Economic modeling of agricultural land-use patterns in forest frontier areas: Theory, empirical assessment and policy implications for Central Sulawesi, Indonesia. dissertation.de - Verlag im Internet, Berlin, Germany

Meeusen W, van den Broeck J (1977) Efficiency estimation from Cobb-Douglas production functions with composed error. International Economic Review 18: 435-444

Pitt MM, Lee LF (1981) Measurement and sources of technical inefficiency in the Indonesian weaving industry. Journal of Development Economics 9: 43-64

Reifschneider D, Stevenson R (1991) Systematic departures from the frontier: A framework for the analysis of firm inefficiency. International Economic Review 32: 715-723

Ruf F (2001) Tree crops as deforestation and reforestation agents: the case of cocoa in Côte d'Ivoire and Sulawesi. In: Angelsen A and Kaimowitz D (eds) Agricultural technologies and tropical deforestation. CABI Publishing, in association with Center for International Forestry Research (CIFOR), Wallingford, UK, pp 291-315

Seiford LM (1996) Data Envelopment Analysis: The evolution of the state of the art (1978-1995). Journal of Productivity Analysis 7: 99-138

Stevenson RE (1980) Likelihood functions for generalized stochastic frontier estimation. Journal of Econometrics 13: 57-66

Thiam A, Bravo-Ureta BE, Rivas TE (2001) Technical efficiency in developing country agriculture: A meta analysis. Agricultural Economics 25: 235-243

von Thünen JH (1826) Der isolierte Staat in Beziehung auf Landwirtschaft und Nationaloekonomie. Reprint in 5th edition (1990). Scientia Verlag, Aalen, Germany

Weber R, Faust H, Schippers B, Mamar S, Sutarto E, Kreisel W (this volume) Migration and ethnicity as cultural impact factors on land use change in the rainforest margins of Central Sulawesi, Indonesia. In: Tscharntke T, Leuschner C, Zeller M, Guhardja E, Bidin A (eds) The stability of tropical rainforest margins, linking ecological, economic and social constraints of land use and conservation. Springer, Berlin Heidelberg New York

WWF (1981) Lore Lindu National Park Management Plan 1981-1986. A World Wildlife Fund report for the Directorate of Nature Conservation. Bogor, Indonesia

Zeller M, Schrieder G, von Braun J, Heidhues F (1997) Rural finance for food security for the poor: Implications for policy and research. International Food Policy Research Institute (IFPRI), Washington, D.C.

Zellner A, Kmenta J, Drèze J (1966) Specification and estimation of Cobb-Douglas production function models. Econometrica 34: 784-795

Migration and ethnicity as cultural impact factors on land use change in the rainforest margins of Central Sulawesi, Indonesia

Robert Weber[1*], Heiko Faust[1], Björn Schippers[1], Sulaiman Mamar[2], Endriatmo Sutarto[3], and Werner Kreisel[1]

[1] Department of Cultural and Social Geography, Institute of Geography, Georg-August-University, Goldschmidtstr. 5, D-37077 Göttingen, Germany
[2] Department of Anthropology, Universitas Tadulako, Kampus Bumi Tondo, Palu 94118, Indonesia
[3] Department of Agricultural Socio-Economic Sciences, Institut Pertanian Bogor, Jl. Raya Pajajaran, Bogor 16143, Indonesia

*corresponding author: Robert Weber, phone: +49 (0)551 398082, Email: rweber@gwdg.de

Summary

Human activity endangers tropical forests in different parts of the world. The conflicting interests of nature conservation on the one hand, and the livelihood of farmers living at the forest margins, on the other, clash noticeably in so-called hotspots of biodiversity, such as the Lore Lindu region of Central Sulawesi, Indonesia. Biodiversity generally decreases along a land use gradient from natural forest to agroforestry and annual crop systems. Thus, before solutions for a sustainable balance between conservation and the needs of people living at the forest margins can be sought, changing land use strategies and the factors that influence them must be analyzed. While similar studies often concentrate on economic indicators of land use change only, this chapter highlights the importance of two cultural realities, namely migration and ethnicity. We will demonstrate the great influence of these two factors on land use decisions and on the accessibility of land in the Lore Lindu region.

Based on results from previous qualitative studies, we selected three upland villages, which represent a continuum that leads, in terms of migration history, from a pre-transition village, to a transitional village, to a post-transition village. We used quantitative and qualitative methods for our analysis.

The results show a general land use change in the region, ranging from a strategy that places food first (e.g., rice) to one that places cash first, especially by the cultivation of cacao. This change cannot be observed to the same

Tscharntke T, Leuschner C, Zeller M, Guhardja E, Bidin A (eds), The stability of tropical rainforest margins, linking ecological, economic and social constraints of land use and conservation, Springer Berlin 2007, pp 417-436

extent in the study region as a whole. It varies from village to village, depending on the share of households belonging to the Bugis migrants. Bugis as an ethnic group and as migrants have an enormous effect on the land use decisions of local ethnic groups in their respective villages. Strong ethnic networks among Bugis migrants play an important role not only with respect to their village preferences, but also with respect to the transfer of knowledge regarding agricultural management. A lack of interaction between local and migrant ethnic groups results in differences in the efficiency of cacao cultivation. While the Bugis migrants mainly buy their agricultural land from local farmers or, depending on the village's migration history, from other Bugis households, members of the autochthonous ethnic group usually clear the primary forest. The findings of the present study reveal that an analysis of migration and ethnicity is a crucial precondition to finding sustainable solutions for ensuring that the margins of the rainforest in the Lore Lindu region remain stable. This chapter will emphasize that, in addition to economic considerations, an analysis of the cultural forces that influence land use changes cannot be neglected.

Keywords: migration, ethnicity, cultural orientation system, land use change, access to land, Central Sulawesi, Indonesia

1 Introduction

Located within the bio-geographic region of Wallacea, the island of Sulawesi, Indonesia, represents one of the world's most important hotspots of biodiversity. Within this region, the Lore Lindu National Park of Central Sulawesi plays an essential role in the conservation of endemic fauna and flora (Waltert et al. 2004, Schulze et al. 2004a). As elsewhere in the tropics, however, this forest reserve is endangered by human activities (for a general overview of deforestation in the tropics, see Achard et al. 2002, DeFries et al. 2000, Turner et al. 1990, Watson et al. 2001). Between 1971 and 2002, the area of primary forest in the Lore Lindu region decreased by 18 percent, resulting in an annual rate of deforestation of 0.6 percent (Erasmi et al. 2004). Forest encroachments along with agricultural land conversion contribute greatly to causing severe disturbances of the region's ecosystems. As demonstrated by Schulze et al. (2004b) for the region covered in this chapter, in most cases diversity of bird, plant, and insect species decreases along a land use gradient from natural forest to agroforestry and annual cropping systems (for the diversity of beetles, see Bos et al. in this volume). Moreover, the risk of soil erosion on mountain slopes increases from natural forest to fallow land, to perennial crop systems (e.g., cacao, *Theobroma cacao*), and annual crops systems (e.g., maize; Kleinhans and Gerold 2004). Population growth and increasing commercialization, along with a growing need for further agricultural land, especially for cash crop and agroforestry use, aggravate this situation.

Faced with this problem of conflicting demands for land use, one is confronted with the following question: How is stability along the rainforest margins possible? Stability in this connection signifies striking a sustainable balance between forest conservation on the one hand and the livelihood of the people living in the forest's vicinity on the other hand. Without taking into account the socio-cultural and economic needs of these people, sustainable stability at rainforest margins cannot be achieved. As the vast majority of people living in areas such as the Lore Lindu region are farmers, an understanding of their decisions with regard to natural resource management is crucial. Therefore, before finding solutions to this question, an investigation of the factors that influence farmers' decision-making on land use is necessary, a point that is underlined as well by Perz et al. (2005), Turner et al. (1995), and Wood (1992). While other studies often regard this topic from an economic point of view (see e.g. Bann 2000, Maertens 2003, Schwarze 2004, Yaron 2001), this paper focuses on the cultural forces that influence land use decisions. We will concentrate on migration patterns and ethnicity, two cultural factors that we regard as crucial for determining land use changes in the Lore Lindu region, as well as in other tropical frontier zones.

2 Conceptual framework

Studies by Gumbo et al. (2000) and Duram et al. (2004) state that to understand global environmental change one must focus on a local area to provide an in-depth view of the components of land use change. Therefore, an analysis of local variations of land use decisions is important as a base for more general conclusions concerning the factors that cause changes in natural resource use and management.

For this study, which focuses on migration and ethnicity, we partly adopt the concept of culture as an orientation system (Thomas 1993, 1999). Central attributes of this orientation system can be defined as culture standards, i.e., the many varieties of perceiving, thinking, judging, and acting that the members of a group take for typical, self-evident, and mandatory. While Thomas's concept supports the view of culture as homogeneous for a certain group of individuals, Hofstede (1994) suggests that culture defined as an orientation system is not necessarily shared by all individuals who live or have lived within the same social environment. Nevertheless, both approaches draw a too static portrait of culture and neglect the underlying dynamism. Allolio-Näcke et al. (2002) complain that Thomas's and Hofstede's concepts lack the necessary awareness of the reciprocal influences that cultures have on one another. With regard to this, they refer to recent theories that emphasize the hybrid nature of cultural orientation systems (Bhabba 1994, Hannerz 1987, Pieterse 1994, Welsh 1999).

Such reinterpretations of culture are important and essential, but they are also liable to tend to an 'anything goes' discourse. As culture should be regarded as a multi-dimensional system, one may conclude that culture is scientifically unmanageable because everything is culture (see Mitchell 2003, Smith 2000) and respective cultures are too different from each other. Although we agree that this critique of Thomas's concept is justified and that culture is a "complex whole" (Tylor 1871), we nevertheless need to find a conceptual solution for the scientific treatment of aspects of a given culture.

For this study, we choose migration and ethnicity as the two main components that shape the cultural orientation system, which in turn influences decision-making concerning land use (see Figure 1). Recent studies in other world regions affirm the importance of migration and ethnicity (see e.g. Breusers 2001, Joseph 2002). Prior qualitative investigations in the Lore Lindu region refer to the great impact of migration and ethnicity on land use change. So far, however, a quantitative demonstration is largely missing. While migration, in general, is influenced by one or more different factors, such as economic, political, or social ones, it has strong implications for the orientation systems of both the migrant and the people in the target region. Thus, in this context we regard migration as a cultural factor that shapes the orientation system(s). In this chapter, we focus mainly on spontaneous rural migration from areas within and outside of the study region to a given village. The terminus ethnicity, in our study, does not only include the peoples' perception of belonging to a certain ethnic group, but also the knowledge and skills that characterize the majority of an ethnic group. Such special characteristics probably influence farmers' land use decisions.

With regard to land use decisions, access regulations to land play a predominant role. They determine the possible land use options on a particular piece of land and are strongly linked to migration and ethnicity (see Figure 1). Citing Gramsci (1971) and Foucault (1984), Peluso (2005) points out that power relations in local territorial resource management may involve the exercise of discipline through coercion and consent, but that they do not necessarily involve the authority of national governments. The examples of local changes in land allocation since the decline of the authoritarian Suharto regime in 1998 and the consequent power decentralization in Indonesia (see Abdulkadir-Sunito and Sitorus, in this volume) demonstrate how local policies (i.e., at the village level) influence access to land. In addition to their causal links to political changes on the national level, however, cases of different local land allocation policies, which are not directly linked to national developments, also exist (Weber 2006). Migration has an impact on access to land, depending on who is migrating and when migration occurs (see e.g. Doevenspeck 2004, Rodgers 1991). Ethnic affiliations play an important role in such regions as our study area and influence the access to land (see e.g. Bonte 1999, Sitorus 2002, Venema and Mguild 2003). Especially if land allocation is managed by local authorities exercising customary law rather than by national authorities, belonging or not belonging to the ethnic group of local rulers can influence one's access to land.

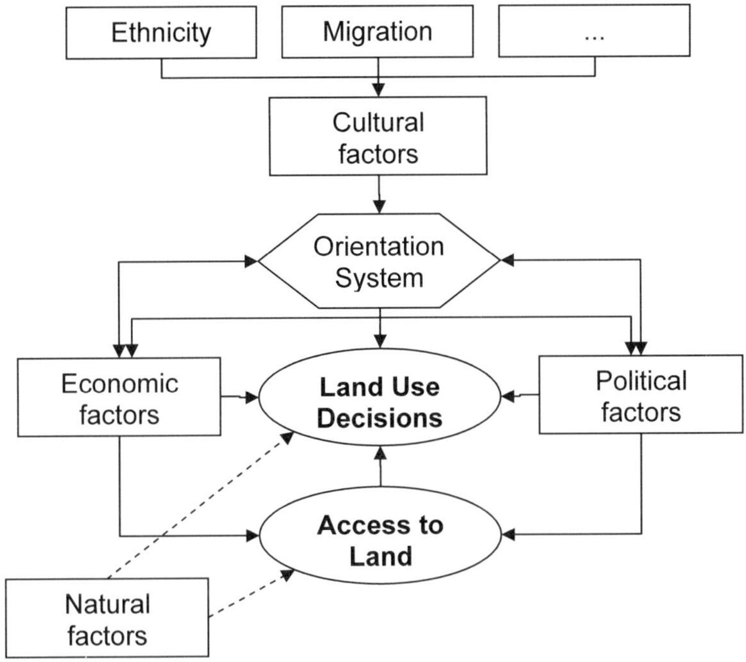

Fig. 1. Conceptual framework

3 Methods

The study in this paper uses a multi-method approach, combining quantitative and qualitative research tools. The quantitative part is covered by a census that was conducted in three villages within the study region in 2004. We selected the villages based on our previous qualitative studies regarding cultural change in the Lore Lindu region during the 20th century (see Weber et al. 2003, Faust et al. 2003, Hoppe and Faust 2004, Kreisel et al. 2004, Weber 2006). The three villages, named A, B, and C, were assumed to represent different stages on a transition continuum, a transition which is for the most part initiated by the villagers' migration patterns. This continuum encompasses a relatively static type of village (A), where the number of new migrants decreased steadily during the last 15 years, to a quite dynamic or post-transitional village (C). The latter is characterized by a high share of local migrants who were among the first settlers in the 1970s, and a high share of regional migrants from South Sulawesi who settled down during a relatively short period (1980-2000). Village B is situated between the two extremes and represents a village in transition, as 30 percent of all migrants arrived during just the past four years (see Figure 2). All three villages are located in the uplands of the Lore Lindu region and have similar access to roadways.

For the village census we interviewed all those household heads who responded using standardized questionnaires. For different reasons, only very few household heads could not be interviewed. In total, 898 households (hh) were interviewed (521 hh in village A, 148 in village B, and 229 in village C). The whole data set provides land use information for 1, 803 agricultural plots (1, 198 plots in village A, 298 plots in village B, and 308 plots in village C). The census data cover the following topics: demography of household members, migration, land access, land use, input/output per plot, use of forest products, labor division, knowledge/information transfer, labor, market access, livestock, income, socio-cultural interrelations, change of perceptions, consumption, dwelling conditions, assets, and environmental perception. Conducting a census is advantageous in that a very realistic picture of the village can be painted, for not only a *sample* of village households, but *every* village household is interviewed.

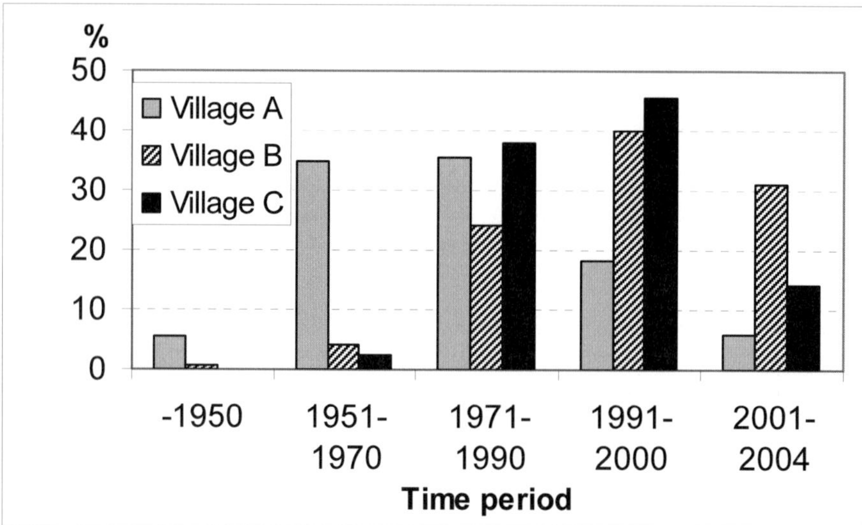

Fig. 2. Temporal distribution of the arrival of migrants in the three study villages (Source: STORMA project A1 village census 2004; number of observations = 557)

In addition to the quantitative census data set, qualitative data are indispensable for interpreting quantitative results, for verifying causal relations, and for revealing why people make the decisions they do. Our qualitative methods included formal, semi-structured in-depth interviews with the heads of selected migrant and non-migrant households belonging to different ethnic groups, and with village representatives. We also conducted informal talks, discussions, and participant observations. Besides the studies in the three census villages, we benefited from our previous qualitative investigations, as noted earlier in this paper. The integration of our research within the interdisciplinary STORMA project also enables us to benefit from the results of other social and natural science research within the study region.

4 Land use changes in the Lore Lindu region

For the past 100 years, the Lore Lindu region has experienced a variety of land use changes, most of which were brought about by the respective political rulers of the central or regional governments. First modifications to land use, beginning in the period from 1904 to 1908, were enforced by the Dutch colonial leaders (Weber et al. 2003). At that time, the then widespread practice of slash-and-burn cultivation was curbed by the expansion of wet rice fields in the plains of the lowlands and the uplands – an expansion that was accompanied by forced local resettlements from the mountain areas to the valleys. Also, indigenous noble families learned about the coffee plant, and coconut trees, an important cash crop, were planted. During their three years of occupation (1942-1945), the Japanese colonial rulers forced the people to plant cotton, a practice that put subsistence cultivation at risk of being neglected. One result was that numerous bouts of famine occurred in the region.

Table 1. Crop preferences in the three study villages

	Main crops cultivated (in %)					
	Cacao	Coffee	Wet rice	Maize	Others	Total
Village A	55.6	3.1	36.6	1.4	3.3	100.0
Village B	87.6	0.7	0.0	8.1	3.7	100.1*
Village C	96.1	0.0	0.3	2.3	1.3	100.0

Source: STORMA project A1 village census 2004
Number of observations = 1804
*Deviation from 100.0% due to rounding

Major land use changes followed the colonial period (Weber 2006). Cotton cultivation was abandoned and the former land use mix of wet rice cultivation, dry land rice, and mixed vegetable crop cultivation (*palawija*), along with coffee and coconut tree cultivation was re-established. By the late 1970s, the Indonesian government officially prohibited shifting cultivation and supported an increase of wet rice production by the continuation of local resettlements from the hills to the plains, as the Dutch had done during their rule. At the same time, world market prices increasingly influenced local farmers' decisions to augment their cash crop production. Besides coffee, clove became an important cash crop. While wet rice was planted in the plains, the former shifting cultivation fields were permanently replaced with cash crops and mixed food crops. The cultivation of the latter, however, remained the first choice for the vast majority of farmers, as they recognized the importance of an independent supply of subsistence crops.

The introduction and spread of the cacao plant in the early 1980s marks the largest and most far-reaching land use transformation in the Lore Lindu region of recent times. Today, many fields formerly planted with annual crops, such as rice, maize, vegetables, and fruits, or with perennial crops, such as coffee and clove, are being converted into cacao fields. Furthermore, recently cleared forest plots are planted with cacao. Cacao is usually grown as a monoculture or cultivated with coffee. For the STORMA project's research, three shading types were identified: (1) cacao planted under natural forest cover; (2) under highly diversified planted shade trees; and (3), under planted shade trees, the majority of which belong to one species.

5 The impact of migration and ethnicity on land use decisions in the Lore Lindu region

Since the mid-1990s (at the latest), cacao has become economically the most important crop in the study region and, thus, one might assume that the vast majority of farmers have changed their land use decisions in favor of this profitable plant. As Table 1 shows, however, this agricultural change did not occur to the same extent in all villages. While the share of cacao fields does not exceed 56 percent in pre-transition village A, it increases to almost 87 percent in transitional village B, and it reaches more than 96 percent in post-transition village C. Correspondingly, the share of food crop fields (e.g. rice and maize) decreases from village A to village C.

Table 2. Share of local and migrant households

	Share of local and migrant households (in %)		
	Local	Migrant	Total
Village A	58.2	41.8	100.0
Village B	28.4	71.6	100.0
Village C	7.0	93.0	100.0

Source: STORMA project A1 village census 2004
Number of observations = 1804

In comparing these results with the share of migrants living in the respective villages (see Table 2), which shows the smallest value in village A and the highest one in village C, one could argue that a direct causal relation exists between the share of migrants and the share of cacao fields in a given village. The Pearson correlation between the share of cacao and the share of migrants in the three villages even shows a significant result ($p < 0.1$, $R^2 = 0.95$). Thus, this outcome would support the simple equation: more migrants = more cacao. A detailed look at the ethnic affiliation of the migrants, however, presents a more complicated picture of this relation. Regarding the ethnic composition of village communities (see Table 3), we observe that the share of cacao fields increases with the share of Bugis migrants living in a village.

Table 3. Distribution of the different ethnic groups in the three study villages

	Distribution of ethnic groups (in %)					
	Local groups	Regional groups	South Sulawesi groups	Bugis	Others	Total
Village A	59.8	13.6	21.5	0.6	4.6	100.1*
Village B	65.3	18.4	1.4	15.0	0.0	100.1*
Village C	53.3	0.9	1.3	43.7	0.9	100.1*

Source: STORMA project A1 village census 2004
Number of observations = 1804
* Deviation from 100.0 percent due to rounding

This result reinforces the findings of previous qualitative studies in this region (see Li 2002, Sitorus 2002, Weber 2006). Introduced by a Bugis trader around 1980, cacao cultivation started to spread in the Lore Lindu region. Bugis farmers from South Sulawesi, some of which gained the necessary knowledge about the cultivation of cacao during their stay as contract workers at cacao plantations in Malaysia during the late 1970s (Akiyama and Nishio 1996), moved to Central Sulawesi during the 1980s. As they faced increasing land scarcity, the Bugis migrants were attracted to the availability of suitable land for cacao cultivation at upland frontier zones such as those of Central Sulawesi. A boom in the Indonesian market price for cacao in the mid-1990s resulted in an enormous population increase of immigrants from South Sulawesi. A historical comparison of cacao production in other regions of the world (Hill 1953, 1963; Jamal and Pomp 1976; Ruf 1995; Ruf and Ehret 1993) shows that most cases of cacao boom are linked to major migration flows. But why is it that, in our case especially, it is the Bugis who play a predominant role in migration to the Lore Lindu region? Answers to this question are found in the history of this ethnic group. The Bugis, to whom Pelras (1998:27) assigns "a tradition specifically open to change" are known to be descendents of seafarers and traders who expanded their sphere of influence within Southeast Asia for centuries. They were interested in the establishment of strategic alliances with newly opened territories, and they cemented these alliances through marriage and the foundation of new settlements (Pelras 1996). Although nowadays most Bugis are not concerned with seafaring anymore, the tradition of opening up new regions seems to have been preserved until today.

Although the Bugis migration led to an obvious land use change in many areas of the Lore Lindu region, not all villages were affected to the same extent. Besides limiting factors such as transportation infrastructure or climate conditions, social ethnic information networks and village policies represent two major determinants of the Bugis' migration decisions and patterns (Weber 2006).

As can be interpreted from Table 1 and 3 and as shown in Figure 3, cacao cultivation did not remain limited to Bugis farmers only. Nonetheless, the population of this migrant ethnic group in a village strongly affects the spread of cacao cultivation among the local ethnic groups.

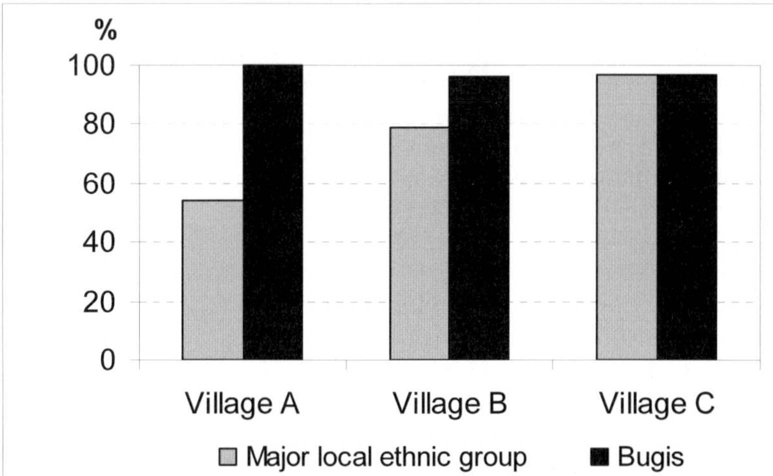

Fig. 3. Share of cacao fields of the major local ethnic group and the Bugis group in the three study villages (in percent of the total number of fields cultivated by the respective group; Source: STORMA project A1 village census 2004; number of observations = 1268)

Figure 3 shows that the share of cacao among the main local ethnic groups' land use decisions is linked to the share of Bugis migrants living in the respective village. The proportion increases from 54 percent in pre-transition village A to 97 percent in post-transition village C.

Among all ethnic groups in the study region, ethnic networks do not only influence migration patterns, but also have an impact on spatial residential patterns and social and economic interactions within the villages. The networking within the largest autochthonous ethnic group of a village influences the village administration in its residential policies. Some villages in the research area practice spatial residential segregation based on such policies. More frequently, this segregation occurs as a result of ethnic networking within the migrant groups. Migrants often prefer living in neighborhoods composed of the same ethnic group. In most cases, family or friendship ties already exist before the arrival of a given migrant, and the information networking within these relationships often influences the decisions of migrants when they select the respective village as a migration target. Many respondents report that being close to people of the same ethnicity eases them into the new village. Given that religion plays an important role in the study region, different religious affiliations of the autochthonous and the migrant group, especially if it is Christian or Muslim, enforce residential segregation. In contrast, if locals and migrants have the same religion, then inter-ethnic marriages are more likely and thus reduce segregation.

Social interaction between local ethnic groups and Bugis is, apart from some exceptions, relatively low. Most local farmers gained their knowledge about cacao cultivation by observing the Bugis farmers, while only few of them directly asked the migrants from South Sulawesi about cultivation techniques. In accordance with the general statement of Laine (2004), which argues that land use decisions also depend on those of neighboring owners, it is understandable that the share of local farmers' investment in cacao cultivation increases in villages where more Bugis have settled. This does not, however, fully explain why the local farmers in village A plant more than 50 percent of their fields with cacao, although the share of Bugis migrants in this village is very low. The respondents report that their widespread family relations for each household are linked to a large network of villages within and outside the research area. These networks allow for information exchanges among members. When asked about what led to an increase in cacao cultivation, many respondents with local ethnic affiliation told that their relatives or friends, who live in other villages, which host a larger number of Bugis migrants, had reported the advantages of planting cacao. Thus, knowledge about cacao is transferred from villages with a higher Bugis population to villages with a lower one.

6 The impact of migration and ethnicity on access to land

In addition to land use decisions, land access plays an important role in the communities that occupy the rainforest margins, especially when it comes to stability and destabilization of these margins. Cacao is mainly grown on the slopes within or close to the forest. In other words, this crop dominates the area along the sensitive forest borders. According to a survey, which included 301 households in the Lore Lindu region, the area of rain-fed farming acquired by clearing away primary forest increased from less than 50 ha during the period extending from 1951 to 1990 to more than 60 ha during the comparably much shorter period of 1996 to 2001 (van Rheenen et al. 2004). This growth in forest conversion has to be seen within the context of the migrants-driven cacao boom in the study region. Although we do not agree with the general argument of Carr (2004:598), namely, that immigration is a "prerequisite to frontier forest conversion", which he explains with examples derived from South America (Pichón and Bilsborrow 1999, Southgate et al. 1991, Wood and Perz 1996, Wood et al. 1996), immigration certainly can play a crucial role in forest conversion (see e.g. Faust et al. 2003).Our census data stand in contrast the general argument of Carr (2004) that immigration directly leads to forest clearing, which supports the assumption that it is predominantly the migrants who encroach into the forests. Figure 4 reveals that mainly ethnic groups, which call themselves the local group in the respective villages, clear the primary forest. Almost 20 percent of the plots owned by the main local ethnic group in village B were acquired in this way. In village C this type of land acquisition by the local group even reaches 25.0 percent. In both villages in contrast, Bugis migrants usually bought their land, while forest clearance makes up for less than one percent of all Bugis land acquisitions in each of the two villages.

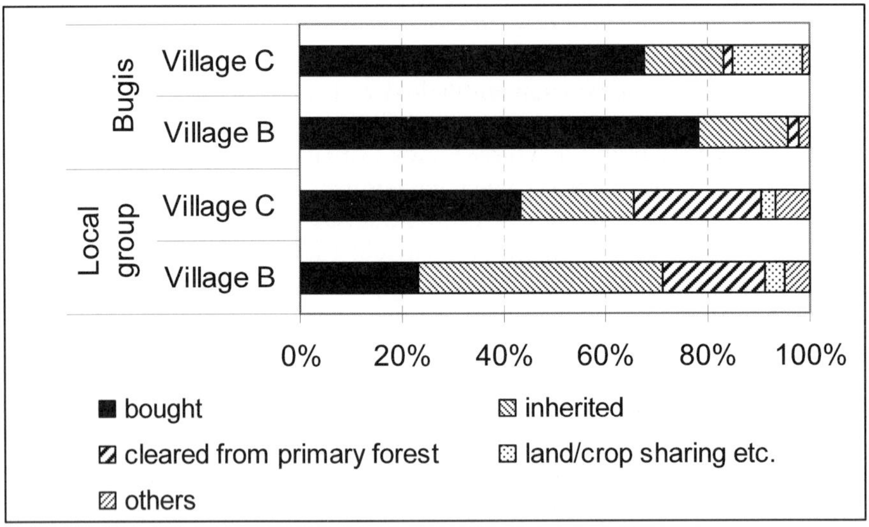

Fig. 4. Land acquisition by the major local ethnic group and the Bugis in the villages B and C (Source: STORMA project A1 village census 2004; number of observations = 543)

When it comes to the Bugis migrants, it is, in contrast to the case study findings of Carr (2002) in Guatemala, not the exceptionally poor that are most likely to migrate to the forest frontiers. Our census data reveal that the majority of the Bugis farmers are found in the highest income groups. They have the financial means to purchase land. The local ethnic groups, which regard themselves as the owners of their village's territory, do not deem it necessary to buy land, but they do realize the opportunity of generating cash by selling parts of their land. The forest, which represents their customary land reserve, is cleared by the local groups to create a balance with the land sold outside the forest or to sell it in order to increase their cash income. This additional income is mainly used for buying status symbols (e.g., motorcycles, TV sets, furniture), purchases that are the results of the improved living standards displayed by the affluent Bugis migrants and the modern lifestyles portrayed on television programs. Additional income is also used for paying medical care and children's education. A quantitative comparison of the ethnic affiliations of buyers and sellers of land in transition village B and post-transition village C (see Table 4) demonstrates the aforementioned relation between local ethnic groups and Bugis. 55.6 percent of the land bought by Bugis in village B and the 38.9 percent bought in village C was sold by members of the main local ethnic group, whereas 41.6 percent in village B and 51.9 percent in village C were sold by other Bugis households.

The different proportion of land in the villages B and C that is sold within the Bugis group can be explained by the respective migration histories of the two villages. Many Bugis migrants in village C arrived earlier than those in village B, where we find a comparably high immigration rate during the past few years. This is one reason we made a distinction between the transitional and post-transitional village. Owing to the fact that more Bugis families have lived in village C for a longer time, more of them were able to increase their land holdings, portions of which they then were able to sell to Bugis migrants arriving later.

7 Conclusions

Our findings pertaining to aspects of the impact of migration and ethnicity on changes in land use contrast with the results of Casey and Caviglia (2000), who state that, in general, information is an important factor in decision-making in the tropical forest regions of Brazil and Mexico regardless of cultural and socio-economic variations. Our study's results reveal that migration patterns and ethnicity do play an important role in land use changes, especially if these changes are investigated at smaller scales (e.g., at the village level). Moreover, intra-ethnic information networks are, in this particular context, highly relevant. In the Lore Lindu region in general, one can observe a change from a 'food crops first' strategy to a 'cash crops first' strategy, dominated by the cultivation of cacao. This transitional change, however, does not occur to

Table 4. Land acquisition by the Bugis in the villages B and C

	How acquired?	From whom acquired?							
		same ethnic group		other ethnic group (local)		other ethnic group (migrant)		village government etc.	
		number	%	number	%	number	%	number	%
Village B	bought	15	41.7	20	55.6	1	2.8	0	0
	inherited	7	87.5	1	12.5	0	0	0	0
	cleared from primary forest	0	0	0	0	0	0	1	100.0
	land/crop sharing etc	0	0	0	0	0	0	0	0
	other	0	0	0	0	0	0	0	0
Village C	bought	56	51.9	42	38.9	9	8.3	1	0.9
	inherited	24	96.0	1	4.0	0	0	0	0
	cleared from primary forest	0	0	0	0	0	0	3	100.0
	land/crop sharing etc.	18	81.8	4	18.2	0	0	0	0
	others	1	50.0	0	0	0	0	1	50.0

Source: STORMA project A1 village census 2004
Number of observations = 205

the same extent within the whole study region, but rather depends largely on migration and ethnicity, as well as natural conditions that already exist in the immediate area. Our study villages represent three stages along a gradient, which includes the two above-mentioned land- use strategies.

The farmers' land use decision favoring the cash crop cacao is generally an economic one, but it is also related to cultural forces. As presented in our conceptual framework, the system that forms the basis for such a decision is influenced by a given individual's cultural background. The ethnic group's historical and cultural traditions, which are linked with territorial expansion, migration, and trade, shape the cultural orientation systems of many Bugis migrants. When they move to the study region, they are drawn to the prospect of economic success in the cultivation (and trade) of cacao. They are, like many entrepreneurs, well prepared for achieving their goals. On arrival at the new destination, they receive the necessary funds to allow them to purchase land, and they are already skilled in the cultivation of the crop. After the Bugis's arrival, the village's autochthonous groups adopt the cultivation of this new crop. The initial cause for this land use change is, once again, economic. Local ethnic farmers realize the financial advantages of engaging in the cultivation of cacao as they observe the rapid improvement of the Buginese cacao farmer's economic status. The majority of the autochthonous farmers, however, do not succeed in the same way as many Bugis because of their different cultural orientation systems. They conform to tradition, for example, by cultivating cacao on much smaller fields than the Buginese farmers. While most Bugis farmers in the study region concentrate their agricultural activities from the beginning on the cash crop cacao only, the cultivation of subsistence crops, especially rice, represents an important aspect of the autochthonous farmers' decision-making. Rice traditionally represents the basis of their livelihoods. Market-orientated crop production was traditionally valued as a way to generate additional income surplus, not as a total replacement of crop cultivation for food. The size of the Bugis community in a given village correlates to the speed of the autochthonous farmers' adoption of cacao cultivation. Here again, migration and ethnicity have an important influence on land use decisions. The Bugis migrants' selection of their target village, strongly depends on their intra-ethnic information networks. These allow for the exchange of information about the target village, including the availability of purchasable land, the suitability of the soil for farming, climate conditions for cacao cultivation, and the target village's immigration policy. In addition, the local ethnic groups' networks support the spread of cacao cultivation to villages where less Bugis migrants have settled. Based on our findings, we predict that if the economic conditions for cacao agroforestry do not deteriorate significantly, the change in land use strategy from food first to cash first will continue, even if one takes into account such factors as migration and ethnicity, both of which we discussed above.

Acknowledgements
This research is part of Project A1, "Demographic change and its impact on land use", of the collaborative research center SFB 552- STORMA "Stability of Rainforest Margins in Indonesia", which is funded by the German Research Council (DFG). We are deeply grateful to the communities of our three study villages for their support. Many thanks to Muhamad Akib, Muhamad Yusup, and Ibnu Saad, as well as all local enumerators, for their research assistance. We are indebted to the coordination staff of STORMA in Göttingen, Bogor, and Palu, and especially to Rina Yusuf, Arman, and the drivers.

References

Abdulkadir-Sunito M, Sitorus MTF (this volume) From ecological to political buffer zone: ethnic politics and forest encroachment in Upland Central Sulawesi. In: Tscharntke T, Leuschner C, Zeller M, Guhardja E, Bidin A (eds) The stability of tropical rainforest margins, linking ecological, economic and social constraints of land use and conservation. Springer, Berlin Heidelberg New York

Achard F, Eva HD, Stibig H, Mayaux P, Gallego J, Richards T, Malingreau J (2002) Determination of Deforestation Rates of the World's Humid Tropical Forests. Science 297: 999-1002

Akiyama T, Nishio A (1996) Indonesia'a cocoa boom: hands-off policy encourages smallholder dynamism. Policy research working papers, no 1580. The World Bank, Washington D.C.

Bann C (2002) Economic analysis of tropical forest land use options in Cambodia. In: Pearce DW (ed) Valuing the environment in developing countries: case studies. Elgar, Cheltenham Glos, pp 536-569

Bhabba HK (1994) The Location of Culture. Routledge, London

Bonte P (1999) Ethnicity and land tenure in the Sahel. In: Reyna SP (ed) Deadly developments: capitalism, states and war. Gordon & Breach, Amsterdam, pp 213-234

Bos M, Höhn P, Saleh S, Büche B, Buchori D, Steffan-Dewenter I, Tscharntke T (this volume) Insect diversity responses to forest conversion and agroforestry management. In: Tscharntke T, Leuschner C, Zeller M, Guhardja E, Bidin A (eds) The stability of tropical rainforest margins, linking ecological, economic and social constraints of land use and conservation. Springer, Berlin Heidelberg New York

Breusers M (2001) Searching for Livelihood Security: Land and Mobility in Burkina Faso. The journal of development studies 37(4): 49-80

Carr DL (2002) Rural frontier-migration and deforestation in the Sierra De Lacandon National Park, Guatemala. Ph.D. thesis, University of North Carolina

Carr DL (2004) Proximate Population Factors and Deforestation in Tropical Agricultural Frontiers. Popul Environ 25(6): 585-612

Casey JF, Caviglia JL (2000) Deforestation and Agroforestry Adoption in Tropical Forests: Can We Generalize? Some Results from Campeche, Mexico and Rondonia, Barzil. Paper presented at the Western Agricultural Economics Association Annual Meetings, Vancouver, British Columbia, June 29 – July 1, 2000.
Electronic resource: http://home.wlu.edu/~caseyj/vancouver.pdf

DeFries R, Hansen M, Towsend JRG, Janetos AC, Loveland TR (2000) A new global 1 km data set of percent tree cover derived from remote sensing. Global Change Biology 6: 247-254

Doevenspeck M (2004) Rural migration and interethnic relationship in land-tenure issues. The example of Borgu (Benin). Afrika-Spectrum 39(3): 359-380

Duram LA, Bathgate J, Ray C (2004) A Local Example of Land-Use Change: Southern llinois – 1807, 1938, and 1993. The Professional Geographer 56(1): 127-140

Erasmi S, Twele A, Ardiansyah M, Malik A, Kappas M (2004): Mapping deforestation and land cover conversion at the rainforest margins in Central Sulawesi, Indonesia. EARSeL eProceedings 3(3): 288-297

Faust H, Maertens M, Weber R, Nuryartono N, van Rheenen T, Birner R (2003) Does migration lead to destabilization of forest margins? – Evidence from an interdisciplinary field study in Central Sulawesi. STORMA Discussion Paper Series Sub-Program A, no 11, Bogor Göttingen Kassel

Foucault M (1976) Überwachen und Strafen: die Geburt des Gefängnisses. Suhrkamp, Frankfurt am Main

Gumbo DJ, Eastman R, Toledano J, Snel M, Haan N (2000) It takes a village – Assessing social causes of environmental change in Africa. GeoInfo Systems 10(1): 22-27

Gramsci A (1991-2002) Gefängnishefte, 10 vol. Argument, Hamburg

Hannerz U (1987) The World in Creolization. Africa 57(4): 546-559

Hill P (1956) The Gold Coast cocoa farmer: a priliminary survey. Oxford University Press, London.

Hill P (1963) The migrant cocoa-farmers of Southern Ghana: a study in rural capitalism. University Press, Cambridge

Hofstede G (1994) Cultures and Organizations. Intercultural Cooperation and its Importance for Survival. Harper Collins Publishers, London

Hoppe M, Faust H (2004) Transmigration and integration in Indonesia – impacts on resource use in the Napu Valley, Central Sulawesi. STORMA Discussion Paper Series Sub-Program A, no 13, Bogor, Göttingen, Kassel

Jamal S, Pomp M (1976) Smallholder adoption of tree crops: A case study of cocoa in Sulawesi. Bull Indonesian Econ Stud 29(3): 69-94

Joseph KV (2002) Migration and changing pattern of land use in Malabar. Journal of Indian School of Political Economy 14(1): 63-81

Kleinhans A, Gerold G (2004) The effects of rainforest conversion on water balance, water yield and seasonal flows in a small tropical catchment in Central Sulawesi. In: Gerold G, Fremerey M, Guhardja E (eds) Land use, nature conservation and the stability of rainforest margins in Southeast Asia. Springer, Berlin Heidelberg New York, pp 353-365

Kreisel W, Weber R, Faust H (2004) Historical impacts on use and management of natural resources in the rainforest margins of Central Sulawesi. In: Gerold G, Fremerey, M, Guhardja E (eds) Land use, nature conservation and the stability of rainforest margins in Southeast Asia. Springer, Berlin Heidelberg New York, pp 39-65

Laine T (2004) Methodology for Comparing Agent-based Models of Land-use Decisions. In: Lovett MC, Schunn CD, Lebiere C, Munro P (eds) Proceedings of the Sixth International Conference on Cognitive Modeling: 6thICCM 2004, integrating models, July 30 – August 1, 2004. Lawrence Erlbaum, Mahwa, pp 410-411

Li TM (2002) Local Histories, Global Markets: Cocoa and Class in Upland Sulawesi. Development and Change 33(3): 415-437

Maertens M (2003) Economic modelling of agricultural land-use patterns in forest frontier areas: theory, empirical assessment and policy implications for Central Sulawesi, Indonesia. dissertation.de, Berlin.

Mitchell D (2003) Cultural Geography. Blackwell, Oxford

Pelras C (1996) The Bugis. Blackwell, Oxford

Pelras C (1998) Bugis culture: a tradition of modernity. In: Robinson K, Paeni M (eds) Living through histories: culture, history and social life in South Sulawesi. Department of Anthropology, Research School of Pacific and Asian Studies, The Australian National University, Canberra, pp 19-28

Peluso NL (2005) Seeing property in land use: Local territorializations in West Kalimantan, Indonesia. Danish Journal of Geography 105(1): 1-15

Perz SG, Aramburú C, Bremner J (2005) Population, land use and deforestation in the Pan Amazon Basin: a comparison of Brazil, Bolivia, Colombia, Ecuador, Perú and Venezuela. Environment, Development and Sustainability 7(1): 23-49

Pichón FJ, Bilsborrow R (1999) Land use systems, deforestation, and demographic factors in the humid tropics: Farm-level evidence from Ecuador. In: Bilsborrow R, Hogan D (eds) Population and deforestation in the humid tropics. IUSSP, Liege, pp 175-207

Pieterse JN (1994) Globalisation as Hybridisation. International Sociology 9(2): 161-184

Rodgers WM (1991) The significance of access to land as a determinant of Kenya's interregional migration. World Development 19(7): 921-926

Ruf F (1995) Booms et Crises du Cacao: Les Vertiges de l'Or Brun. Karthala, Paris

Ruf F, Ehret P (1993) Compétitivité et cycles du cacao: vrais et faux problèmes sous éclairage Indonésien. In: Etienne G, Griffon M, Guillaumont P (eds) Afrique-Asie: Performances agricoles comparées. Edition de la Revue française d'Economie, Paris, pp 255-301

Schulze CH, Steffan-Dewenter I, Tscharntke T (2004a) Effects of land use on butterfly communities at the rain forest margin: a case study from Central

Sulawesi. In: Gerold G, Fremerey M, Guhardja E (eds) Land use, nature conservation and the stability of rainforest margins in Southeast Asia. Springer, Berlin Heidelberg New York, pp 282-297

Schulze CH, Waltert M, Kessler PJA., Pitopang R, Shahabuddin, Veddeler D, Mühlenberg M, Gradstein SR, Leuschner C, Steffan-Dewenter I, Tscharntke T (2004b): Biodiversity indicator groups of different land-use systems: comparing plants, birds, and insects. Ecol Appl 14(5): 1321-1333

Schwarze S (2004) Determinants of income generating activities of rural households: a quantitative study in the vicinity of the Lindu-Lindu National Park in Central Sulawesi/Indonesia. Niedersächsische Staats- und Universitätsbibliothek, Göttingen

Sitorus MTF (2002) Land, ethnicity and the competing power: agrarian dynamics in forest margin communities in Central Celebes, Indonesia. STORMA Discussion Paper Series Sub-Program A, no 5, Bogor, Göttingen, Kassel

Smith N (2000) Socializing culture, radicalizing the social. Sozial & Cultural Geography 1: 25-28

Southgate D, Sierra R, Brown L (1991) The causes of tropical deforestation in Ecuador: A statistical analysis. World Development 19(9): 1145-1151

Thomas A (1993) Psychologie interkulturellen Lernens und Handelns. In: Thomas A (ed) Kulturvergleichende Psychologie. Hogrefe Verlag für Psychologie, Göttingen

Thomas A (1999) Kultur als Orientierungssystem und Kulturstandards als Bauteile. IMIS-Beiträge 10: 91-130

Turner BL, Skole D, Sanderson S, Fischer G, Fresco L, Leemans R (1995) Land-use and Land-cover Change Science/Research Plan.IGBP/HDP, Stockholm Geneva

Turner BI, Turner BC, Clark RW, Kates JF, Richards JT, Meyer WB (eds, 1990) The Earth as Transformed by Human Action. Cambridge University Press, Cambridge

Tylor EB (1871) Primitive culture: researches into the development of mythology, philosophy, religion, language, art and custom: in two volumes. Murray, London

Van Rheenen T, Elbel C, Schwarze S, Nuryartono N, Zeller M, Sanim B (2004) Encroachments on Primary Forests: Are They Really Driven by Dispair? In: Gerold G, Fremerey M, Guhardja E (eds) Land use, nature conservation and the stability of rainforest margins in Southeast Asia. Springer, Berlin Heidelberg New York, pp 199-213

Venema B, Mguild A (2003) Access to Land and Berber Ethnicity in the Middle Atlas, Morocco. Middle Eastern Studies 39(4): 35-53

Waltert M, Lankau M, Maertens M, Härtel M, Erasmi S, Mühlenberg M (2004) Predicting the loss of bird species from deforestation in Central Sulawesi. In: Gerold G, Fremerey, M, Guhardja E (eds) Land use, nature conservation and the stability of rainforest margins in Southeast Asia. Springer, Berlin Heidelberg New York, pp 327-349

Watson RT, Noble IR, Bolin B, Ravindranath NH, Verado DJ, Dokken DJ (eds, 2001) Land Use, Land Use Change, and Forestry. Cambridge University Press, Cambridge

Weber R (2006) Kulturlandschaftswandel in Zentralsulawesi: Historisch-geographische Analyse einer indonesischen Bergregenwaldregion. Universitätsverlag Göttingen, Göttingen

Weber R, Kreisel W, Faust H (2003) Colonial interventions on cultural landscape of Central Sulawesi by Ethical Policy – Impacts of the Dutch rule in Palu and Kulawi valley 1905-1942. Asian Journal of Social Science 31(3): 398-434

Welsh W (1999) Tranculturality: The Puzzling Form of Cultures Today. In: Featherstone M, Lash S (eds) Spaces of Culture. City – Nation – World. Sage, London, pp 194-213

Wood CH (1992) Demographic perspectives on ecological disorder in Bazilian Amazonia. In: Kosinski L (ed) Ecological Disorder in Amazonia: Social aspects. ISSC, Rio de Janeiro, pp 129-148

Wood C, Perz S (1996) Population and land-use changes in the Bazilian Amazon. In: Ramphal S, Sinding SW (eds) Population growth and environmental issues. Praeger, Westport, pp 95-108

Wood C, Skole D, Perz S, Caetano A (1996) Population and deforestation in the Brazilian Amazon. PAA, New Orleans

Yaron G (2001) Forest, Plantation Crops or Small-scale Agriculture? An Economic Analysis of Alternative Land Use Options in the Mount Cameroon Area. Journal of Environmental Planning and Management 44(1): 85-108

From global to regional scale: Remote sensing-based concepts and methods for mapping land-cover and land-cover change in tropical regions

Stefan Erasmi[1*], Martin Kappas[1], André Twele[1], and Muhammad Ardiansyah[2]

[1] Institute of Geography, Department of Cartography, GIS and Remote Sensing, University of Göttingen, Goldschmidtstr. 5, 37077 Göttingen, Germany
[2] Institute of Soil Sciences, Bogor Agricultural University (IPB), Bogor, Indonesia

*corresponding author: Stefan Erasmi, phone: +49 (0)551 398003, fax: +49 (0)551 398020, Email: serasmi@uni-goettingen.de

Summary

A great number of studies have been dealing with land-cover mapping of tropical regions using earth remote sensing technology recently. This is partly due to a growing number of operational sensor systems for both scientific and commercial use and also because of an increasing demand for land-cover information relevant to global environmental issues and international policy instruments (e.g. the Kyoto protocol). Within this context, the present article discusses the state of the art of data processing and analysis for the assessment of broad scale land-cover and land-cover change in tropical regions. Current global scale land-cover maps are compared with regional satellite mapping products (Landsat/ETM+) for a test region in the humid tropics of Central Sulawesi, Indonesia.

The paper suggests the land cover classification system (LCCS) to be used as the conceptual basis for future land-cover analysis in Sulawesi because it delivers a consistent and comparable scale-independent class structure for satellite image-based land-cover mapping and monitoring. The results of the comparative analysis of land-cover and land-cover change document the inhomogeneity, inconsistency and hence high uncertainty of existing estimates. The outcome of the harmonized and generalized land-cover products for two base years (1992 and 2000) indicates considerable disagreements in area estimates and spatial distributions of land-cover classes for a single date that in some cases exceed the detectable changes between years.

Tscharntke T, Leuschner C, Zeller M, Guhardja E, Bidin A (eds), The stability of tropical rainforest margins, linking ecological, economic and social constraints of land use and conservation, Springer Berlin 2007, pp 437-462

Future work aiming at a long-scale operational land-cover mapping of tropical environments has to account for (a) a further harmonization of existing and planned land-cover definitions and products, (b) the regional validation of products and (c) the implementation of a multi-level standardized technical and conceptual classification workflow for ecosystem mapping and monitoring in tropical regions.

Keywords: land-cover, change detection, tropical ecosystems, MODIS, AVHRR, VEGETATION, LCCS, GLC2000

1 Introduction

Since the early 1980ies, information about land-cover has been recognized as being of utmost importance for the description and study of the environment (Watson et al. 2000). The main variable defining primary productivity and biodiversity of a land surface is the type of land-cover (Turner et al. 2005). Land-cover is the easiest detectable indicator of human interventions on the land and is a critical parameter for environmental databases of any kind.

Tremendous advances were made during the 1980ies in the use of remote sensing data to support research related to global change and sustainability. Land-use and land-cover change (LUCC) has become a key topic within global change research programs. The first international research program concerned with this topic was the International Geosphere-Biosphere Programme (IGBP). IGBP and ISCC (International Social Science Council) started up a core project on LUCC (Turner et al. 1995) that combined environmental and social science expertise in land-cover and land-use change analysis. After that, other programs funding research on land-use and land-cover such as the IAI (Inter-American Institute for Global Change Research), APN (Asian-Pacific Network for Global Change Research), START (Global Change System for Analysis, Research and Training), and GCTE (Global Change in Terrestrial Ecosystems) were initiated. Complementary to LUCC and GCTE, NASA developed its own land-cover and land-use change research program (NASA-LCLUC). Each of these programmes identifies land-use and land-cover change as the prior research topic (Townsend et al. 1994). The LUCC and GCTE core projects ended in 2005 but have been continued within the integrated "Global Land Project (GLP)" of the IGBP which aims at the identification, analysis and modelling of the character, causes and consequences of land system change (GLP 2005).

Tropical land-cover, especially the extent of tropical forests and the rate of deforestation, are of particular interest in environmental and socio-economical studies dealing with land-cover and land-use change. This can be documented by three major facts. 1) The tropical land surface is home to more than 55 % of the global human population (UN 2005) and hence human activities related to land-use, commercial structure and social structure constitute the major driving forces of tropical deforestation; 2) Tropical ecosystems harbour a high biodiversity and thus deforestation and land-cover conversion contribute greatly to genetic erosion in tropical regions; 3) Tropical forests comprise a major terrestrial carbon sink and thus tropical deforestation contributes to global warming issues. In summary, the determination of reliable land-cover information constitutes a key input to any kind of spatially explicit analysis and scenario contributing to of one of these topics.

Several efforts to map land-cover in the tropics and monitor changes of forest cover have been undertaken in the past. Besides the problem of accurate assessment, these studies are not directly comparable due to differences in processing and analysis methods, land-cover class definitions and classification concepts. Another disadvantage of recent global land-cover maps (GLC2000, MODIS-IGBP) is the coarse resolution of 1 km^2 / pixel. This restricts the minimum detectable area of change and leads to mixed land-cover types resulting in a significant information loss especially in small-scaled heterogeneous landscapes that are common in Southeast Asia.

This paper deals with the problem of reliable and up-to-date spatially explicit information on land-cover and land-cover change in tropical regions. The introductory chapters provide an overview of the broad range of existing concepts for land-cover change analysis. Furthermore, a review of the most widely used and implemented techniques of image pre-processing, land-cover mapping and change analysis is given. In chapter 6 we present results of a regional study of land-cover monitoring in Central Sulawesi, Indonesia. Within this study different land-cover products are compared with regard to area estimates of land-cover and cover change for the base years 1992 and 2000. The comparison is based on area statistics and spatial distribution of harmonized and aggregated land-cover types. The quality of the global products is investigated by cross comparison with higher level land-cover products. The comparison delivers valuable information about the potential of coarse-scale mapping products for ecological and socioeconomic modeling and scenario analysis in tropical forest margin areas.

2 Basic principles of land-cover mapping and change detection

2.1 Concepts of land-cover characterization

Land-cover is defined as "the observed physical and biological cover of the Earth's land, as vegetation or man-made features" (Watson et al. 2000). In contrast, land-use describes the activities that people undertake in a certain land cover type (FAO 2003) and thus comprises a major driving factor of land-cover change (Meyer and Turner 1994).

In principle, land-cover mapping from satellite data is related to the amount of radiation that is recorded by a sensor on a platform in space for a

Table 1: Land-cover classes of IGBP and LCCS legends for insular Southeast Asia

IGBP class	LCCS class
Evergreen Needleleaf Forest	Forest cover, broadleaved, evergreen, closed and closed to open
Evergreen Broadleaf Forest	Forest cover, broadleaved, deciduous, mainly open (incl. Dry Dipterocarpus)
Deciduous Needleleaf Forest	–
Deciduous Broadleaf Forest	–
Mixed Forests	Mosaic: Forest cover / Other nat. vegetation or Cropland (incl.very degraded and open tree cover)
Closed Shrublands	Mosaics and Shrub Cover, shrub component dominant, mainly evergreen
Open Shrublands	Mosaics and Shrub Cover, shrub component dominant, mainly deciduous
Woody Savannas	Shrub cover, mainly deciduous, (Dry or burnt)
Savannas	Shrub cover, mainly deciduous, (Dry or burnt)
Grasslands	Herbaceous Cover (incl.alpine grassland)
	Sparse herbaceous cover > 3000m
Permanent Wetlands	Tree cover, regularly flooded, Mangrove
	Tree cover, regularly flooded, Swamp
Croplands	Cultivated and managed, non irrigated (mixed)
	Cultivated and managed, irrigated (flooded, rice, shrimp farms)
Cropland/Natural Vegetation Mosaic	Mosaics of Cropland / Other natural vegetation
Barren or Sparsely Vegetated	Bare Areas (Rock: Lime stone)
Urban and Built-Up	Artificial surfaces
Water Bodies	Water Bodies

certain area on the ground. Data acquisition is followed by processing, analysis and product generation (e.g. land-cover maps). However, each of these steps differs fundamentally between different environmental studies, even when they are carried out on the same region and / or topic.

The first major efforts on land-cover mapping at the global to regional scale using earth observing data were initiated by the IGBP in 1990. These activities were driven by the need for information on the distribution of vegetation types and the disturbance of vegetation cover for climate and carbon cycle dynamics studies. The IGBP land-cover mapping studies were based on data from the Advanced Very High Resolution Radiometer (AVHRR) with a nominal resolution of 1 km and covered the period between 1992 and 1993. These data were processed to a global land-cover map and are available to the public (IGBP DISCover). A globally consistent land-cover legend was first defined by Running et al. (1994) and subsequently refined by IGBP core projects (Loveland et al. 2000). It consists of 17 classes and is currently used for the DISCover dataset as well as for present large scale land-cover maps such as the products of the Moderate Resolution Imaging Spectroradiometer (MODIS) land discipline group (Friedl et al. 2002).

Another approach to land-cover description is the Land-cover Classification System (LCCS) of the UN Food and Agricultural Organization (FAO) (Di Gregorio and Jansen 2000). LCCS is not based on a predefined list of class names, but instead uses a simple set of attributes and classifiers that are organized in a hierarchical manner (Bartholomé and Belward 2005). In the first stage of class characterization LCCS uses classifiers that describe the presence of vegetation, the edaphic condition and the nature of ground cover. The results of this "dichotomous phase" are eight land-cover classes (Di Gregorio and Jansen 2000). In the second stage, the modular-hierarchical phase of LCCS, the classes are further subdivided, based on environmental attributes (e.g. life form, land form, climate) and technical specifications (e.g. crop type). In this manner, the LCCS offers high flexibility to the user to describe and classify land-cover features at any desired scale and level of detail. Each of the classifiers and attributes that are used for the generation of a single land-cover class is associated with a unique code and thus every class is explicitly identifiable and is assigned a standard name by the classification system. The system offers the possibility to describe land-cover categories at a high level of detail, but also allows for the generalization and upscaling of complex land-cover maps to national or global levels.

LCCS has been applied to the maps of the Global Land-cover project (GLC2000) and also forms the conceptual basis for the class descriptions of the forthcoming Medium Resolution Imaging Spectroradiometer (MERIS) land-cover product GLOBCOVER. The GLC2000 legend is scale-dependent. At the global scale, it consists of 22 general land-cover types that are defined at the biome level. The legend for Southeast Asia contains 16 classes that inherit features of the major classes while classes that are not prevalent in the tropical biome are neglected. According to Bartholomé and Belward (2005), the classes at the global level provide compatibility with maps that are based on the IGBP classification system. This means that any regional product that is based on the LCCS system can be compared to IGBP-based land-cover maps. In practice, this compatibility is not fully accomplished due to some inconsistencies in class definitions (see table 1 for a comparison of LCCS and IGBP legends). A workaround for the comparison of land-cover products that are based on different classification systems is the aggregation of the land-cover classes in two systems to a set of major land-cover types at the biome level (Giri et al. 2005).

Table 2 provides an overview of current and planned land-cover products covering insular Southeast Asia. These maps can be grouped into two categories: 1) maps based on global AVHRR data that have been generated during the 1990ies and use 1992 as the base year and 2) products for the base year 2000 from MODIS and VEGETATION using different classification concepts. In addition, table 2 includes the forthcoming GLOBCOVER product that is based on Envisat/MERIS. The year 2000 has been chosen as a benchmark for environmental, ecological and economic long-term analysis. The comparison of the land-cover product characteristics points to the general problem of inconsistency and thus incomparability between products that are generated by different workgroups. In addition to these products, the land-cover of insular Southeast Asia or its sub-regions are subject to a number of studies aiming at the characterization and delineation of land-cover types at the global to regional level (Sato and Tateishi 2004, Mayaux et al. 1998, Stibig et al. 2003)

2.2 Concepts of land-cover change analysis

All over the world terrestrial ecosystems are changing at multiple spatial and temporal dimensions. In the context of land-cover, change is defined as the "alteration of the physical or biotic nature of a site". As an example, the transformation of forest to cropland represents a change in land-cover whereas land-use change involves "alterations in the human management of land, including settlement, cultivation, pasture, rangeland and recreation" (Meyer and Turner 1994). Land-use induced land-cover change is one important factor affecting terrestrial ecosystems and human livelihoods and thus comprises a major driving factor of global change. Remote sensing studies are generally restrained to land-cover change analysis. Change can be either abrupt or gradual and thus can be described in categories (classes) or continuous variables. The complete replacement of one land-cover type by another is referred to as land-cover conversion, whereas slight changes affecting the attributes of a land-cover without changing the nature of the land-cover type are defined as land-cover modifications (Coppin et al. 2004).

Considering tropical environments, land-cover change is frequently caused by human-induced deforestation activities (Lambin 1997, Lu et al. 2005). The most general definition of deforestation is given in Watson et al. (2000) as the "sum of both, permanent and temporary human-induced removal of forest

Table 2. Current and planned global land-cover products covering insular Southeast Asia

Product	GLCC	UMD land-cover	GLC2000	MODIS land-cover	TREES	GLOBCOVER
Sensor	AVHRR	AVHRR	SPOT Vegetation	MODIS	SPOT Vegetation	MERIS
Timeframe of data collection	April 1992 - March 1993	April 1992 - March 1993	Nov. 1999 - Dec. 2000	Oct. 2000 - Oct. 2001	March - Sept. 1998 March - Nov. 1999	Jan. 2005 - Dec. 2005
Input data	Monthly NDVI composites	Monthly NDVI composites and metrics	Spot VGT1 S1	MOD43 (nadir BRDF- adjusted reflectance	Spot-VGT1 S10	MERIS Level 2
Spatial Resolution	1km	1km	1km	1km	1km	300m
Map Projection	Goode's Homolosine	Goode's Homolosine	Geographic	Sinusoidal	Geographic	Geographic
Classification technique	Unsupervised clustering	Supervised classification (decision tree)	flexible classification system depending on partner institution	Supervised classification (decision tree)	N/A	N/A
Classification scheme (legend)	IGBP (17 classes)	Simplified IGBP (14 classes)	LCCS (16 classes for SE Asia)	IGBP (and others)	TREES 9 classes	LCCS
Intended update schedule	Annual*	Annual*	2005 (GLOBCOVER)	Annual*	–	N/A
Detailed description of product	Loveland et al. (2000)	Hansen et al. (2000)	Bartholomé and Belward (2005)	Strahler et al. (1999) Friedl et al. (2002)	Stibig et al. (2003)	N/A
Data access	http://edcdaac.usgs.gov/glcc/glcc.asp	http://glcf.umiacs.umd.edu/data/landcover/	http://www.gvm.jrc.it.glc2000	http://edcimswww.cr.usgs.gov/pub/imswelcome/	http://www.gvm.sai.jrc.it/Forest/defaultForest.htm	http://www.gvm.jrc.it/tem/products.complete.htm

* no updates available at time of paper submission + Version 1.2 of GLCC has been updated by version 2.0 in 2004 but is based on the same data set (1992-1993)

cover". This definition includes short-term removals by shifting-cultivation as well as selective and clear-cut logging activities and all other kinds of conversion or modification of forest land. Lund (1999) restricts the term deforestation to "the long-term or permanent removal of forest cover and conversion to a non-forested land-use". Besides human induced alteration, land-cover change can also be caused by long-term changes that are related to climate change, geomorphological (soil erosion) and ecological processes (vegetation succession) as well as inter-annual climate variability (Lu et al. 2004). Within the present study, deforestation is only addressed in terms of short- to long-term forest conversion (e.g. changes in forest or land-cover type).

Detailed descriptions of change detection concepts can be found in Coppin et al. (2004), Lu et al. (2004) and Singh (1989). The most important point in determining the appropriate concept is the decision for a spatial and temporal scale for the intended investigation. Vegetation cover changes seasonally, inter-annually or between any other points in time (directional change). The change concept can either rely on two different images (bi-temporal) or on a set of images covering a complete time trajectory (time series analysis). Bi-temporal change detection concepts have to take into account the dates of image acquisition as the most important factor that may bias the resulting change. This is especially of interest in areas showing significant phenological differences in vegetation status or seasonal / intra-seasonal land-use patterns. Hence, bi-temporal change analysis is only meaningful if acquisition dates are comparable, phenological cycles follow distinct seasonal patterns and if they cover the interval length of the change process that is observed. In practice, bi-temporal change analysis is the dominating concept for change analysis in tropical areas (Achard et al. 2002, Alves et al. 2003, Castro et al. 2003, Erasmi et al. 2004). In most cases, the choice of a data pair for bi-temporal analysis is more related to the availability of appropriate satellite image data than to the above mentioned factors. To overcome the limitations of bi-temporal analysis, recent approaches in satellite based ecosystem analysis focused on the analysis of temporal profiles of image data covering seasonal, annual and inter-annual observation periods. These change concepts require a time series of remote sensing data that are transformed into indicators of ecosystem characteristics (e. g. spectral vegetation indices). Time series-based change detection has recently been successfully implemented for regional and global studies of climate induced land-cover changes (Friedl et al. 2002, Zhang et al. 2004) and for the assessment of the variability of inter-annual net primary productivity of tropical ecosystems (Turner et al. 2005).

Overall, the most important challenges for remote sensing-based ecosystem change analysis are (1) the detection of land-cover modifications in addition to conversions; (2) the detection of abrupt as well as gradual changes in land-cover; (3) the detection and separation of spontaneous, seasonal and annual changes from inter-annual changes and long-term trends; (4) the scale dependency of change estimates derived from satellite images at different spatial resolutions; (5) the precise geometric co-registration of multi-temporal images and time series of satellite data; (6) the development of an appropriate change detection method; (7) the adoption of a consistent and hierarchical classification concept.

2.3 Remote sensing sensors and systems for land-cover mapping

As described in the previous section, the success of land-cover studies is directly related to the availability of satellite imagery at the desired spatial and temporal resolution. Concerning the spatial resolution of existing satellite systems, three main categories of sensors can be defined: (a) coarse to medium-resolution sensors (> 250 m); (b) high-resolution sensors (> 10 m); (c) very high-resolution sensors (< 10 m). These spatial resolutions can be linked to studies at different level of detail: (a) global to landscape level (< 1:200.000); (b) regional level (< 1:25.000); (c) local level (> 1:25.000). In general, the temporal resolution of optical satellite sensors is a function of the spatial resolution, due to sensor limitations (instantaneous field of view) and repetition rate of a sensor system. This means that an increase in the temporal resolution will most likely reduce the spatial resolution available for analysis. In addition to these characteristics, the choice for a satellite system also de-

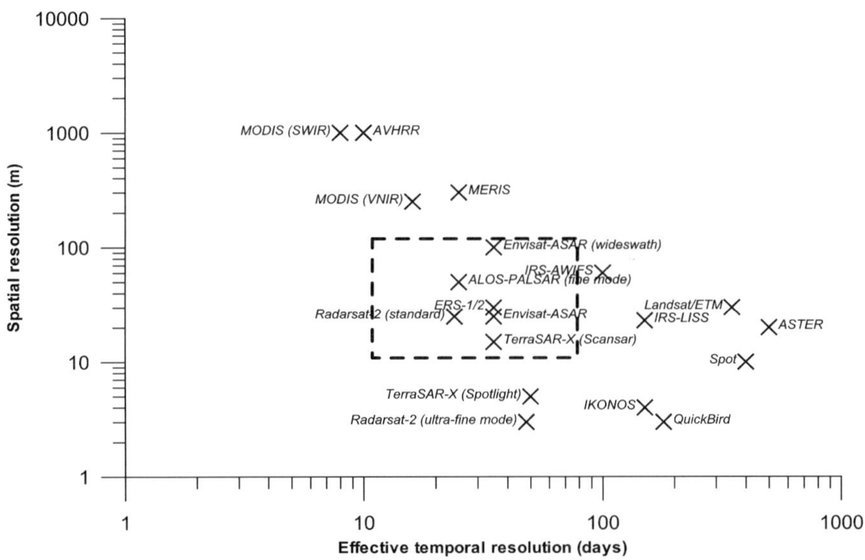

Fig. 1. Effective temporal and spatial characteristics of current and future multispectral and SAR earth observation satellite sensors for humid tropical regions (dotted rectangle highlights desired resolution configuration for operational tropical land-cover monitoring

pends on the spectral resolution (bandwidth, number of channels) and nature of radiation source (optical, radar, lidar). Figure 1 provides a broad overview of selected multi-spectral optical and SAR (synthetic aperture radar) satellite sensors that are used for earth observation purposes at present. The shown temporal coverage for optical sensor systems does not represent the nominal coverage of the observing system based on its repetition rate, but accounts for the effective mean availability of imagery for humid tropical areas due to atmospheric disturbances and cloud coverage. In contrast to this, the temporal coverage of SAR satellite sensors is identical with its repetition rate due to the ability of radar waves to penetrate clouds. The summary shows that, in general, regional to local level mapping and monitoring is mainly related to observation systems with less frequent coverage whereas frequent observations of seasonal, annual and inter-annual phenomena (time series analysis) often require medium to coarse-resolution satellite systems. The lack of available systems in the intermediate space (high acquisition frequency, high spatial resolution) may be accounted for by using spaceborne imaging SAR systems. As illustrated in figure 1, satellite SAR sensors may comprise a key data source for land-cover observations in tropical areas. So far, only few studies have accounted for this opportunity and revealed the potential of the synergetic use of optical and radar data for the mapping and monitoring of land-cover in the tropics (Sano et al. 2005, Siegert et al. 2001). An explanation for this fact is that despite the operational availability of SAR satellite data, still too little is known about the relation of SAR signals to vegetation properties.

3 Pre-Processing of satellite data

3.1 Single-time and bi-temporal mapping

Land-cover studies that are carried out based on single images mainly require accurate geometric registration to a reference system or co-registration to other spatially explicit data sources for geospatial analysis purposes in a geographical information system (GIS). Geometric accuracy of (co-) registration is of utmost importance to any form of land-cover mapping or change analysis and should be accounted for at the sub-pixel level. Mis-registration at the pixel level may lead to significant differences in area or change estimates and, hence, may cause pseudo-events in change detection or geospatial modelling. Radiometric consistency is not needed in the case of single image analysis if isolated empirical studies are intended, because land-cover maps are created individually for a single area and a single date. The challenges for radiometric image pre-processing increase with the number of images that are being processed and analysed.

In the case of bi-temporal analysis, major causes for the alteration of the sensor signals that are recorded and converted into digital numbers (DNs) are the variability of atmospheric constituents and topography. Radiometric processing always aims at the normalisation of these effects by the use of models that in general invert the signal to DN conversion and eliminate the noise that is apparent in the signal or, in other words, maximize the signal to noise ratio (SNR). In general, there are two types of radiometric pre-processing: (a) relative radiometric calibration and (b) absolute radiometric calibration. Detailed summaries of pre-processing techniques for land-cover mapping and change detection are available from Cihlar (2000) and Coppin et al. (2004). Simple relative radiometric normalisation approaches are linear regression algorithms, or pseudo-invariant feature-based techniques. These methods rely on image based statistics and perform a relative normalisation of spectral image layers. Most of these methods use invariant features that are assumed to retain their spectral characteristics over long periods of time (e. g. empirical line calibration). More sophisticated radiometric calibration approaches aim at the computation of the absolute intensity of radiance reflected or emitted by a surface. These models are grouped in image-based atmospheric correction methods (Chavez 1996, Twele and Erasmi 2005) and physically based methods (radiative transfer models). A study by Songh et al. (2001) revealed that the complexity of a radiometric calibration technique does not necessarily improve the accuracy of a classification or change detection analysis. This only implies qualitative analysis of the land surface. If further quantitative analysis (biophysical ecosystem variables) from calibrated remote sensing data is intended, an accurate and robust absolute atmospheric correction method should be applied.

3.2 Time series computation

Satellite image analysis that is based on a time series of observation dates generally requires the same pre-processing workflow as has been explained in the previous section. Adequate geometric accuracy is obligatory for change analysis in time profiles. Current sensor systems already account for this necessity and provide precise geo-located products to the user community (Terra/Aqua-MODIS, Envisat-MERIS, Spot-VEGETATION). The radiometric calibration of time series data has mainly to deal with the removal of clouds and other atmospheric effects (e. g. water vapour content, aerosols) that are inherent in a single image. Since the existence of clouds constitutes a major problem in tropical land surface remote sensing, the development of methods to minimize or remove these effects is imperative for land-cover analysis. A simple but robust approach to reduce the amount of cloudy and atmospherically contaminated image pixels is the generation of temporal composites based on a constant number observation dates. The processing and archiving facilities of the currently available satellite systems that deliver daily image data already, developed and implemented such compositing techniques in standard pre-processing workflows and provide the compositing results as an operational data product (Terra/Aqua-MODIS, Spot-VEGETATION, NOAA-AVHRR). In most cases, these temporal composites rely on the maximum vegetation index value within a given period of time (several days or weeks). Maximum value composites (MVC's) are very simple to use and have been applied to early temporal profiling satellite data sets (e. g. AVHRR Pathfinder; Holben 1986). Newer approaches further account for the influence of the viewing geometry (panoramic distortions, sensor off-nadir angle) (Huete et al. 2002, Duchemin et al. 2002). Global imaging sensor systems are operated at a wide field of view, which results in signal alterations in off-nadir areas. These effects are mainly caused by the directionality of the reflectance behaviour of natural objects and are described for every land surface object by its bi-directional reflectance distribution function (BRDF).

Table 3 summarizes the temporal composite products that are currently available for insular Southeast Asia. All land-cover products that are summarized in Table 2 are based on one of these compositing techniques or use enhanced techniques that have been developed complementary to the basic methods. These recent approaches are mainly driven by the fact that even the BRDF-adjusted composite for MODIS and VEGETATION products (Schaaf et al. 2002, Duchemin et al. 2002) does not yield uniform results for the globe and can further be improved for specific regions, especially the tropics, where cloud coverage and atmospheric disturbances are eminent (Cabral et al. 2003, Carreiras and Perreira 2005, Hagolle et al. 2004).

4 Techniques of land-cover mapping and change detection

Satellite image classification, the assignment of a discrete object attribute (e.g. land-cover type) to a single image unit (pixel or object), is one of the earliest application fields of digital image processing and analysis in remote sensing. Two types of approaches have evolved and are used to this day as the basic options for digital image classification. These types are based on different general concepts about the previous knowledge of the area that is to be classified (Cihlar 2000). Supervised classification uses a priori knowledge about the surface types that is used to create representative signatures which are applied to the entire area of interest. The unsupervised classification concept does not take into account prior information about cover type and distribution. The assignment of clusters to land-cover types is done a posteriori, based on available ground truth information. A number of enhanced techniques have evolved from these two classification techniques, including decision tree classifiers, neural networks and fuzzy logic classification (Coppin et al. 2004).

In addition to the analysis of the spectral dimension, the spatial domain contains also valuable information about land-cover features, especially when high spatial resolution satellite data are available. The most utilized approaches comprise texture analysis and other spatial measures based on image segments.

Table 3. Vegetation index composite products available for Southeast Asia

Product	MODIS MOD13	SPOT-VGT S10	SPOT-VGT D10	AVHRR 1km	AVHRR GIMMS
Sensor	Terra/Aqua MODIS	SPOT-Vegetation-1/2	SPOT-Vegetation-1/2	AVHRR	AVHRR
Temporal coverage	April 2000 - present	April 1998 - present	July 2001 - present	April 1992 - May 1996	July 1981 - Dec. 2003
Spatial resolution	250/500/1000m	1km	1km	1km	8km
Map projection	Sinusoidal	Equirectangular/ Plate Carée (Lat/Lon)	Equirectangular/ Plate Carée (Lat/Lon)	Goode's Homolosine	Geographic
Compositing period	16 days	10 days	10 days	10 days	1 month
Compositing algorithm	MODIS VI compositing algorithm	Maximum value composite (MVC)	Bidirectional composite (BDC)	Maximum value composite (MVC)	Maximum value composite (MVC)
Detailed product description	Huete et al. (2002) Schaaf et al. (2002)	Holben (1986)	Duchemin et al. (2002)	Holben (1986)	Tucker et al. (2005)
Data access	http://edcimswww.cr.usgs.gov/pub/imswelcome/	http://free.vgt.vito.be/index.php	http://free.vgt.vito.be/index.php	http://edcdaac.usgs.gov/1KM/1kmhomepage.asp	http://glcf.umiacs.umd.edu/data/gimms/

If area distributions of land-cover classes are to be compared over time, a classification of all image pixels for each point in time is obligatory. Consequently, the multi-temporal dataset is classified separately and a pixel-by-pixel comparison is employed. Subsequently, changes in area distribution of all land-cover classes can be calculated and the history of land-cover development can be analysed and interpreted under different (e.g. social, ecological) perspectives.

However, land-cover studies frequently deal with single, sometimes drastic change phenomena such as deforestation, forest fires, landslides and floods. In this case, mapping all image pixels is neither necessary nor feasible with respect to time requirements. Therefore, it is possible to use special techniques which are developed to highlight the specific change phenomena. The most popular of these methods utilize image algebra or image transformations to convert multi-temporal spectral reflectance to a feature space that highlights changes. Comprehensive reviews of change detection techniques can be found in Coppin et al. (2004), Lu et al. (2004) and Lu et al. (2005).

If land-cover mapping aims at the detection of gradual changes in land-cover (e.g. forest thinning, selective logging or reforestation), mapping land-cover modifications in a continuous manner is desirable. The most common approach to derive such information from multispectral remote sensing data is the computation of fractions of land-cover types within a pixel (e.g. fraction of vegetation) or of biophysical attributes over time (Lambin 1999).

5 Land-cover and change analysis in Sulawesi

It is evident from numerous studies (Achard et al. 2002, Erasmi et al. 2004, FAO 2003, FWI & GFW 2002) that substantial land degradation and land-cover conversion at the rainforest margin in Southeast Asia has taken place during the past decades and significantly increased during recent years. Regional studies based on satellite data state an annual loss of forest cover of 0.6 % for the time period 1972 to 2002 for Central Sulawesi (Erasmi et al. 2004). National estimates for Indonesia assume a loss of forest of 1.2 % to 1.7 % per year (FAO 2003, FWI & GFW 2002).

These discrepancies in deforestation estimates allude to the problem of data comparability and scale dependency for socio-economic and ecological modelling. Within this context, a study has been carried out for a test region in Central Sulawesi that evaluated the satellite data products and land-cover mapping concepts that have been explained above.

5.1 The land-cover classification system for Sulawesi

The comparison of the IGBP and LCCS classification system for insular Southeast Asia revealed some problems that are mainly caused by the different class definition concepts.

In general, these concepts follow the same idea that land-cover stratification is based on the nature of the surface cover. In addition, LCCS offers the opportunity to further describe the surface cover based on environmental attributes (topography, geology and soils, life form and cover, etc.). The consequence is that the LCCS system can be better adapted to regional land-cover characteristics. Considering the natural vegetation classes, the inconsistencies are generally related to differences in thresholds that are used in both systems to define stages of vegetation coverage (e. g. percent of vegetation or tree cover per pixel). Two examples of class definitions are used here to explain this fact. First, the class evergreen broadleaved forest is employed in both systems. The IGBP legend uses a threshold of 60 % tree cover whereas the LCCS definition is based on a 15 % threshold. The second case of shrubland definitions shows similar inconsistencies in threshold designations (IGBP=60%; LCCS=15%). In addition to this, the class descriptions for shrubland and for evergreen broadleaved forest use different height thresholds (shrubland: IGBP=3 m, LCCS=2 m; forest: IGBP=5 m, LCCS=2 m).

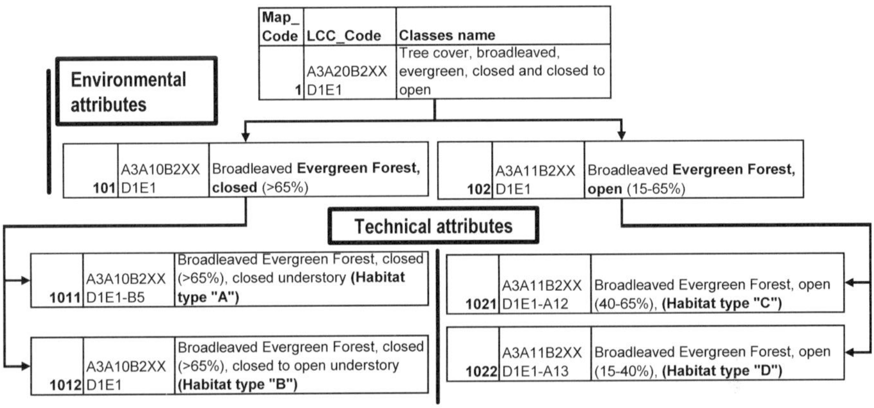

Fig. 2. Subsequent land-cover specification for tropical forest areas in Sulawesi using the modular hierarchical classification system (LCCS)

The flexibility and hierarchical concept of LCCS were the major criteria for the decision to adopt this system as a fundament to create a consistent land-cover classification system for Sulawesi. The first and lowest level of the hierarchical multi-level system comprises the land-cover classes that have been defined within the GLC2000 project for insular Southeast Asia. These class definitions already account for the specifics of tropical land-cover and are tailored to general tropical vegetation types (Table 1). Land-cover analysis at this level is limited to large area observations, temporal trajectory analysis and comparison of results from different regions. At the next and more detailed level, the initial classes are subdivided into more detailed descriptions of the land-cover that is dominant in the area under investigation. Other classes that are less important within the area remain unchanged. At this level regional analysis of land-cover is performed, aiming at a more complex characterization of tropical landscape patterns. The third and highest observation level again contains more detailed descriptions of land-cover units. At this stage, a description of local land-cover phenomena is carried out with regard to the characterization of land-use intensity gradients at the rainforest margin. An example of the specification of land-cover classes within this modular concept is given in Figure 2 for broadleaved evergreen forest areas.

The concept has been adapted to satellite-based mapping approaches for the investigation area in Central Sulawesi. Land-cover analysis based on this classification system guarantees consistent and comparable land-cover information at multiple scales. Land-cover mapping results at higher levels can be transferred to lower levels by means of aggregation of land-cover classes at any time without class overlap or inconsistencies.

5.2 Land-cover area comparison based on global products

Despite the inconsistencies of the IGBP and LCCS classes, the IGBP classes have been translated into the LCCS system for insular Southeast Asia as shown in Table 1. This made a direct quantitative comparison of the MODIS and the GLC2000 land-cover products possible. Figure 3 shows the resulting area fractions of the LCCS classes for the MODIS land-cover compared to the GLC2000 land-cover classification. Most obvious discrepancies are apparent for the classes "broadleaved evergreen forest" and "mosaics of cropland / other natural vegetation". Other findings by Hansen and Reed (2000) and Giri et al. (2005) also reported major differences in forests as well as in shrublands and savannas that are mostly related to the different class definitions of forest, resulting in an overlap of forests in the LCCS system with savannas / shrublands of the IGBP land-cover classes. In the specific case of Sulawesi, the discrepancies cannot solely be explained by these class inconsistencies. In contrast to other reported results (Hansen and Reed 2000, Giri et al. 2005) the forest area in the MODIS land-cover product is 20 percent above the estimates of the GLC2000 product (MODIS: 81.5 %; GLC2000: 61.5 %). The second major difference in area fractions is found for the "mosaics of cropland / other natural vegetation" class. The amount (in percent of total area) nearly compensates for the shift of forest cover rates between the two maps (MODIS: 7.9 %; GLC2000: 24.0 %). The area statistics also reveal some noticeable differences between the estimates for "cultivated areas" as well as "tree cover and other natural vegetation mosaics". The fractions and differences for the remaining classes that are found in the investigation area are below 1 % and neglected in this discussion.

The large overestimation of forest cover in the MODIS land-cover classification compared to the GLC2000 product is surprising, because the class definition of the IGBP legend for forest is based on a threshold of 60 % tree

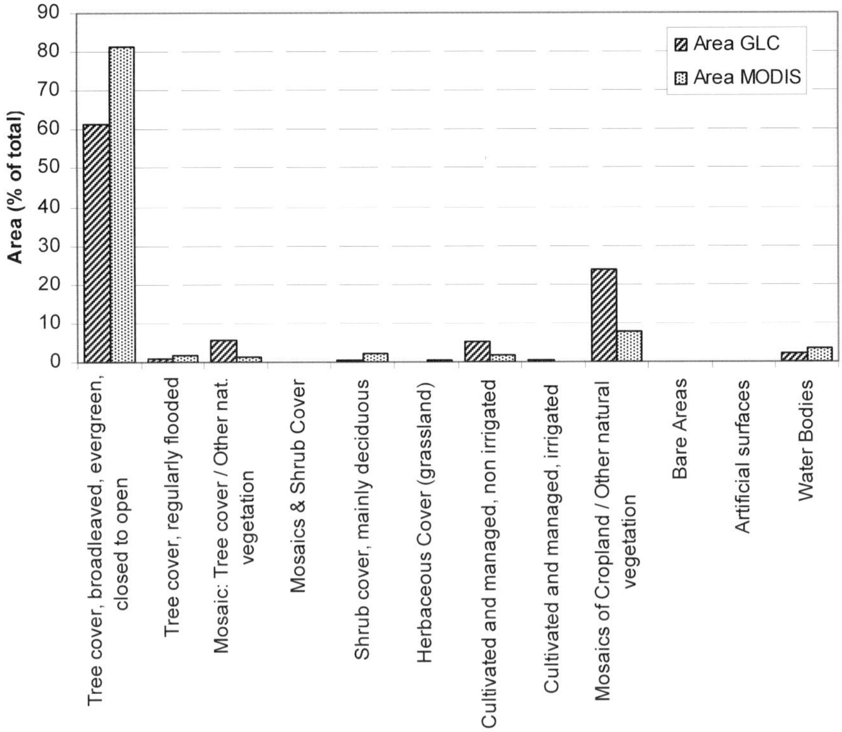

Fig. 3. Occurrence of LCCS land-cover classes based on the GLC2000 and MODIS land-cover products for Sulawesi

cover in contrast to only 15 % for the LCCS. Based only on this fact, the simple translation of the IGBP forest class to the LCCS system that has been accomplished here must lead to an overestimation of forest area in the GLC2000 product. Since the opposite is observed in this study, it cannot be argued that the different area estimates are simply based on different class definitions or inconsistencies in legend translations. This has been further approved by the comparison of the two classifications at the biome level, using only seven aggregated land-cover classes based on the LCCS class definition scheme (Figure 4). Additional investigation thus concentrated on the validation of the two products for a regional study site in Central Sulawesi.

5.3 Comparison of global products with regional studies

There are several options to validate satellite-based land-cover products. The approach for the GLC2000 data sets was based on the participation of international, national and regional experts from more than 30 countries. Within this consortium, the land-cover maps were produced at the regional level and

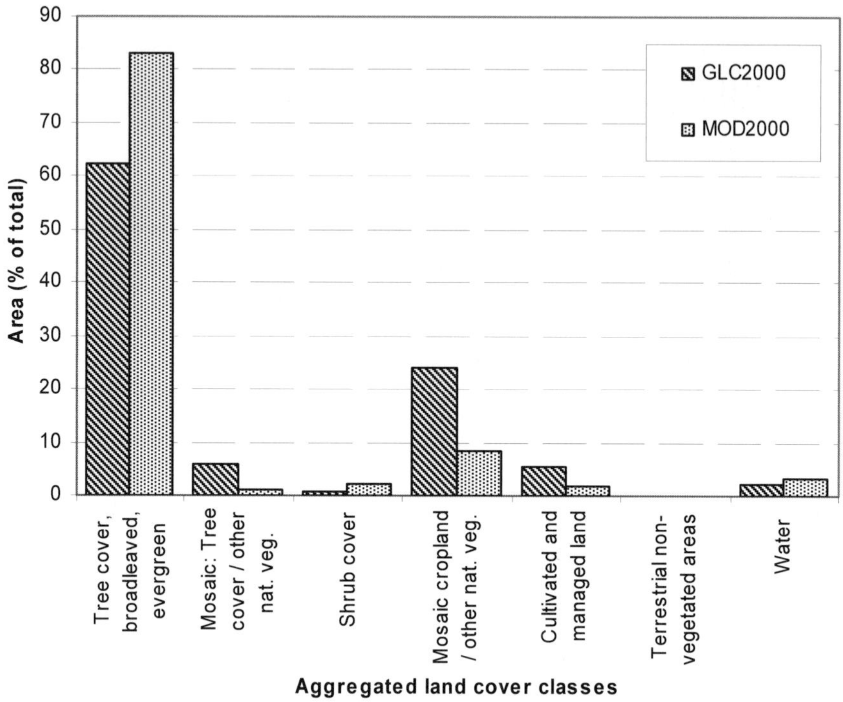

Fig. 4. Area statistics based on aggregated LCCS land-cover classes for the GLC2000 and MODIS land-cover classifications of Sulawesi

later transferred to the continental and global scale using the hierarchical LCCS classification scheme. This bottom-up classification concept already accounted for a regional validation by a team of experts that created maps at higher levels. All input data sources, the classification scheme and methodology were adapted to the needs of the participating institutions and optimized for the land-cover types that are found in the respective regions. In contrast to this, the MODIS product uses a top-down approach to characterize land-cover. The MODIS land-cover product is based on a consistent classification methodology that uses a supervised decision tree classifier and is adopted for the entire terrestrial surface. A validation using results from regional studies and high-resolution satellite data has not been carried out within the project.

The two products (MODIS land-cover, GLC2000) have been validated within the current study using a land-cover map of a test site in Central Sulawesi, Indonesia. The site covers approximately 7.200 square kilometres and is dominated by broadleaved evergreen natural forest and land cultivation of different intensities at the forest margins. The validation dataset consists of a Landsat/ETM+ based land-cover map that has been validated by local

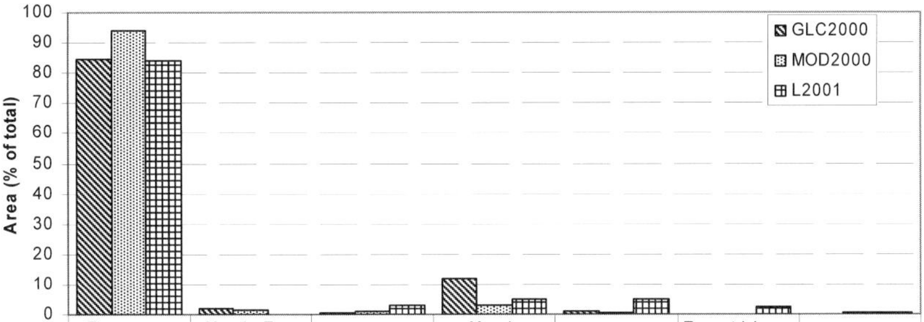

	Tree cover, broadleaved, evergreen	Mosaic: Tree cover / other nat. veg.	Shrub cover	Mosaic cropland / other nat. veg.	Cultivated and managed land	Terrestrial non-vegetated areas	Water
GLC2000	84.4	1.9	0.3	12.1	1.0	0.0	0.2
MOD2000	94.0	1.4	0.8	2.8	0.4	0.2	0.4
L2001	84.3	0.0	3.1	4.8	5.0	2.3	0.6

Aggregated land cover classes

Fig. 5. Area statistics for aggregated classes of Landsat-based classification compared to GLC2000 and MODIS land-cover products (STORMA project area, Central Sulawesi)

and regional experts (producer's accuracy: 87,7 %). All data products have been created using the LCCS legend or translated from other systems and have been aggregated to the major land-cover types that are prevalent in the study area. Figure 5 shows the results of the area statistics calculations for the three land-cover maps. It is apparent that the GLC2000 map better reflects the results of the regional study. The forest areas can be estimated with a total difference of only 10 square kilometres (0.1 % of total area). The "cropland / natural vegetation mosaic" areas are larger in the GLC2000 land-cover map whereas the cropland area is larger in the regional land-cover map. This can be explained by the small-scaled landscape pattern of annual and perennial crops, grassland and agroforestry systems that are mixed up when using lower resolution satellite images. Considering the MODIS regional validation, forest areas are strongly overestimated (94 % of the study area). The remaining area is shared by the other six classes, which leaves no space for interpretation of agreement or disagreement with the regional map. A comparison of the spatial distribution of the forest areas from GLC2000 and MODIS clearly shows the areas of disagreement between these two products (Figure 6). Inconsistencies mainly occur at the transition zones between natural forest and cultivated land and hence in the areas that are of primary interest in most studies with regard to the stability of rainforest margins.

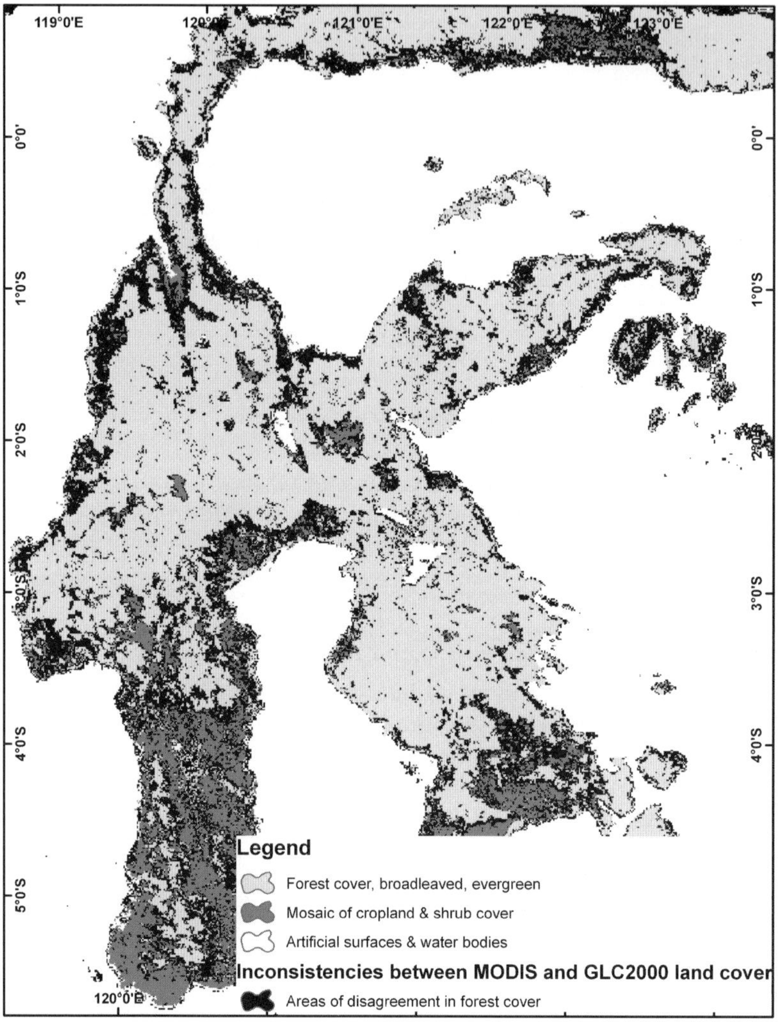

Fig. 6. GLC2000 land-cover map for Sulawesi including areas of disagreement between GLC2000 and MODIS land-cover (in magenta)

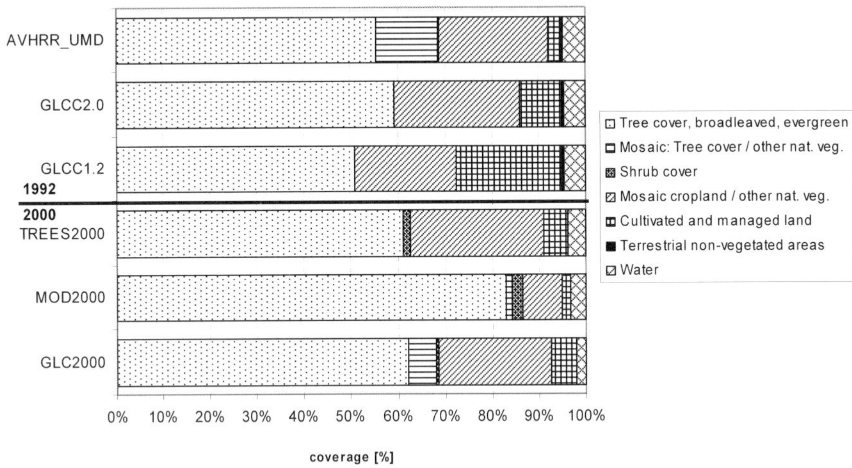

Fig. 7. Area statistics of land-cover change for Sulawesi based on global products for the years of 1992 and 2000

5.4 Change analysis based on global products

The change analysis was based on a comparative post-classification analysis of the existing global land-cover products for insular Southeast Asia. The computation of change statistics and the results of change analysis are strongly dependent on reliable and consistent data for at least two base years. In this study five land-cover products have been compared (see Table 2 for product details) with regard to feasibility and consistency of area estimates for the tropical land-cover of Sulawesi. The results of the area calculations based on aggregated classes are shown in Figure 7. A comparison of the area distribution reveals some remarkable aspects. The revision of the GLCC product (version 2.0) leads to a considerable increase in forest cover. At the same time, cultivated land decreases to less than half of the previous estimate for the GLCC version 1.2. The AVHRR_UMD map includes the "Mosaic: Tree cover / other natural vegetation" class while this class is not apparent in the other AVHRR-based products for the base year 1992. In general, the area estimates for forest cover show an increase in forests during the time period of 1992 to 2000 even if the strong overestimation of the MODIS product is neglected in the comparison. Moreover, there seems to be no increase in land cultivation. The consequence of these results would be that tropical ecosystems in Sulawesi regenerated during the past decade. This finding is contradictory to all assessments and field observations at the local, regional and global level and is abandoned from further interpretation. The reason for the implausible area estimates is rather to be seen in the inconsistencies of data sources, methodologies and classification concepts of the different studies.

As a consequence of these findings, further change analysis based on the existing land-cover products has not been conducted within this study.

6 Discussion and conclusions

A comparative analysis of the global land-cover products for Sulawesi has been carried out. The data set for the comparison consisted of five land-cover maps covering two base years (1992 and 2000). The analysis for the base year 2000 included the MODIS land-cover and the GLC2000 product. Area statistics for these two products show a strong overestimation of forest areas in the MODIS land-cover classification. In comparison, the present study underlines the high consistency of forest cover estimates from GLC2000 land-cover with regional land-cover mapping from high-resolution satellite data for Central Sulawesi. Moreover, cropland is mixed up with natural forest cover in the MODIS product, whereas the GLC2000 map better delineated cultivated land from natural forest areas. The spatial analysis of area distributions for the main land-cover classes reveals large areas of disagreement between MODIS and GLC2000, especially near the rainforest margins. Overall, the GLC2000 is superior to the MODIS product considering the investigation area in Sulawesi. This conclusion is not surprising because of the lack of regional validation studies for the MODIS land-cover data. The data were not intended for regional or local applications, thus the data have not been optimized for those purposes. On the other hand, the GLC2000 data demonstrate the feasibility of integrating global land-cover data sets into regional studies and the importance of regional studies for the verification of global land-cover products.

The GLC2000 concept will be adapted to the forthcoming GLOBCOVER data set in a similar manner. This scheme presumes that GLOBCOVER data will provide a land-cover map that is comparable to the GLC2000 and may deliver the basis for a consistent change analysis in tropical regions. Nevertheless, the classification and validation strategy of GLC/GLOBCOVER is not operational due to the large number of involved partners and institutions at different operating levels. In contrast, the MODIS classification methodology is consistent and repeatable. It aims at the provision of a global data set every six months, but the intended update schedule has not been accomplished up to now.

The MODIS and GLC2000 data sets have also been compared with the global data sets for the base year 1992 (GLCC, AVHRR_UMD). The comparative analysis for the Sulawesi area confirms the problem of inconsistency with regard to data sources, classification methods and concepts. Even if the data source of a land-cover product is identical, the processing and analysis differs fundamentally, thus making it difficult to compare the results (Hansen and Reed 2000, Giri et al. 2005). The results of the present study for Central Sulawesi conclude that the area disagreements between products for a single year can be of the same dimension as estimated changes of land-cover between years.

Considering land-cover classification systems, the LCCS comprises a basis for the definition of land-cover legends at multiple levels, using a hierarchical tree concept and a combination of environmental and technical attributes of land-cover. The concept has been successfully adapted to the land-cover types for the tropical environment of Central Sulawesi. The advantage of the LCCS system is that it is flexible and adaptable to biome specific characteristics while the IGBP consists of an irreversible number of major land-cover classes. Hence the LCCS is more useful for multi-level investigations from local to global scale. In contrast, the IGBP class definitions are restricted to global monitoring issues.

Future work will aim at the generation of consistent data products for change analysis. These attempts have to account for a robust pre-processing chain focussing on data harmonization (spectral, spatial and temporal fitting). If data harmonization is achieved at a certain level, standardized classification workflows can be established and adapted to the different data sets. The overall scope of future studies of land-cover assessment within the project will be on the routine mapping of tropical ecosystems which comprise a hot spot region with regard to global change. It is obvious that future work should therefore focus on the integration of single isolated mapping approaches and aim at developing a comprehensive and reliable operational monitoring concept for tropical environments.

Acknowledgements

We would like to thank the German Research Foundation (DFG) for funding parts of this work within the collaborative research centre SFB-552 (Stability of Rainforest Margins in Indonesia).

References:

Achard F, Eva HD, Stibig HJ, Mayaux P, Gallego J, Richards T, Malingreau JP (2002) Determination of Deforestation Rates of the World's Humid Tropical Forests. Science 297 (9): 999-1002

Alves DS, Escada MIS, Pereira JLG, De Albuquerque Linhares C (2003) Land-use intensification and abandonment in Rondonia, Brazilian Amazonia. Int J Remote Sens 24 (4): 899-903

Bartholomé E, Belward AS (2005) GLC2000: a new approach to global land-cover mapping from Earth observation data. Int J Remote Sens 26 (9): 1959-1977

Cabral A, De Vasconcelos MJP, Pereira JMC, Bartholomé E, Mayaux P (2003) Multi-temporal compositing approaches for SPOT-4 VEGETATION. Int J Remote Sens 24: 3343–3350

Carreiras JMB, Pereira JMC (2005) SPOT-4 VEGETATION multi-temporal compositing for land-cover change studies over tropical regions. Int J Remote Sens 26 (7): 1323-1346

Castro KL, Sanchez-Azofeifa GA, Rivard B (2003) Monitoring secondary tropical forests using space-borne data: implications for Central America. Int J Remote Sens 24 (9): 1853-1894

Chavez PS (1996) Image-based atmospheric corrections. Revisited and improved. Photogramm Eng Rem S 62: 1025–1036

Cihlar J (2000) Land-cover mapping of large areas from satellites: status and research priorities. Int J Remote Sens 21 (6&7): 1093-1114

Coppin P, Jonckheere I, Nackaerts K, Muys B, Lambin E (2004) Digital change detection methods in ecosystem monitoring: a review. Int J Remote Sens 25 (9): 1565-1596

Di Gregorio A, Jansen LJM (2000) Land-cover Classification System (LCCS): Classification Concepts and User Manual. FAO, Rome, 194 pp

Duchemin B, Berthelot B, Dedieu G, Leroy M, Maisongrande P (2002) Normalisation of directional effects in 10-day global syntheses derived from VEGETATION/SPOT: II. Validation of an operational method on actual data sets. Remote Sens Environ 81: 101-113.

Erasmi S, Twele A, Ardiansyah M, Malik A, Kappas M (2004) Mapping deforestation and land-cover conversion at the rainforest margin in Central Sulawesi, Indonesia. EARSeL eProceedings, 3(3): 388-397

FAO (Food and Agriculture Organisation of the UN) (2003) The State of the World's Forests. (FAO, Rome) 151 pp

Forest Watch Indonesia (FWI), Global Forest Watch (GFW) (2002) The State of the Forest: Indonesia. Bogor, Indonesia

Friedl MA, McIver DK, Hodges JCF, Zhang XY, Muchoney D, Strahler AH, Woodcock CE, Gopal S, Schneider A, Cooper A, Baccini A, Gao F, Schaaf C (2002) Global land-cover mapping from MODIS: algorithms and early results. Remote Sens Environ 83 (1): 287-302

Giri C, Zhub Z, Reed B (2005) A comparative analysis of the Global Land-cover 2000 and MODIS land-cover data sets. Remote Sens Environ 94: 123-132

GLP (2005): Science Plan and Implementation Strategy. IGBP Report No. 53/IHDP Report No. 19.Stockholm

Hagolle O, Lobo A, Maisongrande P, Cabot F, Duchemin B, De Pereyra A (2004) Quality assessment and improvement of temporally composited products of remotely sensed imagery by combination of VEGETATION 1 and 2 images. Remote Sens Environ 94 (2004) 172–186

Hansen MC, Reed R (2000) A comparison of the IGBP DISCover and University of Maryland 1 km global land-cover products. Int J Remote Sens 21 (6&7): 1365-1373

Hansen M, DeFries R, Townshend JRG, Sohlberg R (2000) Global land-cover classification at 1km resolution using a decision tree classifier. Int J Remote Sens 21: 1331-1365

Holben BN (1986) Characteristics of maximum-value composite images from temporal AVHRR data. Int J Remote Sens 7: 1417–1434

Huete A, Didan K, Miura T, Rodriguez EP, Gao X, Ferreira LG (2002) Overview of the radiometric and biophysical performance of the MODIS vegetation indices. Remote Sens Environ 83: 195-213

Lambin, EF (1997) Modelling and monitoring land-cover change processes in tropical regions. Prog Phys Geog 21 (3): 375-393

Lambin EF (1999) Monitoring forest degradation in tropical regions by remote sensing: some methodological issues. Global Ecol and Biogeogr 8: 191–198

Loveland TR, Reed BC, Brown JF, Ohlen DO, Zhu Z, Yang L, Merchant JW (2000) Development of a global land-cover characteristics database and IGBP DISCover from 1 km AVHRR data. Int J Remote Sens 21(6+7): 1303-1330

Lu D, Mausel P, Batistella M, Moran E (2005) Land-cover binary change detection methods for use in the moist tropical region of the Amazon: a comparative study. Int J Remote Sens 26(1): 101-114

Lu D, Mausel P, Brondizio E, Moran E (2004) Change detection techniques. Int J Remote Sens 25 (12): 2365-2407

Lund G (1999) A 'forest' by any other name. Environmental Science and Policy 2(2): 125-134

Mayaux P, Achard F, Malingreau JP (1998) Global tropical forest area measurements derived from coarse resolution satellite imagery: a comparison with other approaches. Environ Conserv 25: 37–52

Meyer WB, Turner BL (1994) Changes in Land-use and Land-cover: A Global Perspective. Cambridge University Press, Cambridge

Running SW, Loveland TR, Pierce LL (1994) A vegetation classification logic based on remote sensing for use in global scale biogeochemical models. Ambio 23: 77–81

Sano EE, Laerte GF, Huete AR (2005) Synthetic Aperture Radar (L band) and Optical Vegetation Indices for Discriminating the Brazilian Savanna Physiognomies: A Comparative Analysis. Earth Interactions 9: 1-15

Sato HP, Tateishi R (2004) Land-cover classification in SE Asia using near and short wave infrared bands. Int J Remote Sens 25 (14): 2821-2832

Schaaf CB, Gao F, Strahler AH, Lucht W, Li X, Tsang T, Strugnell NC, Zhang X, Jin Y, Muller JP, Lewis P, Barnsley M, Hobson P, Disney M, Roberts G, Dunderdale M, Doll C, D'Entremont RP, Hu B, Liang S, Privette JL, Roy D (2002) First operational BRDF, albedo nadir reflectance products from MODIS. Remote Sens Environ 83: 135-148

Siegert F, Rueker G, Hinrichs A, Hoffmann AA (2001): Increased damage from fires in logged forests during droughts caused by El Nino. Nature 414(6862): 437-440

Singh A (1989) Digital change detection techniques using remotely sensed data. Int J Remote Sens 10: 989–1003

Songh C, Woodcock CE, Seto KC, Lenney MP, Macomber SA (2001) Classification and change detection using Landsat TM data: when and how to correct atmospheric effects? Remote Sens Environ 75: 230–244

Stibig HJ, Beuchle R, Achard F (2003) Mapping of the tropical forest cover of insular Southeast Asia from SPOT4-Vegetation images. Int J Remote Sens 24 (18): 3651-3662

Strahler A, Muchoney D, Borak J, Friedl M, Gopal S, Lambin E, Moody A (1999) MODIS Land-cover Product Algorithm Theoretical Basis Document (ATBD) Version 5.0: MODIS Land-cover and Land-Cover Change

Townsend JRG, Justice CO, Skole D, Malingreau JP, Cihlar J, Teillet P, Sadowski F, Ruttenberg S (1994) The 1 km resolution global data set – Needs of the Int. Geosphere Biosphere Program. Int J Remote Sens 15 (17): 3417-3441

Tucker CJ, Pinzon JE, Brown ME, Slayback D, Pak EW, Mahoney R, Vermote E, El Saleous N (2005) An Extended AVHRR 8-km NDVI Data Set Compatible with MODIS and SPOT Vegetation NDVI Data. Int J Remote Sens (in press)

Turner II BL, Skole DL, Sanderson S, Fischer G, Fresco L, Leemans R (1995) Land-use and Land-cover Change Science/Research Plan. IGBP Report No. 35 and HDP Report No. 7. Stockholm

Turner DP, Ritts WD, Cohen WB, Maeirsperger TK, Gower ST, Kirschbaum AA, Running SW, Zhao M, Wofsky SC, Dunn AL, Law BE, Campbell JE, Oechel WC, Kwon HJ, Meyers TP, Small EE, Kurc SA, Gamon JA (2005) Site-level evaluation of satellite-based global terrestrial gross primary production and net primary production monitoring. Global Change Biol 11: 666-684

Twele A, Erasmi S (2005) Optimierung der topographischen Normalisierung optischer Satellitendaten durch Einbeziehung von Kohärenzinformation. Photogrammetrie Fernerkundung Geoinformation (PFG) 2005 (3): 227-234

UN (united Nations) (2005) World Population Prospects: The 2004 Revision, vol. I, Comprehensive Tables (United Nations publication, Sales No. E.05.XIII.5), New York 2005

Watson R, Noble I, Bolin B, Ravindranath NH, Verardo D, Dokken DJ (eds) (2000) Land-use, Land-Use Change, and Forestry – a special report of the IPCC. Cambridge University Press, Cambridge

Zhang P, Anderson B, Barlow M, Tan B, Myneni RB (2004) Climate-related vegetation characteristics derived from Moderate Resolution Imaging Spectroradiometer (MODIS) leaf area index and normalized difference vegetation index. J Geophys Res 109: D20105

Effects of land-use change on matter and energy exchange between ecosystems in the rain forest margin and the atmosphere

Andreas Ibrom[1,4*], Alexander Oltchev[1,6], Tania June[2,5], Thomas Ross[1], Heiner Kreilein[1], Ulrike Falk[1], Johannes Merklein[1], André Twele[7], Golam Rakkibu[1], Stefan Grote[1], Abdul Rauf[3], and Gode Gravenhorst[1]

[1] Institute of Bioclimatology, Georg-August-University Göttingen, Büsgenweg 2, Göttingen, Germany
[2] Southeast Asian Ministers of Education Organization Regional Centre for Tropical Biology (SEAMEO BIOTROP), BTIC Building, Jl. Raya Tajur Km.6, Bogor, Indonesia
[3] Tadulaku University (UNTAD), Kampus Bumi Tadulako Tondo, Palu - 94118, Indonesia
[4] Risø National Laboratory, Frederiksborgvej 399, DK-4000 Roskilde, Denmark
[5] Faculty of Mathematics and Natural Resources, Bogor Agricultural University/SEAMEO BIOTROP, Bogor, Indonesia
[6] A.N.Severtsov Institute of Ecology and Evolution of the Russiona Acadamy of Sciences (RAS), 33 Leninskiy Prosp., Moscow 119071, Russia
[7] Institute of Geography, Department of Cartography, GIS and Remote Sensing, University of Göttingen, Goldschmidtstr. 5, 37077 Göttingen, Germany

*corresponding author: Andreas Ibrom, Risø National Laboratory, Frederiksborgvej 399, DK-4000 Roskilde, Denmark, Email: Andreas.Ibrom@risoe.dk

Summary

Greenhouse gas and energy fluxes between the land surface and the atmosphere are important aspects for the evaluation of land-use options in tropical areas. Changes in vegetation cover alter the capacity to absorb carbon dioxide and solar radiation from the atmosphere and influence the magnitudes of latent and sensible heat flows to the atmosphere. If happening at a larger spatial scale, land-use change can lead to significant local feedbacks like drought, flooding, soil erosion or shifts in local climate.

Up to now only little was known about how typical ecosystems of the rain forest margin areas interact with the atmosphere. We present here results from a sub-program of STORMA (Stability of rain forest margins in Indonesia,

Tscharntke T, Leuschner C, Zeller M, Guhardja E, Bidin A (eds), The stability of tropical rainforest margins, linking ecological, economic and social constraints of land use and conservation, Springer Berlin 2007, pp 463-492

SFB-552-B1), to describe the energy, CO_2 and water fluxes between a larger tropical area and the atmosphere.

Field measurements with the eddy correlation method showed that a Cacao agroforestry system (AFS) was a small source of CO_2, whereas a tropical rain forest was a strong CO_2 sink during the one year observation period. In addition, the rain forest evaporated much more water compared to the Cacao AFS.

We applied a series of models to extrapolate the results from the field investigations to the region, including so-called SVAT models of differing complexity, and a simple PAR efficiency model to predict net primary production from regional meteorological and remote sensing data. In all cases, we discovered a large spatial variability in CO_2 and water vapour fluxes. We investigated the sensitivity of the models to changes in climatic drivers or land-use parameters from the current to a more intensified agriculture. The selected land-use scenarios reduced carbon dioxide sequestration and total evaporation and increased sensible heat fluxes and thus surface temperatures.

We conclude that there is still a lack of field observations to better understand and simulate the behaviour of tropical land-use systems. If the necessary detail on model parameter values, which characterise the differences between the land-use systems, is known, the existing model approaches are suitable to investigate consequences of land-use change to regional biogeochemical cycling, water utilisation and climate change.

Keywords: land-use change, carbon dioxide fluxes, energy fluxes, montane tropical rain forest, Cacao agroforestry system, eddy correlation, regional modelling, SVAT model, remote sensing

1 Introduction

Exchange processes of CO_2, water and energy between land surfaces and the atmosphere are highly relevant to the issue of land-use and land-use change. These fluxes are key processes for productivity, have impacts on regional water resources and climate, and are relevant for global climate change.

Matter and energy fluxes are highly variable in space, because of the different biophysical traits of cultivated and natural land surfaces, ranging from bare soil to, e.g., old growth forests. Due to their close connection with atmospheric processes, fluxes vary with weather and climate from hourly to multi annual time scales. This makes them also susceptible to climate anomalies like the El Niño Southern Oscillation (ENSO). It is both a necessity and a challenge to include the spatial and temporal variability with the necessary degree of accuracy in regional modelling.

Matter and energy fluxes in the horizontal direction (e.g. water run-off) on the earth surface are the result of vertical exchange between land surface and the atmosphere. Because the processes in landscapes are highly interconnected, the dynamic consequences within and across ecosystem boundaries are also important results of land-use and land-use change modelling. Small-scale processes may have large effects when accumulated over a long time and across larger areas. Thus, unforeseeable effects of land-use change are likely to occur.

In the tropical rain forest margin zone of Central Sulawesi we meet the challenge to quantify such effects of land-use change. Process based models have been developed, and are going to be tested and later applied as a scientific method to be able to guide land-use decisions in order to avoid negative unforeseen consequences. Intensive field investigations are necessary to collect the necessary information on how relevant ecosystems control the matter and energy exchange with the atmosphere, because only little is known about the functioning and dynamics of the different natural ecosystems, especially in South East Asia (SEA). The information from field observations is used to parameterise simulation models. By simulation of the relevant biophysical processes in soil, vegetation and atmosphere it becomes possible to quantify the matter and energy exchange with the atmosphere at the local scale, and thereby to extrapolate the available information to larger areas.

The objectives of this Chapter are therefore (*i*) to describe observations of biosphere - atmosphere exchange in the field and (*ii*) to outline possible ways to generalise this knowledge in mathematical process models of land surface - atmosphere interaction for prediction of possible impacts of land-use changes at regional and global scales. The results of intensive, continuous, multi-annual flux measurements on micrometeorological towers are compared with findings from other investigations. Modelling case studies are presented and discussed, in which carbon sequestration and regional fluxes of water vapour and carbon dioxide are simulated for the study region in Central Sulawesi.

2 Measurements of biosphere – atmosphere interactions

2.1 Flux methodology and networks

Energy and matter fluxes between the terrestrial biosphere and the atmosphere are in principle well understood. To predict potential impacts of land-use changes on local and regional greenhouse gas budgets and water availability, the fluxes need to be quantified at the regional scale. Finding the scientific basis for such estimations is a major scientific challenge. This is especially true for the tropics, where information, e.g., on canopy structure and physiology, root distribution, soil biological processes, soil water availability and microclimate is scarce. The different types of rainforests are among the most complex examples of land surfaces, because they host a large number of different tree species, with different physiology and architecture. Usually available information is not sufficient to find all relevant parameters for a model simulation and, therefore, the results remain speculative unless tested against independent measures. For both the local and regional scales, micrometeorological methods have been developed that can measure surface fluxes of matter and energy.

Scientific progress during the past 15 years has made it possible to measure fluxes at ecosystems scale in practically all kinds of ecosystems, with high temporal resolution over longer periods, e.g. decades, using the so-called eddy correlation or eddy covariance method (EC, Aubinet et al. 2000). This micrometeorological method measures the fluxes through the atmospheric boundary layer above the ecosystems and does not require any kind of enclosed control volume. Therefore, it does not influence the fluxes directly or indirectly by altering, e.g., the radiation budget, temperature distribution, and, most vital for the energy transports, the turbulent exchange conditions. However, the eddy correlation method requires horizontal homogeneity and stationarity of fluxes over the measurement period.

Provided horizontal homogeneity, the exchange processes between the surface and the atmosphere at 10 to 20 meters distance from the surface can be treated one-dimensionally, in the vertical. For chemically inert gasses and even other atmospheric properties, like energy or momentum, every flux through the surfaces must lead to accumulation or depletion of the transported property in a layer of air that is directly in contact with the surface. Atmospheric vertical transport, however, counterbalances these differences, which leads to steady state vertical concentration profiles. In these conditions the vertical flux is constant with height and approaches the spatial average of all fluxes at the homogeneous surface over one to several km^2. If the surface flux is changing with time, the relation between the net biotic flux at the surface, F_{b}, and the measured flux in the boundary layer above the surface, F_{c}, is

$$F_b = F_c + S_c, \tag{1}$$

with S_{c} being the rate of storage change of the property c in the air column underneath the measurement point.

S_{c} can be estimated with a vertical array of measurements of the transported property. In cases of intensive turbulent mixing it can be approximated by using the rate of concentration change at the measurement point.

Measurement of F_{c} is more difficult, but in absence of transports with the mean flow, which is confined to special meteorological situations or sites where horizontal homogeneity is not fulfilled, F_{c}, equals the vertical turbulent flux (upward positive),

$$F_c = \overline{w'c'}, \tag{2}$$

the covariance between the instantaneous vertical wind velocity and the instantaneous density of the transported entity c. Further information on flux estimations and their background can be found elsewhere (Stull 1988, Kaimal and Finnigan 1994, Aubinet et al. 2000, Lee et al. 2004). We do not report here on methods that assess fluxes at regional scale, like flux measurements at the top of very high towers (see, e.g., Baker et al. 2003), aircraft measurements (Crawford et al. 1996) or boundary layer budgeting methods (Levy et al. 1999, Styles et al. 2002).

The flux methodology has been widely standardized and applied by the so-called flux communities. Scientific networks have been established in America (AMERIFLUX) and Europe (EUROFLUX) and other continents that collect results from more than 200 flux towers (Valentini et al. 2000; Baldocchi et al. 2001). Only a few of these sites are situated in tropical areas, and here mainly in Latin America. Classical work has been done in international projects like ABRACOS (Gash et al. 1998) and the ongoing LBA project (Schimel 2004).

2.2 A case study of flux measurements in Central Sulawesi

As part of the interdisciplinary research on the "Stability of Rainforest Margin Areas in Indonesia" (STORMA), a sub-project, B1, was dedicated to investigate the impact of land –use change on vertical CO_2, water vapour, and energy fluxes. The approach followed the idea to assess fluxes and relevant system parameters in the most relevant land-use types in the investigated area, the Lore Lindu National Park, including its margin zone to farmed area in the surrounding valleys. Typical land-use types in this area are the natural montane tropical rainforest, different forms of forest use, slash and burn cultivation, perennial crops, mainly Cacao plantations, wetland and dry rice cultivation and fallow land. Lowland tropical rain forests are already almost entirely converted to farmland; thus, the montane tropical rainforest is the only remainder of natural vegetation.

The hilly area and the patchy land-use pattern made it very difficult to find suitable sites for micrometeorological flux measurements. Up to now, two sites have been investigated, a mature Cacao agroforestry system (AFS) in the Palolo valley near the village Nopu at 640 m a.s.l. (Falk et al. 2005) and a montane rainforest near the village Bariri in 1420 m elevation (Ibrom et al. 2006).

In both cases, fluxes of CO_2, water vapour and energy have been measured continuously together with other meteorological variables for more than one year. The investigation of the Cacao AFS started in December 2001 and was completed in April 2003; the measurements at the Bariri site began in October 2003 and are still going on.

In both cases the same eddy correlation system was used. It consisted of a three dimensional sonic anemometer (USA-1, Metek, Elmshorn, Germany) and an open-path CO_2 – H_2O sensor (LI-7500, Li-Cor, Lincoln, Nebraska, USA), both of which connected to a field computer that logged the 10 Hz raw data. Measurement heights were 5.3 m and 12 m above the canopies of the Cacao AFS and the tropical rainforest, respectively. Additionally meteorological variables like short and long wave radiation (R_s and R_l), air temperature (T_a), water vapour pressure, horizontal wind speed (u) and precipitation rate

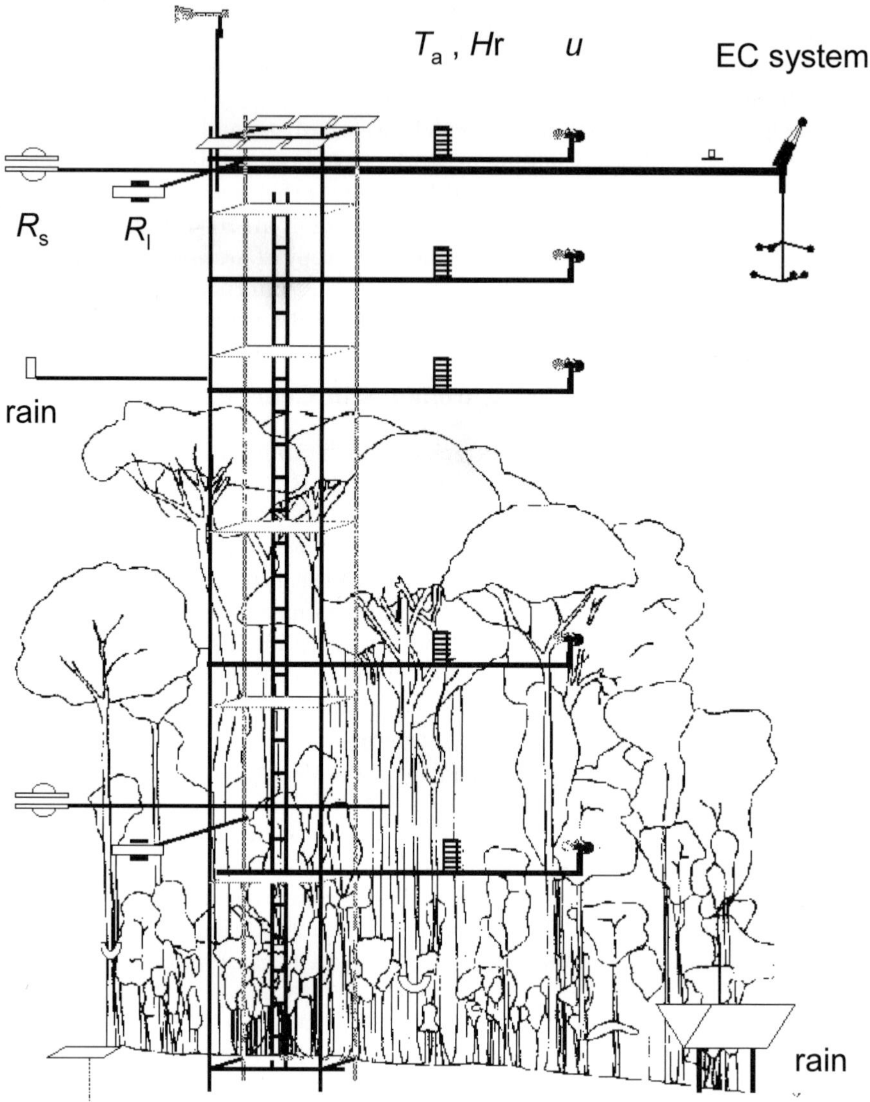

Fig. 1. Schematic diagram of the flux tower setup. The abbreviations are: R_s and R_l short and long wave radiation, T_a air temperature, H_r relative humidity, u horizontal wind speed, EC eddy correlation.

were measured either at the 10 m mast (Cacao AFS) or the 70 m steel tower (montane rainforest). A typical setup of a flux tower is sketched in Figure 1.

In order to be able to explain the fluxes from physical and biological boundary conditions, a set of additional structural and physiological parameters was measured. In case of the Cacao AFS we investigated exact locations of all Cacao trees within two subplots in the source area of the flux measurements and determined parameter values for a physiological leaf gas exchange model (unpublished, Merklein 2003). Such investigations are much more difficult in the tropical rainforest, where the number of tree species with a tree diameter of more than 10 cm was 88 ha^{-1} and the height distribution ranged from 4 m to 41 m. The most relevant 5 tree species could be characterized by ecophysiological measurements.

Table 1. Canopy traits of the montane tropical rain forest and the Cacao agroforestry system

	montane tropical rain forest	Cacao AFS
Age	old	13 yr
Tree species	88 ha^{-1}	3 ha^{-1}
Dominating tree species	*Castanopsis accuminatissima* (29%) *Canarium vulgare* (18%) *Ficus spec.* (9.5%) *NN* (4.8%)	*Theobroma Cacao* L. > 98 %
Tree density	567 trees ha^{-1}	1054 trees ha^{-1}
Basal area	53 m^2 ha^{-1}	10.8 m^2/ha
LAI [1,2]	7.2	7.1
Average tree height	21 m	5.1 m
Maximum height	36 m	6.2 m

1 unpublished, A. Twele and J. Dietz, personal communication
2 unpublished data, (Merklein 2003)

Table 1 characterises the two vegetations in greater detail. It can be clearly seen that there are some fundamental differences between the two canopies, e.g. with regard to tree sizes and density. However, interestingly, one of the most important parameters, leaf area index, LAI, was similar. The LAI of the Cacao AFS was surprisingly high, and comparably accurate, because it had been calculated from direct leaf counting and measurements at ten entire representative trees. The LAI of the rain forest was derived indirectly from hemispherical photography, which introduces some uncertainty to this estimate.

Due to the different elevation, climate was different between the two sites. The montane rainforest site was colder (annual mean T_a=19.1 °C as compared to 24.5 °C), precipitation was higher (2800 mm yr^{-1} compared to 1970 mm

yr^{-1}) and short wave radiation was slightly higher (19.0 MJ m^{-2} d^{-1} compared to 17.3 MJ m^{-2} d^{-1}) compared to the Cacao AFS site near Nopu.

In the following we use the currently existing results from these two investigations to highlight the measurable different biometeorological characteristics between two different, but not essentially contrasting, tropical land-use types.

Energy fluxes

On a global scale, the ground absorbs on average a little more than half of the extraterrestrial solar radiation flux; in the tropics this flux is twice as high as the global average. The amount of absorbed shortwave radiation depends on incoming solar radiation R_s and the albedo α, i.e. the short wave surface reflectance. α changes with the optical leaf and soil properties and with canopy structure. Thus the radiation budget, R_n, of different surfaces can be very different and undergoes changes if the land-use changes.

The energy, which is gained from absorbed radiation, is transformed into other energy forms, thereby inducing strong heat fluxes. Storage of biochemical energy in the system, which is important for plant production, is only small compared to R_n, which in turn mainly feeds the sensible and the latent heat fluxes, H and λE, respectively. Biological regulation and soil water availability steer the energy partitioning either into warming up the boundary layer, high H, or humidifying it, high λE. In addition to its role for the climate near the ground, the water use has also a crucial effect on the surface itself. Depletion of soil water bears the risk for the vegetation to run out of water in dryer seasons. On the other hand, loss of water enables the soil to store rainwater again, thereby buffering the dynamics of water flows. In an area with usually heavy rainfalls this mechanism can prevent flooding and soil erosion. Heating up the atmospheric boundary layer does, beyond the surface temperature effect, increase the saturation deficit, i.e. potential evaporation becomes larger. A consequence for plant cultures or natural vegetation would be that either water use increases or productivity might drop, because of stomatal limitation of photosynthesis.

Our radiation measurements showed that the Cacao canopy behaved differently from the natural forest system. The Cacao AFS reflected 15 % of the incident R_s (Falk et al. 2005); α of the forest in Bariri was only two thirds of that, 10 %.

In accordance with the higher α, the ratio between R_n and incident R_s at daytime was higher in the rain forest (87 %) than in the Cacao AFS (79 %). These differences were larger than expected from the measured short wave radiation budgets (85 % and 90 % of R_s, for rainforest and Cacao AFS, respectively). The long wave radiation budget caused this difference. Due to the higher surface temperatures the long wave emission of the Cacao AFS exceeded that of the montane rainforest. Thus, the long wave radiation budget became be more negative. During the nighttime, the tropical rain forest lost somewhat more thermal radiation than the Cacao AFS, leading to a daily R_n

of 14.1 MJ m^{-2} d^{-1} (74.1 % of daily R_s) as compared to 11.1 MJ m^{-2} d^{-1} (70.4 % of daily R_s).

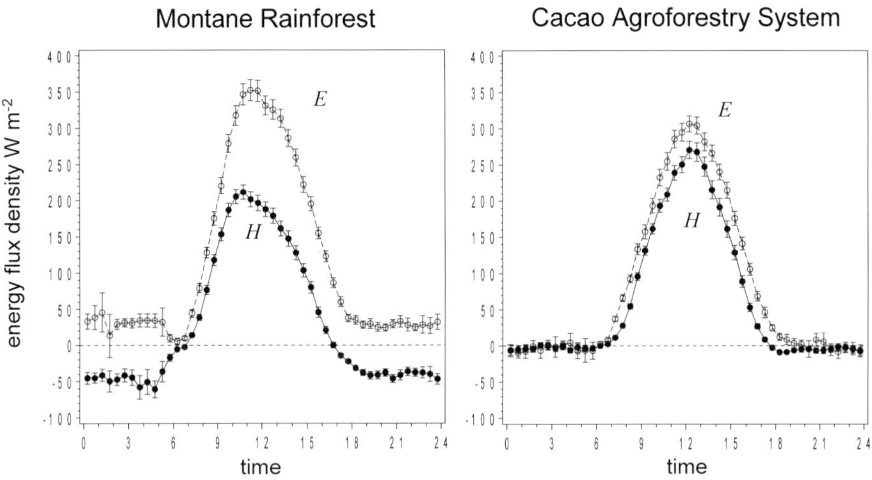

Fig. 2. Turbulent energy fluxes above the montane rainforest and the Cacao AFS. Data have been selected for turbulent night time conditions (rainforest friction velocity > 0.3 m s^{-1}, Cacao AFS friction velocity > 0.2 m s^{-1}). Error bars denote standard errors of the mean due to seasonal variability. Standard errors are particularly higher at the Cacao AFS during night, because of the strong reduction of data due to low wind speeds.

Figure 2 shows how R_n was transformed into latent and sensible heat at the two sites. At night we observed only very small fluxes at the Cacao AFS, which can be partly due to a methodical problem but can also be caused by regular cloud formation in the valley.

The methodical problem arose from the very calm wind conditions we met at night in the Nopu valley, when non turbulent energy fluxes might have occurred, which can not be measured with the eddy correlation system (see discussion in Aubinet et al. 2000, Falk et al. 2005, Ibrom et al. 2006). At no time did R_n reach strong negative values during nights because clouds formed in the valley and these emitted thermal radiation towards the ground. Hence, H was never as negative as at the rain forest site. Thus, the forest on the upper slopes cooled the air around the canopy more during night than the Cacao AFS in the valley did. During daytime the forest evaporated much more than the Cacao AFS, and, accordingly, H was also lower, as during the night.

The above-described differences in energy exchange were most likely caused both by differences between the two vegetations covers and by the different microclimates, as imposed by different topography. During the daytime, the different surface properties were more important for the different behaviour; during the night, the microclimatic differences were more important. Thus the daytime patterns are more relevant in the context of the discussion on potential effects of land-use change on energy fluxes. Our findings suggest that at daytime the climate near the ground will become warmer and drier after rain forest conversion to Cacao AFS, whereas at night the effects on atmospheric properties will be small.

Water balances

The water balances of the two systems, montane rainforest and Cacao AFS, are also affected by differences between the surface parameters, i.e. the different vegetations, and by local microclimatic effects.

The rainfall was higher at the rainforest site. This should have induced increased evaporation of intercepted rainwater and potentially also transpiration, if soil water availability was maintained. The more turbulent conditions at the forest site fostered evapotranspiration (E), but lesser cloudiness at night decreased the surface temperatures and therefore also the rates of evaporation of intercepted rainwater. Thus it is difficult to directly conclude from our measurements, whether the observed differences in water fluxes were due to the different vegetation or due to the different atmospheric conditions. In total, E reached 4 mm d^{-1} (1450 mm yr^{-1}) at the rain forest site and 2.9 mm d^{-1} (1045 mm yr^{-1}) at the Cacao site. If differences in water storage (that means in our cases water gain of the land-use system) can be neglected, the water balance of the rain forest was 1450 mm yr^{-1} (48 % of local precipitation) and 845 mm yr^{-1} (43 % of local precipitation) at the Cacao AFS. In relation to the precipitation, the figures are relative close to each other. Future investigations should help to explain these differences, i.e. whether different vegetation traits or the different weather conditions at different sites are more influential. If it turns out that differences in water fluxes are mainly caused by surface characteristics, they should be considered more carefully in the modelling of regional effects of land-use change on water resources.

CO_2 fluxes

Before the first CO_2 flux measurements above a pristine tropical rain forest were published (Grace et al. 1995) it could be assumed that the undisturbed old growth ecosystems were carbon neutral, i.e. they reached equilibrium between CO_2 assimilation and ecosystem respiration. Since then, different investigations in Latin America showed rain forests either to be taking up considerable amounts of carbon dioxide, e.g. 500 – 600 g (C) m^{-2} yr^{-1} (Malhi et al. 1999, Loescher et al. 2003), or moderate amounts (ca. 100 g(C) m^{-2} yr^{-1} Grace et al. 1995) or they were carbon neutral or even a small sources of CO_2 (Loescher et al. 2003, Saleska et al. 2003, Miller et al. 2004). Longer observation periods revealed that the carbon dioxide budgets of tropical forests can vary considerably from year to year (Loescher et al. 2003).

There is also an ongoing debate on how accurate the micrometeorological flux estimates can and may be, especially in tropical areas, where the above mentioned "night time problem" is more severe than in other places with higher wind speeds. So far, our investigations went on for only a little more than one year, so it cannot yet be assured that the carbon budget estimates represent a typical year.

The flux measurements in the Nopu valley suffered from a technical problem with the first generation of the open-path analyser, where direct radiation affected the CO_2 concentration measurements. Therefore, the flux estimations were corrected by band-pass filtering and extrapolation to the entire spectral domain with the help of undisturbed temperature spectra (Falk 2004). The potential loss of information by this methodical problem is not yet quantified. At night, when the source area of the flux measurements became large, the flux estimate might include CO_2 fluxes from farms that are part of the plantation. This factor has yet to be quantified by source area modelling.

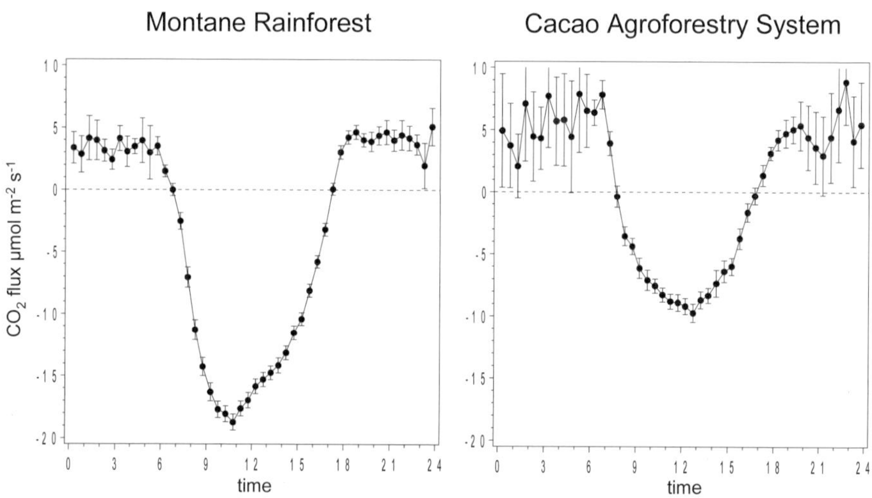

Fig. 3. Averaged diurnal net carbon dioxide fluxes. Night time flux data have been filtered for turbulent conditions, for the rainforest if friction velocity > 0.3 m s^{-1} and for the Cacao AFS if friction velocity > 0.2 m s^{-1}. Upward fluxes are positive, thus net release is positive and net uptake is negative. Error bars denote standard errors of the mean due to seasonal variability.

The carbon dioxide fluxes of the two investigated sites are presented in Figure 3. They differed fundamentally from each other. Gross photosynthesis (P_g) of the rain forest was more than twice as high as that in the Cacao AFS. Ecosystem respiration (R_E) was, however, smaller in the rain forest than in the Cacao AFS, where the exclusion of many data from calm nights led to a generally larger standard error. We discuss the relatively small R_E of the montane rainforest in Ibrom et al. (2006).

The combination of a high P_g with low respiration rates made the montane rainforest a very strong CO_2 sink, -970 g (C) m^{-2} yr^{-1} (Ibrom et al. 2006), whereas the Cacao AFS was a small CO_2 source. According to our measurements, the Cacao AFS released about 170 g (C) m^{-2} yr^{-1}.

Despite some open questions on the exact flux values and the need for longer observation periods, our measurements indicate fundamental differences in the carbon budgets between the two investigated tropical ecosystems. Compared to the other intensively managed land-use systems in the tropical rain forest margin, such as annual crops or rice paddies, we expected the perennial Cacao AFS to be relatively close in structure and function to the rain forest. This means that we can expect even stronger differences in CO_2 budgets, if we take other agricultural systems into account.

3 Simulation of biosphere – atmosphere interactions

Mathematical models of land-surface – atmosphere interactions are effective tools in meteorology, hydrology, geochemistry, and ecology. They are indispensable to quantify fluxes at local, regional and global scales, to extrapolate fluxes from point measurements to areal information, and to predict possible changes of surface fluxes under environmental (climatic) and land-use changes. These models are usually based on aggregated descriptions of the physical, chemical and biological transport processes in soil, vegetation and atmospheric surface layer and are thus called "Soil - Vegetation - Atmosphere - Transfer" (SVAT) schemes.

SVAT models have a varying degree of complexity depending on at which spatial and temporal scales they describe energy, water, carbon dioxide and other atmospheric trace compound fluxes between the ground surface and the atmosphere (e.g. Sellers et al. 1997, Pitman and Henderson-Sellers 1998, Falge et al. 2005).

It is obvious, that the complexity of any SVAT model should depend on the main goals of the modelling study, hence on the required spatial and temporal resolutions, on the properties of land surface and vegetation canopy and, especially, on availability of input parameters for model calculations. Generally, the coarser the scale the more simplification is needed in the model approaches. Thus global and regional scale models are generally much more simplified than those acting at ecosystems scale. However, there are also differences between the different ecosystems. The interaction between short vegetation (e.g. grassland, some agricultural crops) and the atmosphere can be simulated successfully using simplified modelling approaches, e.g. one-layer "Big-leaf" models (e.g. Falge et al. 2005). On the other hand, mono- or multi-specific forest stands need in most cases more sophisticated approaches, which allow for taking into account the spatial heterogeneity of the canopy and describing both the total fluxes and flux partitioning among different layers and species in the canopy and in the soil (e.g. Raupach and Finnigan 1986).

This sub-chapter focuses on the description of three different models for the prediction of energy, H_2O and CO_2 fluxes at ecosystem (local) and at regional scales, and on different applications for the analysis of the effects of land-use change on biophysical processes in a tropical rain forest margin area.

A sophisticated process-based model Mixfor-SVAT (Oltchev et al. 2002, Falge et al. 2005), is applied to describe possible changes of surface fluxes after deforestation and environmental changes at local ecosystem scale. The SVAT-Regio model (Oltchev 2003) is used for spatial interpolation of meteorological data, for the description of regional fluxes of energy, H_2O and CO_2 in the rainforest margin zone, and to study the influence of land-use change on surface fluxes. A third model, NetPro, reduces the process description to a very minimum and allows to quantify the net primary production in a larger area using remote sensing and regional meteorological data.

3.1 Local scales

Description of the model Mixfor-SVAT

Mixfor-SVAT is a one-dimensional (1D) process-based SVAT model for the parameterisation of the energy, H_2O and CO_2 exchange between multi-specific and vertically structured vegetation and the atmosphere (Oltchev et al. 2002, Falge et al. 2005). It describes a forest stand taking into account its three-dimensional features, such as multi-specific structure, and different tree heights, leaf area index (LAI), crown shapes and foliage distributions of different tree species. It is assumed that all trees of the different species are evenly distributed over the homogeneous ground surface area (at least 1 ha) and that trees of the same species are identical. Horizontal advection of energy, water and CO_2 through the boundaries of the modelled forest plot is not taken into account.

Mixfor-SVAT consists of several closely coupled sub-models describing:

1. two-directional solar radiation transport within the forest canopy;
2. vertical turbulent exchange of momentum, sensible heat, H_2O, CO_2 within and above the forest canopy;
3. transpiration of different tree species in different sub-layers of overstorey and understorey;
4. water uptake by roots and water transport through the roots, trunks and branches to the leaves of different tree species and understorey plants;
5. interception, storage and evaporation of rain water and dew by different sub-layers;
6. evaporation and CO_2 efflux from bare soil;
7. photosynthesis of leaves, autotrophic and heterotrophic respiration;

8. water and heat transport in soil.

The different sub-models are coupled using the aggregated numerical solution to close the energy and water balance equations for individual tree species, different canopy layers, and, finally, the entire forest stand. The energy, water vapour and CO_2 fluxes for both the entire forest ecosystem and for each canopy sub-layer are determined by iteration procedures for each intermediate (30-60 seconds) and main time step (15 to 60 minutes). Additionally, Mixfor-SVAT predicts vertical profiles of various concentrations, temperatures, liquid water contents in soil, vegetation and atmospheric boundary layer.

The required meteorological input information is measured or predicted air temperature, air specific humidity, wind speed, global radiation, precipitation rate and CO_2 concentration at a reference height within the atmospheric surface layer above a forest stand.

The main concept of Mixfor-SVAT is the aggregated description of physical and biological processes in a forest ecosystem on different spatial scales: individual leaf, canopy and soil sub-layers, individual tree (plant) and entire ecosystem. To integrate the information from different scales in multi-species forest ecosystems Mixfor-SVAT uses both species specific and averaged input parameters. For the description of the processes occurring inside an individual leaf or tree the model uses individual species specific input parameters. On the other hand, for parameterisation of exchange processes between different tree species within each sub-layer, as well as for description of the processes on ecosystem scale (e.g. turbulent exchange, radiation transport) the model uses species averaged parameters of vegetation cover (e.g. morphological and physical properties of the canopy). These effective parameters are calculated for each individual canopy sub-layer from individual properties of tree species taking into account the vertical distributions of living and dead biomass of the trees.

Model validation

Mixfor-SVAT was tested using EC flux measurements in a Cacao AFS in Nopu (from December 2001 to April 2003) and in natural tropical rain forest in Bariri (from October 2003 to February 2005, see Sub-chapter 2). Meteorological input variables were measured from meteorological towers above the plant canopies (Figure 1). The parameters describing the biological and physical properties of the different plant and tree species, as well as the physical properties of different soil layers, were either determined from field measurements or obtained from literature.

In order to exclude uncertainties of EC flux estimations for the model test, the flux data were critically screened for technical errors and statistical stationarity of the time series (Ibrom et al. 2006). Only those night time flux data were used when turbulence was well developed, i.e. when friction velocity $(u_*) > 0.3$ m/s.

Three statistical parameters, the determination coefficient (r^2), the absolute fractional bias (AFB) and Willmott's index of agreement (IA, see also Falge et al. 2005) were used to quantify the agreement between modelled and measured fluxes.

Because of weak turbulent conditions during the nights, a large number of measured nocturnal fluxes were excluded from model validation procedures for both land-use types. Lack of EC data with turbulent conditions, sensor wetness and spikes in the flux pattern due to likely impacts of horizontal and vertical advections resulted in only moderately high r^2 values between modelled and measured fluxes for tropical rain forest ($r^2 = 0.62$ for CO_2 and $r^2 = 0.64$ for H_2O fluxes) and for the Cacao AFS ($r^2 = 0.66$ for CO_2 and $r^2 = 0.71$ for H_2O fluxes).

At the same time, results of model comparisons using values of AFB and IA show a good performance of the MixFor-SVAT, which means that especially the averaged daily dynamics of H_2O and CO_2 fluxes were described with sufficient accuracy. For CO_2 fluxes of the tropical rain forest AFB was about 0.52 and for H_2O fluxes 0.54. AFB for the Cacao AFS were somewhat higher: 0.66 for CO_2 and 0.65 for H_2O fluxes. All AFB values stayed within a critical interval of AFB (between 0 and 0.67) that indicates a sufficiently good agreement of the model predictions and field measurements.

Willmott's index of agreement, which ranges between 0 and 1 and indicates the degree to which model predictions are error free (Willmott 1981), was 0.87 for CO_2 and 0.86 for H_2O fluxes for the tropical rain forest, and 0.88 for CO_2 and 0.87 for H_2O fluxes for the Cacao AFS. These results indicate a good performance of Mixfor-SVAT, too.

Modelling effects of land-use changes on energy, H_2O and CO_2 surface fluxes

The meteorological data obtained from measurements at the forest tower in Bariri from January to December 2004 were used to predict possible impacts of deforestation on surface energy, H_2O and CO_2 fluxes. Three different land-use change scenarios were selected for this analysis (Figure 4); the substitution of tropical rain forest by Cacao AFS, by grassland and by corn crop. The third scenario assumes two harvests of corn per year, mainly during periods with moderate precipitation in summer and in winter.

Meteorological conditions for the modelling period were characterised by the small variation of air temperature, high variation of relative humidity and very high precipitation (2511 mm yr^{-1}) with the maximum in April.

Modelling results show that all deforestation scenarios result in a decrease of annual E and F_b. Very small differences were obtained between modelled F_b of tropical rainforest and Cacao AFS. The annual trend of F_b for the corn field scenario is characterised by a very high seasonal F_b during periods of active vegetation, i.e. for the selected scenario in summer and in winter. However, annual F_b for the corn field was significantly less than for tropical

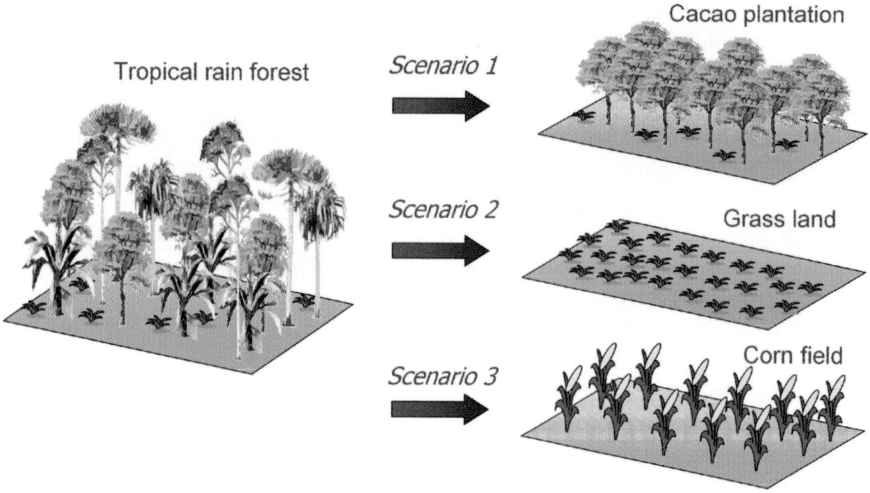

Fig. 4. Scenarios of land-use changes that were used in the simulations.

rain forest. Maximal decrease of annual E and transpiration was simulated for the substitution of tropical rain forest by grassland and by corn crop. Transpiration of the Cacao AFS was about 20 % smaller than transpiration of the tropical rain forest despite similar LAI for both vegetation types. However, evaporation of intercepted water was about 3% higher than for tropical rain forest. This can probably be explained by differences in canopy structure and the vertical wind profile.

3.2 Regional scales

Description of SVAT-Regio and its application for the estimation of regional energy, H_2O and CO_2 fluxes

The SVAT-Regio model was developed to describe the energy, H_2O and CO_2 fluxes between heterogeneous landscapes and the atmosphere (Oltchev 2003). It is a quasi three-dimensional process-based SVAT model, which uses a simplified multi-layer representation of the vegetation and the soil to quantify both the total fluxes and the contributions of the individual canopy and soil layers. Neighbouring grid cells of the model domain are connected via lateral water flows, thus, the approach is effective to describe the regional dynamics of soil moisture, E, F_b, R_E and P_g of the vegetation cover.

The modelling procedure consists of several consecutive steps including (Figure 5):

1. Spatial interpolation of daily meteorological data from the nearest meteorological stations into individual grid cells into which the entire study area

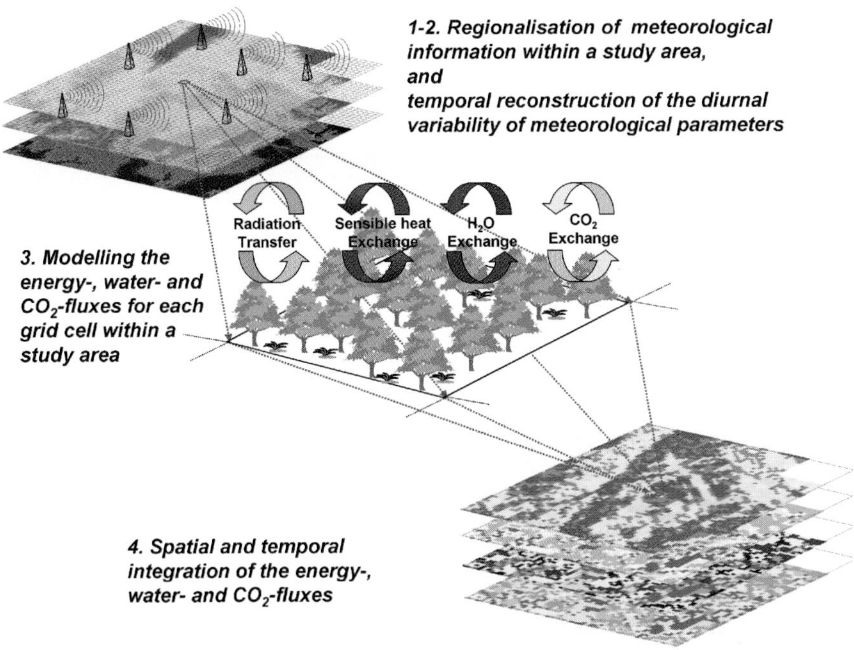

Fig. 5. The main steps used by the SVAT-Regio model for simulation of the energy, water and carbon dioxide fluxes in regional scale.

is subdivided. The interpolation algorithms take into account relief, local atmospheric circulation and land-use pattern of the study area. The land surface, soil and vegetation properties within each grid cell are assumed to be spatially uniform.

2. Temporal reconstruction of the daily course of meteorological parameters for each grid cell with a time step of 1 hour from regionalised daily meteorological data.
3. Simulation of energy, H_2O and CO_2 fluxes, canopy microclimate, canopy and soil hydrology for each grid cell including the impact of horizontal advection between neighbouring grid cells.
4. Summation of energy, H_2O and CO_2 fluxes over time and area, e.g., ecosystems, catchment areas, entire study area.

SVAT-Regio uses mean daily air temperature and humidity, wind speed and direction, amount of precipitation, R_s and digital maps of relief, land-use, vegetation and soil types. The vegetation of each grid cell is described by mean LAI, mean canopy height, tree density, mean root depth and physiological parameters such as stomatal conductance, photosynthesis and dark respiration rates for each grid cell. Based on the soil types, structure and texture of different soil layers, hydraulic and heat conductivities are determined. SVAT-

Regio can be parameterised for all relevant land-use and vegetation types, such as forests, bush land, meadows, agricultural crops (e.g. corn, wheat, rice fields) and water bodies (e.g. rivers, lakes).

The main outputs of SVAT-Regio are: R_n, water flows, E and transpiration rates, as well as F_b for the canopy overstorey and understorey and for the soil for every grid cell. Additionally, SVAT-Regio outputs the vertical stratification of physical and biological state variables for every grid cell.

The choice of spatial resolution in SVAT-Regio is determined mainly by the size of the target area and by the spatial heterogeneity of relief, land-use, vegetation and soil patterns. Furthermore, it depends on available input data, and on the main purposes of a modelling study. The current version of the SVAT-Regio model uses a regular grid consisting of 500 $\times$ 500 grid cells. The number of grid cells can be adapted to the needs of the modelling tasks and available computer capacities. However, the size of an individual grid cell should not be smaller than 100 m $\times$ 100 m in order to minimise boundary effects between neighbouring grid cells.

E and F_b were modelled for a mountainous landscape (about 26 $\times$ 26 km^2) located in the northern part of the LLNP for September 2004 (Figure 6 a). The target area is very heterogeneous and is mainly covered by tropical rain forests. Lake Lore Lindu lies in its southern part. The spatial resolution of the simulation was 300 m $\times$ 300 m.

The regional pattern of meteorological parameters was derived from measurements provided by 10 meteorological stations. The mean air temperature was about 21.8 C°, mean relative humidity 83.6 %, total solar radiation about 590 MJ m^{-2} month^{-1}, and total precipitation amount on average 114 mm month^{-1}. Soil moisture was close to field capacity in the entire study area during almost all of the time.

Figures 6 b and c show the simulated regional patterns of monthly E and F_b. The spatial pattern of E is very heterogeneous. It ranges from 54 to 152 mm per month. Minimal values are observed in urban and grassland areas and maximal in the area of Lake Lore Lindu. Modelling results show that given the high soil water availability during September 2004 there were no significant differences between E of tropical rain forests and other land-use types in lower elevation parts of the model domain. Despite larger precipitation rates, E from montane tropical rain forests in the target area is lower than that of lowland rain forests. These differences result from lower air and surface temperatures, higher relative humidity and lower incoming solar radiation in these higher elevation areas.

Monthly F_b ranged from 41 to -184 g (C) m^{-2} month^{-1}. Maximal CO_2 uptake was observed in areas covered by tropical rain forest and CO_2 losses in urban areas. The lake was not simulated but instead F_b was set to 0 g (C) m^{-2} month^{-1}.

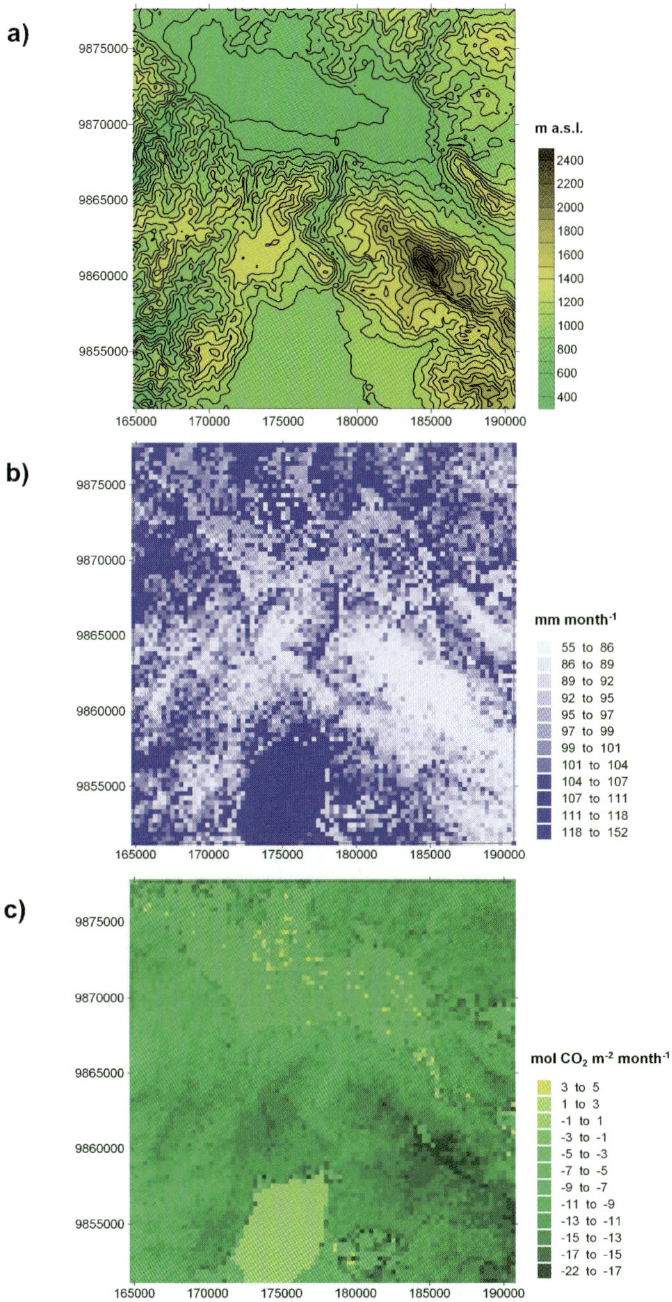

Fig. 6. (a) Relief of the study area in Central Sulawesi and modelled patterns of (b) monthly evapotranspiration in mm month^{-1} and (c) net ecosystem carbon dioxide exchange for September 2004.

Prediction of regional net primary productivity with the model NetPro

Plants assimilate carbon dioxide from the atmosphere by photosynthesis. The sum of the carbon fixed during photosynthesis (P_g) is often measured as gross primary production in terms of rate, for example, tonnes of carbon per hectare per year. A large part of the fixed carbon is used by the plants for metabolic processes, largely respiration (R_A), thereby returning carbon dioxide to the atmosphere. The net carbon balance between P_g and R_A is called net primary production (NPP) and represents the net new carbon stored in plants as biomass in growing stems, leaves or roots, or lost as litter or herbivore consumption. In the following we describe our attempts to estimate NPP of a larger tropical area with a simple model approach.

Our model, NetPro, uses remote sensing data, and assumes that NPP of a vegetation cover is basically proportional to the photosynthetically active radiation (PAR) that is absorbed by the canopy (f_{APAR}, Monteith 1972, 1977). The proportionality factor is called PAR use efficiency for NPP (e_{NPP}, e.g., Running et al. 2000, Xiao et al. 2004, June 2004):

$$\mathrm{NPP} = e_{\mathrm{NPP}} \cdot f_{\mathrm{APAR}} \cdot \mathrm{PAR} \tag{3}$$

e_{NPP} is derived using an established leaf scale C_3 photosynthesis model. Up-scaling to canopy NPP is done with a model that treats sunlit and shaded leaf fractions separately (De Pury and Farquhar 1997, June 2002). The responses of C_3 leaf photosynthesis to PAR, temperature and CO_2 concentration are described by the biochemical properties of just two steps in the process, the carboxylation reaction (constrained by V_{cmax}) and the regeneration of the acceptor for carboxylation (constrained by J_{max}). This mechanistic model has been widely used as an accurate predictor of photosynthetic carbon uptake by leaves with variation in environmental conditions. In the simulation, supply of CO_2 (c_i) into the leaf is modelled as 0.7 of ambient CO_2 (c_a). This is a condition where water is not a limiting factor and vapour pressure deficit is set around 12.5 hPa. J_{max} is taken as 2.1 V_{cmax}. V_{cmax} and R_d are taken from field measurements and related to the leaf nitrogen content, an easily measurable parameter.

To estimate e_{NPP} we used generalised photosynthetic parameters (Table 2) but individual LAI for each homogeneous region in the investigation area. LAI was derived from remotely sensed Normalized Difference Vegetation Index (NDVI) using the linear equation published by Ibrahim (2001).

$$\mathrm{LAI} = 12.74 \cdot \mathrm{NDVI} + 1.34 \tag{4}$$

The second term in (3), the fraction of PAR absorption, f_{APAR}, is also derived from NDVI data using an empirical relationship that has been suggested by Ochi and Shibasaki (1999),

Table 2. Parameters for the photosynthesis model used in the estimation of the PAR use efficiency

Parameters		Value	Sources
Γ^*	CO_2 compensation partial pressure in the absence of dark respiration (μmol mol^{-1}) (at 25 °C)	42.75	Bernacchi et al. (2001)
$\boldsymbol{E}$	activation energy for carboxylation, oxygenation, respiration, rubisco activity and CO_2 compensation point (J mol^{-1})	79430, 36380, 46390, 65330, 37830	Bernacchi et al. (2001)
$\boldsymbol{K_c}$	Michaelis-Menten constant for carboxylation by Rubisco (μmol mol^{-1})	40.49	Bernacchi et al. (2001)
$\boldsymbol{K_o}$	Michaelis-Menten constant for oxygenation by Rubisco (μmol mol^{-1})	278.40	Bernacchi et al. (2001)
$\boldsymbol{R}$	universal gas constant (J mol^{-1} K^{-1})	8.3144	Goudriaan (1977)
$\mathbf{P_{cd}}$	canopy reflection coefficient for diffuse PAR	0.036	Goudriaan (1977)
$\mathbf{K_{d'}}$	diffuse and scattered diffuse PAR extinction coefficient	0.715	Goudriaan (1977)
$\varrho\mathbf{H}$	reflection coefficient of a canopy with horizontal leaves	0.041	Goudriaan (1977)
$\mathbf{k_{b'}}$	beam and scattered beam PAR extinction coefficient (for random orientation of leaves) at solar elevation β	$0.69/\sin\beta^{-1}$	June (2002)
$\mathbf{N_b}$	base level of nitrogen not associated with photosynthesis (mmol (N) m^{-2})	29	Anten et al. (1995)
χ_n	ratio of rubisco capacity to leaf nitrogen content μmol (C) mmol (N)$^{-1}$ s^{-1}	1.63	June (2002)
ϱ_1	the leaf reflection coefficient for PAR	0.10	De Pury&Farquhar (1997)
τ_1	the leaf transmissivity to PAR	0.05	De Pury&Farquhar (1997)
θ	curvature factor of the PAR response curve	0.7	June (2002)

$$f_{\mathrm{APAR}} = 1.075 \cdot \mathrm{NDVI} - 0.08 \tag{5}$$

To be able to calculate regional NPP with (3) the regional field of PAR data is obtained from global radiation measured by 10 meteorological stations around LLNP.

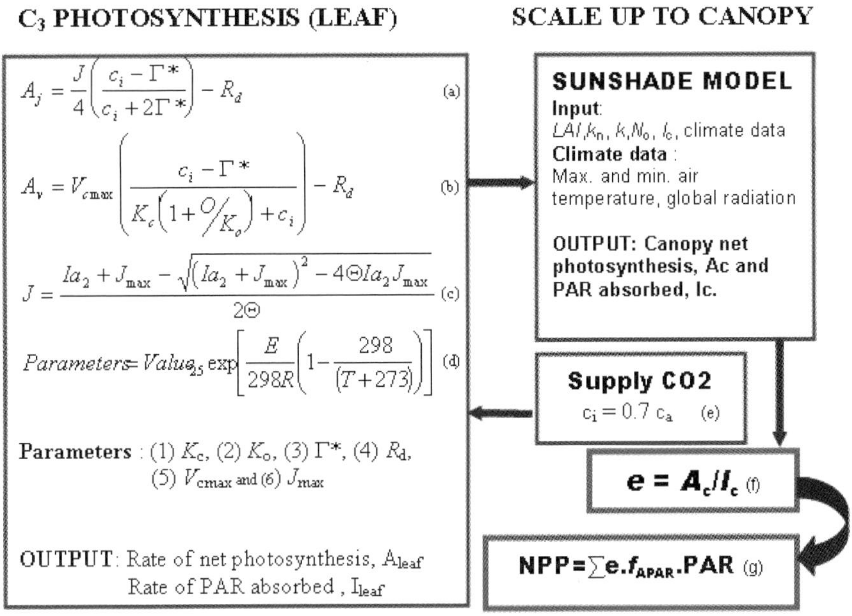

Fig. 7. Outline of the C_3 photosynthesis model in NetPro v. 1.0 to estimate canopy PAR use efficiency (e_{NPP}) and net primary production (NPP): (a) and (b) describe the effect of CO_2 concentration on the leaf assimilation rate; (c) describes the photosynthetic PAR response; (d) describes the temperature effects on the model parameters; (e) is the supply function of CO_2 and (f) and (g) define e_{NPP} and relate NPP to it.

Definitions of symbols in Figure 7: (see also Table 2)

Θ curvature factor of the PAR response curve

A net CO_2 assimilation rate (μmol m^{-2} s^{-1}), subscript ***c*** represents canopy and subscript ***leaf*** represent leaf

a_2 quantum yield (in terms of incident PAR) of electron transport at low PAR

A_j RuBP-regeneration limited rate of CO_2 assimilation (μmol m^{-2} s^{-1})

A_v Rubisco-limited rate of CO_2 assimilation (μmol m^{-2} s^{-1})

c_i partial pressure of CO_2 in the leaf (μmol mol^{-1})

I PAR photon flux density incident on leaf surface (μmol m^{-2} s^{-1})

J rate of actual electron transport (μmol m^{-2} s^{-1})

J_{max} maximum electron transport rate (μmol m^{-2} s^{-1})

O ambient partial pressure of oxygen (mmol mol^{-1})

R universal gas constant, 8.3144 J mol^{-1} K^{-1}

R_d dark respiration of leaf, which continues in the light (μmol m^{-2} s^{-1})

T leaf temperature (°C)

V_{cmax} maximum rate of Rubisco activity in the leaf (μmol m^{-2} s^{-1})

k_n nitrogen extinction coefficient

k light extinction coefficient
N_o leaf nitrogen concentration on top of canopy (mmol m^{-2})
$I_{c,leaf}$ total PAR absorbed by the canopy and leaf (μmol m^{-2} s^{-1})

NetPro makes it possible to easily acquire the LAI and f_{APAR} distribution from a regional NDVI data base, to estimate the regional distribution of e_{NPP}, to calculate the regional radiation and air temperature field from measured daily data and to, finally, estimate NPP with (3). The model structure is summarised in Figure 7.

NetPro V. 1.0 was applied to predict the NPP of the Lore Lindu National Park. The whole national park is located at 1°10' – 1°50' S and 119°50' – 120°20' E. NDVI was derived from a Landsat TM 7 picture dated from the 21. August 2001. Climate data were measured with 10 meteorological stations in the investigation area during the years 2001-2005.

NDVI values were sub-divided into 4 classes, which corresponded to LAI ranges from 0 to 10. f_{APAR} ranged from 0 to 65 %. Classes of average NDVI, air temperature and R_s were overlaid to form 17 polygons with different characteristics of climate and NDVI as input to NetPro v 1.0.

Under the present conditions, averaged NPP reached 14.5 t (C) ha^{-1} yr^{-1}. Given the total area of 217,991 ha the NPP of the entire LLNP was estimated to take up 3.2 Mt (C) yr^{-1}.

Table 3. Simulated NPP values for the study area in Lore Lindu National Park under altered leaf nitrogen contents and climate. The simulation used parameters obtained from field measurement as the following: Leaf nitrogen content 108.5 ± 44.1 mmol m^{-2}, R_d -1.20 μmol m^{-2} s^{-1}; quantum yield = 0.02 mol mol^{-1} and V_{cmax} = 33 ± 11.29 μmol m^{-2} s^{-1} (n=15)

CO_2 (ppm)	LAI	Temperature (°C)	Leaf Nitrogen concentration (mmol (N) m^{-2})	NPP (mg (C) ha^{-1} year^{-1})	NPP_{700}-NPP_{350} (%)
350	6.1	0	95	12.1	
			145	15.9	
	6.1	4	95	12.9	
			145	15.4	
700	6.1	0	95	14.5	20
			145	19.2	27
	6.1	4	95	16.2	26
			145	19.8	28

In a sensitivity study the leaf nitrogen content and the driving variables, CO_2 concentration and air temperature were changed to give four different scenarios. Results are presented in Table 3. It is shown, that the model responds sensitive to the changed environmental conditions and parameterizations within a reasonable range.

Further comparisons of the model with more sophisticated model approaches or even with measured flux data are yet lacking. This is, however, required to gain confidence in the simulations. At the time being, the accuracy of the NPP estimates can not yet be determined. There are also two aspects of the model approach that need to be improved for a more general applicability in the discussion of land-use change with respect to climate protection.

1. The neglect of possible drought effects on photosynthesis may hold for our investigation area, because it is consistent with our other SVAT model simulations in the investigation area. However, this assumption is very likely an oversimplification if NPP should be simulated under drier weather conditions, e.g. during climate anomalies.

2. Another aspect of how NetPro should be further developed, is to simulate not only NPP, but also the net ecosystem productivity, i.e. the net CO_2 budgets (F_b), which are more relevant in the discussion of climate protection. This requires the inclusion of R_E in the simulation.

Despite these deficiencies and improvement options, we could demonstrate with this case study that NetPro is an easy to use operational tool to integrate remote sensing data, regional meteorological data and measured plant ecophysiological parameters for the quantitative estimation of regional carbon fluxes in the tropics.

4 Status and outlook

We presented various ways of looking at the effects of land-use and land-use change on biosphere – atmosphere interactions in tropical areas, which, we believe, are an essential aspect in evaluating land-use options in tropical areas for two reasons. Given the high atmospheric water and energy fluxes in the humid tropics, surface fluxes of energy and water are much more important for natural resources, like water availability, soil stability and also local climate, than in other climate zones. A second reason is the dominant role of agriculture and forest use for the production of subsidence and income in many poorer tropical countries, which strongly depend on the efficient and sustainable use of the limited natural resources.

Despite this necessity, the initial situation prior to the presented studies was still mainly coined by a sobering lack of knowledge with respect to the quantities of driving variables, system properties, and, the magnitude of matter and energy fluxes.

In three case studies taken from an ongoing project, we presented and discussed advanced scientific methods to describe and forecast the CO_2, water and energy fluxes between ecosystems in the tropical rainforest margin zone of Central Sulawesi and the atmosphere.

Micrometeorological flux observations at local scale showed a surprisingly large CO_2 uptake by an almost undisturbed montane rainforest. These high uptake rates have still to be critically examined, but at the time being this result is the best available knowledge on the flux magnitude. Such unusual results would certainly have been overlooked in modelling studies, if parameters from other tropical sites would have been used.

Comparisons between a montane rain forest and a Cacao AFS showed significant differences both in some relevant vegetation parameters and in the measured magnitudes of energy and matter fluxes. Perennial farming systems are currently being discussed as a sustainable land-use option (van Noordwijk et al. 2003), which might be correct if compared to alternatives based on annual cropping systems. However, our results suggest that conversion of tropical rain forests to such Cacao AFS will alter the fluxes and may lead to foreseeable consequences on regional greenhouse gas budgets, water resources and local climate. Such regional feed back mechanisms have to be considered in the evaluation of the sustainability of land-use, even if the change of land-use seems to be relatively moderate.

The necessary step to scale up from local scales to regional scales was taken in two very different modelling approaches, both making use of the information collected in the investigated area. A sophisticated SVAT model was applied to describe fluxes and to interpret the driving variables and system parameters that infer the fluxes. From these kinds of studies, it could be derived, that certain process parameterisations in the model were in fact indispensable but others became redundant, when moving from local to regional scales. A slightly simpler SVAT model version, driven by regionally available land surface parameters and the regional field of meteorological variables, could be constructed and applied to estimate both regional fluxes between the atmosphere and the land surfaces and also the consequences for water flows in the investigated area.

A third model used the well established PAR use efficiency approach to calculate the NPP in the area based on remotely sensed data and the measured and interpolated temperature and radiation field. The innovation compared to the traditional approach is that PAR use efficiency was calculated with a process-based model, thereby introducing a prognostic model feature that helps to forecast carbon dioxide sequestration into biomass under a changed climate.

Although we discussed all methods self-critically, we are aware that more efforts are needed to critically estimate the accuracy of both the modelling and the measuring approaches. The flux measurements above the montane rain forest have clearly to be continued for longer time periods, because carbon fluxes are known to vary from year to year. Both modelling strategies still lack an atmospheric component that would enable us to forecast the feedback of land-use change to local climate. A good example that these approaches exist and can yield reasonable results has been given by Van der Molen (2002), who investigated the consequences of deforestation in the comparable situation of Puerto Rico.

With this study, we sketched how advanced scientific methods can be used, finally, to help people foreseeing the biophysical consequences of their future land-use planning. It shows, how the best available knowledge can be included in the analysis using operational model approaches that can be designed with regard to the needs of the application. One general problem, however, was identified, which limited the potential of our modelling approaches. The detail of knowledge on biophysical processes in the tropics is still much poorer than in other climate zones. Thus, we advocate placing much more efforts into biometeorological field research in tropical areas to overcome the still strong meridional knowledge gradient.

Acknowledgements

The investigations in Sulawesi were mainly funded by the DFG within the German Indonesian collaborated, interdisciplinary research project "Stability of Rain Forest Margins in Indonesia" (SFB 552). Andreas Ibrom was funded by a Marie-Curie Action, project MEIF-CT-2005-008354. We are grateful for the enthusiastic help and support by the local people and the administrations of Central Sulawesi. We thank Mike C. Courtney, Risø National Laboratory, Denmark, for correcting the manuscript and for his useful scientific comments.

References

Anten NPR, Schieving F, Werger MJA (1995) Patterns of light and nitrogen distribution in relation to whole canopy carbon gain in C_3 and C_4 mono- and dicotyledonous species. Oecologia 101: 504-513.

Aubinet M., Grelle A, Ibrom A, Rannik Ü, Moncrieff J, Foken T, Kowalski AS, Martin PH, Berbigier P, Bernhofer C, Clement R, Elbers J, Granier A, Grünwald T, Morgenstern K, Pilegaard K, Rebmann C, Snijders W, Valentini R, Vesala T. (2000) Estimates of the annual net carbon and water exchange of European forests: The EUROFLUX Methodology. Adv Ecol Res 30: 113-175.

Baker I, Denning AS, Hanan N, Prihodko L, Uliasz M, Vidale P-L, Davis K, Bakwin P (2003) Simulated and observed fluxes of sensible and latent heat and CO_2 at the WLEF-TV tower using SiB2.5. Global Change Biol 9: 1262-1277

Baldocchi D, Falge E, Liahong G, Olson R, Hollinger D, Running S, Anthoni P, Bernhofer C, Davis K, Evans R, Fuentes J, Goldstein A, Katul G, Law B, Lee X, Malhi Y, Meyers T, Munger W, Oechel W, Paw U KT, Pilegaard K, Schmid HP, Valentini R, Verma S, Vesala T, Wilson K, Wofsy S (2001) FLUXNET: A new tool to study the temporal and spatial variability of ecosystem-scale carbon dioxide, water vapor and energy flux densities. Bull Am Meteorol Soc 82: 2415-2435

Bernacchi CJ, Singsaas EL, Pimentel C, Portis Jr. AR, Long SP (2001) Improved temperature response functions for models of Rubisco-limited photosynthesis. Plant Cell and Environ 24: 253-260

Crawford TL, Dobosy RJ, McMillen RT, Vogel CA, Hicks BB (1996) Air-surface exchange measurement in heterogeneous regions ± extending tower observations with spatial structure observed from small aircraft. Global Change Biol 2: 275-285

De Pury DGG, Farquhar GD (1997) Simple scaling of photosynthesis from leaves to canopies without the errors of big-leaf models. Plant, Cell and Environment 20: 537-557

Falge E, Reth S, Bruggemann N, Butterbach-Bahl K, Goldberg V, Oltchev A, Schaaf S, Spindler G, Stiller B, Queck R (2005) Comparison of surface energy exchange models with eddy flux data in forest and grassland ecosystems of Germany. Ecol Mod 188: 174-216.

Falk U (2004) Turbulent fluxes of CO_2, H_2O and energy in the atmospheric boundary layer above tropical vegetation investigated by Eddy-Covariance measurements. Ph.D. thesis Institute of Bioclimatology. Georg-August-University, Goettingen, Germany p 116

Falk U, Ibrom A, Oltchev A, Kreilein H, June T, Rauf A, Merklein J, Gravenhorst G (2005) Energy and water fluxes above a Cacao agroforestry system in Central Sulawesi, Indonesia, indicate effects of land use change on local climate. Z. Meteorol. 14: 219-225

Gash J, Nobre CA, Roberts JM, Victoria RL (1998) Amazonian Deforestation and Climate. John Wiley and Sons, Chichester

Grace J, Lloyd J, McIntyre J, Miranda AC, Meir P, Miranda HS, Nobre C, Moncrieff J, Massheder J, Malhi Y, Wright I, Gash J (1995) Carbon dioxide uptake by an undisturbed tropical rain forest in southwest Amazonia, 1992 to 1993. Science 270: 778-780

Goudriaan J (1977). Crop micrometeorology: a simulation study. PUDOC, Center for Agricultural Publishing and Documentation, Wageningen

Ibrahim (2001). Environment and Development in Coastal Region and in Small Island. Assessing Mangrove Leaf Area Index and Canopy Closure.

Ibrom A, Ross T, Kreilein H, June T, Oltchev A, Falk U, Rauf A, Gravenhorst G (2006) Large net CO2 uptake by a tropical upland rain forest in Central Sulawesi, Indonesia. submitted to Agric For Meteorol, 52 p

June T (2002). Environmental Effects on photosynthesis of C_3 plants: scaling up from electron transport to the canopy (Study case: Glycine max L. Merr). PhD Thesis, Environmental Biology, Research School of Biological Sciences. Australian National University, Canberra.

Kaimal JC, Finnigan J (1994) Atmospheric Boundary Layer Flows : Their Structure and Measurement . Oxford University Press, New York, Oxford, p 289

Lee X, Massman W, Law B (2004) Handbook of Micrometeorology: A guide for Surface Flux Measurements. Kluwer Academic Publishers, Dordrecht

Levy PE, Grelle A, Lindroth A, Mölder M, Jarvis PG, Kruijt B, Moncrieff JB (1999) Regional-scale fluxes over central Sweden by a boundary layer budget method. Agric For Meteorol 98-99: 169-180

Loescher H, Oberbauer SH, Gholz HL, Clark DA (2003) Environmental control on net ecosystem-level carbon exchange and productivity in a Central American tropical wet forest. Global Change Biol 9: 396-412

Malhi Y, Baldocchi DD, Jarvis PG (1999) The carbon balance of tropical, temperate and boreal forests. Plant, Cell and Environ 22: 1-26

Merklein J (2003) Simulation des CO_2-und H_2O-Gaswechsels einer Kakaoplantage in Indonesien(Sulawesi). MSc. Thesis, Institut für Bioklimatologie. Georg-August-Universität, Göttingen (*in German*).

Miller SD, Goulden ML, Menton MC, Da Rocha HR, De Freitas HC, Michela ESFA, Dias De Sousa CA (2004) Biometric and micrometeorological measurements of tropical forest Carbon balance. Ecol Appl 14: 114-126

Monteith JL (1972) Solar radiation and productivity in tropical ecosystems. J Appl Ecol 9: 747-766.

Monteith JL. (1977) Climate and the efficiency of crop production in Britain Phil Trans R Soc Lond B 281: 277-294.

Oltchev A (2003) Concept of a model of evaporation for non-uniform forest landscape. J Rus For Soc 2 (in Russian): 55-64

Oltchev A, Cermak J, Nadezhdina N, Tatarinov F, Tishenko A, Ibrom A, Gravenhorst G (2002) Transpiration of a mixed forest stand: field measurements and simulation using SVAT models. Boreal Environ Res 7: 389-397

Ochi S, Shibasaki R (1999) Estimation of NPP based agricultural production for Asian countries using remote sensing data and GIS, Proceedings of the 20th Asian Conference on Remote Sensing, Hong Kong, 7-10

Pitman AJ, Henderson-Sellers A (1998) Recent progress and results from the project for the intercomparison of landsurface parameterization schemes. J Hydrol 128-135

Raupach MR, Finnigan JJ (1988) Single-layer models of evaporation from plant canopies are incorrect but useful, whereas multilayer models are correct but useless: discuss. Austral J Plant Physiol 15: 705-716

Sellers PJ, Dickinson RE, Randall DA, Betts AK, Hall FG, Berry JA, Collatz GJ, Denning AS, Mooney HA, Nobre CA, Sato N, Field CB, Henderson-Sellers A (1997) Modeling the exchanges of energy, water, and carbon between continents and the atmosphere. Science 275: 502-509

Saleska SR, Miller SD, Matross DM, Goulden ML, Wofsy SC, da Rocha HR, de Camargo PB, Crill P, Daube BC, de Freitas HC, Hutyra L, Keller M, Kirchhoff V, Menton M, Munger JW, Pyle EH, Rice AH, Silva H (2003) Carbon in Amazon Forests: Unexpected Seasonal Fluxes and Disturbance-Induced Losses. Science 302: 1554-1557.

Schimel D (2004) The Large-Scale Biosphere–Atmosphere Experiment in the Amazon. Ecol Appl 14: 1-2

Stull R (1988) An Introduction to Boundary Layer Meteorology. Kluwer Academic Publisher, Dordrecht, The Netherlands.

Styles M, Lloyd J, Zolotukhin D, Lawton KA, Tchebakova N, Francey RJ, Arneth A, Salamakho D, Kolle O, Schulze E-D (2002) Estimates of regional surface carbon dioxide exchange and carbon and oxygen isotope discrimination during photosynthesis from concentration profiles in the atmospheric boundary layer. Tellus B 54: 768-783

Valentini R, Matteucci G, Dolman AJ, Schulze E-D, Rebmann C, Moors EJ, Granier A, Gross P, Jensen NO, Pilegaard K, Lindroth A, Grelle A, Bernhofer C, Grünwald T, Aubinet M, Ceulemans R, Kowalski AS, Vesala T, Rannik Ü, Berbigier P, Loustau D, Gudmundsson J, Thorgeirsson H, Ibrom A, Morgenstern K, Clement R, Moncrieff J, Montagnani L, Minerbi S, Jarvis PG (2000) Respiration as the main determinant of carbon balance in European Forests. Nature 404: 861-865

Willmott CJ (1981) On the validations of models. Phys Geogr 2: 184-194

Science and technology and sustainable development in Brazilian Amazon

Diógenes S. Alves

Instituto Nacional de Pesquisas Espaciais (INPE), Av dos Astronautas 1758, CEP 12227-010 São José dos Campos SP, Brazil, Email: dalves@dpi.inpe.br

Summary

Deforestation in the Brazilian Amazon increased from 10 million hectares in the 1970s to more than 60 million hectares at the turn of the century, resulting in growing awareness about deforestation impacts like greenhouse-gas emissions, loss of biodiversity, and motivating a number of initiatives involving the science and technology (S&T) field to address the issues of deforestation and sustainable development in the Amazon. The present work summarizes part of the large-scale land cover-use changes that occurred in the region and then analyses the organization of four S&T programs carried out in the Amazon in the context of alarming deforestation rates and rapidly changing land use in the Amazonian frontier. The four programs include two major research programs - the Large-Scale Biosphere-Atmosphere Experiment in the Amazon (LBA) and the Science & Technology Subprogram (S&T) of the Pilot Program to Conserve the Brazilian Rain Forest (PPG7) - and two examples of environmental monitoring and management - the Ecological-Economic Zoning (ZEE) and deforestation monitoring programs. In the context of high rates of forest loss and generally very weak institutions, the initiatives organized with the concourse of the S&T field may have significantly contributed to advance the discussions of sustainable development and sustainable land use in the frontier, and to mature some ideas about the participation of civil society, national environmental policy and, also, international cooperation. These exercises also suggest that the tasks of reducing and mitigating deforestation impacts and fostering sustainable land use are not to be engineered but, rather, negotiated, and that understanding how to contribute to such negotiations seems to be a major challenge for the science and technology field in Brazil.

Tscharntke T, Leuschner C, Zeller M, Guhardja E, Bidin A (eds), The stability of tropical rainforest margins, linking ecological, economic and social constraints of land use and conservation, Springer Berlin 2007, pp 493-512

Keywords: Amazon , deforestation monitoring, ecologic-economic zoning, land zoning, pilot program to conserve the Amazon rain forest, PPG7, large-scale biosphere-atmosphere experiment in the Amazon, LBA

1 Introduction

A comparison of STORMA with similar initiatives in other tropical regions can be important to discuss the multiple questions of sustainable land use and stability of forest margins. The present work attempts to contribute to such an exercise by examining four programs carried out by the science and technology (S&T) field in Brazil in the context of alarming deforestation rates and precarious living conditions in the Amazonian frontier. These initiatives – including deforestation monitoring, ecological-economic zoning (ZEE), the S&T Subprogram of the Pilot Program to Conserve the Brazilian Rain Forest (PPG7/S&T) and the Large-Scale Biosphere-Atmosphere Experiment in the Amazon (LBA) – are of interest because of the variety of technical approaches and multi-lateral arrangements they include and some existing reflections about their workings under conditions of persistently expanding deforestation.

The chapter starts with a review of the rates and the spatial distribution of forest clearing and some large-scale land use changes in Brazilian Amazon, attempting to point out at some traits of frontier expansion into the Amazon that seem critical to discuss the S&T programs. The text then proceeds by reviewing the major goals and the implementation of each of the four programs, trying to find out how they addressed specific problems associated with the frontier expansion, and, whenever possible, discussing how they dealt with the different actors and stakeholders that are part of the frontier. Here, a major concern was that reducing and mitigating deforestation impacts should require a major involvement of frontier actors.

It would be fair to see the present reflections as preliminary, in particular, because of the recentness of all four examined programs and an almost complete absence of substantive studies of the way science and technology addresses the problems of frontier expansion – with the most remarkable exception of PPG7 and its S&T subprogram. It also seems appropriate to remember that the very notion of frontier expansion can be less meaningful outside of the New World. However, as some of the programs have already pleaded, understanding the capabilities and limitations of the S&T field is a critical part of discussing the very question of sustainable land use. This work is an attempt to contribute to such discussions, too.

2 Changes in forest extent and land use in the Amazonian frontier

2.1 Land cover changes and spatial patterns of forest clearing

The debate about deforestation in Brazilian Amazon gained increasing importance in the 1980s, when it was widely publicized that Government colonization and development programs accelerated road building, migration and the conversion of large areas of forest to pasture and agriculture in the region. At the time, deforestation rates became a controversial issue (Fearnside 1993 and Faminow 1998), up to the point that a high level Congressional Panel was appointed to investigate them (Brasil 1990).

Table 1. Changes in forest cover according to National Institute for Space Research (INPE) surveys

Annual forest loss		Accumulated forest loss		
Period	ha x 10^6yr^{-1}	Year	ha x 10^6 (6)	% forest (7)
1978-1988 (1)	2.1	1978	15.2	3.8
1988-1989 (2)	1.8	1988	37.7	9.3
1989-1990 (2)	1.4	1989	40.1	9.9
1990-1991 (2)	1.1	1990	41.5	10.3
1991-1992 (2)	1.4	1991	42.6	10.5
1992-1993 (2,3)	1.5	1992	44.0	10.9
1993-1994 (2,3)	1.5	1993	45.5	11.2
1994-1995 (4)	2.9	1994	47.0	11.6
1995-1996 (4)	1.8	1995	49.7	12.3
1996-1997 (4)	1.3	1996	51.7	12.8
1997-1998 (4)	1.7	1997	53.2	13.1
1998-1999 (4)	1.7	1998	55.2	13.6
1999-2000 (4)	1.8	1999	56.9	14.1
2000-2001 (5)	1.8	2000	58.7	14.5
2001-2002 (5)	2.3	2001	61.1	15.1
2002-2003 (5)	2.5	2002	63.6	15.7
2003-2004 (5)	2.7	2003	66.3	16.3

(1) - 10-year average
(2) - based on visual interpretation and manual digitizing
(3) - 2-year average
(4) - based on visual interpretation and electronic scanning
(5) - based on image segmentation and classification
(6) - includes 97.6 x 10^6 ha of old deforestation (Fearnside et al 1990); after 2000, estimated by adding annual forest loss
(7) - following Fearnside (1993) estimate of percent forest loss by 1988; after 2000, estimated by adding annual forest loss
Sources: Fearnside et al (1990), INPE (1992), Fearnside (1993), INPE (2002), and http://www.inpe.br

Table 1 summarizes deforestation statistics from the monitoring program carried out by the National Institute for Space Research (INPE 2002, http://www.inpe.br), based on a number of methodologies. The estimates for 1978 were produced after partial reanalysis of the black-and-white Landsat MSS 1:500,000-scale images and maps by Tardin et al (1980) to deal with some inconsistencies between the 1980 pioneer study and later surveys (Fearnside et al 1990); statistics for the 1987-2000 period are based on visual interpretation of color composites of Landsat TM red, NIR and MIR channels at the 1:250,000-scale (Alves 2001); more recent statistics are based on digital processing of Landsat TM images with detection of clearly cut areas above 6.25 ha (http://www.inpe.br).

Estimates from Table 1 show a relatively large year-to-year variation in annual rates, which I believe remains to be better investigated. Although the minimum annualized rate estimated by INPE (nearly 1.110^6ha yr $^{-1}$ in 1990-1991) could be easily associated with a huge economic depression, I think the mechanisms behind the maximum rate (nearly 2.910^6ha yr $^{-1}$ in 1994-1995) and the persistently high rates in recent years are not really well understood, and would require a methodical examination of geographic differences of occupation, of the evolution of policy and economy at the national and regional levels, and, last but not least, of the flaws in the enforcement of forest regulations and land appropriation rules.

Analysis of INPE deforestation maps shows that forest clearing has been highly concentrated near roads and colonization and development zones established after the 1960s. Eighty-seven percent of the 1991-1996 deforested area was observed within 25 kilometers from clearings observed by 1978 (Alves 2002b), while 90% of the 1991-1997 deforestation was found within 100 kilometers from major roads, (Alves 2002a), where most colonization projects were located (Machado 1998).

Concentration of forest clearing has persistently led to clearing beyond the limits established by Brazilian forest regulations (Brasil 1965, 2001). In fact, Alves et al (2003) showed that the fraction of cleared forest in Central Rondônia tended to increase continually, largely exceeding the 1965 Forest Code 50%-limits to clear cut (Brasil 1965), not to mention the more restrictive 20%-limits established more recently (Brasil 2001). Preliminary analysis of the forest clearings for the periods 1978-1991, 1991-2000, 1991-1994, 1994-1997 and 1997-2000 suggests that the continual clearing of forest beyond the limits established by forest regulation is characteristic of the areas of deforestation concentration.

It must be noted that concentration of forest clearing does not exclude frontier expansion into new areas. Furthermore, whereas concentration of clearing has been widely associated with disregard of forest regulations in private land, the spreading of forest clearing into new areas can be preceded by illegal appropriation of public land, a recurrent problem since the very foundation of the country (Silva 1996). Here, a recent study by the Federal land agency INCRA found some 100 million hectares of land suspected of illegal/irregular appropriation, the majority of it in the Amazon (Sayago and Machado 2004; data refer only to estates of 10,000 hectares or more).

2.2 Land use changes in the Amazon frontier

The dynamics of land use are still poorly understood in the rapidly changing Amazon, where a number of issues remain to be explored further. For example, understanding the different pathways from forest clearing to use intensification to land degradation and abandonment, identifying the different driving forces and actors of land use change and the processes behind the movement of the frontier, and searching for alternatives of land use that might prevent deforestation have been recognized as relevant research questions (see for example, Faminow 1998, Margulis 2004 and Sayago et al 2004, for discussions of some topical important questions; Velho (1976), Silva (1996) and Machado (1998), for a broader discussion of the questions about the frontier).

Here, I will limit myself to present a summary of large-scale land use changes by comparing data from the two last Agricultural Censuses (1985 and 1995/96) and will refer to a few other work that can help to delineate how land cover change has evolved in the last two decades. Before starting, let us keep in mind the limits imposed to our inferences and generalization due to the lack of up-to-date, reliable land-use data in an area of rapid changes and to the important regional differences existing in the Amazon.

Table 2. Agricultural Census average land use statistics for the states of Mato Grosso, Pará and Rondônia

	Fraction (% of total used land) of land used by Census category					
Year	Permanent crops	Temporary crops	Pasture (native/planted)	Tree plantations	Unused land	Cattle density head ha^{-1}
1985	2.1	10.3	82.4 (42/40.4)	0.4	14.9	0.44
1995	1.7	9.3	85.7 (21.9/63.8)	0.6	7.7	0.76

Sources: 1985 (Waniez et al. 1997) and 1995/96 Agricultural Censuses (http://www.ibge.gov.br)
Notes:
a. Municipal units after discarding those with detectable inconsistencies accounting for 90% of total use in 1985 (n1=128) and 99% of total use in 1995/96 (n2=236)
b. Unused land for more than 4 years

Table 2 summarizes changes in land use in the states of Mato Grosso, Pará and Rondônia, the 3 states that showed the largest increases in deforested area in the last decades, accounting for 80% of forest clearing (INPE 2002) (the table reports only fractions of land under different categories of use and cattle densities because of the difficulties in comparing the absolute numbers from the two Censuses due to changes in sampling strategy and other problems; see also Andersen et al 2002 for a further discussion about using data from these Censuses).

One first thing to notice from Table 2 refers to changes in the fraction of pasture, the predominant category of use reported in both Censuses. While the fraction of land used as pasture slightly increased from 1985 to 1995, the share of natural pastures - generally developed in areas of *cerrado* (a savannah-type vegetation common in central parts of Brazil) – decreased from 42% to 22% of the used land, suggesting that pastures and cattle ranching may have expanded largely outside the savannah biomes, i.e. at the expense of forest.

Cattle ranching - the predominant land use in the Amazon - has been a critical topic of investigation during the last decades. To start with, cattle ranching has been shown to be unproductive and highly dependent on Government subsidies, to thrive on land speculation, and to lead to land conflicts and marginalization of the poor (see, for example, Hecht et al 1988). Chomitz and Thomas (2001) considered that stocking rates were systematically low in 1995/96, in particular in areas of *cerrado* and under high precipitation, although they would increase within 25 km of major good roads and within 200 km from areas cleared by 1976, suggesting that this would be "an encouraging sign that pasture use intensifies over time" (Chomitz and Thomas (2001) p.13).

More recently, the issues of intensification of pasture use and changes in the cattle market chains started to receive more attention. Margulis (2004) attributed the considerable growth of the Amazon cattle herds in the 1970-1996 period to both deforestation and intensification; Alves (in press), investigating the increase in cattle densities shown in Table 2, found evidence of intensification of pasture use between 1985 and 1995/96 in municipal units with smaller fractions of remaining forest. Faminow (1997) argued that a fundamental cause of the growth of cattle herds in the Amazon was the considerable expansion of regional demand associated with urban growth that induced more private investments in cattle ranching; Andersen et al (2002) and Margulis (2004) developed a broader discussion of these trends, reviewing the many motivations for cattle ranching and intensification of pasture use, concluding that this activity became lucrative and no longer dependent on subsidies thanks to the growth of urban demand and increased productivity. Veiga et al (2004) analyzed the market chains for cattle products finding "local sub-chains, aimed at the markets within the pioneer fronts; regional sub-chains aimed at the large Northern urban centers; and the export sub-chains that fulfill the demands outside the Amazon" (Veiga et al (2004) p.67).

Changes in the economy of cattle ranching did not prevent, however, many of problems associated with the activity in the Amazon. As noted by Margulis (2004 p xi) "the fact that cattle ranching is viable from the private perspective does not mean that the activity is socially desirable or environmentally sustainable. Private gain needs to be contrasted with the environmental (social) costs associated with cattle ranching and deforestation. From the social perspective, it is legitimate to argue that the private benefits from large-scale cattle ranching are largely exclusive, having contributed little to alleviate social and economic inequalities". In general, it might be argued that it remains crucial to investigate further the social, economic and ecologic connections of cattle ranching, and, in particular, to recognize the role of this sector's stakeholders in policy formulation (Mahar 2002, Margulis 2004). It is also critical to note that the functioning of the cattle sector shows important geographic differences in respect to many factors, a question that would deserve more attention from researchers.

The second most important land use in the last two Censuses was crops (Table 2), even if both temporary and permanent crops have shown a reduction in their share of total land use. As with cattle, agriculture has shown important geographical differences, some of which are reflected in the discussions below.

A most remarkable change in agriculture was due to the introduction of soybean in the Amazon in the 1980s, resulting in this crop becoming by far the most important by harvest area and production value. Soybean cultivation has been mostly limited to *cerrado* areas of Mato Grosso, where soil correction and mechanization are generally uncomplicated, and rainfall in the 1,600-2,000-mm range does not impede seeding and harvesting (Chomitz and Thomas 2001, Andersen et al 2002, Margulis 2004). Some authors have defended soybean cultivation as profitable and highly competitive in the Amazon, arguing that it is an alternative "to reduce wasteful land uses like extensive cattle ranching" (Andersen et al 2002 p.207), "represent[ing] only 10% of the converted area in 2000" (Margulis 2004 p.10). At the same time, the impacts of this crop have been a controversial topic: earlier soybean introduction in other regions of Brazil is believed to have "push[ed] land-extensive activities like cattle-ranching away from the South and Center-West towards the cheaper land in the Amazon" (Andersen et al 2002 p.76); also, an assessment of its social and environmental impacts, including concentration of land tenure and income, expulsion of rural populations, and loss of biodiversity, is still do be done (Fearnside 2001).

The occurrence of highly mechanized crops like soybeans appears to have followed a different trajectory than other subsistence or cash crops grown in the frontier between the 1985 and the 1995 Censuses. In general, Census data suggest the development of specialized forms of capital-intensive agriculture in a few regions following the conversion of most of their original forests and savannahs, while in most of the Amazon crops tended to be important in the original phases of colonization – probably in the form of subsistence or cash crops – largely loosing its importance to cattle ranching as more forest is converted (Alves in press) (despite this general trend, some crops like rice, maize, sugar cane, oil palm and other can represent a large share of land use in a few areas of older colonization where larger fractions of forest have been converted).

The last category of land use analyzed here – tree plantations – represents a small fraction of the total land (Table 2). Margulis (2004) reported an increase in tree plantation area in the 1970-1995 but emphasized that this area is 2 to 3 orders of magnitude smaller than that covered by natural forests in private land. Alves (in press) suggested that the modest increase in the relative importance of tree plantations between the last two Censuses may be due to their relative growth in municipal units with less deforested area, although the occurrence of this category of land use may be too limited to characterize it as a typical activity of recently occupied or more preserved regions.

Unused – deforested and abandoned for more than 4 years – land has received particular attention from research, because of its potential contribution to mitigate deforestation impacts by sequestering carbon, restoring soil properties and other potential benefits (see, for example, Brown and Lugo 1990). At the same time, abandoned land is a sign of low productivity and land degradation in some studies (see, for example, Chomitz and Thomas 2001). It is important to analyze temporal and spatial changes in unused land. Based on 1970-1995 Census data, Margulis (2004) reported an increase in the fraction of unused and fallow land in 1970-1985 that then decreased after 1985, as shown in Table 2. Alves et al (2003) found a pattern of decreasing secondary vegetation associated with increasing forest clearing in Rondônia; furthermore, Alves (in press) found that the decline in the fraction of unused land in Table 2 can be attributed to a significant reduction of this figure in municipal units with larger fractions of cleared forest, suggesting a scenario where both primary forest and secondary vegetation are simultaneously cleared as land use intensifies.

Decreasing rates of unused land in areas of more intensive forest clearing has at least two interesting implications. Firstly, it may lead to reduced carbon sequestration by regrowing vegetation that would not compensate for larger fractions of gross CO_2 emissions from deforestation. Secondly, the pathways leading to land degradation, generally believed to intervene after a short period of use rapidly exhausting soil nutrients, seem to be more complex than initially thought, in particular if considering the large-scale predominance of pastures, the prevailing low level of pasture management, and the intensification of land use in some areas.

To conclude this section, it seems appropriate to emphasize three major points. First of all, both forest conversion and land use (cattle ranching and agriculture) show a very dynamic behavior: it is critical to be aware of temporal and geographical differences in the expansion of the frontier, specially in a context where categories as distinct as land use intensification, land degradation, and social and environmental impacts are investigated in a still unformed territory. Secondly, discussing alternatives for sustainable land use and reduction and mitigation of deforestation impacts requires understanding the actors and stakeholders involved in the frontier, their motivations, dealings, articulations, and conflicts. Finally, to address the social, economical and ecological dimensions of the frontier expansion, a major challenge is that frontier actors are often alienated from any robust institutional framework to deal with their problems as it has recurrently been the case in the past (Velho 1976, Silva 1996, Franco 1997, Machado 1998, World Bank 2000a, Martins 2003, Candido 2003).

3 The S&T field and the questions of sustainable development

The four programs described in this section – the deforestation monitoring program, the Ecological-Economic Zoning, the Science & Technology Subprogram under the Pilot Program to Conserve the Brazilian Rain Forest, and the Large-Scale Biosphere-Atmosphere Experiment in the Amazon – were carried out mainly at S&T centers and universities, and have focused on a variety of questions ranging from mapping deforestation to land zoning to some level of explicit discussion of the very notion of sustainable development. It should be noted that they do not represent the entire range of initiatives involving science and technology in the Amazon, but their progresses and difficulties can be helpful in analyzing how the science and technology field has evolved both in dealing with monitoring and research questions, and in articulating its works with the rest of society. After discussing each program, the text presents a short reflection on possible lessons learnt from these initiatives.

3.1 Deforestation monitoring in Brazilian Amazon

Deforestation monitoring based on satellite imagery started by a few concept-proving projects resulting in a pioneer Amazon-wide mapping of deforested areas by Tardin et al (1980). It is worth noting that the complexity of the first survey – including more than 200 Landsat MSS scenes and a variety of biomes with different spectral signatures - was managed by the close collaboration of the Brazilian Institute of Forestry Development (IBDF) and the Space Research Institute (INPE), bringing together various teams to work on the then novel field of remote sensing. After 1988, a new monitoring program was established at INPE to follow the deforestation process. The results of this program are summarized in Table 1. More importantly, several federal and state organizations (among others, the Amazonian Surveillance System (SIVAM/SIPAM) at the federal level, and major initiatives by the states of Mato Grosso, Pará and Rondônia) accomplished significant progresses in the 1990s, improving the ability to track deforestation, direct enforcement operations and contribute to land use planning.

Because of the complex institutional framework required to control deforestation, it seems appropriate to analyze the S&T field's role in relation to other organizations. First of all, it appears relevant to mention that policy makers have frequently requested organizations from the S&T field to provide reliable information about forest clearing in the Amazon (see, for example, the report of the Congress Panel (CPI) on the Amazon, Brasil 1990). Secondly, some level of cooperation was established between S&T centers and organizations with more operational mandates, although a lack of articulation has been pointed out by the national press (see, for example, Folha de São Paulo 2004). More to the point, the many technical progresses achieved in monitoring deforestation may have masked the fact that the very few operational initiatives to track forest clearing at the level of individual properties were erratic and short-lived, resulting in no effective enforcement of forest regulations; also, no consistent efforts to demarcate and control public lands have been put into place, making it practically impossible to prevent illegal appropriation of public land (see, for example, Silva 1996 and Sayago and Machado 2004 for a broader discussion of this long-lasting problem).

In a way, zoning might be expected to partially address some of these problems, so it is worth looking at how this other initiative has been carried out.

3.2 Ecological-economic zoning (ZEE)

Ecological-economic zoning (ZEE) is closely anchored in the ideas about the role of the State in promoting socio-economic development and territorial ordering inscribed in the Brazilian 1988 Federal Constitution (Art. 21 § IX; see also Art. 3, § 1 of Federal Decree 99,540 of 21 September 1990). Although the origins and the character of these ideas might be the object of other studies it is interesting to note the association between the notion of socio-economic development and the need to "establish order over the territory" ("*ordenação/ordenamento territorial*") in the context of the frontier. In fact, Federal Decree 99,540 of 21 September 1990 designates the Amazon as priority area for zoning, in response to widely publicized deforestation and conflicts; likewise, the two major zoning exercises attempted in Rondônia and Mato Grosso recognized the need to create adequate institutional frameworks in these frontier States (Mahar 2002, Diewald and Chavez 2003). Of special interest to our discussion are the demands put before the S&T field and the role played by this field in ZEE.

The importance of scientific and technical support for ZEE seems to have originated at its foundation. Federal Decree 99,540/1990 stated that macro and regional zoning "were to be conducted according to ... principles [of] ... inter-disciplinary ... [and] systemic vision" (Art. 3, § 2), but, in contrast, includes no major political directives except in connection with associations to state-level ZEE initiatives (Art. 1). In this vision, one of the major challenges of zoning – the negotiation among stakeholders with conflicting interests – receives no explicit mention, while science and technology appear to be expected to provide objective, maybe undisputable criteria for the delimitation of "zones".

The analyses of the Rondônia and the Mato Grosso cases by Mahar (2002) and Diewald and Chaves (2003) may help to discuss this further. These studies recognized the need to involve sectors of the civil society in the discussion of zoning; the programs in both states also required zoning to be based on sound technical recommendations and accomplished the development of very elaborate (even if "extremely costly") datasets. However, the studies found critical difficulties in recognizing and involving major actors in these zoning exercises, and in articulating ZEE technical concepts with society. More to the point, a major challenge was considered the fact that "some of the strongest and most influential support ... for zoning has come from beyond the state's borders ... because many of the potential benefits of zoning ... are seen to accrue to groups outside the state, and even outside Brazil" (Mahar 2002 p.118). In the case of Rondônia, the difficulties lead to open conflicts between the Executive and Legislative and to considerable social tensions in some regions of the State, despite all technical efforts employed in land surveys, soil mappings etc., forcing "the state government [to develop] transparent and participatory administrative rules and processes" (Mahar 2002 p.125). Furthermore, both the Mato Grosso and the Rondonia projects appear to have contributed to advance the debate of forest regulations, even if such regulations were a major source of disagreement during ZEE proceedings. Also, the inclusion of the ZEE among wider efforts to create protected areas and indigenous reserves subtracted part of the public land from the "common pool" of land subjected to fraud, indirectly contributing to restrain the problem of illegal appropriation of land.

As noted before, the macro and regional initiatives articulated by the Federal Government established a prominent role of the S&T field, resulting in its active participation in all zoning phases. First of all, the development of a set of methodological guidelines for zoning and the articulation of some level of public discussion indicates that the S&T field was extremely capable in working out at least one contribution to ZEE (see, for example, Brasil 2003). At the same time, the nature and the pace of the ZEE under the Federal sphere may have subjected its technical and theoretical concepts to a moderate level of pressure from society in comparison to State-level exercises. In fact, these zoning methodological guidelines (Brasil 2003) are centered on discussions about the concepts of "territory", "systemic approaches", "multi-disciplinarity" and "information systems", leading one to wonder, for example, why no S&T contributions are to be found (in the guidelines or elsewhere in these zoning exercises) in the fields of studying political articulation, addressing the issues of conflicts, or the evident difficulties created by the weakness of institutions in the frontier.

To conclude, it is necessary to stress that the ZEE, and the role of S&T within it, would merit far more detailed analyses that are still to be done. However, it already seems clear that achieving ZEE goals to intervene in the rapidly changing Amazon would require wide-ranging contributions extending beyond the formulation of objective, undisputable technical proposals for zoning. Instead, the S&T field might be needed to understand a multitude of interests, conflicts and stakeholders, and to adjust its methods and proposals to the pressures of the public arena.

3.3 The Pilot Program to Conserve the Brazilian Rain Forest (PPG7) and its Science & Technology subprogram (S&T)

The PPG7 is an international cooperation program supported by donations from G7 countries and counterparts from the Brazilian government with the objective of protecting Brazilian rain forests and promoting sustainable development. As characterized in the Mid Term Review of the Pilot Program "the main lines of action in the Pilot Program implementation have been (i) demonstration and experimentation, (ii) conservation, (iii) institutional strengthening, (iv) scientific research, and (v) lessons and dissemination" (World Bank 2000b: Internet pages).

The experiences from most of its projects are very useful, in particular because of its numerous systematic assessments and revisions that provide valuable, probably unprecedented, insights about the questions of promoting conservation and sustainable development in the Amazon. Here, the discussion starts with a review of the PPG7 program that defined the context in which all subprograms were required to operate and then proceeds to review its S&T component. It should be noted that this is not intended to be a comprehensive analysis of PPG7, being as it is mostly based on its initial, more comprehensively reviewed phases.

The Pilot Program is considered to have been very valuable and relatively successful "in developing innovative mechanisms for international cooperation … within a complex national administrative structure", in helping the establishment of relatively extensive protected areas (indigenous territories, extractive and conservation reserves, and other categories), in capacity building and strengthening Brazilian institutions (particularly in the environmental field), and in involving government and civil society in discussing environmental issues (World Bank 2000b: Internet pages). Adjustments were continually sought by a series of mechanisms propitiating discussion of the program's goals and strategies, among which PPG7's International Advisory Group (IAG), whose contributions included "shifting the original Program emphasis on conservation towards the notion of sustainable development, creation of a monitoring and analysis component, and initiation of the mid-term review" (World Bank 2000b: Internet pages).

The program execution revealed serious difficulties in addressing key issues (World Bank 2000a, 2000b, 2001), among which could be mentioned a lack of long-term strategic visions and programmatic approach, as well as a lack of geographic focus; the perception of conservation and development as opposing alternatives by some stakeholders in combination with the program's disregard of fundamental economic issues, which contributed to alienate the private sector and continually engendered conflicts; the wide-varying abilities of the diverse segments of society and government in effectively taking part in participatory exercises. Here the assessment presented in the Mid-Term Review might be cited verbatim: "the Program was externally conceived and, at first, mostly driven by donor policies and priorities" [without] "sufficient trust by the donor agencies in the capacity, transparency and accountability of the Brazilian ministries and institutions to efficiently manage the Program" (World Bank 2000b: Internet pages).

Furthermore, the very statement of PPG7 objectives by the World Bank and by the Brazilian Government (see, for example, the 1992 Rain Forest Trust 1992 resolution and Federal Decree 563 of 05 June 1992) suggests that these parts had very different conceptions of PPG7 priorities that came to surface when questions about the definition of the program's agenda and funding allocation actually started (see also Egler 2001). However the particularities of this problem, it emphasizes the delicate nature of international cooperation, and reveals that more trust between PPG7 partners was achieved only after increasing transparency and persistent negotiations.

In relation to the S&T subprogram, the contribution of which to PPG7 objectives had been generally perceived as modest (World Bank 2000b, Egler 2001), the Mid-Term Review report suggests that the Pilot Programs general objectives in fact provided limited guidance for the strategic focus of research, and that there were poorly conceived linkages and feedback mechanisms with the other subprograms and projects (World Bank 2000b: Internet pages). The analysis by Egler (2001) provides further insights into these issues. Recognizing the Pilot Program as a successful example of mutually-beneficial cooperation between the international community, the Brazilian government and the Brazilian society, this author also claimed that the initial negotiations were jeopardized by insufficient trust on the Brazilian counterpart, and that scientific research did not fulfill its original expectations, despite of the quality of the projects; more interestingly, he develops the debate on the role of the S&T field in the Amazon by proposing that the PPG7 research should have contributed to the exploitation of the Amazon natural resources, concluding that to fulfill such role "the creation of mechanisms that would allow to integrate research ... with the possible users of its results" would have been required (Egler 2001 p.324).

3.4 The Large-Scale Biosphere-Atmosphere Experiment in the Amazon (LBA)

LBA is a multi-national, multi-disciplinary research effort built to investigate the biological, biogeochemical and physical functioning of the Amazon, and the effects of land-cover/use and climate changes on such functioning. LBA Science Plan and research strategy integrate several scientific disciplines from climate to hydrology to biogeochemistry to land-cover/use changes (LBA 1996a, 1996b).

Although the experiment is still concluding its first phase, it has been recognized as a successful example of science planning and implementation, in which the periodic revisions of LBA Scientific Steering Committee (SSC) played a crucial role (see, for example, Schor 2005). My personal experience shows that peer discussions within the SSC have been fundamental to establish an unprecedented level of communication and cooperation between disciplines studying the Amazon, allowing the formulation of scientific questions and helping in specific topics like establishing and managing field equipment, experiment design, and model design and parameterization.

Other LBA recognized successes include the training of more than 640 undergraduate and graduate students by 2004 (Folha Amazonica 2004), many of them from the Amazonian states; a hands-on attitude to discuss land cover/use changes that ended up in establishing a few linkages between Earth scientists and social and human scientists (Alves et al 2004); and increasing the ability to model climate and biogeochemistry, especially in Brazil, where building capacity in this kind of models is recognized as important (Lahsen 2002). Also, LBA could be considered a very interesting experience of partnership among scientists from different countries, with many notable progresses in a field where differences in institutional arrangements and in scientific capacity generally create barriers as challenging as dealing with problems across disciplinary borders.

Beyond the scientific achievements of LBA it is interesting to analyze how it addressed its own expectations of contributing to sustainable land use in the Amazon. Here, it is generally recognized that more "reflections about the role of science for sustainable development of the Amazon" are needed (LBA 2004). It has been noted that the "discussion and acceptance [of LBA findings] by society are still incipient, even [within] agro-ecological and ecological-economic zoning programmes ... or ... international climate change negotiations" and that closer collaborative work with social scientists would be required "to make LBA results more accessible for policy-makers, stakeholders and to society ... by helping to recognize political and societal implications of research" (Alves et al 2004 p.5). New forms of institutional engagement and organization within the Brazilian S&T field may be required to achieve this goal; the inexistence of cooperation between LBA and ZEE is noteworthy, despite the formal commitment of S&T to both initiatives and the evident superimposition of the agenda of both projects in some issues, ranging from agro-ecology to land use to the interest on systems theory inherent to both initiatives.

To conclude, it could be mentioned that international cooperation was a very challenging issue throughout the Experiment (see, for example, Schor 2005), and the S&T field may have failed to properly recognize and address it as a sensitive matter. More specifically, discussions of foreign and national participation in LBA planning and implementation, foreign and national funding, and the close association between LBA research and topics of Brazilian national policy (e.g. greenhouse gas emissions) do not appear to have been properly faced, as was the case, for example, under PPG7. My personal belief is that if these issues would have been addressed more clearly from the beginning, the LBA Science Plan might have included more topics of research at the interface with other S&T initiatives in the Amazon, Brazil might have benefited more from capacity building, and LBA planning and execution would have gone more smoothly. At any rate, a more profound discussion of this issue appears to be crucial to understand the actual capabilities and limitations of the science and technology field in Brazil.

3.5 What can be learnt?

Although the four selected S&T programs represent only part of the S&T actions involving the Amazon and my analyses are not intended to be comprehensive and final, there are a few lessons that can be abstracted from these experiences.

Looking at the programs from the strict perspective of their objectives and implementation, a major lesson is the need to distinguish between the different ecological and socio-economic situations that can vary significantly from region to region. This has been particularly pointed out within PPG7 and its subprograms that recognized "the need for differentiated intervention strategies and priorities in addressing the problem of deforestation [according to geographic differences]" (World Bank 2000b: Internet pages) and LBA, where the issues of spatial variability and scale presented a permanent challenge during all phases of the Experiment (LBA 1996a,b, Keller et al 2004).

Another important lesson that I'd like to point at is that these experiments show that a very open attitude is required to identify and examine problems like ecosystem functioning, land use dynamics, deforestation impacts and alternatives of sustainable land use. Here, it might be recommended to develop a more detailed analysis of LBA, which succeeded in carrying out a continuous reassessment of its scientific questions and observational methods but nevertheless was challenged in addressing its human dimensions (Alves et al 2004, Keller et al 2004, Schor 2005). A permanently open attitude will certainly require closer interactions between social and natural sciences that none of the four analyzed programs have achieved so far.

It could be also recommended to examine some particularities of science and technology in Brazil, in particular, the inexistent or inadequate funding and infrastructure in key areas, the small number of specialists in some disciplines and a strong tradition of secluded disciplines that inhibits environmental trans-disciplinary studies. Designing education and capacity building programs, as those promoted under LBA and PPG7 (Folha Amazônica 2004, World bank 2000b), and sponsoring multi-institutional projects like in LBA and problem-oriented discussions as attempted under PPG7 (Egler 2001), may be good ideas to overcome those particularities.

Looking at the programs from the perspective of their impact on society, it might be argued that the most notable progresses have been achieved when S&T initiatives were exposed to closer interaction with policy makers and the actors and stakeholders of the frontier. Here, it might be recommended to further analyze the experiences of PPG7, its S&T subprogram, and ZEE, with their programmatic efforts to learn how to involve civil society in their actions (World Bank 2001, Egler 2001 and Mahar 2001), and, also, to scrutinize the initiatives and the struggles of all four programs to deal with the question of interactions with society.

In relation to this, the major lesson that I personally learned may seem naïve but, hopefully, it can help in addressing the question of interactions with society. It tells that the S&T field in Brazil should not be expected to disclose the final, undisputable solutions for the questions of sustainable land use and reducing deforestation, which indeed are problems of the entire society. I also went as far as suspecting the need to examine interactions between society and the S&T field but could not be sure about the methods to do this. It appeared necessary to look at the ways the S&T field works, with its puzzling possibilities of autonomy. To understand these, we may be warned against attempts to separate the "purely" technical or scientific constructions from the social conditions in which these constructions are conceived and proposed (Bourdieu 2003, Mack 1990, Collins 1992, Mackenzie 1993). It also may look reasonable to identify conditions that might be more specific to Brazil, including old linkages of Brazilian technocracy and of some intellectuals with authoritarianism that Velho (1976) related to the themes of the frontier, the personal stakes that many Brazilian scientists have progressively but idealistically taken in environmental problems (Alonso and Costa 2002), and, in particular, the astonishing contradictions of the deforestation debate in Brazil. Here, it seems important to recognize the opposition between development and conservation by influent sectors inside and outside the Amazon, the active pro-conservation voice of some groups frequently based outside the region, the intriguing precedence that deforestation in the remote Amazon can assume in public opinion in Brazil in comparison to other serious environmental problems in Brazil, like critical sanitation and pollution (World Bank 2000b, Mahar 2001, Tesh and Paes-Machado 2004), and a still unexplored symbiosis between social and environmental questions that has developed in the Amazon frontier in the last decades. To me, this complex setting only showed the diversity of challenges for science, technology, and society, warning that we should be conscious about them.

4 Conclusions and outlook

Frontier expansion in the Brazilian Amazon and the clearing of large areas of forest in the last decades have resulted in growing awareness about deforestation at the national and international levels, and triggered widespread claims for forest conservation and reduction of CO_2 emissions due to biomass burning. The implications of this state of affairs were many, including the setting up of initiatives to monitor land cover changes and planning land use, and the search for new frameworks of international cooperation, putting the questions of improving our understanding of the nature and the impacts of land cover/use changes, and of discussing the many aspects of sustainable land use before the S&T field.

After some initiatives more strictly focused on forest conservation, a new framework started to emerge, delineating the concept of sustainable land use in the frontier, while some questions of international cooperation, national policy and participation of civil society have matured. As a result, the complexities of addressing the frontier problems began to be perceived more widely and the discussion of forest conservation and sustainable land use have ripened.

To be sure, these progresses did not solve our problems. In fact, the fraction of forest lost doubled in the last two decades, key stakeholders perceive conservation as opposing development, tense and anomic conditions persist as an intrinsic trait of the frontier. My perception is that this situation shows that science and technology alone should not be expected to propose objective, undisputable answers to the frontier problems, but rather, to be one of the parts of a larger context of policy making, negotiations, and conflictive situations. It also seems sensible to regard the very concept of sustainable land use as uncharted waters for science and technology as it is for society, and recognize the difficulties in addressing conflicts related to environmental issues, that often assume very contradictory forms in Brazil.

Amid conflicts and contradictions, the tasks of reducing and mitigating deforestation impacts and fostering sustainable land use are not to be engineered but, rather, negotiated. In this case, understanding how to contribute to such negotiations seems to be a major challenge for the science and technology field. This is probably one of the major motivations for analyzing existing experiences.

Acknowledgements
I sincerely thank the organizers of the 2005 STORMA Symposium for the opportunity that inspired these reflections and allowed me to learn more about STORMA. I also thank for comments and suggestions from one anonymous reviewer and from Eric Lambin. I'd like to acknowledge also that this work benefited from the long-standing stance of my colleagues inside the LBA Scientific Steering Committee to raise and discuss some of the issues addressed here.

References

Alonso A, Costa V (2002) Por uma sociologia dos conflitos ambientais no Brasil. In: Ecologia Política. CLACSO, 115-135

Alves DS (2001) O processo de desmatamento na Amazônia. Parcerias Estratégicas 12. 259-275

Alves DS (2002a) An analysis of the geographical patterns of deforestation in Brazilian Amazon in the period 1991-1996. In: Wood CH, Porro R (eds)

Deforestation and land use in the Amazon, University Press of Florida, Gainesville, pp 95-106

Alves DS (2002b) Space-time dynamics of deforestation in Brazilian Amazônia. International Journal of Remote Sensing 23, 2903-2908

Alves DS (2004) Quantificar e compreender o desmatamento na Amazônia. Available from http://www.dpi.inpe.br/dalves/sumarios/dalves-prelba-17ago04.pdf

Alves DS (in press) Cenários de Cobertura e Uso da Terra e Dimensões Humanas no LBA. In: Becker BK, Alves DS, ca Costa WM (eds) Dimensões Humanas da Atmosfera-Biosfera da Amazônia. EDUSP, Sao Paulo, pp 39-64

Alves DS, Escada MIS, Pereira JLG, Linhares CA (2003) Land use intensification and abandonment in Rondônia, Brazilian Amazônia. International Journal of Remote Sensing 24, 899-903

Alves DS, Becker BK, Batistella M (2004) Land Cover/Land Use Change and Human Dimensions in the Large Scale Biosphere-Atmosphere Experiment in Amazonia (LBA). LUCC Newsletter 10, Dec 2004, 4-5. Available from http://www.geo.ucl.ac.be/LUCC/pdf/news_lucc10.pdf

Andersen LE, Granger CWJ, Reis EJ, Weinhold D, Wunder S (2002) The Economics of Deforestation: Dynamic Modeling of Amazonia. Cambridge University Press, Cambridge

Bourdieu P (2003) O campo científico. In: A sociologia de Pierre Bourdieu (org. R Ortiz), Olho d'Agua São Paulo, pp 112-143 (*originally published as Bourdieu P, Le champ cientifique, Actes de la recherche em Sciences Sociales, n 2, 1976, 88-104*)

Brasil (1965) Lei No 4.771. Diário Oficial da União. 15 de setembro de 1965

Brasil (1990) CPI da Amazônia. Relator Senador Jarbas Passarinho. Senado Federal, Brasília.

Brasil (2001) M.P. 2.080-63 Diário Oficial da União. 18 de maio de 2001

Brasil (2003) Diretrizes metodológicas e artigos selecionados. Programa Zoneamento Ecológico-Econômico. CDROM. Secretaria de Políticas para o Desenvolvimento Sustentável. Ministério do Meio Ambiente, Brasília

Brown S. And Lugo A. (1990) Tropical secondary forests. Journal of tropical Ecology 6, 1-32

Candido A (2003) Parceiros do Rio Bonito. Livraria Duas Cidades, São Paulo

Chomitz KM, Thomas TS (2001) Geographic Patterns of Land Use and Land Intensity in the Brazilian Amazon. Policy Research Working Paper 2687. World Bank, Washington DC. Available from http://econ.worldbank.org

Collins HM (1992) Experts artificiels. Seuil, Paris

Diewald C, Chaves F(2003) Planafloro e Prodeagro – Dez anos perdidos ? Uma avaliação de duas tentativas de mudança no padrão de ocupação da Amazônia. Available from http://www.worldbank.org/rfpp/news/debates/debates.htm

Egler PCG (2001) Capacitação para pesquisa e desenvolvimento em ciência tecnologia e inovação na Amazônia In: P. Faulhaber P, de Toledo PM (eds) Conhecimento e fronteira: História da Ciência na Amazônia. Museu Paraense Emilio Goeldi, Belém, pp 317-342

Faminow MD (1997) Spatial economics of local demand for cattle products in Amazon development. Agriculture, Ecosystems and Environment 62, 1-11

Faminow MD (1998) Cattle deforestation and development in the Amazon. CAB International, New York

Fearnside PM (1993) Deforestation in Brazilian Amazonia: the effect of population and land tenure. Ambio 22(8), 537-545

Fearnside PM, Tardin AT, Meira Filho LG (1990) Deforestation rate in Brazilian Amazonia. Reprint. Instituto de Pesquisas Espaciais, São José dos Campos

Folha Amazônica (Boletim do Experimento de Grande Escala da Biosfera-Atmosfera na Amazônia (LBA)) (2004). LBA em Números, Ano 6, N 12, Julho de 2004

Folha de São Paulo (2004) Ministérios da Casa Civil e da Ciência e Tecnologia duplicam esforços para produzir sistemas de monitoramento – Governo paga dois vigias para Amazônia. Folha Ciência 17 November 2004

Franco MSC (1997) Homens livres na ordem escravocrata. 4a edição. Editora UNESP São Paulo

Hecht SB, Norgaard RB and Possio G (1988) The economics of cattle ranching in Eastern Amazonia. Interciencia 13(5), 233-240

Instituto Nacional de Pesquisas Espaciais (INPE) (1992) Deforestation in Brazilian Amazonia. Reprint. Instituto Nacional de Pesquisas Espaciais, São José dos Campos

Instituto Nacional de Pesquisas Espaciais (INPE) (2002) Monitoramento da Floresta Amazônica por Satélite 1999-2000. Reprint. INPE, São José dos Campos

Keller M, Silva-Dias MA, Nepstad DC, Andreae MO (2004) The Large-Scale Biosphere-Atmosphere Experiment in Amazonia: Analyzing regional land use change effects. In: DeFries RS, Asner GP, Houghton RA (eds) Ecosystems and Land Use Change. American Geophysical Union, Washington, DC, pp 321-334

Lahsen M (2002) Brazilian Climate Epistemers' Multiple Epistemes: An Exploration of Shared Meaning, Diverse Identities and Geopolitics in Global Change Science. Available from http://environment.harvard.edu/gea/pubs/ 2002-01.html

LBA (1996a) Concise Science Plan. Available from http://lba.inpa.gov.br/lba

LBA (1996b) Plano Científico Extendido. Available from http://lba.cptec.inpe. br/lba/port/extplan/capap.html

LBA (2004) Painel Pré-Conferencia: O conhecimento científico e a formulação de políticas públicas para a Amazônia: a experiência do programa LBA. Available from http://www.lbaconferencia.org/port/assoc_meetings_res1. htm

Machado L (1998) A fronteira agrícola na Amazônia. In: Becker BK, Christofoletti A, Davidovich FR, Geiger PP (eds) Geografia e Meio Ambiente no Brasil, 2ª edição, Hucitec, São Paulo, pp 181-217

Mack PE (1990) Viewing the Earth. The MIT Press, Cambridge, Massachusetts

Mackenzie PE (1993) Inventing accuracy. The MIT Press, Cambridge, Massachusetts

Mahar D (2002) Agro-Ecological Zoning in Rondônia, Brazil: What are the lessons. In: Hall A (ed) Amazonia at the crossroads. Institute of Latin American Studies, London, pp 115-128

Margulis S (2004) Causes of deforestation in Brazilian Amazon. World Bank, Washington. Available from http://www-wds.worldbank.org

Martins JS (2003) O sujeito oculto. Vozes, Petrópolis

Sayago D, Machado L (2004) O pulo do grilo: o INCRA e a questão fundiária na Amazônia. In: Amazônia: Cenas e cenários. Org. D. Sayago, JF Tourrand, M Bursztyn, 217-235. Editora da Universidade de Brasília

Sayago D, Tourrand JF, Bursztyn M (orgs) (2004) Amazônia: Cenas e cenários. Editora da Universidade de Brasília, Brasilia, pp 217-235

Schor T (2005) Ciência e Tecnologia: Uma interpretação da pesquisa na Amazônia – o caso do Experimento de Grande Escala da Biosfera-Atmosfera na Amazônia (LBA). Doctoral Thesis, University of São Paulo, São Paulo

Silva LO (1996) Terras devolutas e latifndio Efeitos da lei de 1850. Editora da UNICAMP, Campinas

Tardin AT, Lee DCL, Santos RJR, Assis OR, Barbosa MPS, Moreira ML, Pereira MT, Silva D, Santos Filio CP (1980) Subprojeto desmatamento: convnio IBDF/CNPq - INPE. Technical Report. INPE-1649-RPE/103. Instituto de Pesquisas Espaciais, São José dos Campos

Tesh SN, Paes-Machado E (2004) Sewers, Garbage, and Environmentalism in Brazil. Journal of Environment & Development 13, 42-72

Veiga JB, Tourrand JF, Piketty MG, Poccard-Chapuis R, Alves AM, Thales MC (2004) Expansão e trajetórias da pecuária na Amazônia: Pará, Brasil. Editora Universidade de Brasilia, Brasília

Velho OG (1976) Capitalismo Autoritário e campesinato. DIFEL, São Paulo

Waniez P, Brustlein V, O'Neill M (1997) SAMBA 2000, CD-ROM (Rio de Janeiro/Paris: IBGE/ORSTOM/CREDAL)

World Bank (2000a) Land issues and the G7 Pilot Programme to conserve the Brazilian Rain Forest. Available from http://www.worldbank.org/rfpp/iag/ landissues.htm

World Bank (2000b) Mid-term review of the pilot program to conserve the Brazilian Rain Forest. Executive Summary. Available from http://www.worldbank.org/rfpp/docs/mtrexsum.htm

World Bank (2001) Civil society participation in the Pilot Program to conserve the Brazilian rain forest. Executive Summary. October 2000. Available from http://www.worldbank.org

Index of keywords

Printing: Krips bv, Meppel
Binding: Stürtz, Würzburg